Lecture Notes in Artificial Intelligence   4573

Edited by J. G. Carbonell and J. Siekmann

Subseries of Lecture Notes in Computer Science

T0223158

Manuel Kauers   Manfred Kerber
Robert Miner   Wolfgang Windsteiger (Eds.)

Towards Mechanized Mathematical Assistants

14th Symposium, Calculemus 2007
6th International Conference, MKM 2007
Hagenberg, Austria, June 27-30, 2007
Proceedings

 Springer

Series Editors
Jaime G. Carbonell, Carnegie Mellon University, Pittsburgh, PA, USA
Jörg Siekmann, University of Saarland, Saarbrücken, Germany

Volume Editors

Manuel Kauers
Wolfgang Windsteiger

Johannes Kepler University
Research Institute for Symbolic Computation, Linz, Austria
E-mail: {Manuel.Kauers, Wolfgang.Windsteiger}@risc.uni-linz.ac.at

Manfred Kerber
The University of Birmingham
School of Computer Science, Birmingham B15 2TT, England
E-mail: M.Kerber@cs.bham.ac.uk

Robert Miner
Design Science, Inc., St. Paul, Minnesota, USA
E-mail: robertm@dessci.com

Library of Congress Control Number: 2007928735

CR Subject Classification (1998):
I.2.2, I.2, H.3, H.2.8, I.7.2, F.4.1, H.4, C.2.4, G.4, I.1

LNCS Sublibrary: SL 7 – Artificial Intelligence

ISSN   0302-9743
ISBN-10   3-540-73083-4 Springer Berlin Heidelberg New York
ISBN-13   978-3-540-73083-5 Springer Berlin Heidelberg New York

Springer is a part of Springer Science+Business Media

springer.com

© Springer-Verlag Berlin Heidelberg 2007
Printed in Germany

Typesetting: Camera-ready by author, data conversion by Scientific Publishing Services, Chennai, India
Printed on acid-free paper   SPIN: 12077151   06/3180   5 4 3 2 1 0

Preface

This volume contains the collected contributions of two conferences, Calculemus 2007 and MKM 2007. Calculemus 2007 was the 14th in a series of conferences dedicated to the integration of computer algebra systems (CAS) and automated deduction systems (ADS). MKM 2007 was the sixth International Conference on Mathematical Knowledge Management, an emerging interdisciplinary field of research in the intersection of mathematics, computer science, library science, and scientific publishing. Both conferences aimed to provide mechanized mathematical assistants.

Although the two conferences have separate communities and separate foci, there is a significant overlap in the interests in building mechanized mathematical assistants. For this reason it was decided to collocate the two events in 2007 for the first time, at RISC in Hagenberg, Austria. The number and quality of the submissions show that this was a good decision. While the proceedings are shared, the submission process was separate. The responsibility for acceptance/rejection rests completely with the two separate Program Committees.

By this collocation we made a contribution against the fragmentation of communities which work on different aspects of different independent branches, traditional branches (e.g., computer algebra and theorem proving), as well as newly emerging ones (on user interfaces, knowledge management, theory exploration, etc.). This will also facilitate the development of integrated mechanized mathematical assistants that will be routinely used by mathematicians, computer scientists, and engineers in their every-day business.

In total, 23 papers were submitted to Calculemus. For each paper there were three reviews and, finally, ten papers were accepted for publication in these proceedings. MKM received 52 submissions (more than double last year's number). For each paper there were at least two reviews; if the evaluation was not uniform we had three and in some cases four reviews. After discussions, we accepted 19 high-quality papers for these proceedings. In the preparation of these proceedings and in managing the whole discussion process, Andrei Voronkov's EasyChair conference management system proved itself an excellent tool. In addition to the contributed papers, abstracts of the invited speakers of MKM are found in these proceedings.

April 2007

Manuel Kauers
Manfred Kerber
Robert Miner
Wolfgang Windsteiger

Preface

Calculemus & MKM Organization

Conference Chair   Wolfgang Windsteiger (RISC, Linz, Austria)
Local Arrangements   Laura Kovács (RISC, Linz, Austria)
WWW: http://www.risc.uni-linz.ac.at/about/conferences/
                       summer2007/

Sponsors

Calculemus and MKM greatfully acknowledge the financial support of the following institutions:

- Bundesministerium für Wissenschaft und Bildung, Österreich
  (Federal Minister of Science and Education, Austria)
- Land Oberösterreich (Upper Austrian Government)
- Raiffaisen Landesbank Oberösterreich
- Johannes Kepler University
- Linzer Hochschulfond
- Spezialforschungsbereich SFB F013
  "Numerical and Symbolic Scientific Computing"
- Research Institute for Symbolic Computation (RISC)
- RISC Software GmbH
- uni software plus, Mathematica reseller, Austria

Calculemus 2007 Organization

Program Committee

Alessandro Armando	University of Genova, Italy
Christoph Benzmüller	University of Cambridge, UK
Olga Caprotti	University of Helsinki, Finland
Jacques Carette	McMaster University, Canada
Timothy Daly	Carnegie Mellon University, USA
William Farmer	McMaster University, Canada
Keith Geddes	University of Waterloo, Canada
Tom Hales	University of Pittsburgh, USA
Hoon Hong	North Carolina State University, USA
Deepak Kapur	University of New Mexico, USA
Manuel Kauers	RISC-Linz, Austria (*Co-chair*)
Laura Kovacs	RISC-Linz, Austria
Petr Lisonek	Simon Fraser University, Canada
Renaud Rioboo	Universtite Pierre et Marie Curie, France

Volker Sorge   University of Birmingham, UK
Thomas Sturm   Carnegie Mellon University, USA
Klaus Sutner   University of Passau, Germany
Wolfgang Windsteiger   RISC-Linz, Austria (*Co-chair*)

External Reviewers

Chad Brown   Martin Kreuzer   Wolfgang Schreiner
Stephen Forrest   Aless Lasaruk   Dongming Wang
Therese Hardin   Piotr Rudnicki   Tetsu Yamaguchi

MKM 2007 Organization

Program Committee

Andrea Asperti   University of Bologna, Italy
Laurent Bernardin   Maplesoft, Canada
Jonathan Borwein   Dalhousie University, Halifax, Canada
Thierry Bouche   Université de Grenoble I, France
Bruno Buchberger   Johannes Kepler University, Linz, Austria
Paul Cairns   University College London, UK
Olga Caprotti   University of Helsinki, Finland
Bruce Char   Drexel University, Philadelphia, USA
Simon Colton   Imperial College, London, UK
Mike Dewar   Numerical Algorithms Group, Oxford, UK
William Farmer   McMaster University, Hamilton, Canada
Herman Geuvers   Radboud Univ. Nijmegen, The Netherlands
Tetsuo Ida   University of Tsukuba, Japan
Mateja Jamnik   University of Cambridge, UK
Fairouz Kamareddine   Heriot-Watt University, UK
Manfred Kerber   University of Birmingham, UK (*Co-chair*)
Michael Kohlhase   International University Bremen, Germany
Paul Libbrecht   DFKI Saarbrücken, Germany
Robert Miner   Design Science, Inc., USA (*Co-chair*)
Bengt Nordström   Chalmers University of Technology, Göteborg,
                            Sweden
Ross Reedstrom   Rice University, USA
Eugénio Rocha   University of Aveiro, Portugal
Alan Sexton   University of Birmingham, UK
Andrzej Trybulec   University of Białystok, Poland
Stephen Watt   The University of Western Ontario, Canada
Abdou Youssef   George Washington University, Washington, DC,
                            USA

External Reviewers

Pierre Corbineau
Cezary Kaliszyk
Pouya Larjani
Mircea Marin
Russell O'Connor
Florian Rabe
Freek Wiedijk

Claudio Sacerdoti Coen
Robert Lamar
Lionel Mamane
Normen Mueller
Martijn Oostdijk
Krzysztof Retel
Jian Xu

Jeremy Gow
Christoph Lange
Manuel Maarek
Christine Mueller
Matti Pauna
Clare So

Table of Contents

Executing in Common Lisp, Proving in ACL2*

Mirian Andrés, Laureano Lambán, and Julio Rubio

Departamento de Matemáticas y Computación, Universidad de La Rioja,
Edificio Vives. Calle Luis de Ulloa s/n, E-26004 Logroño, Spain
{mirian.andres,lalamban,julio.rubio}@unirioja.es

Abstract. In this paper, an approach to integrate an already-written
Common Lisp program for algebraic manipulation with ACL2 proofs
of properties of that program is presented. We report on a particular
property called "cancellation theorem", which has been proved in ACL2,
and could be applied to several problems in the field of Computational
Algebraic Topology.

1 Introduction

Kenzo is a Common Lisp program [10] designed by Sergeraert, implementing
his ideas on *Constructive Algebraic Topology* [19]. Kenzo, and its predecessor
EAT [21], were capable of computing homology groups unknown by any other
means. Kenzo continues to evolve and has been recently released as an open
source computer algebra system [10] and extended with new modules on Koszul
Homology [20], Spectral Sequences [18] and Coalgebras [4].

Several years ago a project was launched to analyze the Kenzo system by
means of formal methods. The objective of the project is twofold. Better knowledge of the internal processes and structures in Kenzo is intented, thus increasing
the reliability of the system. Besides, Kenzo is also a good "laboratory" (due to
its structural richness and to the presence of challenging results which have been
obtained using it) to experiment with different tools and approaches in the field
of formal methods in Software Engineering, allowing the analyst to compare
them, to evaluate them and, hopefully, to apply them to other fields unrelated
to Algebraic Topology or Computer Algebra.

The first efforts were devoted to the Algebraic Specification of EAT [13] and
Kenzo [8,9]. After that, these rather theoretical results were put into practice
through *theorem provers*. The tactical assistant Isabelle [17] was chosen for the
first studies [1,2] on the application of automated theorem proving in the area
of Algebraic Topology. These preliminary works led to the recent Isabelle mechanized proof of the Basic Perturbation Lemma [3], one of the central results in
Algorithmic Homological Algebra. Other lines of research include modeling and
proving with Coq [5], and programming and proving with the system FoCaL [6].

In this paper we report on a relative approach, by using the theorem prover
ACL2 [12]. The limitations of this prover with respect to Isabelle or Coq are

---

* Partially supported by Comunidad Autónoma de La Rioja, project ANGI-2005/19,
and by Ministerio de Educación y Ciencia, project MTM2006-06513.

M. Kauers et al. (Eds.): MKM/Calculemus 2007, LNAI 4573, pp. 1–12, 2007.

well-known and are essentially related to the underlying logics. ACL2 is based on a weak form of first order logic, while both Coq and Isabelle can work with higher order logic. On the positive side, ACL2 is based on Common Lisp (as Kenzo itself) and is very suitable when linking proofs and running programs. In addition, the treatment of Symbolic Computation problems with the help of ACL2 has obtained important successes in recent years (see, for instance, [15]).

The organization of the paper is as follows. The next section introduces our methodological approach to relate an already-written program with the proofs of properties in ACL2. Section 3 and 4 are devoted to introduce, respectively, our motivating examples from Homological Algebra and the basic data structures and proofs in ACL2. Section 5 presents the main contribution of the paper, reporting on the automated proof of a "cancellation theorem". This theorem is applied in Section 6 to the proof of an algebraic property of our programs. The paper ends with the section of conclusions and future work, and the bibliography.

2   Proving and Then... Testing

There are many ways in which Symbolic Computation (or programming, more generally) can interplay with theorem proving. For instance, Computer Algebra programs can be used as oracles for theorem provers. In the other direction, theorem provers can be used to ensure the correctness of Computer Algebra programs. In this paper we will introduce a third manner of interaction: theorem provers can be used for automated-testing of programs. Although it is usually considered that testing is easier than proving, and so that testing should occur in early stages of the quality control cycle, our proposal is the reversal (in a sense which will be clear later on): *first proving and then... testing*. Of course, the complete picture of our view is more complex than indicated by that simplistic phrase. Let us explore it in a concrete situation.

Let us assume that someone gave us a Common Lisp `program1` with the following characteristics:

- it is difficult to test, perhaps because it produces results difficult to interpret, or, even worse, some of its results are unknown by any other means, and
- the program correctness is difficult to prove, perhaps due to being logically complex, based on higher-order constructions, for instance.

An example of such a `program1` could be the Kenzo system, which has been developed in Common Lisp and has been successfully tested for more than fifteen years, but ... not always: some of the results found with the help of Kenzo continue to be unverifiable by any other means at this moment (homology groups of some iterated loop spaces, for instance; see [10]). In addition, Kenzo is based on both object-orientation and higher-order functional programming, in such a way that its formal specification is challenging (see [13,8,9]), and therefore its verification with theorem provers poses problems far from trivial. The formal specification and verification of some of the `algorithms` appearing in Kenzo have been carried out with the Isabelle assistant [17], and were explained in

[1] and [2]. The most relevant result in this line is the recent Aransay's proof in Isabelle/HOL of the BPL, the Basic Perturbation Lemma [3]. The BPL is one of the most important theorems and algorithms used to build Kenzo. But, independently of the merits of this mechanized proof of the BPL, the distance with respect to the *programs* implementing the BPL in Kenzo, continues to be quite large.

Since our goal is to verify *real* Common Lisp programs, a sensible idea should be to use the ACL2 system to devise proofs (instead of Isabelle or Coq). ACL2 [12] is both a programming language and an environment to produce proofs of properties of programs. As programming language, ACL2 is an extension of a sub-language of Common Lisp. The extensions added to Common Lisp in ACL2 are not relevant for our work. On the contrary, the features erased from Common Lisp in ACL2 are very important with respect to Kenzo. In particular, ACL2 does not allow the programmer to use higher-order functionals, a tool intensively employed in Kenzo. Thus, in order to study a Common Lisp program1 within ACL2, we are proposing to write a new Common Lisp program2 emulating the behavior of program1, but programmed this time in ACL2.

Let us enumerate the characteristics of this situation:

- program1 is
  - already written
  - in Common Lisp (not necessarily in ACL2);
  - efficient;
  - tested;
  - unproved.
- program2 is
  - specially designed to be proved;
  - programmed in ACL2 (and Common Lisp);
  - efficient or not: irrelevant;
  - tested;
  - proved in ACL2.

In our approach, program2 is *supposed to be equivalent* to program1. But we do not pretend to prove this equivalence: this option would lead us to a form of ill-founded recursion. Our aim should be to use the *highly reliable* program2 to perform automated testing of the *efficient* program1.

The following toy program will illustrate this idea:

```
(defun automated-testing ()
    (let ((case (generate-test-case)))
     (if (not (equal (program1 case)
                     (program2 case)))
         (report-on-failure case))))
```

Note that it is an (unverified!) Common Lisp program, but not an ACL2 one (at least, if program1 is not).

The relationship of these ideas with *Model Checking* is appealing. Even if the field of application (reactive systems modeled as state machines) and the formal methods used (temporal logics) are different from ours, at least in the standard literature on Model Checking [7], the underlying philosophy is the same. In our case, the system (an already written `program1`) is abstracted into a model (`program2`). Then, formal methods (theorem proving in our case) are used to get theoretical properties of the model (the correctness of `program2`, proved in ACL2). The final step is to interpret the results obtained from the model with respect to reality (automated testing of the `program1` against `program2`).

As in Model Checking, one of the important bottlenecks of the method is to build a model which is an accurate representation of the system to be modeled. In Model Checking one such difficult step occurs when an infinite system (that is to say, a system with an unbounded number of possible reachable states) is modeled by means of a finite graph (the condition of finiteness is mandatory, because the checking of properties is done by exhaustive traversal of state spaces).

In our context, it is hopeless to apply our method to the whole Kenzo system. The most important constraint is that we must restrict our ACL2 study to the parts of Kenzo which are *first-order*[1]. This excludes large (and interesting!) fragments of Kenzo, that should be analyzed by using tools such as Isabelle (as in [1], [2] or [3]) or Coq.

Once a part of Kenzo with this characteristic has been chosen (let us call it `program1`), the (heuristic) transformations we apply to construct the model `program2` are the following:

- iterations and loops are replaced by recursive functions (this step could be automated);
- first-order functional programming is replaced by standard functions[2];
- data structures are "flattened" to lists: objects, structs and arrays are replaced by convenient nested lists;
- destructive operations are replaced by the corresponding constructive ones (this is a problematic point, but destructive updates appear in very precisely located Kenzo fragments, and so this task is quite relaxed).

With these cautions, it is hoped that `program2` accurately models `program1`, and then our strategy could be safely applied.

3   Homological Algebra

A first application of the ideas presented in the previous section arises from two different on-going projects devoted to analyze formally Kenzo [10], the system for computing in Algebraic Topology.

---

[1] Interestingly enough, this constraint seems related, in some sense, with the finite/infinite dichotomy evoked previously on Model Checking.

[2] For instance, an occurrence of (`mapcar #'cadr l`) should be replaced by (`mapcadar l`) where the new function `mapcadar` is simply:
(`defun mapcadar (l) (if (endp l) l (cons (cadar l) (mapcadar (cdr l))))`)

One of these projects is related to *Koszul Homology*, a module included recently in the Kenzo system (see [20]). The *Koszul complex* (see [14]) is defined as a free chain complex which can be presented in the following way. The generators are expressions such as $x_1^{a_1} \ldots x_n^{a_n} \otimes b_1 \wedge \ldots \wedge b_n$, where $x_1^{a_1} \ldots x_n^{a_n}$ is called the *monomial* part and $b_1 \wedge \ldots \wedge b_n$ is called the *exterior* part. In the monomial part, the x_i are undeterminates and the exponents a_i are simply integer numbers. In the exterior part, each b_i is or 0 or 1. In other words, as a Lisp datum a generator can be represented as a pair ((a1 ... aN) (b1 ... bN)), where the elements of the first part are integers and the elements of the second one are bits. The free abelian group generated by these pairs is graded, by means of the degree $\sum_{i=1}^{n} b_i$. Then the Koszul differential on this graded free abelian group is defined by the following formulae:

$$d(x_1^{a_1} \ldots x_n^{a_n} \otimes b_1 \wedge \ldots \wedge b_n) := \sum_{i=1}^{n} (-1)^{sg(i)} d_i(x_1^{a_1} \ldots x_n^{a_n} \otimes b_1 \wedge \ldots \wedge b_n)$$

where

$sg(i)$ denotes the number of 1's in $b_1 \wedge \ldots \wedge b_{i-1}$,
$d_i(x_1^{a_1} \ldots x_i^{a_i} \ldots x_n^{a_n} \otimes b_1 \wedge \ldots \wedge b_i \wedge \ldots \wedge b_n) := 0$, if $b_i = 0$

and

$d_i(x_1^{a_1} \ldots x_i^{a_i} \ldots x_n^{a_n} \otimes b_1 \wedge \ldots \wedge b_i \wedge \ldots \wedge b_n) :=$
$:= x_1^{a_1} \ldots x_i^{a_i+1} \ldots x_n^{a_n} \otimes b_1 \wedge \ldots \wedge 0 \wedge \ldots \wedge b_n$, if $b_i = 1$.

The first fact we are trying to verify with the help of ACL2 is that the morphism defined by d is actually a differential; in other words, we are trying to give a mechanized proof of $dd = 0$. (To be more precise, following the guidelines explained in Section 2: we are trying to verify this property on an ACL2 function d which has been obtained from the real Common Lisp encoding done in Kenzo.)

The second project we are dealing with is related to the automated proof of the *Eilenberg-Zilber theorem* [11]. The statement of this result is too complex to be presented here. For our purposes, it is enough to know that we are planning to prove it on the chain complex associated to a universal simplicial set called in the literature Δ [16]. The generators of the free chain complex are lists $(z_0, \ldots, z_n)$, and the degree of such a list is n. The differential is then defined by the following formulae.

$$d(z_0, \ldots, z_n) := \sum_{i=0}^{n} (-1)^i d_i(z_0, \ldots, z_n)$$

where

$$d_i(z_0, \ldots, z_{i-1}, z_i, z_{i+1}, \ldots, z_n) := (z_0, \ldots, z_{i-1}, z_{i+1}, \ldots, z_n).$$

Here, again, one of the first tasks to be done is to verify $dd = 0$.

The (pencil & paper) proofs of both instances of $dd = 0$ lie in the same kind of argument:

- in the case of the Koszul complex, $d_i d_j = -d_j d_i$, if $i < j$, and $d_i d_i = 0$;
- in the case of the chain complex of Δ, $d_i d_j = -d_{j-1} d_i$, if $i < j$.

That is to say, in both cases the morphism *dd* applied on any generator will give, if no intermediary reduction is achieved, a combination where each term is matched with its inverse. This idea led us to the statement (and ACL2 proof) of a so-called "cancellation theorem" which could be applied, later on, in these two case studies.

4   Canonical Forms

Before starting the formal specification of the cancellation theorem, let us first establish the elementary data structures involved in our computations and proofs. Let us call any ACL2 object a *generator* (or *gnrt*, in an abbreviated manner). Then a *term* (*trm*) is a pair (coef gnrt) where coef is a non-null integer and gnrt is a generator. In Lisp terminology, extracting the coefficient of a term consists in applying the car operator (the *first* element of a list), and extracting its generator consists in applying the cadr operator (the *second* element of a list).

A *list of terms* (*lot*) is, obviously, a list of terms ((coef0 gnrt0) ...(coefN gnrtN)). And, finally, a *combination* is a list of terms strictly increasingly ordered with respect to the ACL2 total order (see lexorder in [12]) *on generators*. In other words, combinations are a representation of the elements on the *free abelian group generated by* **all** *the ACL2 objects*.

Usually, it is more convenient to work with canonical forms, that is to say with *combinations* instead of with *lots*, both from the computing and the proving points of view. Nevertheless, when trying to prove that a combination is null, things can be slightly different: to work with sorted structures means to have a very tight control of the order in which cancellations occur. On the contrary, to have an unordered list of terms can allow the prover to organize the proof in a more convenient way. This point will be made clearer in the following section.

With this idea in mind, it is a well-known fact that combinations can be added in two different ways. Terms of the first combination can be added one-by-one to the second combination. Or the two combinations can be concatenated (without simplification) and then the resulting list of terms can be reduced to canonical form. Let us assume that two ACL2 functions, called a2c and c-f, have been programmed (the actual programming of both is a very simple exercise). The first one, a2c, adds two combinations following the first strategy previously mentioned. The second function, c-f takes as argument a list of terms and constructs the combination being the canonical form of its argument.

Then the first relevant ACL2 theorem can be presented in the following way:

```
(defthm a2c-equivalence
    (implies (and (cmbnp l1) (cmbnp l2))
             (equal (a2c l1 l2)
                    (c-f (append l1 l2)))))
```

where cmbnp is the predicate checking whether an ACL2 object is a combination.

It is worth noting that ACL2, with the help of the natural lemmas on adding a term, is capable of finding the right induction schemes without any human aid.

5   The Cancellation Theorem in ACL2

Now, we can prepare the statement of our main result: the cancellation theorem. Its informal wording is as follows: "if in a list of terms each element has its inverse, and the corresponding generator appears exactly twice, then the canonical form of the list is the null combination".

Due to the representation of terms, the *inverse* of a term is read simply as:

```
(defun inverse (trm)
  (cons (- (car trm)) (cdr trm)))
```

The `almost-zerop` function checks if in a list 1 each element has its inverse in 1 itself. In order to define it, we first consider the following generalization, which checks if each element in a list 11 has its inverse in a second list 12.

```
(defun almost-zeropi (11 12)
  (if (endp 11)
      T
    (and (member (inverse (car 11)) 12)
         (almost-zeropi (cdr 11) 12))))
```

So, the `almost-zerop` function is given by:

```
(defun almost-zerop (1)
  (almost-zeropi 1 1))
```

This property is not enough to ensure that the canonical form of 1 is null, because an element could occur repeated in 1. Then the key observation is that in the applications considered (Koszul homology and chain complex associated to the universal simplicial set Δ) the generators appear exactly twice in the composition of the differential. Thus, the `exactly-two` function checks if each element in a list appears exactly twice. Using the same strategy as above, we first consider the natural embedding:

```
(defun eti (11 12)
  (or (endp 11)
      (and (= 2 (count-times (car 11) 12))
           (eti (cdr 11) 12))))
```

And then we define:

```
(defun exactly-two (1)
  (eti 1 1))
```

The hypotheses for the cancellation theorem are then collected in the following function `azp`, where `mapcadar` extracts the list of generators of a given list of terms.

```
(defun azp (1)
  (and (almost-zerop 1)
       (exactly-two (mapcadar 1))))
```

The statement of the cancellation theorem looks as follows

```
(defthm cancellation-theorem
  (implies (and (lotp l) (azp l))
           (endp (c-f l))))
```

where `lotp` is the predicate checking if a given ACL2 object is a list of terms.

As usual in ACL2 (and, in fact, in any other theorem prover) the important point in order to build a proof is to find the right lemmas. In the case of the cancellation theorem the most important lemma is the following one.

```
(defthm essential-lemma
  (implies (and (lotp l)
                (member x l)
                (member (inverse x) l)
                (= (count-times (cadr x) (mapcadar l)) 2))
           (not (member (cadr x) (mapcadar (c-f l))))))
```

This `essential-lemma` explains that if a term and its inverse belong to a list of terms, and the generator of the term appears twice in the list, then the generator does not appear in the canonical form of the list. Since generators in the canonical form of a list l must proceed from generators in l, the cancellation theorem follows from the `essential-lemma`.

In order to prove this `essential-lemma` a brick is the following lemma. It particularizes the `essential-lemma` when the term x is the `car` of the list of terms l.

```
(defthm essential-lemma-x
  (implies (and (lotp l) (not (endp l))
                (equal (car l) x)
                (member x l)
                (member (inverse x) l)
                (= (count-times (cadr x) (mapcadar l)) 2))
           (not (member (cadr x) (mapcadar (c-f l))))))
```

Once this lemma is proved, the following one is a simple corollary.

```
(defthm essential-lemma-inverse-x
  (implies (and (lotp l) (not (endp l))
                (equal (car l) (inverse x))
                (member x l)
                (member (inverse x) l)
                (= (count-times (cadr x) (mapcadar l)) 2))
           (not (member (cadr x) (mapcadar (c-f l)))))
  :hints (("Goal" :do-not-induct t
           :use (:instance essential-lemma-x (x (inverse x))))))
```

Now, the idea to prove the `essential-lemma` is to apply induction: a non-empty list either has as first element x or (inverse x) or its rest (cdr in Lisp terminology) has the same properties as above. Nevertheless, ACL2 is unable to find the convenient induction schema. We must aid it by defining the following recursive function:

```
(defun induct-essential-lemma (x l)
  (if (endp l)
      nil
    (if (equal (car l) x)
        T
      (if (equal (car l) (inverse x))
          T
        (induct-essential-lemma x (cdr l))))))
```

and then giving it to the prover as an induction schema:

```
(defthm essential-lemma
  (implies (and (lotp l)
                (member x l)
                (member (inverse x) l)
                (= (count-times (cadr x) (mapcadar l)) 2))
           (not (member (cadr x) (mapcadar (c-f l)))))
  :hints (("Goal" :induct (induct-essential-lemma x l))))
```

From this, the `cancellation-theorem` can be easily deduced in ACL2.

6   An Application of the Cancellation Theorem

Since the applications to Algebraic Topology are far complex to be explained in detail, we prefer to present an elementary example of using the cancellation theorem (together with the `a2c-equivalence` theorem introduced at the end of Section 3). The example is also chosen to illustrate another way in which we can help ACL2, by avoiding unsuitable induction schemes (the first way is to define our own induction schemes as in Section 5).

The inverse of a combination (or the opposite, in additive terminology) is simply programmed as:

```
(defun invc (l)
  (if (endp l)
      l
    (cons (inverse (car l)) (invc (cdr l)))))
```

Now, one of the algebraic properties needed to ensure that we are really working with a free abelian *group* is that this `invc` acts actually as an inverse for combinations. In other words, we must prove the following result.

```
(defthm invc-is-inverse
  (implies (cmbnp l)
           (endp (a2c l (invc l)))))
```

To this aim, the cancellation theorem can be used by simply proving:

```
(defthm invcIsInverse-lemma-preparation-lemma
   (implies (cmbnp l)
            (and (lotp (append l (invc l)))
                 (azp (append l (invc l))))))
```

together with the a2c-equivalence theorem.

Now, the unique technical problem when trying to build this proof in ACL2 is the following elementary property.

```
(defthm member-append-twice-lemma
   (implies (member x (append l l))
            (member x l)))
```

ACL2 is unable to prove it directly, since any recursion schema would be run on the list l, obtaining a property which is not true anymore. The solution in this case is to state the right generalization lemma (this generalization is required due to our way of working in ACL2, but it seems unnecessary in a traditional by hand proof):

```
(defthm member-append-my-lemma
   (implies (and (member x (append l1 l2))
                 (not (member x l1)))
            (member x l2)))
```

In this case, ACL2 finds automatically the right induction, and so the complete proof. Going back to our technical lemma, a simple hint gives the solution:

```
(defthm member-append-twice-lemma
   (implies (member x (append l l))
            (member x l))
   :hints (("Goal" :use
               (:instance member-append-my-lemma (l1 l) (l2 l)))))
```

A final word on this small proof. It is worth noting that the application of member-append-my-lemma amounts to an instance of the *exclude-middle principle*. This application poses no problem with respect to constructiveness, since every predicate test involved is decidable.

7   Conclusions and Further Work

The last technical sections show how ACL2 is a very good tool when dealing with inductive reasoning, as it is usual in the field of algebraic manipulation. ACL2 finds by itself the natural induction schemes in many cases, and it allows the prover, in this particular aspect, to automate much more tasks than other tools as Isabelle or Coq. This does not imply, by no means, that ACL2 allows the user a higher-level or abstract view on proofs: one stops frequently on frustrating technical details and proofs that must be accomplished before ACL2 deploys its very powerful heuristics. As seen in Section 5, even when ACL2 does not find the right induction scheme, the prover can be helped by explicit hints from the user. In Section 6, we showed another way of guiding ACL2: by introducing convenient generalization lemmas.

Going back to Section 2, and to the initial proposal of the paper, the automated testing of Kenzo fragments could be applied to the case of the differential in the Koszul complex, whose implementation is tricky and efficient in Kenzo, and could be matched with the much more explicit (and verbose) version we are trying to verify in ACL2.

With respect to future work, it is clear that many pieces of the puzzle are still missing. It is necessary to finish the proofs in cases of the Koszul complex and of the chain complex of Δ, allowing us to deduce $dd = 0$ by applying our cancellation theorem. Even then, much effort should be employed to conclude the formalization of both the Koszul Homology [14] package and the Eilenberg-Zilber theorem [11]. Another critical issue would be to find a good strategy for test generation, in order to be able to give a smart implementation of the function `generate-test-case` evoked in Section 2.

Another line of research consists in modeling this same example (the cancellation theorem) in Isabelle [17], Coq [5] and FoCaL [6], in order to evaluate and compare them, and with respect to ACL2 [12], focusing on our particular area of interest: verified Computer Algebra.

References

1. Aransay, J., Ballarin, C., Rubio, J.: Towards a Higher Reasoning Level in Formalized Homological Algebra. In: Proceedings Calculemus 2003, Aracné Éditrice, pp. 84–88 (2003)
2. Aransay, J., Ballarin, C., Rubio, J.: Four Approaches to Automated Reasoning with Differential Algebraic Structures. In: Proceedings Artificial Intelligence and Symbolic Computation, AISC 2004, Lecture Notes in Artificial Intelligence, vol. 2349, pp. 221–234 (2004)
3. Aransay, J., Ballarin, C., Rubio, J.: A Mechanized Proof of the Basic Perturbation Lemma. Preprint
4. Berciano, A., Sergeraert, F.: Software to compute A_∞-(co)algebras: Araia & Craic (2006) http://www.ehu.es/aba/araia-craic.htm
5. Bertot, Y., Castéran, P.: Interactive Theorem Proving and Program Development. In: Coq'Art: The Calculus of Inductive Constructions, Springer, Heidelberg (2004)
6. Boulmé, S., Hardin, T., Hirschkoff, D., Ménissier-Morain, V., Rioboo, R.: On the way to certify Computer Algebra systems. In: Proceedings Calculemus 1999, Electronic Notes in Theoretical Computer Science, vol. 23 (1999)
7. Clarke, E.M., Emerson, E.A., Sistla, A.P.: Automatic verification of finite-state concurrent systems using temporal logic specifications. ACM Transactions on Programming Languages and Systems 8(2), 244–263 (1986)
8. Domínguez, C., Rubio, J., Sergeraert, F.: Modeling Inheritance as Coercion in the Kenzo System. Journal of Universal Computer Science 12(12), 1701–1730 (2006)

9. Domínguez, C., Lambán, L., Rubio, J.: Object-Oriented Institutions to Specify Symbolic Computation Systems. To appear in Rairo Theoretical Informatics and Applications
10. Dousson, X., Sergeraert, F., Siret, Y.: The Kenzo program (1999-2007) http://www-fourier.ujf-grenoble.fr/~sergerar/Kenzo/
11. Eilenberg, S., Zilber, J.A.: On products of complexes. American Journal of Mathematics 75, 200–204 (1953)
12. Kaufmann, M., Manolios, P., Moore, J.: Computer-Aided Reasoning: An Approach. Kluwer Academic Press, Boston (2000)
13. Lambán, L., Pascual, V., Rubio, J.: An Object-Oriented Interpretation of the EAT System, Applicable Algebra in Engineering. Communication and Computing 14(3), 187–215 (2003)
14. Mac Lane, S.: Homology. Springer, Heidelberg (1994)
15. Martín, F.J., Alonso, J.A., Hidalgo, M.J., Ruiz-Reina, J.L.: Formal verification of a generic framework to synthesize SAT-provers. Journal of Automated Reasoning 32, 287–313 (2004)
16. May, P.: Simplicial Objects in Algebraic Topology, Van Nostrand (1967)
17. Nipkow, T., Paulson, L.C., Wenzel, M.: Isabelle/HOL. LNCS, vol. 2283. Springer, Heidelberg (2002)
18. Romero, A., Rubio, J., Sergeraert, F.: Computing Spectral Sequences. Journal of Symbolic Computation 41, 1059–1079 (2006)
19. Rubio, J., Sergeraert, F.: Constructive Algebraic Topology. Bulletin Sciences Mathématiques 126, 389–412 (2002)
20. Rubio, J., Sergeraert, F.: Constructive Homological Algebra, Lectures Notes MAP Summer School (2006) http://map.disi.unige.it/summer_school/index.php?page=2
21. Rubio, J., Sergeraert, F., Siret, Y.: EAT: Symbolic software for effective homology computation, Institut Fourier (1997) ftp://ftp-fourier.ujf-grenoble.fr/pub/EAT

A Rational Reconstruction of a System for Experimental Mathematics

Jacques Carette[1], William M. Farmer[1], and Volker Sorge[2]

[1] Department of Computing and Software
McMaster University, Hamilton, Ontario, Canada
{carette,wmfarmer}@mcmaster.ca
http://www.cas.mcmaster.ca{~carette,~wmfarmer}
[2] School of Computer Science, University of Birmingham, UK
V.Sorge@cs.bham.ac.uk
http://www.cs.bham.ac.uk/~vxs

Abstract. In previous papers we described the implementation of a system which combines mathematical object generation, transformation and filtering, conjecture generation, proving and disproving for mathematical discovery in non-associative algebra. While the system has generated novel, fully verified theorems, their construction involved a lot of ad hoc communication between disparate systems. In this paper we carefully reconstruct a specification of a sub-process of the original system in a framework for trustable communication between mathematics systems put forth by us. It employs the concept of biform theories that enables the combined formalisation of the axiomatic and algorithmic theories behind the generation process. This allows us to gain a much better understanding of the original system, and exposes clear generalisation opportunities.

1 Introduction

Over the last decade several environments and formalisms for the combination and integration of mathematical software systems [2, 15] have been proposed. Many of these systems aim at a traditional automated theorem proving approach, in which a given conjecture is to be proved or refuted by the cooperation of different reasoning engines. However, they offer little support for experimental mathematics in which new conjectures are constructed by an interleaved process of model computation, model inspection, property conjecture and verification. And while for example the Theorema system [4] supports many of these activities, there are currently no systems available that provide, in an easy to use environment, the flexible combination of diverse reasoning systems in a plug-and-play fashion via a high level specification of experiments, despite some previous research in that direction [1, 3].

[8, 14] presents an integration of more than a dozen different reasoning systems — first order theorem provers, SAT solvers, SMT solvers, model generators, computer algebra, and machine learning systems — in a general bootstrapping algorithm to generate novel theorems in the specialised algebraic domain

M. Kauers et al. (Eds.): MKM/Calculemus 2007, LNAI 4573, pp. 13–26, 2007.

of quasigroups and loops. While the integration leads to provably correct results, the integration itself was achieved in an ad-hoc manner, i.e., systems were combined and recombined in an experimental fashion with a set of custom-built bridges that not only perform syntax translations but also certain filtering functions.

In this paper we report on a rational reconstruction of an interesting sub-process of the bootstrapping algorithm, namely the generation of isotopy invariants (presented in Sec. 2), using the framework for trustable communication between mathematics systems that was put forth in [6]. It employs the concept of biform theories (Sec. 3) that enables the combined formalisation of the axiomatic and algorithmic theories behind the generation process. It turns out that it is surprisingly difficult to separate the syntactic, semantic, and algorithmic level of the current implementation. We present the formalisation in terms of the necessary semantic and syntactic concepts in Sec. 4 and of biform theories describing the actual computations in Sec. 5. The aim of this work is to expose the general principles behind the combination and communications of the single systems. It is currently only a purely theoretical reconstruction of the current implementation, and we do not have or even plan an implementation of the process in the framework of biform theories. However, the work should ultimately be used in the design of a flexible environment for experimental mathematics that enables a user to specify complex experiments on a high level without the need of detailed knowledge of the underlying logical relations and the particularities of the integrated systems.

Note that, as in [6], we abstract out the details of the idiosyncratic syntax of each of the systems. We use a uniform abstract syntax (in this case, we need 4 separate languages, each embedded in the other) for the specification. This allows us to abstract out the tedious engineering of transformations in and out of the actual systems. On the other hand, any transformation beyond trivial parsing and pretty-printing is explicitly specified.

The specification we present involves (at least) 3 levels of mathematical discourse: using the language of mathematics, we are specifying (syntactically) the semantics of a computer system which manipulates (the syntax of) mathematical theories, which are themselves inhabited by mathematical objects represented syntactically. Each level also possesses semantic models, which is ultimately what we want, but for computer manipulation, must be handled syntactically. Separating these languages cleanly is a difficult task – as the reader will soon witness.

2 Problem

The particular problem we are concerned with is the process of generating isotopy invariants for loops, which is part of the overall classification procedure presented in [14]. We give a brief, high level description of the procedure here. The more formal, mathematical details and definitions are presented in Sec. 4.

A loop is a quasigroup with identity, i.e., an algebraic structure $(L, \circ)$ satisfying the following two axioms: $\forall a, b. \ (\exists x. x \circ a = b) \wedge (\exists y. a \circ y = b)$ and

$\forall x . x \circ e = e \circ x = x$. We say two loops $(L, \circ)$ and $(M, *)$ are *isotopic* to each other—or L is an *isotope* of M—if there are bijective mappings α, β, γ from L to M such that, for all $x, y \in L$, $\alpha(x) * \beta(y) = \gamma(x \circ y)$ holds. A property P, that is preserved under the isotopy mapping (i.e., if P holds for L then it also holds for all its isotopes) is called an *isotopy invariant*. In our approach we generate universal identities, a particular type of isotopy invariants, presented by Falconer [9]. To find universal identities we have implemented the procedure presented in Fig 1.

The basic idea of our procedure is to find identities (i.e., universally quantified equations) that hold for some loop, by first generating identities and then checking, which identity has a non-trivial loop satisfying it, using a model generator. All identities for which a loop exists are then transformed into derived identities (see Sec. 4 Def. 6). All derived identities for which we can prove, by means of a first order automated theorem prover, that they are invariant under isotopy are universal identities. Note that, for each universal identity, we show that it is an invariant under isotopy independently of the size of a loop. We can therefore reuse these universal identities in different classifications. Consequently, we collect universal identities in a pool of confirmed isotopy invariants, which we use in the overall bootstrapping algorithm. That is, during the classification of loops of a particular size n, we draw on the pool of invariants by first filtering them again using another model generator to generate loops of size n that satisfy the invariant. We then extract those invariants for which at least one loop of order n exists, and we use only these as potential discriminants. Note that the filter discards any invariants which cannot solve any discrimination problem, as no loop of size n satisfies the invariant property.

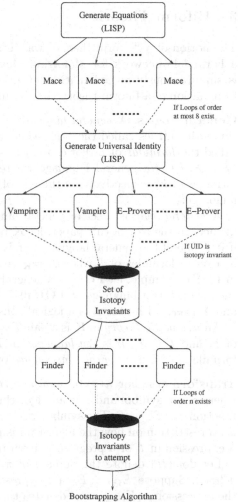

Fig. 1. Current implementation

So far we have generated and verified 8, 530 isotopy invariants. These have been employed as one means, among others, to generate two novel classifications for loops of order 6 and 7 with respect to isomorphism. Observe, that we currently concentrate only on the subprocess for generating isotopy invariants and that the general bootstrapping algorithm is not the subject of the paper. For more details and results of the classification of quasigroups and loops with respect to both isomorphism and isotopism we refer the reader to [8, 14].

3   Biform Theories

The notion of a "biform theory" was introduced in [13] as the basis for FFMM, a Formal Framework for Managing Mathematics. Informally, a biform theory is simultaneously an axiomatic and an algorithmic theory. We present here a formulation of a biform theory that is simpler than the one given in [13].

General Logics. A *general language* is a pair $L = (\mathcal{E}, \mathcal{F})$ where $\mathcal{E}$ is a set of syntactic entities called the *expressions* of L and $\mathcal{F} \subseteq \mathcal{E}$ is a set of expressions called the *formulas* of L. For example, if F is a first order language, then $L_F = (\mathcal{T} \cup \mathcal{F}, \mathcal{F})$ is a general language where $\mathcal{T}$ and $\mathcal{F}$ are the sets of terms and formulas of F, respectively. In the rest of this paper, let $L = (\mathcal{E}, \mathcal{F})$ be a general language.

A *general logic* is a set of general languages with a notion of logical consequence. In the rest of this paper, let $\mathbf{K}$ be a general logic. L is a *language* of $\mathbf{K}$ if it is one of the general languages of $\mathbf{K}$. If L is a language of $\mathbf{K}$ and $\Sigma \cup \{A\}$ is a set of formulas of L, then $\Sigma \models_{\mathbf{K}} A$ means A is a logical consequence of Σ in $\mathbf{K}$. For example, let $\mathbf{FOL}$ be a general logic representation of first order logic such that L is a language of $\mathbf{FOL}$ iff $L = L_F$ for some first order language F and $\Sigma \models_{\mathbf{FOL}} A$ means A is a logical consequence of Σ in first order logic.

An *axiomatic theory* in $\mathbf{K}$ is a pair $T = (L, \Gamma)$ where $L = (\mathcal{E}, \mathcal{F})$ is a language of $\mathbf{K}$ and $\Gamma \subseteq \mathcal{F}$. L is the *language* of T, and Γ is the set of *axioms* of T. A formula A of L is a *logical consequence* of T if $\Gamma \models_{\mathbf{K}} A$.

Transformers. For $n \geq 0$, an *n-ary transformer in L* is a pair $\Pi = (\pi, \hat{\pi})$ where π is a symbol and $\hat{\pi}$ is an algorithm that implements a (possibly partial) function $f_{\hat{\pi}} : \mathcal{E}^n \to \mathcal{E}$. The symbol π serves as a name for the algorithm $\hat{\pi}$. There is no restriction on how the algorithm is presented. For example, it could be a a λ-expression in L or a program written in Lisp or Haskell (or even C or Java).

Let $\mathsf{dom}(\Pi)$ denote the domain of $\hat{\pi}$, i.e., the subset of $\mathcal{E}^n$ on which $f_{\hat{\pi}}$ is defined. Suppose $E_1, \ldots, E_n$ are expressions in $\mathcal{E}$. If $(E_1, \ldots, E_n) \in \mathsf{dom}(\Pi)$, the expression $\pi(E_1, \ldots, E_n)$ denotes the output of $\hat{\pi}$ when given $E_1, \ldots, E_n$ as input, i.e., it denotes $f_{\hat{\pi}}(E_1, \ldots, E_n) \in \mathcal{E}$ (and is thus *defined*). If $(E_1, \ldots, E_n) \notin \mathsf{dom}(\Pi)$, $\pi(E_1, \ldots, E_n)$ does not denote anything (and is thus *undefined*). The expression $\pi(E_1, \ldots, E_n)$ is not required to be in $\mathcal{E}$; it will usually be an expression of the metalanguage of L but not of L itself.

Example 1. Suppose $L_F = \{\mathcal{E}_F, \mathcal{F}_F\}$ is the general language corresponding to a first order language F. Let $\Pi = (\pi, \hat{\pi})$ be a unary transformer in L_F such that:

1. $\pi(E)$ is defined iff $E \in \mathcal{F}_F$.
2. If $\pi(A)$ is defined, it denotes a formula $B \in \mathcal{F}_F$ that is in prenex normal form and is logically equivalent to A.

That is, the algorithm $\hat{\pi}$ transforms any formula of L_F into a logically equivalent formula in prenex normal form. The expression $\pi(E)$ cannot be an expression in L_F (without some mechanism, such as Gödel numbering, for formalising the syntax of L_F in L_F itself).   □

Example 2. Suppose $L_F = \{\mathcal{E}_F, \mathcal{F}_F\}$ is again the general language corresponding to a first order language F. Let $\Pi = (\pi, \hat{\pi})$ be a ternary transformer in L_F such that:

1. $\pi(E_1, E_2, E_3)$ is defined iff E_1 is a term of F, E_2 is a variable of F, and E_3 is a formula of F.
2. If $\pi(t, x, A)$ is defined, it denotes the result of simultaneously substituting t for each free occurrence of x in A.

That is, given t, x, A, the algorithm $\hat{\pi}$ transforms the formula A into the formula $A[x \mapsto t]$. Again the expression $\pi(E_1, E_2, E_3)$ cannot be an expression in $\mathcal{E}_F$. □

Example 3. Let **STT** be a general logic representation of simple type theory. Suppose $T = (L, \Gamma)$ is an axiomatic theory of a complete ordered field in **STT** and that we have defined in T a type real of real numbers and the basic concepts of calculus such as limits, continuity, derivatives, etc. Let $\Pi = (\pi, \hat{\pi})$ be a unary transformer in L such that:

1. $\pi(E)$ is defined iff E is an expression of L of type real $\rightarrow$ real.
2. If $\pi(E)$ is defined, it is an expression of L of type real $\rightarrow$ real that denotes the derivative of the function denoted by E.

That is, $\hat{\pi}$ is an algorithm that differentiates expressions that denote functions on the real numbers.   □

An *algorithmic theory* is a pair $T = (L, \Delta)$ where L is a general language and Δ is a set of transformers in L. L is called the *language* of T, and Δ is the set of *algorithms* of T. For more on transformers, see [12, 13].

Rules. A *rule* in L is a pair $R = (\Pi, M)$ where:

1. $\Pi = (\pi, \hat{\pi})$ is an *n*-ary transformer in L.
2. M is a formula that uses π to relate the values of the inputs to $\hat{\pi}$ to the value of the output of $\hat{\pi}$.

The *transformer* of R, written trans(R), is Π, and the *meaning formula* of R, written mean(R), is M. The meaning formula M, which specifies the semantic relationship between the tuple of inputs and the output of the algorithm $\hat{\pi}$, will usually be an expression of the metalanguage of L but not of L itself. For each *n*-tuple $I = (E_1, \ldots, E_n)$ of inputs to $\hat{\pi}$, we assume that M reduces to a formula M_I of L which is called the *instance* of M with respect to I. An instance of M specifies the relationship between the values of a given tuple of input expressions and the value of the resulting output expression. Let $\text{inst}(R)$ be the set of instances of M. M can often be conveniently expressed as a formula schema.

Example 4. Let $R = (\Pi, M)$ where:

1. $\Pi = (\pi, \hat{\pi})$ is the transformer in L_F given in Example 1.
2. M is the formula schema $A \equiv \pi(A)$ where A is a formula of L_F.

If A is the formula $p(c) \supset \forall x.q(x)$ (where c is a constant) and the result of applying $\hat{\pi}$ to (A) is $\forall x.p(c) \supset q(x)$, then $(p(c) \supset \forall x.q(x)) \equiv (\forall x.p(c) \supset q(x))$ is the instance of M with respect to (A). $\qquad\square$

Example 5. Let $R = (\Pi, M)$ where:

1. $\Pi = (\pi, \hat{\pi})$ is the transformer in L_F given in Example 2.
2. M is the formula schema $(x = t \wedge A) \supset \pi(t, x, A)$ where t is a term, x is a variable, and A is a formula of L_F and t is free for x in A.

If t is a term, x is a variable, and A is $f(x, y) = g(x)$, then $(x = t \wedge f(x, y) = g(x)) \supset f(t, y) = g(t)$ is the instance of M with respect to (t, x, A). $\qquad\square$

Example 6. Let $R = (\Pi, M)$ where:

1. $\Pi = (\pi, \hat{\pi})$ is the transformer in the language L of the theory T given in Example 3.
2. M is the formula schema $\text{derivative}(E) = \pi(E)$ where E is of type real $\to$ real. derivative is an expression of L of type (real $\to$ real) $\to$ (real $\to$ real) that maps a function to its derivative. M thus asserts that the derivative of the function denoted by E is the function denoted by $\pi(E)$.

If E is $\lambda x : \text{real}.x^2$, then $\text{derivative}(\lambda x : \text{real}.x^2) = (\lambda x : \text{real}.2 \cdot x)$ is the instance of M with respect to (E). $\qquad\square$

For the sake of convenience, we will view a formula A of L as a (transformer-less) rule in L and assume that $\text{trans}(A)$ is undefined, $\text{mean}(A) = A$, and $\text{inst}(A) = \{A\}$.

Biform Theories. A *biform theory* in **K** is a pair $T = (L, \Omega)$ where L is a language of **K** and Ω is a set of rules in L. (Ω may include formulas of L viewed as transformer-less rules.) L is the *language* of T, and Ω is the set of *axioms* of T.

T can be viewed as simultaneously both an *axiomatic theory* and an *algorithmic theory*. The axiomatic theory *of* T is the axiomatic theory $T_{\text{axm}} = (L, \Gamma)$ in **K** where $\Gamma = \bigcup_{R \in \Omega} \text{inst}(R)$, while the *algorithmic theory of* T is the algorithmic theory $T_{\text{alg}} = (L, \Delta)$ where $\Delta = \{\text{trans}(R) \mid R \in \Omega \text{ and } \text{trans}(R) \text{ is defined}\}$.

The axioms of T—which are formulas and rules—are the background assumptions of T in an implicit form. The axioms of T_{axm}—which are formulas alone—are the background assumptions of T in an explicit form. A rule R in L is a *logical consequence* of T if, for all formulas $A \in \text{inst}(R)$, A is a logical consequence of T_{axm}. Thus, the axioms of T are trivially logical consequences of T. Notice also that, since we are assuming that the formulas of L are rules in L, every logical consequence of T_{axm} is also a logical consequence of T.

4 Definitions

We now render the problem from Sec 2 precisely by giving the relevant formal definitions. To facilitate the formal specification as biform theories in Sec 5 we painstakingly distinguish between the semantics of the mathematical concepts, the languages necessary to express them, and the purely syntactic expression.

4.1 Semantic Concepts

Definition 1. *A* loop *is a non-empty set G together with a binary operation $\circ$ and a distinguished element $e \in G$ such that*

$$\forall a, b \in G.\, (\exists x \in G.x \circ a = b) \wedge (\exists y \in G.a \circ y = b) \text{ and } \forall x \in G.x \circ e = e \circ x = x.$$

Definition 2. *Let G be a loop with binary operation $\circ$, then we can define two additional binary operations $/$ and $\backslash$ by*
1. *$\forall x, y \in G.x \circ (x \backslash y) = y$ and $\forall x, y \in G.x \backslash (x \circ y) = y$*
2. *$\forall x, y \in G.(y/x) \circ x = y$ and $\forall x, y \in G.(y \circ x)/x = y$*

Definition 3. *Let G, H be two loops with respective binary operation $\circ$ and $*$. We say G is* isotopic *to H if there are bijective mappings α, β, γ from G to H such that for all $x, y \in G$, $\alpha(x) * \beta(y) = \gamma(x \circ y)$ holds.*

Definition 4. *Let G be a loop and let P be a property that holds for G. We call P an* isotopy invariant *if P is preserved under the isotopy mapping (i.e., if P holds for G than it also holds for all its isotopes).*

Definition 5. *Let G be a loop with binary operation $\circ$. w is a* word *of G if it is a combination of elements of G with respect to the loop operation $\circ$. Let w_1, w_2 be two words in G, then $w_1 = w_2$ defines a* loop identity *if it holds for all elements of G.*

Definition 6. *Let G be a loop with binary operations $\circ, \backslash, /$. Given a word w in G, we define its corresponding* derived word *$\overline{w}$ by*
1. *if $w = x, x \in G$, then $\overline{w} = x$;*
2. *if $w = u \circ v$, then $\overline{w} = (\overline{u}/y) \circ (z \backslash \overline{v})$, where $u, v, y, z \in G$.*
Given a loop identity $w_1 = w_2$ of G, we define its corresponding derived identity *as $\overline{w_1} = \overline{w_2}$.*

Definition 7. *Let G be a loop with binary operations $\circ, \backslash, /$ and $\overline{w_1} = \overline{w_2}$ be a derived identity in G. Then $\overline{w_1} = \overline{w_2}$ is a* universal identity *if it is an isotopy invariant.*

4.2 Languages

In order to express the definitions of the syntactic concepts we give the necessary languages by stepwise extending the basic language of first order logic with equality (**FOL+EQ**). This will later enable us to define biform theories with minimal languages. Generally, we need a fair bit more machinery to define our various meaning functions; thus we will freely use Simple Type Theory (**STT**) [7, 10] as our general environment.

As a general typographical convention, we will underline all the symbols when we refer to the syntactic version of a symbol we already have in our semantics. We will not however underline variables, to ease (somewhat) the readability of the results. We assume that the reader is proficient enough in **FOL+EQ** and **STT** so that we do not need to repeat their syntactic definition here.

We first extend **FOL+EQ** to the language **Loop** by adding a binary function $\underline{\circ}$ and a constant $\underline{e}$. In a next step we add the two binary operations $\underline{\backslash}$ and $\underline{/}$ to **Loop** to obtain **Loop'**. In fact, in **Loop'**, we need *two* loops, so we in fact add $e_1, \circ_1, \backslash_1, /_1$ and $e_2, \circ_2, \backslash_2, /_2$ to **Loop'**. While these languages are sufficient to express the syntactic objects manipulated during the computation, we also need to express the meaning formulas for transformers in the language of a biform theory. This will necessarily be (at least) a second order logic, as we are quantifying over loops. It also is much easier to specify if we have access to a bit more machinery, such as lambda expressions and unique choice. We therefore use **STT**, as a superset of **Loop'**, for this purpose. Thus altogether we have the following sequence of languages: **FOL+EQ** $\subset$ **Loop** $\subset$ **Loop'** $\subset$ **STT**

4.3   Syntactic Concepts

Let $\mathcal{V}$ be a set of variables, with $x_i \in \mathcal{V}$. Let y, z be two new symbols not in $\mathcal{V}$ and let $\mathcal{V}' = \mathcal{V} \cup \{z, y\}$.

Word ::= $x_i \mid \underline{e} \mid$ Word $\underline{\circ}$ Word
Identity ::= $\underline{\forall} x.$Word $\underline{=}$ Word
Word' ::= $x_i \mid \underline{e} \mid$ (Word' $\underline{\backslash}$ y) $\underline{\circ}$ (z $\underline{/}$ Word')
DerivedIdentity ::= $\underline{\forall} x.$Word' $\underline{=}$ Word'

A Word is a word in the language of loops, composed of variables, an identity element $\underline{e}$ and an operation $- \underline{\circ} -$, where all variables of $\mathcal{V}$ are understood to be universally quantified. A Word' is a word in the extended language of loops, composed of variables, and identity element, operation and two new operators, $/$ and $\backslash$, where again variables $(\mathcal{V}')$ are universally quantified. Then Identity and DerivedIdentity are identities over the respective languages.

We also need syntactic representations of various axioms. For example, we have that

CircAxm $\hat{=}$ $\underline{\forall} a, b.$ $(\underline{\exists} x.x \underline{\circ} a \underline{=} b) \underline{\wedge} (\underline{\exists} y.x \underline{\circ} y \underline{=} b)$ in **Loop**.
IdAxm $\hat{=}$ $(\underline{\forall} x.x \underline{\circ} e \underline{=} x) \underline{\wedge} (\underline{\forall} x.e \underline{\circ} x \underline{=} x)$ in **Loop**.
DivLAxm $\hat{=}$ $(\underline{\forall} x, y.x \underline{\circ} (x \underline{\backslash} y) \underline{=} y) \underline{\wedge} (\underline{\forall} x, y.x \underline{\backslash} (x \underline{\circ} y) \underline{=} y)$ in **Loop'**
DivRAxm $\hat{=}$ $(\underline{\forall} x, y.(y \underline{/} x) \underline{\circ} x \underline{=} y) \underline{\wedge} (\underline{\forall} x, y.(y \underline{\circ} x) \underline{/} x \underline{=} y)$ in **Loop'**.

These express respectively the axiom for $\circ$, the identity e, the left division $\backslash$ and the right division $/$.

Actually, to describe the full semantics, we need *two* copies of the above, for two different loops, whose components we'll naturally denote $(\underline{e_1}, \underline{\circ_1}, \underline{\backslash_1}, \underline{/_1})$

and $(e_2, \circ_2, \backslash_2, /_2)$ respectively. Since we are in **STT**, we could have easily have written the above as functions from syntax to syntax, but that would have made the presentation too opaque. We also need to represent a *finite domain* syntactically. But this essentially amounts to creating n unique names.

We can then continue thus, for all the various concepts defined semantically in the previous section, which we use syntactically later (like bijective). We should also define the full syntax for a language of proofs (as the language for the output of one of our intermediate transformers below), but since the actual implementation ignores these proofs, it suffices to posit that this language exists.

5 Specification

We can view the generation of isotopy invariants from derived identities and their selection as possible discriminants for loops of a given size n as a sequence of single computational processes as displayed in Fig. 2. Each process accomplishes a different function in the overall computational process, e.g., is a source of equations, transforms expressions, or filters with respect to different criteria.

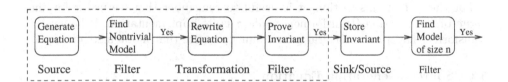

Fig. 2. Abstract view of the process

For each module, we have a background (biform) theory that expresses the language and axioms necessary to describe the rules (and thus the inputs and outputs) encapsulated in that module. Using appropriate translations and interpretations, we can set up a communication channel (a connection in the language of [6]). We give detailed formal specifications for each process and their communications.

We can also give a specification for the global problem for generating universal identities in general, which corresponds to the dashed box in Fig. 2. This results in a transformer that operates on a more abstract level. We start by giving the formal specification for it, before going into the details of the component processes.

Concretely, for the remainder we let **K** be the general logic based on **STT** as general language. We then define the biform theory for the overall process consisting as axiomatic theory $T_{\mathrm{axm}} = (L, \Gamma)$ and algorithmic theory $T_{\mathrm{alg}} = (L, \Delta)$, where $L = \mathbf{STT}$, Γ contains all axioms defined in 4.1, and Δ contains one transformer $\Pi = (\pi, \hat{\pi})$.

1. $\pi() = $ `Generate_Universal_Identity()` is always defined, as it takes no input. It returns a set of syntactical structures DerivedIdentity, as defined in Sec. 4.3

2. $\hat{\pi}$ is the implementation of `Generate_Universal_Identity` as presented in [14].
3. M is the meaning formula

$\forall U \in$ Generate_Universal_Identity$().U \in$ DerivedIdentity

$\wedge Subst(\mathsf{CircAxm}, \underline{e}_1, \underline{\circ}_1, \underline{\backslash}_1, \underline{/}_1) \wedge Subst(\mathsf{IdAxm}, \underline{e}_1, \underline{\circ}_1, \underline{\backslash}_1, \underline{/}_1)$

$\wedge Subst(\mathsf{DivLAxm}, \underline{e}_1, \underline{\circ}_1, \underline{\backslash}_1, \underline{/}_1) \wedge Subst(\mathsf{DivRAxm}, \underline{e}_1, \underline{\circ}_1, \underline{\backslash}_1, \underline{/}_1)$

$\wedge Subst(\mathsf{CircAxm}, \underline{e}_2, \underline{\circ}_2, \underline{\backslash}_2, \underline{/}_2) \wedge Subst(\mathsf{IdAxm}, \underline{e}_2, \underline{\circ}_2, \underline{\backslash}_2, \underline{/}_2)$

$\wedge Subst(\mathsf{DivLAxm}, \underline{e}_2, \underline{\circ}_2, \underline{\backslash}_2, \underline{/}_2) \wedge Subst(\mathsf{DivRAxm}, \underline{e}_2, \underline{\circ}_2, \underline{\backslash}_2, \underline{/}_2)$

$\wedge Subst(U, \underline{e}_1, \underline{\circ}_1, \underline{\backslash}_1, \underline{/}_1) \wedge \exists \alpha, \beta, \gamma.\mathsf{bijective}(\alpha) \wedge \mathsf{bijective}(\beta) \wedge \mathsf{bijective}(\gamma)$

$\wedge \forall x, y.\alpha(x)\underline{\circ}_2\beta(y) = \gamma(x\underline{\circ}_1 y) \supset Subst(U, \underline{e}_2, \underline{\circ}_2, \underline{\backslash}_2, \underline{/}_2).$

In the formula M the predicate $Subst$ is shorthand for a syntactic replacement of the symbols $\underline{e}, \underline{\circ}, , \underline{\backslash} \underline{/}$ in U by their indexed counterpart to obtain two copies of loops. This can be implemented similar to the transformer for substitution in Example 2. However, we omit this level of detail here. Observe also that bijective is shorthand for the formulas representing the bijectivity property.

For the rest of this section, it is useful to remember that we use π for the *name* of the function (transformer) while $\hat{\pi}$ is its implementation. Whenever a transformer has a more natural name than π, we use that name instead.

5.1   Source: Generating Equation

The first step generates a set of identities. The language of the biform theory is $L =$ **Loop** and the axiomatic theory is $T_{\mathrm{axm}} = (L, \{\})$. We do not need any axioms as the computation of this module is only a generation of constructs in our language and, therefore, not based on any logical consequences. In $T_{\mathrm{alg}} = (L, \Delta)$, Δ contains one transformer $\Pi = (\pi, \hat{\pi})$.

1. $\pi() =$ `Generate_Identity`$()$ is always defined, as it takes no input. It returns a set of syntactical structures Identity.
2. $\hat{\pi}$ is the implementation of `Generate_Identity` that corresponds to the grammar given for Identity in Sec. 4.3.
3. M is $\forall i \in$ Generate_Identity.$i \in$ Identity.

Here M simply states that the elements of the generated set are indeed members of the syntactic class of identities.

5.2   Filter: Find Non-trivial Model

The computation defined in this module is more complicated to formalise since we take a set of identities as input, for each one we try to generate a model of some specified size, and only output those identities for which we could successfully generate a model. Its formalisation can be achieved by using one transformer in the context of the other. We have a transformer modelling the general computation of a model generator that operates in the context of a transformer for the particular generation of models of loops satisfying given identities. In addition this latter transformer performs the actual filtering by discarding those identities for which no models exist.

The language of the biform theory is Loop and $T_{\text{axm}} = (L, \Gamma)$, where Γ contains the axioms given in definition 1. The algorithmic theory T_{alg} contains two transformers: $\Pi_1 = (\pi_1, \hat{\pi}_1)$ specifying the computation of the model generator and $\Pi_2 = (\pi_2, \hat{\pi}_2)$ specifying the filtering operation carried out on a set of identities.

1. $\pi_1(S, D) = \texttt{Generate\_Model}(S, D)$ is defined iff S is a set of syntactic formulas and D is a syntactic representation of the domain. It returns set of syntactic representation of models if one exists, i.e., for each constant element in the input a relation in D satisfying E.
2. $\hat{\pi}_1$ is the model generation process performed by some integrated model generator.
3. M is $\forall m \in \texttt{Generate\_Model}(S, D).(m \models S)$. Note that m is a model with interpretations of the symbols of S given as functions represented by sets of ordered pairs.

Given the transformer for general model generation, we now define the actual filter operation as:

1. $\pi_2(E, n) = \texttt{Filter\_Identities}(E, n)$ is defined iff E is a set of Identity n is a positive integer. It also returns a set of syntactical structures Identity.
2. $\hat{\pi}_2$ is the implementation of $\texttt{Filter\_Identities}$ that generates the syntactic representation of the domain of size n given to π_1 and applies π_1 to each element in E.
3. M is $\forall i \in \texttt{Filter\_Identities}(E, n).\texttt{Generate\_Model}(\{i\} \cup \Gamma, Dom(n)) \neq \emptyset$.

Observe that in M above $Dom(n)$ is a schema specifying the set of domain elements computed in $\texttt{Filter\_Identities}$, which could itself be specified using a transformer. Observe also that we have slightly simplified the formalisation, in that it only takes domains of one size, rather than a range of domain sizes.

We note that both transformers are very general and do not depend on the particular biform theory they live in. In particular $\texttt{Generate\_Model}$ works on any first order language regardless of the language of the biform theory. Here the theory serves as a way to parameterise the input to the transformer. $\texttt{Filter\_Identities}$ performs filtering with respect to model existence. Likewise it could filter with respect to non-existence of models or return computed models by interpreting them in the biform theory. Note that $\texttt{Filter\_Identities}$ is a total set-to-set function, and the underlying implementation is total as well; even $\texttt{Generate\_Model}$ is a total function (it always terminates), however it may return an *empty* result[1].

---

[1] In Haskell, we could say that $\texttt{Generate\_Model}$ belongs in the Maybe Monad.

5.3   Transformation: Rewrite Equations

This step rewrites loop identities to derived identities. The language of the biform theory is $L = \textbf{Loop}'$ and the axiomatic theory is $T_{\text{axm}} = (L, \Gamma)$ where Γ contains the axioms given in Definitions 1, 2, and 6. In $T_{\text{alg}} = (L, \Delta)$, Δ contains one transformer $\Pi = (\pi, \hat{\pi})$.

1. $\pi(E) = \texttt{Generate\_Derived\_Identity}(E)$ is defined iff E is a set of Identity. It returns a set of syntactical structures DerivedIdentity.
2. $\hat{\pi}$ is the implementation of the rewrite system given in Def. 6 that performs syntactic rewriting of an Identity into a DerivedIdentity.
3. M is $\forall i \in E. \exists! d \in \texttt{Generate\_Derived\_Identity}(E). \Gamma \models i \equiv d$.

5.4   Filter: Prove Invariant

Similar to the specification of the filter in Sec. 5.2 this process requires the combination of two transformers. One for the general proving process and one that performs the actual filtering.

We define the biform theory for $L = \textbf{STT}$, with $\textbf{Loop}'$ as a distinct sublanguage, $T_{\text{axm}} = (L, \Gamma)$, where Γ contains all axioms defined in 4.1, and T_{alg} containing two transformers: $\Pi_1 = (\pi_1, \hat{\pi}_1)$ specifying the computation of the theorem prover and $\Pi_2 = (\pi_2, \hat{\pi}_2)$ specifying the filtering operation.

1. $\pi_1(A, C) = \texttt{Prove}(A, C)$ is defined iff A is a set of syntactic formulas in $\textbf{Loop}'$ and C is a formula in $\textbf{Loop}'$. It returns a syntactic representation of a proof if one exists.
2. $\hat{\pi}_2$ is the theorem proving process performed by some automated theorem prover. It takes the elements of A as assumptions and C as a conclusion.
3. M is $Proves(\texttt{Prove}(A, C), A \vdash C)$

Here the predicate $Proves$ checks the correctness of the derivation. The formalisation of this predicate depends on the calculus of the integrated prover and is generally very lengthy to formalise. Since we are not interested in examining proofs but only their existence at this point, we do not go into any detail here.

Given the transformer for theorem proving, we now define the actual filter operation as:

1. $\pi_2(E) = \texttt{Filter\_Derived\_Identities}(E)$ is defined iff E is a set of expressions of the form DerivedIdentity. It also returns a set of DerivedIdentity.
2. $\hat{\pi}_2$ is the implementation of $\texttt{Filter\_Derived\_Identities}$ that generates an assumption set A from the syntactic representation of the elements of Γ computes for each element i of E $\texttt{Prove}(A, i)$. It returns a set of all elements in E for which a proof exists.
3. The meaning function M is of the same form as the meaning function of the $\texttt{Generate\_Universal\_Identity}$ transformer, with the exception that we now universally quantify over $U \in \texttt{Filter\_Derived\_Identities}(E)$.

Observe that the *Subst* is the symbol substitution transformer already used at the beginning of this section.

The result of this last transformer, the set of syntactic universal identities, can now be stored for later use. Choosing from this store is again achieved with a filter that uses model generation, which is similar to the transformer `Filter_Identities` above. Due to lack of space we omit the detailed formalisation of these processes here.

6   Conclusions

We have presented a first step towards a rational reconstruction of the classification procedure for finite algebras in [8, 14]. The use of biform theories enables us to express both axiomatic and algorithmic properties of the procedure while clearly distinguishing syntactic, semantic, and algorithmic levels. Although the generation of universal identities is only a small sub-process of the overall classification, its formalisation is surprisingly involved.

This is not really due to the design and current implementation of the algorithm, but rather because some of the operations necessarily intermix syntax and semantics. Keeping straight what has to be in **FOL+EQ**, what is safely in **STT**, and what is in fact in the meta-language is very difficult. For example when generating models with respect to a particular domain, the result is a semantic entity. Nevertheless, models have to be interpreted as syntax not only to express the meaning function but also in case the models are used in further computations and syntactic manipulations. This also has the effect that the given domain elements have to be incorporated into the language of the biform theory, which is currently not possible and subject to future work.

On the other hand, communication between the components is very simple, since it is all done via **FOL+EQ** or conservative extensions such as **Loop** and **Loop'**. The necessary interpretations [6] between these theories are straightforward, unlike the more general case where communication occurs amongst more disparate logics.

The current formalisation already exposes some generalisations. In particular many of the sub-processes can be expressed as a mixture of computation and filtering, where the computation is often independent of the particular theory. It also becomes apparent what is actual input and what has to be specified in the background theory of a process. This information could be exploited to design an environment enabling mathematical experimentation by combining systems on a high level, such that it is only necessary to specify input, output and parts of the background theory without interaction on the actual logical level.

Our formalisation is certainly not the only possible approach to reconstruct the generation process. Indeed we could view the entire process as a sequence of recursive generators and filters. E.g., the first three boxes in Fig. 2 could be combined in a single transformer that acts as a generator for the next filter. Comparing different specifications and combinations of transformers for the same process could expose possible optimisation opportunities for the overall process.

As we have already discussed, biform theories—particularly the meaning formulas of rules—are difficult to formalise in a traditional logic without the means to reason about syntax. The paper [11] illustrates how biform theories can be formalised in Chiron, a derivative of von-Neumann-Bernays-Gödel (NBG) set theory that directly supports reasoning about the syntax of expressions. After an implementation of Chiron is produced, we would like to use Chiron to fully formalize the work presented in this paper.

References

1. Armando, A., Ranise, S.: From integrated reasoning specialists to "plug-and-play" reasoning components. In: Calmet and Plaza [5], pp. 42–54
2. Armando, A., Zini, D.: Towards Interoperable Mechanized Reasoning Systems: the Logic Broker Architecture. In: AI*IA Notizie Anno XIII (2000) Parma, Italy, vol. 3, pp. 70–75 (2000)
3. Bertoli, P.G., Calmet, J., Giunchiglia, F., Homann, K.: Specification and Integration of Theorem Provers and Computer Algebra Systems. In: Calmet and Plaza [5], pp. 94–106
4. Buchberger, B., Craciun, A., Jebelean, T., Kovacs, L., Kutsia, T., Nakagawa, K., Piroi, F., Popov, N., Robu, J., Rosenkranz, M., Windsteiger, W.: Theorema: Towards computer-aided mathematical theory exploration. Journal of Applied Logic 4, 470–504 (2006)
5. Calmet, J., Plaza, J. (eds.): AISC 1998. LNCS (LNAI), vol. 1476. Springer, Heidelberg (1998)
6. Carette, J., Farmer, W.M., Wajs, J.: Trustable communication between mathematical systems. In: Proc. of Calculemus 2003, pp. 58–68 (2003)
7. Church, A.: A Formulation of the Simple Theory of Types. Journal of Symbolic Logic 5, 56–68 (1940)
8. Colton, S., Meier, A., Sorge, V., McCasland, R.: Automatic generation of classification theorems for finite algebras. In: Basin, D., Rusinowitch, M. (eds.) IJCAR 2004. LNCS (LNAI), vol. 3097, pp. 400–414. Springer, Heidelberg (2004)
9. Falconer, E.: Isotopy Invariants in Quasigroups. Transactions of the American Mathematical Society 151(2), 511–526 (1970)
10. Farmer, W.M.: STMM: A set theory for mechanized mathematics. J. Autom. Reasoning 26(3), 269–289 (2001)
11. Farmer, W.M.: Biform theories in Chiron. In: Kauers, M., Windsteiger, W. (eds.) Towards Mechanized Mathematical Assistants. LNCS (LNAI), vol. 4573, Springer, Heidelberg (2007)
12. Farmer, W.M., von Mohrenschildt, M.: Transformers for symbolic computation and formal deduction. In: Colton, S., Martin, U., Sorge, V. (eds.) CADE-17 Workshop on the Role of Automated Deduction in Mathematics, pp. 36–45 (2000)
13. Farmer, W.M., von Mohrenschildt, M.: An overview of a formal framework for managing mathematics. Annals of Mathematics and Artificial Intelligence 38, 165–191 (2003)
14. Sorge, V., Meier, A., McCasland, R., Colton, S.: Automatic construction and verification of isotopy invariants. In: Furbach, U., Shankar, N. (eds.) IJCAR 2006. LNCS (LNAI), vol. 4130, pp. 36–51. Springer, Heidelberg (2006)
15. Zimmer, J., Kohlhase, M.: System Description: The MathWeb Software Bus for Distributed Mathematical Reasoning. In: Voronkov, A. (ed.) Automated Deduction - CADE-18. LNCS (LNAI), vol. 2392, Springer, Heidelberg (2002)

Context Aware Calculation and Deduction
Ring Equalities Via Gröbner Bases in Isabelle

Amine Chaieb and Makarius Wenzel*

Technische Universität München
Institut für Informatik, Boltzmannstraße 3, 85748 Garching, Germany

Abstract. We address some aspects of a system architecture for mathematical assistants that integrates calculations and deductions by common infrastructure within the Isabelle theorem proving environment. Here calculations may refer to arbitrary extra-logical mechanisms, operating on the syntactic structure of logical statements. Deductions are devoid of any computational content, but driven by procedures external to the logic, following to the traditional "LCF system approach". The latter is extended towards explicit dependency on abstract theory contexts, with separate mechanisms to interpret both logical and extra-logical content uniformly. Thus we are able to implement proof methods that operate on abstract theories and a range of particular theory interpretations. Our approach is demonstrated in Isabelle/HOL by a proof-procedure for generic ring equalities via Gröbner Bases.

1 Introduction

The requirements for mathematical assistants come in various and sometimes conflicting flavors: efficient calculations (computer algebra systems have excelled here), deduction with a precise notion of logical correctness (the theorem proving approach), and ability to work relative to abstract theory contexts as in traditional mathematics. In the present paper we address the issue of combining calculations and deductions with theory abstraction and interpretation in the Isabelle theorem proving environment [17]. Since calculations are external to the logic, we may employ arbitrary programming techniques, with full access to the syntactic structure of statements. Deductions are driven by such external procedures, to replay proofs via primitive inferences, following the traditional "LCF system approach" of correctness-by-construction introduced by Milner [10].

Proof methods are typically inhomogeneous because of a natural division of *computing* results vs. *checking* witnesses within the logic. From the algorithmic perspective it is usually harder to produce a result than to check it, e.g. consider long division of polynomials. From the logical point of view, results may emerge spontaneously by an external "oracle", but checking requires tedious inferences.

Implementing proof tools by orchestrating non-trivial calculations and deductions is a difficult task. In providing a link to abstract theory mechanisms we

---

* Supported by BMBF project "Verisoft".

M. Kauers et al. (Eds.): MKM/Calculemus 2007, LNAI 4573, pp. 27–39, 2007.

contribute to systematic development of advanced methods, being organized according to the logical context required at each stage. Thus we transfer principles of modular program and theory development towards proof procedures.

Related work. There are mainly two other approaches in implementing proof procedures in theorem proving environments: *interpretation* and *assimilation*.

Interpretation means that the proof system provides a specific language for describing procedures, being interpreted at run-time and reduced to basic inferences eventually. This language is usually limited to common expressions for combining terms, theorems, tactics etc. Even if it is made computationally complete, there is normally no access to arbitrary libraries and components of the underlying implementation platform. For example, see *Ltac* in Coq [3], or the language of *macetes* in IMPS [9]. The latter is also notable due to its integration with the concept of "little theories", by means of *transportable macetes*.

Assimilation means that powerful computational concepts (such as algebraic datatypes, recursive functions) are integrated into the very logical environment itself. Calculations are then native reductions of the underlying calculus, as in the type-theory of Coq [3], which augments plain syntactic $\alpha\beta$-reduction by $\delta\iota$-reduction (expansion of simple and inductive definitions). Thus calculations may be performed within the main calculus, while proof terms are reduced to mere reflexivity according to the "Poincaré principle". Further advanced implementation techniques for functional programming language are assimilated into the Coq inference kernel, in order to achieve reasonable run-time performance.

This unified approach of calculations and deductions (also called "computational reflection") looks very attractive at first sight: various proof tools have been implemented like this [11, 5, 6]. On the other hand, there is some extra tedium involved, because all manipulations need to be formalized within the logic. First this requires facilities to *quote* (or "reify") logical statements in order to access their syntactic structure. Then fully formal correctness proofs need to be provided, to get manipulated statements interpreted back into the logic.

An important point of the "LCF approach" is to allow arbitrary manipulations without requiring quoting or formally proven programs — correctness is achieved by checking the logically relevant bits at run-time. There are numerous procedures implemented in HOL Light [13]; see also [14] for particular applications involving Computer Algebra Systems as search oracles for HOL proofs. Recent interesting work following this approach include [8, 7].

Basic notation. Keywords like **theorem, locale, interpretation** etc. refer to Isabelle theory elements. Types and terms are embedded here as smaller units, and notation approximates mathematical conventions despite a bias towards λ-calculus. Types τ consist of type variables α, type constants like *nat* or *int*, or function spaces $\tau_1 \Rightarrow \tau_2$. Terms t consist of variables x, constants c, abstractions $\lambda x.\ t$, or applications $t_1\ t_2$. Propositions are terms of type *prop*, with the outermost logical rule structure represented via implication $A \Longrightarrow B$, or quantification $\bigwedge x.\ B$, where outermost quantifiers are implicit.

Mathematical functions $A \to B$ are approximated by computable functions implemented in SML. We use "curried" notation uniformly for iterated applications: $f\ x\ y$ for $f : A \to B \to C$ and $x : A$ and $y : B$. This is particularly important for *partial application*, such as $f\ x : B \to C$, which requires another argument $y : B$ to continue evaluation. Despite some similarities in notation, a plain mathematical function $(x \mapsto f\ x) : A \to B$ should not be confused with a symbolic term $(\lambda x.\ t\ x) : \alpha \Rightarrow \beta$ within the logical framework!

Overview. In §2 we present a simple example of calculating with (semi)ring numerals. In §3 we introduce the general principles of contexts and context declarations in Isabelle. In §4 we review some aspects of Isabelle's Simplifier that use the context infrastructure to implement facilities shown in the example. In §5 we report on a larger application, a proof method for ring equalities via Gröbner Bases.

2 Example: Calculations on Generic Numerals

Our main application (cf. §5) operates on general semiring and ring structures in non-trivial ways. In order to illustrate the integration of proof methods with abstract theory mechanisms of Isabelle, we merely consider the trivial sub-problem of representing constants (abstract numerals) adequately, such that basic calculations (addition, subtraction etc.) can be performed efficiently.

We shall be using Isabelle locales [15, 1, 2] which provide a general concept for abstracting over operations with certain properties. Locales are similar to the "little theories" of IMPS [9], but based on pure predicate definitions together with some infrastructure to manage Isabelle/Isar proof contexts [20].

Existing calculational tools in Isabelle [18] use the more restricted mechanism of axiomatic type classes [19], where the only parameter is the underlying type, but the signature is *fixed*. According to [12], type-classes may be understood as particular locale interpretations, where the variable parameters are replaced by polymorphic constants — fixing everything except the underlying type. Thus our general locale-based proof methods will be able to cover type-classes as well.

2.1 Semirings

A *semiring* provides operations $\oplus$, $\odot$, 0, 1 over type α, with the usual laws for associativity, commutativity, and distributivity etc.

locale *semiring* =
  fixes *add* :: $\alpha \Rightarrow \alpha \Rightarrow \alpha$ (**infixl** $\oplus$ 65)
  and *mul* :: $\alpha \Rightarrow \alpha \Rightarrow \alpha$ (**infixl** $\odot$ 70)
  and *zero* :: α (0) **and** *one* :: α (1)
  assumes *add-a*: $(x \oplus y) \oplus z = x \oplus (y \oplus z)$
  and *add-c*: $x \oplus y = y \oplus x$
  and *add-0*: $0 \oplus x = x$ **and** ...

This locale definition provides a context with hypothetical parameters and assumptions. Results within this abstract theory are specified using the notation "(**in** *semiring*)".

We single out canonical *semiring* expressions as "constants", corresponding to non-negative natural numbers. To achieve a reasonably efficient representation, we postulate another operation *bit* for shifting a least-significant 0 or 1:

locale *semiring-bin* = *semiring* +
 fixes *bit* :: $\alpha \Rightarrow bool \Rightarrow \alpha$
 assumes *bit-def*: *bit x b* = $x \oplus x \oplus$ (*if b then* 1 *else* 0)

We can now represent numerals compactly using binary representation according to the recursive definition *numeral* = 0 | 1 | *bit numeral False* | *bit numeral True* (excluding leading zeros). E.g. 6 is represented as binary *bit* (*bit* 1 *True*) *False* instead of unary $1 \oplus 1 \oplus 1 \oplus 1 \oplus 1 \oplus 1$. The syntax machinery of Isabelle can be modified to support human-readable (decimal) notation; we shall use #*digits* for this purpose. Then the term #6 will stand for the previous binary expression.

Symbolic calculation on numerals means to manipulate canonical expressions such that proven equalities emerge in the end. For example, addition is easily implemented as a strongly-normalizing term rewrite system:

theorem (**in** *semiring-bin*) *binary-add*:
 $x \oplus 0 = x$
 $0 \oplus x = x$
 $1 \oplus 1 = bit\ 1\ False$
 bit x False $\oplus 1 = bit\ x\ True$
 bit x True $\oplus 1 = bit\ (add\ x\ 1)\ False$
 $1 \oplus bit\ x\ False = bit\ x\ True$
 $1 \oplus bit\ x\ True = bit\ (add\ x\ 1)\ False$
 bit x False $\oplus bit\ y\ False = bit\ (x \oplus y)\ False$
 bit x False $\oplus bit\ y\ True = bit\ (x \oplus y)\ True$
 bit x True $\oplus bit\ y\ False = bit\ (x \oplus y)\ True$
 bit x True $\oplus bit\ y\ True = bit\ (x \oplus y \oplus 1)\ False$

Isabelle's Simplifier depends on rewrite rules declared in the context as [*simp*]. The proof method *simp* derives equalities according to a bottom-up strategy, effectively simulating strict functional evaluation with proven results $t = t'$.

declare (**in** *semiring-bin*) *binary-add* [*simp*]

We can now prove equations involving addition of abstract numerals, e.g.:

theorem (**in** *semiring-bin*) #48734 $\oplus$ #3762039758274 = #3762039807008
 by *simp*

The same works for various interpretations of the abstract theory context, e.g. for type *nat* with the usual operations +, *, 0, 1, and *bit-nat* being defined separately:

definition *bit-nat n b* = $n + n +$ (*if b then* (1::*nat*) *else* (0::*nat*))
interpretation *semiring-bin* [(*op* +) (*op* *) 0 1 *bit-nat*] ⟨*proof*⟩

Now the background context contains corresponding instances of the *semi-ring* locale content (e.g. for type *nat*), including *binary-add* with the [*simp*] declaration. This allows to calculate on interpreted *nat* numerals as follows:

theorem #48734 + #3762039758274 = #3762039807008
by *simp*

Multiplication on numerals is handled by the very same technique.

2.2   Ring Numerals

The subsequent locale *ring-bin* extends semirings with subtraction and negation. Our particular axioms are not an abstract algebraic characterization, but express the requirements of the proof method introduced later.

locale *ring-bin* = *semiring-bin* +
  fixes *sub* :: $\alpha \Rightarrow \alpha \Rightarrow \alpha$ (**infixl** $\ominus$ 65)
    and *neg* :: $\alpha \Rightarrow \alpha$ ($\ominus$ - [81] 80)
  assumes *neg-mul*: ($\ominus$ 1) $\odot$ x = $\ominus$ x
    and *sub-add* [*simp*]: $x \ominus y = x \oplus (\ominus y)$
    and *add-neg*: $z \oplus y = x \Longrightarrow x \oplus (\ominus y) = z$

Canonical expressions for signed numerals are defined as *numeral* | $\ominus$ *numeral* (excluding negated zeros). For symbolic computation, subtraction is reduced to negation and addition immediately, and negations are normalized as follows:

theorem (**in** *ring-bin*) *neg-norm* [*simp*]:
  $\ominus$ 0 = 0
  $\ominus$ ($\ominus$ x) = x
  ($\ominus$ x) $\oplus$ ($\ominus$ y) = $\ominus$ ($x \oplus y$)

This reduces signed addition to $x \oplus (\ominus y)$ or $(\ominus y) \oplus x$, where x and y are unsigned *semiring* numerals. The result is either some z or $\ominus z$, depending on the sign of the difference of x and y. These cases are covered by *diff-rules*:

theorem (**in** *ring-bin*) *diff-rules*:
  $z \oplus x = y \Longrightarrow x \oplus (\ominus y) = \ominus z$
  $z \oplus x = y \Longrightarrow (\ominus y) \oplus x = \ominus z$
  $z \oplus y = x \Longrightarrow x \oplus (\ominus y) = z$
  $z \oplus y = x \Longrightarrow (\ominus y) \oplus x = z$

Here z needs to be produced separately in order to apply these rules effectively. Checking the preconditions afterwards works by term rewriting, but computing z the same way would be slightly awkward (requiring auxiliary constructions within the logical context, like twos-complement binary numerals). Instead we shall now use the existing programming environment of SML to work out signed addition (including negation and subtraction). Converting symbolic expressions back and forth to SML integers is straight-forward, by providing functions *dest-binary* and *mk-binary*. The remaining task is to test the result and instruct the logical engine to apply an appropriate *diff-rules* case.

The **simproc-setup** declaration below augments the Simplifier context by a function that proves rewrite rules on the fly (depending on the current redex):

```
simproc-setup (in ring-bin) binary-sub (x ⊕ (⊖ y) | (⊖ y) ⊕ x) =
  (fn φ ↦
    let
      val [rule₁, rule₂, rule₃, rule₄] = φ diff-rules
      val int-of = dest-binary φ
      val of-int = mk-binary φ
      val add = φ (op ⊕)
      val add-conv = Simplifier.rewrite (φ binary-add)
    in
      fn (@{term add t u}) ↦
        let
          val x = int-of t
          val y = int-of u
          val z = x + y
          val v = of-int (abs z)
          val (w, rule) =
            if z < 0 then if y < 0 then (t, rule₁) else (u, rule₂)
            else if y < 0 then (operand-of u, rule₃) else (operand-of t, rule₄)
        in rule OF [add-conv (@{term add v w})] end
  end)
```

Everything depends on a morphism φ that provides the view of abstract entities from *ring-bin* in the particular application context. In the first stage, the procedure applies φ to all required logical entities of *ring-bin*, notably terms and theorems. In the second stage it operates on the particular problem, accepting a term *add t u* and producing a proven result involving another term *add v w*.

The main calculation happens in $z = x + y$, which re-uses the existing implementation of addition on unbounded integers in SML. In more complex applications we could just as well refer to a suitable library or communicate with an external CA system — the full SML programming environment is accessible here. The final deduction is performed in "*rule OF* [*add-conv* ...]", where the conversion function turns the term *add v w* into a proven theorem *add v w = result* (using the Simplifier internally analogous to §2.1); this equality is resolved with the precondition of our selected *rule*.

With the above simproc installed we may now establish equalities for abstract and concrete numerals:

theorem (in *ring-bin*) #48734 ⊖ #3762039758274 = ⊖ #3762039709540
  by *simp*

definition *bit-int i b* = $i + i + (if\ b\ then\ (1::int)\ else\ (0::int))$
interpretation *ring-bin* [(op +) (op *) 0 1 *bit-int* (op −) *uminus*] ⟨*proof*⟩

theorem #48734 − #3762039758274 = − #3762039709540
  by *simp*

The same technique also works for more complex calculations, like *div/mod*.

3   Contexts and Declarations

3.1   Generic Contexts

Isabelle supports two notions of context: *theory* for large-scale library organization, and *proof context* as medium-scale reasoning infrastructure for structured proofs [20] and structured specifications [15, 1, 2, 12]. Here we present a unified view of *generic contexts* that covers the common characteristics of both kinds.

In primitive inferences $\Gamma \vdash A$ means that A is derivable within a context Γ, which typically contains fixed variables and assumptions. We generalize this idea towards a container for arbitrary logical and extra-logical data, including background theory declarations (types, constants, axioms), local parameters and assumptions, definitions and theorems, syntax and type-inference information, hints for automated proof tools (e.g. the Simplifier, arithmetic procedures) etc.

There are two main logical operations on generic contexts:

1. *Context construction* starts with an empty context Γ_0 and proceeds by adding further context elements consecutively. In particular, $\Gamma + \mathbf{fix}\ x$ declares a local variable, and $\Gamma + \mathbf{assume}\ A$ states a local assumption.
2. *Context export* destructs the logical difference of two contexts, by imposing it on local results. The effect of *export* $\Gamma_1\ \Gamma_2$ is to discharge portions of the context on terms and theorems as follows:

$$export\ (\Gamma + \mathbf{fix}\ x)\ \Gamma\ (t\ x) \qquad = (\lambda x.\ t\ x)$$
$$export\ (\Gamma + \mathbf{fix}\ x)\ \Gamma\ (\vdash B\ x) \qquad = (\vdash \bigwedge x.\ B\ x)$$
$$export\ (\Gamma + \mathbf{assume}\ A)\ \Gamma\ (A \vdash B) = (\vdash A \Longrightarrow B)$$

Discharging assumptions on syntactic terms has no effect.

3.2   Context Data

Isabelle belongs to the tradition of "LCF-style" provers, which is centered around formally checked entities being implemented as abstract datatypes. All values of such types being produced at run-time are "correct by construction" — relative to the correctness of a few core modules. For example, certified terms may be composed by syntax primitives to produce again certified terms; likewise theorems are composed by primitive rules to produce new theorems.

We extend this principle towards an abstract type of well-formed proof contexts, with additional support for type-safe additions of context data at compile time. Internally, context data consists of an inhomogeneous record of individual data slots (based on dynamically typed disjoint sums). The external programming interface recovers strong static typing by means of an SML functor, involving dependently-typed modules, functor $Data(ARGS)$: $RESULT$, where:

$$ARGS = \mathsf{sig}\ \mathsf{type}\ T\ \mathsf{val}\ init:\ T\ \mathsf{end}$$
$$RESULT = \mathsf{sig}\ \mathsf{val}\ put:\ T \to context \to context\ \mathsf{val}\ get:\ context \to T\ \mathsf{end}$$

For example, the inference kernel requires a table of axioms, which is declared by $\mathbf{structure}\ Axioms = Data(\mathsf{type}\ T = (name \times term)\ list\ \mathsf{val}\ init = [])$. The resulting operations *Axioms.put*: $(name \times term)$ *list* $\to$ *context* $\to$ *context* and *Axioms.get*: *context* $\to$ $(name \times term)$ *list* are kept private to the kernel implementation, exposing only interfaces for adding axioms strictly monotonically, and for turning named axioms into theorems. Thus abstract datatype integrity can be maintained, according to the "LCF system approach".

Slightly less critical applications of the same mechanism involve automated proof tools that depend on "hints" in the context, such as the Simplifier (cf. §4).

3.3   Morphisms

In general a morphism is just abstract non-sense to organize certain logical operations. The idea is that a morphism determines how results may be transferred into another context, providing a different *view* of previous results.

Formally, a morphism φ is a tuple $(\varphi_{type}, \varphi_{term}, \varphi_{thm})$ of mappings on types, terms, and theorems, respectively. φ_0 refers to the identity morphism. We write uniformly $\varphi \ \tau$ for some type τ, and $\varphi \ t$ for some term t, and $\varphi \ th$ for some theorem th, meaning that the appropriate component is applied. Usually morphisms respect the overall syntactic structure of arguments, by mapping types within terms uniformly etc. Then the full morphism φ is already determined by the φ_{thm} part, since types and terms are sub-structurally included in thm.

A morphism is called *well-formed*, if it preserves logical well-formedness of its arguments. This desirable property may be achieved in practice by composing morphisms from basic operations of the inference kernel. For example, a morphism consisting of primitive inference rules (e.g. for instantiation of variables) will map theorems again to theorems by construction.

The following two kinds of morphism, which resemble abstraction and application in λ-calculus or type-theory, are particularly important in practice.

1. *Export morphism:* the export operation between two contexts (cf. §3.1) determines a well-formed morphism $\varphi = export \ \Gamma_1 \ \Gamma_2$. By this view, local results (with fixed variables and assumptions) appear in generalized form (with arbitrary variables and discharged premises).
2. *Interpretation morphism:* given concrete terms for the fixed variables, and theorems for the assumptions of a context, the substitution operation determines a well-formed morphism $\varphi = interpret \ [t/x] \ [th/A]$. By this view, results of an abstract theory are turned into concrete instances.

3.4   Generic Declarations

Ultimately we would like to view arbitrary context data after morphism application, but doing this directly turns out as unsatisfactory — essentially violating abstract datatype integrity. This could be amended by incorporating morphisms into the generic data interface (cf. §3.2), but then implementations would become more complicated due to additional invariants (e.g. consider efficient term index structures in the presence of arbitrary morphisms).

Instead, we handle morphisms at the level of *data declarations*. Recalling that arbitrary datatypes can be incorporated into the generic *context*, any operation on data is subsumed by *context* → *context*. This motivates the our definition:

$$declaration = morphism \rightarrow context \rightarrow context$$

This means a declaration participates in applying the morphism: being passed some φ as additional argument it is supposed to apply it to any logical parameters (types, terms, theorems) involved in its operation. Note that immediate declaration in the current context works by passing the identity morphism φ_0.

A *fact declaration* is represented as a theorem-attribute pair. This is an important special case; the general declaration is recovered by the *apply* operation:

$$fact\text{-}declaration = thm \times attribute$$
$$attribute = thm \rightarrow context \rightarrow context$$
$$apply\ (th,\ att) = (\varphi \mapsto att\ (\varphi\ th))$$

Observe that *th* is transformed separately before invoking the *attribute*. Here the internal operation does not have to consider morphisms at all.

The Isabelle syntax for (th, att) is "*th* [*att*]", which may appear wherever facts occur, as in "**theorem** *th* [*att*]: *A*" or "**declare** *th* [*att*]". General declarations are written as "**declaration** *d*", for *d* : *morphism* → *context* → *context*.

3.5   Locales

Locales [15] provide a high-level mechanism to organize proof context elements and declarations (cf. §3.4). *Locale expressions* [1] compose existing locales by means of merge and rename operations. *Locale interpretation* [2] transfers results stemming from a locale into another context. *Type-classes* [19] express properties of polymorphic entities within the type-system; there is a canonical interpretation of classes as specific locales [12]. These building-blocks provide a medium-scale module concept on top of the existing Isabelle logic — locale operations are reduced to basic inferences (usually expressed via morphisms).

A locale specification consists of the following two distinctive parts.

1. *Assumptions* refer to fixed types, terms, and hypotheses. The latter two are specified explicitly by the notation "**fixes** *x*" and "**assumes** *A x*"; types are maintained implicitly according to occurrences in the given formulae.

   The assumption part determines the logical meaning of a locale, which is expressed as predicate definition together with a context construction: the definition "**locale** *c* = **fixes** *x* **assumes** *A x*" produces a predicate constant $c = (\lambda x.\ A\ x)$, and a context construction $\Gamma + $ **fix** *x* **assume** *A x*.

2. *Conclusions* are essentially theorems that depend on the locale context and are adjoined to the locale later on, without changing the logical meaning.

   The notation "(**in** *c*)" may be added to various theory elements (notably **theorem**) to indicate that the result is established within the context of *c*, and stored there for later use. Fact declarations may be included, too.

The conclusion part is what really matters in practical use, including declarations of arbitrary extra-logical data. The locale infrastructure maintains a canonical order of declarations $d_1, \ldots, d_n$. Whenever the locale context is reconstructed later, relative to a morphism φ, the context will be augmented by the collective declaration of $d_n \; \varphi \; (\cdots (d_1 \; \varphi \; \Gamma))$. For our purpose, we have generalized the existing fact declaration mechanism towards arbitrary **declaration** functions (cf. §3.4). Thus we may attach arbitrary SML values to a locale, which will be transformed together with the logical content by morphism application.

4   Context-Dependent Simplification

Isabelle's generic Simplifier [17] is already a non-trivial example of integrating automated proof tools into our architecture of context data and declarations (cf. §3). The generic proof method *simp* depends on a *simpset* container being maintained as context data. A *simpset* roughly consists of a set of simplification rules and simplification procedures (of type *simproc*), which are maintained by:

> *add-simp*: *thm* → *simpset* → *simpset*
> *add-simproc*: *simproc* → *simpset* → *simpset*

Declaration of plain simplification *rules* (cf. the example in §2.1) follows the general scheme of *fact-declaration* of §3.4. Here *add-simp* is wrapped into an attribute called *simp*. Morphism application is trivial, since rules are theorems.

Declaration of simplification *procedures* is more involved, cf. the example in §2.2. A *simproc* essentially consists of a function *morphism* → *term* → *thm* that turns a redex t into a theorem $t = t'$. We introduce a derived declaration "**simproc-setup** *name* (*patterns*) = *f*" for *patterns* : *term** and *f* : *morphism* → *simproc*, which roughly abbreviates "**declaration** $(\varphi \mapsto \mathit{add\text{-}simproc}\; (\varphi \; f))$". What does it mean to to apply a morphism to a function? We define:

$$\varphi \, f = (\psi \mapsto f \; (\psi \circ \varphi))$$

Our main observation is that a function transforms itself by a given morphism. The body will apply the morphism to all required logical entities, or pass it on to other functions. To apply φ to f, we compose φ with any morphism ψ being passed in the future. Actual evaluation in the present context is commenced by invoking $f \; \varphi_0$ (the identity morphism). Note that this scheme is reminiscent of programming language semantics presented in "continuation-passing style".

Tool-compliant morphisms. A well-formed morphism (cf. §3.3) guarantees stability of logical entities after transformation. Unfortunately, this is *not* sufficient to achieve stability of arbitrary functions. For example, even with plain simplification rules alone, the Simplifier is not necessarily stable after arbitrary interpretation of its context: although an interpreted rewrite system still represents equational consequences of the original theory, it may cause the rewrite strategy to produce unwanted results or fail to terminate.

A morphism φ is called f-*compliant* iff $\varphi\,f$ preserves the intended operational behavior of $f\,\varphi_0$. E.g. for *binary-sub* defined in §2.2, tool-compliant morphisms include those that replace *bit*, *add*, *neg* by rigid first-order terms, but not arbitrary λ-abstractions that destroy the shape of the term patterns involved.

Morphisms stemming from locale import expressions [1] (type instantiations and renaming) are usually tool-compliant. Substitution of parameters by constants works equally well. Assuming that the tool is able to perform unification of simple types at runtime, types may be even generalized here, e.g. replacing the fixed α of locale *ring-bin* by an arbitrary $\beta::c$ constrained by a suitable type-class c that ensures the required algebraic properties. This means that the canonical interpretation of locales as type-classes [12], where fixed parameters are replaced by polymorphic constants yields morphisms that comply with such tools.

5   Application: Gröbner Bases

We use the techniques of §3 to integrate a proof-tool for (semi)ring equalities. The procedure follows the traditional refutation scheme: negate the input then transform to DNF and disprove each disjunct. Hence we consider formulae P of the form $\bigwedge_{i=1}^{n} p_i = 0 \wedge f \neq 0$, where the p_i's and f are polynomials within *semiring*. Recall that we can turn several inequalities into only one. We use an oracle to find a Nullstellensatz refutation consisting of polynomials $h_1, \ldots, h_n$ and a number k s.t. $\sum_{i=1}^{n} h_i \cdot p_i = f^k$. It is clear that by proving this last equality, $\neg\,P$ follows. The oracle computes a Gröbner basis [4] for the ideal generated by the p_i's and reduces f in that basis. The whole process keeps track on where the S-polynomials originate from in order to express the final result in terms of the p_i's. This implementation has been ported from HOL Light [13], see also [16, §3.4.2]. Note that this procedure also solves many problems in *semiring*, by dealing with $p = q$ as if it were $p - q = 0$ and separating the "negative" and "positive" parts of the Nullstellensatz certificate. This yields a correct yet incomplete procedure, depending only on two axioms stated in *semiringb*:

locale *semiringb* = *semiring* +
  assumes *add-cancel*: $(x \oplus y = x \oplus z) \leftrightarrow y = z$
  and *idom*: $(w \odot y \oplus x \odot z = w \odot z \oplus x \odot y) \leftrightarrow w = x \vee y = z$

We emphasize that our procedure *proves* results by inference: the oracle only finds a certificate, used to derive the conflicting equation. The final equation emerges by normalizing both sides, using genuine inferences. This approach also allows to integrate other tools solving the detachability problem [16, §3.4.2] or provide the construction information of the S-polynomials (as e.g. in AXIOM).

Context data. The method sketched above is essentially parameterized by the operations and axioms of (semi)ring (in order to *prove* the normal form of polynomials and recognize their structure) and by four non-logical functions: *is-const* recognizes canonical ring expressions as "constants", *dest-const* produces SML rational numbers from ring constants, *mk-const* produces ring constants from

SML rational numbers, and *ring-conv*: *morphism* → *term* → *thm* produces *proven* normalization results of numeral expressions.

Following §3.2, the method reserves a data slot in the context to manage all its instances. We emphasize that instances no longer depend on a morphism, since these are applied at declaration-time, where the corresponding instance of our procedure is computed. At run-time, when we have to prove P, we extract a polynomial q occurring in P and try to find a corresponding interpretation installed in our context-data. Afterwards it proceeds as explained above.

Using the proof method. The end-user sees an Isabelle proof method *algebra*, that works for any tool-compliant interpretation of *semiringb*. The user only needs to interpret the *semiringb* locale to activate the method in his context. The default configuration includes interpretations of *semiringb* to "prominent" types in Isabelle/HOL: natural numbers and the type-class *idom*, which covers e.g. integers and real numbers. Here is a simple example:

theorem
  fixes $x :: int$
  shows $x * (x^2 - x - 5) - 3 = 0 \leftrightarrow (x = 3 \vee x = -1)$
    by *algebra*

6   Conclusion

We have presented a system architecture that emphasizes the explicit context-dependency of both logical and non-logical entities, with specific infrastructure to manage generic data, declarations, morphisms etc. An interesting observation about context-aware proof tools is that there are three phases: compile-time (when loading the sources), declaration-time (when applying morphisms to transfer abstract entities to particular interpretations), and run time (when the tool is invoked on concrete problems).

Traditional "LCF system programming" directly uses features of SML to abstract over theory dependencies. Our approach goes beyond this by providing specific means to organize the requirements of a tool. Thus the implementor is enabled to build generic methods that are transferred automatically into the application context later on. The end-user merely needs to perform a logical interpretation of the corresponding tool context, without being exposed to SML.

This integration of abstract theory mechanisms and proof methods is achieved without changing the logical foundations of Isabelle. All additional mechanisms are reduced to existing logical principles. Neither do we require a separate formal model of computation, we merely access the underlying implementation platform (SML) directly.

Apart from these theoretical considerations there remains the important practical issue to implement tools that are actually stable under application of typical theory morphisms. While the LCF approach guarantees logical correctness, proof methods may still *fail* due to unexpected interpretation of abstract theory contexts. This can be addressed by focusing on restricted classes of morphisms, such as those stemming from interpretation of Isabelle locales as axiomatic type-classes or concrete base types.

References

[1] Ballarin, C.: Locales and locale expressions in Isabelle/Isar. In: Berardi, S., Coppo, M., Damiani, F. (eds.) TYPES 2003. LNCS, vol. 3085, Springer, Heidelberg (2004)

[2] Ballarin, C.: Interpretation of locales in Isabelle: Theories and proof contexts. In: Borwein, J.M., Farmer, W.M. (eds.) MKM 2006. LNCS (LNAI), vol. 4108, Springer, Heidelberg (2006)

[3] Barras, B., et al.: The Coq Proof Assistant Reference Manual. INRIA, vol. 8 (2006)

[4] Buchberger, B.: Ein algorithmisches Kriterium für die Lösbarkeit eines algebraischen Gleichungssystems. Aequationes mathematicae 3, 374–383 (1970)

[5] Chaieb, A.: Verifying mixed real-integer quantifier elimination. In: Furbach, U., Shankar, N. (eds.) IJCAR 2006. LNCS (LNAI), vol. 4130, Springer, Heidelberg (2006)

[6] Chaieb, A., Nipkow, T.: Verifying and reflecting quantifier elimination for Presburger arithmetic. In: Stutcliffe, G., Voronkov, A. (eds.) Logic for Programming, Artificial Intelligence, and Reasoning, vol. 3835 (2005)

[7] Delahaye, D., Mayero, M.: Dealing with Algebraic Expressions over a Field in Coq using Maple. In: Journal of Symbolic Computation: special issue on the integration of automated reasoning and computer algebra systems (2004)

[8] Delahaye, D., Mayero, M.: Quantifier Elimination over Algebraically Closed Fields in a Proof Assistant using a Computer Algebra System. In: Proceedings of Calculemus 2005 (2005)

[9] Farmer, W.M., Guttman, J.D., Thayer, F.J.: IMPS: An interactive mathematical proof system. Journal of Automated Reasoning, 11(2) (1993)

[10] Gordon, M.J.C, Milner, R., Wadsworth, C.P.: Edinburgh LCF. LNCS, vol. 78. Springer, Heidelberg (1979)

[11] Grégoire, B., Mahboubi, A.: Proving equalities in a commutative ring done right in Coq. In: Hurd, J., Melham, T. (eds.) TPHOLs 2005. LNCS, vol. 3603, Springer, Heidelberg (2005)

[12] Haftmann, F., Wenzel, M.: Constructive type classes in Isabelle. In: Altenkirch, T., McBride, C. (eds.) Types for Proofs and Programs (TYPES 2006)

[13] Harrison, J.: HOL Light Tutorial (for version 2.20) (September 2006)

[14] Harrison, J., Théry, L.: A skeptic's approach to combining HOL and Maple. Journal of Automated Reasoning 21, 279–294 (1998)

[15] Kammüller, F., Wenzel, M., Paulson, L.C.: Locales: A sectioning concept for Isabelle. In: Bertot, Y., Dowek, G., Hirschowitz, A., Paulin, C., Théry, L. (eds.) TPHOLs 1999. LNCS, vol. 1690, Springer, Heidelberg (1999)

[16] Mishra, B.: Algorithmic algebra. Springer, Heidelberg (1993)

[17] Nipkow, T., Paulson, L.C., Wenzel, M.: Isabelle/HOL. LNCS, vol. 2283. Springer, Heidelberg (2002)

[18] Paulson, L.C.: Organizing numerical theories using axiomatic type classes. Journal of Automated Reasoning, 33(1) (2004)

[19] Wenzel, M.: Type classes and overloading in higher-order logic. In: Gunter, E.L., Felty, A.P. (eds.) TPHOLs 1997. LNCS, vol. 1275, Springer, Heidelberg (1997)

[20] Wenzel, M.: Isabelle/Isar — a versatile environment for human-readable formal proof documents. PhD thesis, Institut für Informatik, TU München (2002)

Towards Constructive Homological Algebra in Type Theory

Thierry Coquand[1] and Arnaud Spiwack[2]

[1] Göteborg University
[2] Ecole Normale Supérieure de Cachan

Abstract. This paper reports on ongoing work on the project of representing the Kenzo system [15] in type theory [11].

Introduction

This paper reports on ongoing work on the project of representing the system Kenzo [15] in type theory. Kenzo is a symbolic computation system in algebraic topology, based on a rich mathematical theory described in [15], and which uses in an essential way ideas from functional programming. Our ultimate goal is to represent the mathematical results of [15] as constructive mathematical results developed in type theory. Using type theory as a functional programming language, this representation should then give a fully specified and checked functional version of Kenzo. Besides the Kenzo system, we can hope to develop in this way a library of reasonably efficient algorithms in homological algebra, similar to [5] but specified and written in type theory.

One of the main mathematical results on which the system Kenzo relies (the Basic Pertubation Lemma) has been already checked in the system Isabelle [1,15]. Our paper complements this work by exploring the formalisation of Kenzo at the level of *preabelian category*. For this, we show in detail how to represent category theory, and in particular preabelian categories, in type theory. We use then this formalisation on the test example suggested in [1], where it is explained why it is quite subtle to represent it formally. We believe that to express reasoning at the level of category theory, in a characteristic "pointfree" way, is perfect for formalisation. Indeed it works well on this test example, and the formal reasoning in type theory follows closely the informal argument. We can then instantiate this abstract argument on the example of the category of abelian groups to get back the statement in [1].

This paper is organised as follows. First we describe in detail the general setting in which we represent the mathematics of Kenzo: dependent type theory with universes, essentially the system [11], with a special universe of propositions. The system Coq [6] is a possible implementation of this system. We explain then how usual mathematical notions (sets, groups, ...) and then the notion of category theory are represented in this setting. The main example is the notion of preabelian category. We show how a test example [1] can be represented as a general property of preabelian categories. This has been done formally in Coq.

M. Kauers et al. (Eds.): MKM/Calculemus 2007, LNAI 4573, pp. 40–54, 2007.

We end by listing the remaining steps for having a representation of Kenzo. An appendix presents the formal statements, that are reasonably close to the informal statements.

The results of this paper are quite preliminary w.r.t. to the general goal of actually running Kenzo program in type theory. They show however that this project should be feasible, and provide already interesting observations on the formal representation of mathematics in type theory.

1   Type Theory

We shall use type theory [11] as a model for the mathematics used in homological algebra. It is an alternative to the system ZF, with closer connection with functional programming (and should be thus a priori well adapted for representing Kenzo). All mathematical notions and proofs are represented as λ-terms. These terms can then be directly computed [10] and there exist now actual efficient implementations of such computations [8].

The terms are untyped lambda terms with constants. We consider terms up to α-conversion. We have a constructor Π of arity 2 and we write $\Pi x{:}A.B$ instead of $\Pi\ A\ (\lambda x.B)$, and $A \to B$ instead of $\Pi\ A\ (\lambda x.B)$ if x is not free in B. We write also $\Pi x_1\ \dots\ x_n : A.B$ for $\Pi x_1 : A.\dots.\Pi x_n : A.B$. We write $B[x = M]$, or even $B[M]$ if x is clear, the substitution of the term M for the variable x in B. We have special constants $U_1, U_2, \dots$ for universes. A *context* is a sequence $x_1 : A_1, \dots, x_n : A_n$.

The minimal type theory we use has three forms of judgements

$$\Gamma \vdash A \qquad \Gamma \vdash M : A \qquad \Gamma \vdash$$

The last judgement $\Gamma \vdash$ expresses that Γ is a well-typed context. We may write $J\ [x : A]$ for $x : A \vdash J$.

The typing rules are as follows.

$$\frac{}{\vdash} \qquad \frac{\Gamma \vdash A}{\Gamma, x : A \vdash}$$

$$\frac{\Gamma \vdash}{\Gamma \vdash U_i} \qquad \frac{\Gamma \vdash A : U_i}{\Gamma \vdash A} \qquad \frac{\Gamma, x : A \vdash B}{\Gamma \vdash \Pi x{:}A.B}$$

$$\frac{(x : A) \in \Gamma \quad \Gamma \vdash}{\Gamma \vdash x : A} \qquad \frac{\Gamma, x : A \vdash M : B}{\Gamma \vdash \lambda x.M : \Pi x{:}A.B} \qquad \frac{\Gamma \vdash N : \Pi x{:}A.B \quad \Gamma \vdash M : A}{\Gamma \vdash N\ M : B[M]}$$

$$\frac{\Gamma \vdash M : A \quad \Gamma \vdash B \quad A =_\beta B}{\Gamma \vdash M : B}$$

We have also

$$\frac{\Gamma \vdash}{\Gamma \vdash U_i : U_{i+1}} \qquad \frac{\Gamma \vdash A : U_i}{\Gamma \vdash A : U_{i+1}}$$

We express finally that each universe U_i is closed under the product operation.

$$\frac{\Gamma \vdash A : U_i \quad \Gamma, x : A \vdash B : U_i}{\Gamma \vdash \Pi x{:}A.B : U_i}$$

These rules have an intuitive interpretation in set theory, where U_i are Grothendieck universes [2]. The dependent product $\Pi x : A.B$ if $B(x)$ is a family of sets over a set A is the set of all families $(b_x)_{x \in A}$ such that $b_x \in B(x)$ for all $x \in A$.

It is convenient to introduce sigma types, with the following rules, and adding the conversion rules $(M_1, M_2).1 =_\beta M_1$, $(M_1, M_2).2 =_\beta M_2$

$$\frac{\Gamma, x : A \vdash B}{\Gamma \vdash \Sigma x : A.B}$$

$$\frac{\Gamma \vdash M : A \quad \Gamma \vdash N : B[M]}{\Gamma \vdash (M, N) : \Sigma x : A.B} \qquad \frac{\Gamma \vdash P : \Sigma x : A.B}{\Gamma \vdash P.1 : A} \qquad \frac{\Gamma \vdash P : \Sigma x : A.B}{\Gamma \vdash P.2 : B[P.1]}$$

$$\frac{\Gamma \vdash A : U_i \quad \Gamma, x : A \vdash B : U_i}{\Gamma \vdash \Sigma x : A.B : U_i}$$

In set theory, $\Sigma x : A.B$ is the set of pairs (x, b_x) with $x \in A$ and $b_x \in B(x)$. We shall write $(x_1, \ldots, x_n)$ for $(\ldots (x_1, x_2), \ldots, x_n)$[1].

2   A Type of Propositions

It is convenient also to introduce a special universe U_0, which in set theory would be the set of truth values, with the special rules

$$\frac{\Gamma \vdash}{\Gamma \vdash U_0 : U_1} \qquad \frac{\Gamma \vdash A : U_0}{\Gamma \vdash A : U_1} \qquad \frac{\Gamma \vdash A \quad \Gamma, x : A \vdash B : U_0}{\Gamma \vdash \forall x : A.B : U_0}$$

$$\frac{\Gamma, x : A \vdash B : U_0 \quad \Gamma, x : A \vdash M : B}{\Gamma \vdash \lambda x.M : \forall x : A.B} \qquad \frac{\Gamma \vdash N : \forall x : A.B \quad \Gamma \vdash M : A}{\Gamma \vdash N\ M : B[M]}$$

We write $A \Rightarrow B$ for $\forall x : A.B$ if x is not free in B, and $A_1 \Rightarrow \ldots \Rightarrow A_n$ denotes $(\ldots (A_1 \Rightarrow A_2) \ldots \Rightarrow A_n$.

We can quantify over any type: $\forall x : A.B : U_0$ for any type A, if $B : U_0$ $[x : A]$. In particular, we have $\forall x : U_0.B : U_0$ if $B : U_0$ $[x : U_0]$. This impredicativity is convenient, but not necessary for representing the reasonings in [15]. We can represent logical connectives as operations of type $U_0 \to U_0 \to U_0$. For instance $A \wedge B$ is defined as

$$\forall P : U_0.(A \Rightarrow B \Rightarrow P) \Rightarrow P$$

We define also $A \Leftrightarrow B$ as $(A \Rightarrow B) \wedge (B \Rightarrow A)$ and $\bot : U_0$ as $\forall A : U_0.A$ and $\top : U_0$ as $\forall A : U_0.A \Rightarrow A$, which has an inhabitant $\lambda A \lambda x.x$.

---

[1] The addition of sigma types is convenient but not strictly necessary. One can work with *vectors* of terms and *telescopes* instead [4]. A telescope is like a context $T = x_1 : A_1, \ldots, x_{n-1} : A_{n-1}, A_n$ and a vector $P_1, \ldots, P_n$ fits this telescope iff $P_1 : A_1, \ldots, P_n : A_n[P_1, \ldots, P_{n-1}]$.

These rules have also a direct interpretation in ZF set theory. The type U_0 is interpreted as the set $\{0, 1\}$, 0 being the empty set and 1 being the set $\{0\}$, and $\forall x : A.B$ is 1 iff $B(x) = 1$ for all x in A and is 0 otherwise. See [13] for a careful presentation of this model.

Notice that if $A : U_1$ and $B : U_0$ $[x : A]$ we have also $B : U_1$ $[x : A]$ and we can also form $\Pi x : A.B : U_1$. There are two maps

$$(\Pi x : A.B) \to \forall x : A.B \qquad (\forall x : A.B) \to \Pi x : A.B$$

but the two types $\forall x : A.B$ and $\Pi x : A.B$ are not convertible. (They are not identical in general in the set theoretical model. Indeed, in this model an element of $\Pi x : A.B$ will be a family, which in set theory has to be the set of pairs $(a, 0)$ with a in A, while an element of $\forall x : A.B$ can only be the empty set 0. See [13] for a general discussion of this point.)

The existence of a set theoretical model entails the *consistency* of our type theory: the type $\perp$ is not inhabited (*i.e.* there is no proof of false). A sharper result is the strong normalisation theorem [12]. It follows from this that if M, A are in normal form then the judgement $\vdash M : A$ is decidable. It follows from this that one can actually checked the correctness of proofs. Furthermore, one can use type theory as a terminating functional programming language.

By analogy with $\forall x : A.B$ it would be natural to add an operation $\exists x : A.B$ with the rules

$$\frac{\Gamma \vdash A \quad \Gamma, x : A \vdash B : U_0}{\Gamma \vdash \exists x : A.B : U_0}$$

$$\frac{\Gamma \vdash M : A \quad \Gamma \vdash N : B[M]}{\Gamma \vdash (M, N) : \exists x : A.B} \qquad \frac{\Gamma \vdash P : \exists x : A.B}{\Gamma \vdash P.1 : A} \qquad \frac{\Gamma \vdash P : \exists x : A.B}{\Gamma \vdash P.2 : B[P.1]}$$

A fundamental result, which plays a role in this paper, is that the addition of these rules is *contradictory*: it is possible to build a proof of $\perp$ (which is then automatically not normalisable) in this extended system [7].

It would be difficult to try and give an intuitive reason why the system with $\exists$ is inconsistent. Unfortunately the proof of inconsistency is not so intuitive. In this particular case, it is possible to encore a type theory with a type of all types, which is a well known case of inconsistent theory [7].

Since these rules are contradictory, we cannot use them to represent mathematics. It is possible however to define $\exists x : A.B : U_0$ as

$$\forall P : U_0.(\forall x : A.B \Rightarrow P) \Rightarrow P$$

Since $B : U_1$ $[x : A]$ we can also form $\Sigma x : A.B : U_1$. We then have a map

$$(\Sigma x : A.B) \to \exists x : A.B$$

but in general there is no map in the other direction.

3   Representation of Bishop Set Theory in Type Theory

3.1   Bishop Sets

Bishop [3] specified the notion of set by stating that a set has to be given by a description of how to build element of this set and by giving a binary relation of equality, which has to be an equivalence relation. A function from a set A to a set B is then given by an *operation*, which is compatible with the equality (*i.e.* two elements which are equal in A are mapped to two elements which are equal in B), and is described as "a finite routine f which assigns an element $f(a)$ of B to each given element a of A". This notion of routine is left informal but must "afford an explicit, finite, mechanical reduction of the procedure for constructing $f(a)$ to the procedure for constructing a." It is direct and natural to represent formally all these notions in our type theory.

A *Bishop set* is defined to be a type $A : \mathsf{U}_1$ together with an *equivalence relation* over A that is an element $R : A \to A \to \mathsf{U}_0$ with a proof of *equiv A R*, where

$$equiv : \Pi A : \mathsf{U}_1.(A \to A \to \mathsf{U}_0) \to \mathsf{U}_0$$

$$equiv\ A\ R = refl\ A\ R \wedge sym\ A\ R \wedge trans\ A\ R$$

$$refl\ A\ R = \forall x : A.R\ x\ x, \quad sym\ A\ R = \forall x\ y : A.\ R\ x\ y \Rightarrow R\ y\ x$$

$$trans\ A\ R = \forall x\ y\ z : A.\ R\ x\ y \wedge R\ y\ z \Rightarrow R\ x\ z$$

It is possible to represent the collection of all Bishop sets as the type

$$\Sigma A : \mathsf{U}_1.\Sigma R : A \to A \to \mathsf{U}_0.equiv\ A\ R$$

which is itself an element of type U_2.

If $X = (A, R, p)$ is a Bishop set, we write $|X| = X.1 = A$ its corresponding type and $=_X$ for its corresponding equivalence relation $R = X.2.1$. If there is no confusion, we shall say simply "set" for "Bishop set". If $Y = (B, S, q)$ is another set, then one can form the set Y^X of *functions* from X to Y by taking

$$|Y^X| = \Sigma f : |X| \to |Y|.\forall x_1\ x_2 : |X|.\ x_1 =_X x_2 \Rightarrow f\ x_1 =_Y f\ x_2$$

and $(f_1, p_1) =_{Y^X} (f_2, p_2)$ is the proposition $\Pi x : A.\ f_1\ x =_Y f_2\ x$.

Bishop sets form a category. One can ask how similar is this category to the category of sets in ZF. We analyse this in the next subsection.

3.2   Truth Values, Properties and Subsets

An important set is the set of truth values Ω such that $|\Omega|$ is U_0 and $=_\Omega$ is $\Leftrightarrow$. A *property* on X is a function from X to Ω (in Bishop's sense).

If $P : A \to \mathsf{U}_0$ is a property on a set $X = (A, =_X, p)$, which means that $x_1 =_X x_2$ implies $P\ x_1 \Leftrightarrow P\ x_2$, it is possible to define the set $|Y| = \Sigma x : A.P\ x$ with the equality $(x_1, p_1) =_Y (x_2, p_2)$ iff $x_1 =_X x_2$. One can then check that the map $m : y \longmapsto y.1$ is a function from the set Y to the set X which is one-to-one: $y_1 =_Y y_2$ iff $m\ y_1 =_X m\ y_2$. Bishop defines a *subset* of X to be such a function $i : Z \to X$ which is one-to-one. We have just seen that any property P on X defines a subset $m : Y \to X$ of X, and it is natural to write $m : \{x : X \mid P\ x\} \to X$ for this subset. This corresponds to the comprehension axiom in systems such as ZF.

Conversely, given a subset Z with $|Z| = C$, $i : C \to A$ it is possible to define a property $P : A \to U_0$ on X by taking $P\ x$ to be $\exists z : C.x =_X i\ z$. This property defines a subset $m : Y \to X$ with $Y = \{x : X \mid P\ x\}$. In ZF set theory the two subsets Y and Z are equivalent and it is possible to find a (unique) map $f : Y \to Z$ such that $i\ f = m$. This is *not* possible in our representation: given $x : A$ and a proof that $\exists z : C.x =_X i\ z$ there is no way in general to extract from this proof an element $z : C$ such that $x =_X i\ z$ holds. In general, we do *not* have the implication

$$(\forall x : A.\exists! z : C.R\ x\ z) \to \exists f : A \to C.\forall x : A.R\ x\ (f\ x) \qquad (*)$$

In set theory, this implication is achieved by reducing *functions* to *functional relations*. However, we want here to be able to use functions as *functional programs* for our representation of Kenzo. Since functional programs do not coincide with functional relations, it is natural that the implication $(*)$ is not valid.

From these remarks, one can see that the category of Bishop sets we have defined is *not* a topos. It thus differs in subtle way from the usual category of sets. We shall see similarly that in this setting the category of abelian groups is *not* an abelian category (but the category of finitely presented abelian groups is). However, and this is an important point, it is not an obstacle in representing Kenzo. (This can be expected since the goal of Kenzo is precisely to obtain functional programs, and not abstractly defined relations.)

3.3 Alternative Representation

Following Curry-Howard, one can represent propositions as types in U_1. We don't need then to introduce the type U_0. Existential propositions are then represented using sigma types, and in this representation, there is a good correspondance between subsets and properties and the implication $(*)$ is valid. The problem there is that we don't have a *set* of truth values any more, since the type of propositions, U_1, is itself of type U_2. (So the category of sets do not form a topos either in this representation.)

We feel that our representation is closer to mathematical practice, and separates more clearly what is the computational part, at level U_1, and the specification part, at level $U_0{}^2$.

4 Category Theory in Type Theory

We can represent the type of "all" Bishop sets, and this itself is of type U_2. It is possible similarly to represent the collection of all groups, the category of all sets, the category of all groups, and these are represented by types that are in

---

[2] For the development of Kenzo however, both approaches seem possible.

U_2. (This corresponds to the notion of locally small categories.) One can then also consider the 2-category of all these categories, and this will be represented by a type in U_3.

In general, a locally small category will be represented by a *type* of objects $Obj : U_2$ (for instance the type of Bishop sets). If $A, B : Obj$ we suppose given a *set Hom A B* of morphisms. Thus, if $A, B : Obj$ we have $Hom\ A\ B : U_1$ and we have an equality $f =_{Hom\ A\ B} g$ which is in U_0 for $f, g : Hom\ A\ B$. We introduce also the identity morphism, the composition operator, and the usual axioms of associativity and identity.

An important instance is provided by the category of abelian groups[3].

4.1   Properties of the Category of Abelian Groups

In classical mathematics, an elegant axiomatisation of the category of abelian groups is provided by the notion of *abelian category*. What are the properties of the category of abelian groups represented in type theory?

It is clear that this category is *preabelian*: it is preadditive, all hom-sets are abelian groups and the composition of morphisms is bilinear, it is additive since we can form finite direct sums and direct products, and finally, every morphism has both a kernel and a cokernel. This can also be quite directly checked formally.

So the category of abelian groups represented in type theory is preabelian. Is it abelian? Surprisingly it is *not* the case that every monomorphism and every epimorphism is normal (that is, a monomorphism for instance is not necessarily the kernel of a map). If we have a map $u : A \to B$ which is mono, that is $u\ x = 0 \to x = 0$ then, in usual set theory, this map is the kernel of the map $s : B \to B/Im\ u$ (which always makes sense since all the groups are abelian). This means that if we have a map $f : X \to B$ such that $s\ f = 0$ then there exists a unique map $g : X \to A$ such that $f = u\ g$. That $s\ f = 0$ means that for all $x \in X$ there exists $a \in A$ (unique) such that $f\ x = u\ a$. But it is not possible in our representation of Bishop sets to deduce from this the existence of a map $g : X \to A$ such that $f\ x = u\ (g\ x)$. One would need for this the implication

$$(\forall x : A.\exists! z : C.R\ x\ z) \to \exists f : A \to C.\forall x : A.R\ x\ (f\ x) \qquad (*)$$

that, as we have seen, does not hold in general.

In any case, we have chosen to axiomatise the algebraic reasoning justifying Kenzo at the level of *preabelian* category and not at the level of *abelian* category. We believe that actually this reflects better the reasonings done in [15][4].

---

[3] The main difference with the treatment in [9] is the following. We use the structure of universes to stratify the categories: locally small categories are represented with a type of object in U_2, 2-categories with a type of objects in U_3, ...

[4] Let us consider as an example the long exact sequence of a short exact sequence (section 2.6 of [15]). This is something that only can be done in an *abelian* category. However, the Kenzo version of this notion requires a further hypothesis. The exactness property of the short exact sequence must be *effective* (definition 80 of [15]). With this extra hypothesis, the reasoning can then be represented at the level of preabelian category.

4.2   Preabelian Category

We represent the notion of general preabelian category in type theory in the following way. (All these axioms are instantiated by the category of abelian groups.)

First we have a type of objects $Obj : U_2$. We have to use the type U_2 since it is the type of the collection of all abelian groups, represented as a sigma type. For any two objects $A, B : Obj$ we have an abelian group of morphisms $Hom\ A\ B$. Thus $Hom\ A\ B : U_1$ and we have an equality on $Hom\ A\ B$ and a group operation $f + g : Hom\ A\ B$ for $f, g : Hom\ A\ B$ with a zero element $0 : Hom\ A\ B$. We have a composition operation $gf : Hom\ A\ C$ for $g : Hom\ B\ C$ and $f : Hom\ A\ B$. We require the equations $(f + g)h = fh + gh$ and $h(f + g) = hf + hg$.

There is a zero object $0 : Obj$ such that $f = 0$ if $f : Hom\ A\ 0$ or $f : Hom\ 0\ A$. We have biproducts, and there is an operation $(+) : Obj \to Obj \to Obj$ with morphisms $i : Hom\ A\ (A+B)$, $j : Hom\ B\ (A+B)$ and $p : Hom\ (A+B)\ A$ and $q : Hom\ (A+B)\ B$ with equations $pi = 1,\ qj = 1,\ pj = 0,\ qi = 0,\ ip + jq = 1$.

For stating the existence of the kernel, it is convenient to use the telescope notation [4]

$$(Ker, inj, pinj) : \Pi A\ B : Obj.\Pi f : Hom\ A\ B.\ (K : Obj,\ i : Hom\ K\ A,\ f\ i = 0)$$

We require also the universal condition

$$(univ, puniv) : \Pi X : Obj.\Pi u : Hom\ X\ A.\ f\ u = 0 \to (v : Hom\ X\ (Ker\ f),\ u = (inj\ f)\ v)$$

and the unicity condition

$$\Pi X : Obj.\Pi u : Hom\ X A.\Pi p : f\ u = 0.\Pi v : Hom\ X (Ker\ f).\ u = (inj\ f)v \to v = univ\ f\ u\ p$$

We state the existence of cokernel in a dual way.

We can define in this way a sigma type $PreAb : U_3$ which represents the collection of all prebalian categories. In particular one can define an element of type $PreAb$ which is the category of all abelian groups. One could also define the notion of additive functors between two elements of type $PreAb$.

4.3   Implicit Arguments

Besides dependent types, we use an important notational facility: *implicit arguments*. We don't need to give explicitly the arguments that can be inferred from the context. For instance the composition operator is of type

$$\Pi\ A\ B\ C : Obj.\ Hom\ A\ B \to Hom\ B\ C \to Hom\ A\ C$$

and expects 5 arguments. Since the 3 first arguments can be inferred from the last 2 arguments, one needs only to give the last 2 arguments and can write the composition (almost) as usual. In this way, we can write gf instead of *comp A B C f g*.

5   Formalisation of Kenzo

The main idea is to represent the reasonings done in [15] in an arbitrary pre-abelian category.

5.1   A Test Example

We have tested this approach on the introductory example Lemma 3.3.1 of [1]. As stated in [1], this example seems to contain the most interesting problems that have been found in the formalisation of proofs in Kenzo. It requires reasoning with homomorphisms and endomorphisms as if they were elements of certain algebraic structures, but also dealing with their functional definition; furthermore the domain conditions on the source or the target of the homomorphisms also are important. We try here to approach these issues by a formalisation at the level of category theory. The domain and codomain are explicit, but can be hidden since they can be inferred from the context.

We work in an arbitrary preabelian category. We suppose that we have $h, d :$ $G \to G$ such that $dd = hh = 0$ and $hdh = h$. We define $p = dh + hd$. We consider then the inclusion $i : K \to G$ where $K = Ker\, p$.

Proposition 1. *We have*

$$pp = p, \ ph = hp = h, \ pd = dp, \ hi = pi = 0$$

It follows that $pdi = dpi = 0$. Hence there exists $d_1 : K \to K$ such that $id_1 = di$. We then have $d_1 d_1 = 0$. Since $p(1 - p) = 0$ there exists $j : G \to K$ such that $ij = 1 - p$ and we have $jh = 0$, $jd = d_1 j$.

Proof. The equality $pp = p$ is proved by computation

$$(hd + dh)(hd + dh) = hdhd + hddh + dhhd + dhdh = hd + 0 + 0 + dh = hd + dh = p$$

Similarly we check $hp = hhd + hdh = 0 + h = h$, $ph = hdh + dhh = h + 0 = h$ and $dp = ddh + dhd = dhd = dhd + hdd = pd$. Since i is mono, $id_1 d_1 = did_1 = ddi = 0$ implies that $d_1 d_1 = 0$.

(The proposition can be stated as the fact that (j, i, h) is a *reduction* from G, d to K, d_1, as defined below.)

5.2   Use of Category Theory

This "pointfree" style, which requires to represent formally some basic notion of category theory, can be compared to the formalisation in [1].

The statement of Proposition 1 is not exactly the same as the one of Lemma 3.3.1 of [1] or even the corresponding informal statement in [15]. The formal representation in [1] allows to consider for instance $1 - p$ both as a map of type $G \to G$ and of type $G \to K$. We do not allow this, and have to distinguish between $j : G \to K$ and $1 - p : G \to G$ such that $ij = 1 - p$. We do not think however that this is a problem in practice.

The point is that the proof is essentially equational, like in a non commutative ring, but with an addition and a multiplication operations that are not always defined (the arity should be compatible). We can furthermore represent it as it is in type theory (see the appendix). We believe that this formalisation is close to a precise informal mathematical reasoning.

5.3   Use of Dependent Types

The extension of higher-order logic to a type system with universes is natural to represent category theory. It seems also *necessary* if one wants to have a system in which one can state general properties of an arbitrary category, and then instantiate it on concrete categories.

Dependent types are also used to facilitate *modular* reasoning. We can state a property about an arbitrary preabelian category, and then instantiate it to the category of abelian groups.

Finally, in this representation, all the terms can directly be seen as functional programs.

5.4   Refinement of the Test Example

In an arbitrary preabelian category, we define a *differential object* to be a pair $G, d : G \to G$ with $dd = 0$. If $G, d : G \to G$ is a differential object, we define the homology $H(G, d)$ of G, d as follows. We consider $m : Ker\ d \to G$. Since $dd = 0$ there exists a map $d' : G \to Ker\ d$ such that $md' = d$. This map has a cokernel $s : Ker\ d \to H(G, d)$.

If $G_1, d_1 : G_1 \to G_1$ is another differential object, and $f : G \to G_1$ satisfies $fd = d_1 f$, we can build a map $H(f) : H(G, d) \to H(G_1, d_1)$ characterised by the condition $s_1 d_1' fm = H(f)s$.

A *reduction* from G, d to G_1, d_1 is a triple f, g, h such that $f : G \to G_1$, $g : G_1 \to G$, $h : G \to G$ such that $d_1 f = fd$, $gd_1 = dg$, $fg = 1$, $gf = 1 - hd - dh$, $hh = fh = hg = 0$.

Proposition 2. *If f, g, h is a reduction from G, d to G_1, d_1 then $H(f)$, $H(g)$ define an isomorphism between $H(G, d)$ and $H(G_1, d_1)$, that is $H(g)H(f) = 1$ and $H(f)H(g) = 1$.*

6   Main Remaining Steps

We believe that the notion of preabelian category gives the right axiomatic level to formally represent what is going on in Kenzo. The next natural step is to represent the previous notion of equivalence and homology group for *chain-complexes*. This is naturally represented in a system with dependent types and we don't expect essential problems here. We hope then to be able to represent formally the Basic Pertubation Lemma as it is formulated in [15] (but for an arbitrary preabelian category). We should then, using [15] and the Pertubation Lemma as an essential tool, develop a library of programs to reduce a chain-complex to a chain-complex of finitely presented modules, for which one can compute explicitly the homology group.

The computationally challenging part is to represent the notion of *finitely presented* abelian group, and the main algorithms on these groups: effective computation of the kernel, cokernel via Smith reduction. This would involve computations on matrices of integers, and may be done by formalising Chapter V 1 of [14]. Using these algorithms one can then compute a canonical representation of the homology groups of a chain-complex of finitely presented modules. This should be a perfect test for the new representation of integers in type theory [16].

References

1. Aransay, J.M.: Razonamiento mecanizado en álgebra homológica. PhD thesis (2006)
2. Artin, M., Grothendieck, A., Verdier, J,-L.: Univers Séminaire de Géométrie Algébrique du Bois Marie -,1963-64 - Théorie des topos et cohomologie étale des schémas - (SGA 4) - vol. 1, LNM 269, 185-217, Springer-Verlag, Berlin, New York (1963)
3. Bishop, E.: Foundations of Constructive Analysis. McGraw-Hill, New York (1967)
4. de Bruijn, N.G.: Telescopic mappings in typed lambda calculus. Inform. and Comput. 91(2), 189–204 (1991)
5. Barakat, M., Robertz, D.: homalg: First steps to an abstract package for homological algebra. In: Proceedings of the X meeting on computational algebra and its applications (EACA 2006), Sevilla, Spain, pp. 29–32 (2006)
6. LogiCal project. The Coq Proof Assistant (2007) http://coq.inria.fr/V8.1/refman/index.html
7. Th. Coquand. Metamathematical investigations of a calculus of constructions. Rapport de recherche de l'INRIA (1989)
8. Gregoire, B., Leroy, X.: A Compiled Implementation of Strong Reduction. International Conference on Functional Programming (2002)
9. Huet, G., Saibi, A.: Constructive category theory. In: Proof, language, and interaction, 239–275, Found. Comput. Ser., pp. 239–275. MIT Press, Cambridge, MA (2000)
10. Landin, P.: The mechanical evaluation of expressions. Comput. J. 6, 308–320 (1964)
11. Martin-Löf, P.: An intuitionistic theory of types. In: Twenty-five years of constructive type theory (Venice, 1995), pp. 127–172, Oxford Logic Guides, 36, Oxford Univ. Press, New York (1998)
12. Melliès, P.-A., Werner, B.: A Generic Normalization Proof for Pure Type Systems. In: Paulin-Mohring, C., Gimenez, E. (eds.) TYPES'96. LNCS, Springer, Heidelberg (1997)
13. Miquel, A., Werner, B.: The Not So Simple Proof-Irrelevant Model of CC. In: Geuvers, H., Wiedijk, F. (eds.) TYPES 2002. LNCS, vol. 2646, Springer, Heidelberg (2003)
14. Mines, R., Richman, F., Ruitenburg, W.: A Course in Constructive Algebra. Springer, Heidelberg (1988)
15. Rubio, J., Sergereart, F.: Constructive Homological Algebra and Applications, Genove Summer School (2006) Available at http://www-fourier.ujf-grenoble.fr/šergerar/Papers/
16. Spiwack, A.: Ajouter des entiers machine à Coq. Master thesis, http://arnaud.spiwack.free.fr/

Appendix: Formal Representation in Coq

We have represented formally the notion of preabelian category, and checked that abelian groups form a preabelian category. We have then formulated and proved Lemma 3.3.1 of [1] in an arbitrary prebalian category.

Sum-Up of the Syntactic Facilities

Coq provides a lot of syntactic sugar to ease the work of the user. Here is a few words about these:

1. Record types: the syntax `Record` *name* : *univ* := *c* {*field* : *type* [;...]} allows to declare dependent tuple-types (*i.e. n*-ary Σ types) with named field. This is actually no extension of the theory shown in Sections 1 and 2. Internally, this syntax declares a regular Σ-type that goes in universe *univ*, and name it *name*. It also declares a *constructor i.e.* a term of type $\Pi f{:}type\dots.name$ this term has a fundamental meaning in Coq, but in a first approximation, one can see it as a tool for tactic-based proofs. Finally it declares a function *field* of type *name* $\rightarrow$ *type* for each field declaration (*field* : *type*). It is rather useful to use this notation when building very large tuples like the definition of a preabelian category.

2. Implicit coercions: in Coq, one can declare a function f as an implicit coercion between two *classes* of types, a class being basically an identifiable type construction. If f is a coercion from type t to type u, then f is automatically inserted whenever a term of type t is given where a term of type u is expected. This allows to consider a category as a type: for a category C there is a type of its object `dom C`. If we declare `dom` as a coercion, then we can "abuse notations" and write simply C instead of `dom C`.

3. Implicit coercions in record definition: as an additional notation we can define fields of a record as implicit coercions during the definition of the record with the syntax —*field* :> *type* instead of the usual —*field* : *type*. An intuitive way to read this syntactic construction would be to consider the new record type as an extension of *type*. For instance :

   `Record preabelian_category : Type := mk_preabcat { preab_cat :> category; ... }`

   reads "a preabelian category is a category with ...".

Preabelian Category

Here are additional notes to read this formalisation :

- The composition of `f` and `g` is written `f!g` (instead of `gf`).
- In Coq, $\Pi x{:}A.B$ and $\forall x{:}A.B$ are both written `forall x:A,B`. The design choices have led to use the dot as the end-of-line symbol, so a coma was use in the `forall` construction instead.
- The definition of the type `category` is not included, but it is worth noticing that both the domain function `dom` and the hom-set `hom` function have been

declared as coercions. Thus we can write C instead of dom C and C X Y instead of hom C X Y. As an example, forall (X Y:C) (f:C X Y), f == f reads "for all objects X and Y of the category C and for all arrow f in the hom-set from X to Y, f is equal to itself"

```
Require Export category.

Record preabelian_category : Type := mk_preabcat {
  preab_cat :> category;
  zero : preab_cat;
  zero_is_zero : zero_object zero;
  zerom : forall (X Y:preab_cat), preab_cat X Y;
  zerom_is_zero : forall (X Y:preab_cat), zero_morphism zero (zerom X Y);
  hom_plus : forall (X Y:preab_cat) (f g:preab_cat X Y), preab_cat X Y;
  hom_plus_morphism : forall (X Y:preab_cat) (f1 f2 g1 g2:preab_cat X Y),
                      f1==f2 -> g1==g2 ->
                      hom_plus f1 g1 == hom_plus f2 g2;
  hom_plus_assoc : forall (X Y:preab_cat) (f g h:preab_cat X Y),
                   hom_plus f (hom_plus g h ) ==
                   hom_plus (hom_plus f g) h;
  hom_plus_comm : forall (X Y:preab_cat) (f g:preab_cat X Y),
                  hom_plus f g == hom_plus g f;
  hom_plus_zero_l : forall (X Y:preab_cat) (f:preab_cat X Y),
                    hom_plus zerom f == f;
  hom_plus_zero_r : forall (X Y:preab_cat) (f:preab_cat X Y),
                    hom_plus f zerom == f;
  hom_minus : forall (X Y:preab_cat) (f:preab_cat X Y), preab_cat X Y;
  hom_minus_morphism : forall (X Y:preab_cat) (f1 f2:preab_cat X Y),
                       f1 == f2 -> hom_minus f1 == hom_minus f2;
  hom_minus_is_minus_l : forall (X Y:preab_cat) (f:preab_cat X Y),
                         (hom_plus f (hom_minus f)) == zerom ;
  hom_minus_is_minus_r : forall (X Y:preab_cat) (f:preab_cat X Y),
                         (hom_plus (hom_minus f) f) == zerom ;
  comp_plus_linear : forall (X Y Z:preab_cat) (f1 f2:preab_cat X Y)
                     (g:preab_cat Y Z),
                     (hom_plus f1 f2)!g ==
                     hom_plus (f1!g) (f2!g);
  comp_plus_colinear : forall (X Y Z:preab_cat) (f:preab_cat X Y)
                       (g1 g2:preab_cat Y Z),
                       f!(hom_plus g1 g2) ==
                       hom_plus (f!g1) (f!g2);
  comp_minus_linear : forall (X Y Z:preab_cat) (f:preab_cat X Y)
                      (g:preab_cat Y Z),
                      (hom_minus f)!g == hom_minus (f!g);
  comp_minus_colinear : forall (X Y Z:preab_cat) (f:preab_cat X Y)
                        (g:preab_cat Y Z),
```

```
                          f!(hom_minus  g) == hom_minus  (f!g);
biproduct :
  forall (A B:preab_cat),
   {A_B:preab_cat &
    {pA:preab_cat A_B A &
     {pB:preab_cat A_B B &
      {iA:preab_cat A A_B &
       {iB:preab_cat B A_B |
        hom_plus  (pA!iA) (pB!iB) == id /\
        iA!pA == id /\ iB!pB == id /\
        iA!pB == zerom  /\ iB!pA == zerom }}}}}
ker_obj : forall (X Y : preab_cat) (f:preab_cat X Y), preab_cat
ker_arr : forall (X Y : preab_cat) (f:preab_cat X Y),
                    preab_cat (ker_obj f) X;
ker_univ_arr : forall (H X Y : preab_cat) (f:preab_cat X Y)
                  (h:preab_cat H X) (p: h!f == zerom ),
                    preab_cat H (ker_obj f);
ker_univ_com : forall (H X Y : preab_cat) (f:preab_cat X Y)
                  (h:preab_cat H X) (p: h!f == zerom ),
                  h == (ker_univ_arr  f h p)!(ker_arr  f);
ker_univ_uniq : forall (X Y:preab_cat) (f:preab_cat X Y),
                  (h:preab_cat H X) (p: h!f == zerom ),
                  (i:preab_cat H (ker_obj  f)),
                  (q: h == i!(ker_arr  f)),
                  i == ker_univ_arr  f h p;
coker_obj : forall (X Y : preab_cat) (f:preab_cat X Y), preab_cat
coker_arr : forall (X Y : preab_cat) (f:preab_cat X Y),
                    preab_cat X (coker_obj  f);
coker_univ_arr : forall (H X Y : preab_cat) (f:preab_cat X Y)
                  (h:preab_cat Y H) (p: f!h == zerom ),
                    preab_cat (coker_obj  f) H;
coker_univ_com : forall (H X Y : preab_cat) (f:preab_cat X Y)
                  (h:preab_cat X H) (p: f!h == zerom ),
                  h == (coker_arr  f)!(coker_univ_arr  f h p);
coker_univ_uniq : forall (X Y:preab_cat) (f:preab_cat X Y),
                  (h:preab_cat Y H) (p: f!h == zerom ),
                  (i:preab_cat (coker_obj  f) H),
                  (q: h == (coker_arr  f)!i),
                  i == coker_univ_arr f h p;
}.

Notation "0" := (zerom) : preabelian_category_scope.
Notation "f + g" := (hom_plus f g) : preabelian_category_scope.
Notation "~ f" := (hom_minus f) : preabelian_category_scope.
Notation "f - g" := (f+~g)%preab : preabelian_category_scope.
```

The Test Example in Type Theory

We use implicit coercions to be able to write write both C for the type obj C of objects of C and C A B for the type hom C A B of morphisms from A to B.

```
Variable C:preabelian_category.

Variable G:C.
Variable h d:C G G.

Variable (dd_zero: d!d == 0) (hh_zero : h!h == 0) (hdh_h : h!d!h == h).

Definition p := d!h+h!d.

Lemma ph_h : p!h == h.

Lemma uv_v_implies_id_minus_u_v_zero :
  forall u v:C G G, u!v == v -> (id-u)!v == 0.

Lemma id_minus_p_h_zero : (id-p)!h == 0.

Lemma hp_h : h!p == h.

Lemma vu_v_implies_v_id_minus_u_zero :
  forall u v:C G G, v!u == v -> v!(id-u) == 0.

Lemma h_id_minus_p_zero : h!(id-p) == 0.

Lemma pp_p : p!p == p.

Lemma id_minus_p_p_zero : (id - p)!p == 0.

Lemma p_id_minus_p_zero : p!(id - p) == 0.

Definition K := ker_obj p.
Definition i : C K G := ker_arr p.

Lemma ip_zero : i!p == 0.

Lemma ih_zero : i!h == 0.

Definition j : C G K := ker_univ_arr p (id - p) id_minus_p_p_zero.

Lemma ji_id_minus_p : j!i == id - p.

Lemma hj_zero : h!j == 0.
```

What Might "Understand a Function" Mean?

J.H. Davenport*

Department of Computer Science, University of Bath, Bath BA2 7AY England
J.H.Davenport@bath.ac.uk
http://staff.bath.ac.uk/masjhd

Abstract. Many functions in classical mathematics are largely defined
in terms of their derivatives, so Bessel's function is "the" solution of
Bessel's equation, etc. For definiteness, we need to add other properties,
such as initial values, branch cuts, etc. What actually makes up "the
definition" of a function in computer algebra? The answer turns out to
be a combination of arithmetic and analytic properties.

1 Introduction

The claim is often made (these days generally informally) that a given computer
algebra system "understands" tan, or some other function, generally a function
defined through some analytic process. Here we ask three questions.

1. What does this mean?
2. What might it mean?
3. How should a system "understand" such a new function?

More generally, to what extent does such an analytic process, or a built-in function, define a function, and what properties does such a function have?

Notation: throughout this paper, the term 'function' means a total[1] function
from $\mathbf{R}$ to $\mathbf{R}$ or $\mathbf{C}$ to $\mathbf{C}$. The principles apply to functions $\mathbf{R}^n$ to $\mathbf{R}$ or $\mathbf{C}^n$ to $\mathbf{C}$,
but we shall not consider such functions here. C will denote an arbitrary field of
constants (of characteristic zero). From the point of view of differential algebra,
x will be the variable of differentiation/integration, i.e. $x' = 1$. From the point of
view of functions, x is the variable being evaluated. Functions such as log have
the meaning given in [1], as refined by [9].

We remind the reader of a couple of definitions from differential algebra.

Definition 1. θ *is said to be* elementary *over a differential field K if one of the
following is true:*

(a) θ *is algebraic over K;*
(b) $\theta' = \eta'/\eta$ *for some $\eta \in K$ (we write $\theta = \log \eta$);*
(c) $\theta' = \eta'\theta$ *for some $\eta \in K$ (we write $\theta = \exp \eta$).*

---

* The author is grateful to the referees, whose thoughtful comments significantly improved the paper. He is also grateful to Dr. Bradford for his comments.
[1] Or at least "total with singularities". We then define equality $f = g$ to mean that, at
all x where f and g are both defined, $f(x) = g(x)$ [12]. A full exposition of removable
singularities would be a paper in itself.

M. Kauers et al. (Eds.): MKM/Calculemus 2007, LNAI 4573, pp. 55–65, 2007.
© Springer-Verlag Berlin Heidelberg 2007

The object f is said to be elementary over K if it can be expressed in some $K(\theta_1, \ldots \theta_n)$ with each θ_i elementary over $K(\theta_i \ldots \theta_{i-1})$. If K is omitted, $C(x)$ is assumed.

Much of the theory of integration [5] is cast in terms of elementary functions. We can generalise the concept as follows.

Definition 2. *θ is said to be Liouvillian over a differential field K if one of the following is true:*

(a) *θ is algebraic over K;*
(b) *$\theta' = \eta$ for some $\eta \in K$ (we write $\theta = \int \eta$);*
(c) *$\theta' = \eta'\theta$ for some $\eta \in K$ (we write $\theta = \exp \eta$).*

The object f is said to be Liouvillian over K if it can be expressed in some $K(\theta_1, \ldots \theta_n)$ with each θ_i Liouvillian over $K(\theta_i \ldots \theta_{i-1})$. If K is omitted, $C(x)$ is assumed.

Note that, even if K is a field of functions embedded in $\mathbf{R} \to \mathbf{R}$ (or $\mathbf{C} \to \mathbf{C}$), there is no requirement that f should be such a function: we have merely stated a property of the *abstract* derivative of f. In practice, we also want each θ_i to be an elementary (resp. Liouvillian) *function* as well, i.e. that its numerical values, as well as its differential properties, be specified.

Definition 3. *Let K be a field of functions in $\mathbf{R} \to \mathbf{R}$ (or $\mathbf{C} \to \mathbf{C}$). $f(x)$, a function from $\mathbf{R} \to \mathbf{R}$ (or $\mathbf{C} \to \mathbf{C}$) is said to be an elementary (resp. Liouvillian) function if it lies in some elementary (resp. Liouvillian) extension $K(\theta_1, \ldots \theta_n)$ of K.*

However, even this is not enough.

Definition 4. *Let K be a field of functions in $\mathbf{R} \to \mathbf{R}$ (or $\mathbf{C} \to \mathbf{C}$). $f(x)$, a function from $\mathbf{R} \to \mathbf{R}$ (or $\mathbf{C} \to \mathbf{C}$) is said to be a proper elementary (resp. Liouvillian) function if it lies in some elementary (resp. Liouvillian) extension $K(\theta_1, \ldots \theta_n)$ of K, where each is θ_i proper elementary (resp. Liouvillian) over $K(\theta_i \ldots \theta_{i-1})$, and, for each x where both are defined,*

$$(f')(x) = \lim_{\epsilon \to 0} \frac{f(x + \epsilon) - f(x)}{\epsilon}. \tag{1}$$

Furthermore, we require that the right-hand side of (1) be defined almost everywhere.

As examples of the various pathologies that can occur, we give the following examples, where K is the field $\mathbf{Q}(x)$ of rational functions $\mathbf{C} \to \mathbf{C}$ equipped with the derivation induced by $x' = 1$.

1. $K(\theta)$ where $\theta' = \frac{1}{x}$. Here θ is merely an abstract symbol, not a function at all.
2. $K(\theta)$ where $\theta' = \frac{1}{x}$ and $\theta : x \mapsto 0$. Here θ is elementary, and a function, but not a proper elementary function since equation (1) is violated.

3. $K(\theta)$ where $\theta' = \frac{1}{x}$ and $\theta : x \mapsto \begin{cases} 1 & x \in \mathbf{Q} \\ 0 & x \notin \mathbf{Q} \end{cases}$. Here equation (1) is satisfied, but only because the right-hand side is nowhere defined, and therefore this falls foul of the last clause in definition 4.
4. $K(\theta = \log(x) + 42)$ where $\theta' = \frac{1}{x}$. This is indeed a proper elementary function in the sense of definition 4, even though it is not "what we all mean by" $\log x$.
5. $K(\theta)$ where $\theta' = \frac{1}{x}$ and $\theta : x \mapsto \log x + \begin{cases} 0 & x > 0 \\ -i\pi & x < 0 \end{cases}$. As a function $\mathbf{R} \to \mathbf{R}$ this is $\log|x|$, and is a proper elementary function in our sense. Whether it is "what we all mean by" $\log x$ has been debated elsewhere [31].

2   What Does It Mean?

1. Numerical evaluation. Generally speaking, if the input is real, this means real evaluation where possible. To do numerical evaluation, one has to choose the branch cuts (if there are any) of the relevant function — see [9].
2. Plotting — generally a consequence of the above, though more can in fact be done [2] if the function is better "understood".
3. Differentiation. This property is generally hard-coded for some functions, with an extension mechanism for others, e.g. defining diff/f for a function f in Maple, or giving a symbol a !*DF property in REDUCE.
4. Integration. This is the difficult one, and is discussed during much of the rest of this paper.
5. Special values. This is not the same as numerical evaluation (though the two can easily be confused): $\sin(\pi)$ is precisely 0, whereas

$$\sin(3.1415926535897932384626433) = 8.32795 \times 10^{-26}$$

(with an appropriate setting of Digits or the equivalent). This is a case where the precise nature, and the adherence [4], of the branch cuts is critical: $\log(-1.0 + \epsilon i)$ might be near either of πi or $-\pi i$, but $\log(-1)$ has (with the standard definitions) to be πi.
6. Simplification. Some of this is built in, e.g. for even/odd functions, as in $\sin(-x)$ and $\cos(-x)$: other simplifications can be invoked via commands such as expand or collect, or by giving functions properties (REDUCE).

3   Defining Functions

There are various ways by which new functions can be defined.

3.1   By Explicit Formulae, Normally Composition

"Let $h(x) = f(g(x))$". Provided that f and g are "understood", and that the system knows the chain rule, this more or less means that the system "understands" h, at least as well as it understands $f(g(x))$. This may be "not at all", as in the case of the real-valued function $\log \log \sin x$, which is nowhere defined.

Numerical and symbolic evaluation and plotting are, at least conceptually, simple. Difficulties can arise, though, if we expect the algebra system to remove removable singularities, i.e.

$$h(x) = \begin{cases} f(g(x)) & g(x) \text{ well-defined} \\ \lim_{y \to x} f(g(y)) & \text{otherwise.} \end{cases} \tag{2}$$

Expecting a system to perform (2) automatically is, in the author's opinion, expecting too much, though possibly systems might provide some help in this direction. Problems ought, where possible, to be signalled at definition time rather than at use time, so an explicit, tool-supported, definition mechanism is probably what should be provided. An example of what can go wrong is provided by $\arctan\left(\frac{1}{1-x}\right)$, where there is a jump discontinuity at $x = 1$ corresponding to the "discontinuity at infinity" of arctan.

3.2   By Indefinite Integration

One might define erf to be the integral[2] of $\exp(-x^2)$, or, more formally,

$$\operatorname{erf}(x) = \int_0^x \exp -t^2 \, dt,$$

in order to fix the constant of integration.

Such a definition tells us explicitly how to evaluate the function numerically[3], and implicitly how to differentiate the new function. Risch's algorithm [27,5] will tell us whether this is a 'new' function or can be defined in terms of previously known ones (though current systems are not always good at getting the constant of integration right).

Indefinite integration is much harder. All integration algorithms for elementary functions rely on Liouville's principle: that the only new elementary functions which can be introduced are logarithms, and that only with constant coefficients. This theorem remains true even if the integrand is not elementary. However, this is not what one actually wants. Just as we added a new logarithm to compute $\int \frac{1}{x \log x} = \log \log x$, we would like to add new error functions, or whatever is in the domain of discourse, and this is not always obvious. The first term in the following integral (taken from [7]) is pretty obvious, but it is far from clear where the last term comes from.

$$\int \operatorname{erf}(ax) \operatorname{erf}(bx) = x \operatorname{erf}(ax) \operatorname{erf}(bx) +$$

$$\frac{e^{-a^2 x^2} \operatorname{erf}(bx)}{\sqrt{\pi} a} + \frac{e^{-b^2 x^2} \operatorname{erf}(ax)}{\sqrt{\pi} b} - \operatorname{erf}\left(\sqrt{a^2 + b^2} x\right) \left(\frac{a}{b} + \frac{b}{a}\right) \frac{1}{\sqrt{\pi}} \frac{1}{\sqrt{a^2 + b^2}}$$

---

[2] In practice, one introduces a multiplicative factor of $2/\sqrt{\pi}$ to keep the statisticians happy, but the principle is the same.

[3] And hence how to plot it. However, there are much better, and more stable, ways of plotting an integral than via a sequence of *de novo* numerical evaluations.

Similarly [8]

$$\int \frac{x}{\log^2 x} = 2\,\mathrm{li}(x^2) - \frac{x^2}{\log x}, \tag{3}$$

where $\mathrm{li}(x) = \int \frac{1}{\log x}$, and one could wonder where the $\mathrm{li}(x^2)$ comes from.

In general, one needs a fresh generalisation of Liouville's Principle for each new function generator introduced. Some such have been proved [3,7,8,19,20], but even the most general [30] is far from complete: it deals with EL-elementary extensions subject to the restriction that, for each H in case (e) below, the degree of the numerator of H does not exceed the degree of the denominator by more than 1.

Definition 5. θ *is said to be EL-elementary over a differential field K if one of the following is true:*

(a) θ *is algebraic over K;*
(b) $\theta' = \eta'/\eta$ *for some $\eta \in K$ (we write $\theta = \log \eta$);*
(c) $\theta' = \eta'\theta$ *for some $\eta \in K$ (we write $\theta = \exp \eta$);*
(d) $\theta' = \zeta' R'(\zeta)\eta G(\eta))$ *and $\eta' = \zeta' R'(\zeta)\eta$ for some $\zeta \in K$ (we might[4] write* $\eta = \exp(R(\zeta))$ *and $\theta = \int G(\exp(R(\zeta))))$;*
(e) $\theta' = \zeta' \frac{S'(\zeta)}{S(\zeta)} H(\eta))$ *and $\eta' = \zeta' \frac{S'(\zeta)}{S(z\eta)}$ for some $\zeta \in K$ (we might[4] write* $\eta = \log(S(\zeta))$ *and $\theta = \int H(\log(S(\zeta))))$,*

where each of G, H, R, S are prescribed[5] rational functions of one variable. The function f is said to be EL-elementary over K if it can be expressed in some $K(\theta_1, \ldots \theta_n)$ with each θ_i EL-elementary over $K(\theta_i \ldots \theta_{i-1})$. If K is omitted, $C(x)$ is assumed.

For example, error functions would be coped with by having $(G, R) = (t \mapsto t, t \mapsto -t^2)$.

Special values are essentially then problems of definite integration. Tricks such as evaluating $\mathrm{erf}(\infty)$ by writing

$$\mathrm{erf}^2(\infty) = \left(\int_0^\infty \frac{2}{\sqrt{\pi}} e^{-x^2}\, dx \right) \left(\int_0^\infty \frac{2}{\sqrt{\pi}} e^{-y^2}\, dy \right)$$
$$= \int_0^\infty \int_0^\infty \frac{4}{\pi} e^{-x^2-y^2}\, dx dy$$
$$= \int_0^\infty \int_0^{\pi/2} \frac{4}{\pi} e^{-r^2} r\, dr d\theta$$
$$= 1$$

are within the scope of heuristics rather than algorithms at the current time.

---

[4] The use of "might" here indicates that the problem of introducing new constants by this formulation of such an extension is a delicate one.
[5] This is the original definition from [30]. In practice the (G, R) and (H, S) are prescribed pairs of rational functions, so that a given G goes with a given R, etc.

Even/odd simplifications are generally possible, but deducing further rules is again a matter for heuristics. If $F = \int_0 f$, then

$$F(a+b) = \int_0^a f + \int_a^{a+b} f = F(a) + \int_a^{a+b} f,$$

and if the last term can be transformed into $\int_0^c f$, then a simplification can be deduced.

3.3   By First Order Linear Differential Equations

In general, one would consider a y defined by

$$y' + fy = g, \tag{4}$$

with an initial condition equivalent to the constant of integration discussed above. Let $F = \int f$ and $y = z \exp(-F)$. Then (4) becomes

$$z' \exp(-F) - fz \exp(-F) + fz \exp(-F) = g, \tag{5}$$

i.e. $z' = g \exp(F)$. Hence

$$y = \exp(-F) \int (g \exp(F)), \tag{6}$$

and the problem is reduced to the previous one, i.e. the solution is Liouvillian over the field generated by f and g.

However, there are some caveats here [11]. The first is that any logarithms with rational coefficients in F have to be expressed explicitly, and the exponentiation has to perform the "simplification" $\exp \log(h)) \mapsto h$, thereby possibly adding radicals to the mix. The second is that, if F has any components other than logarithms with rational coefficients, then $\int (g \exp(F)) = G \exp(F)$, and then $\exp(-F)$ and $\exp(F)$ cancel, and the integration is in fact the solution of the original differential equation.

3.4   By Higher Order Linear Differential Equations

If we take second-order linear differential equations with coefficients in $C(x)$, there are four possibilities [22] (generalised in [29] to Liouvillian coefficients).

(1) There is a solution of the form $e^{\int f}$, where f is a rational function. In this case, the differential operator factorises, and we get a first order equation, the solutions of which are always Liouvillian.

(2) The first case is not satisfied, but there is a solution of the form $e^{\int f}$, where f satisfies a quadratic equation with rational functions as coefficients. In this case, the differential operator factorises and we get a first order equation, the solutions of which are always Liouvillian.

(3) The first two cases are not satisfied, but there is a non-zero Liouvillian solution. In this case, every solution is an algebraic function.

(4) The non-zero solutions are not Liouvillian.

We could extend the definition of "Liouvillian" to allow solutions of second-order differential equations (normally called "Eulerian"), and ask whether differential equations can be solved in terms of Eulerian functions [28], and so on, but the underlying differential Galois theory becomes intractable.

Is this function "new"?. A more fundamental question might be: "can special function g, defined as a solution of equation (g), be expressed in terms of special function f, defined as a solution of equation (f)?" Assuming that (f) and (g) have order greater than one, this would be more precisely defined as "can special function g, defined as a solution of equation (g) with given initial conditions, be defined in terms of a basis $f_1, \ldots, f_n$ of solutions of (f)?"

An example is given by the Bessel functions. [1, chap. 9] defines J_ν as solutions of

$$x^2 y'' + xy' + (x^2 - \nu^2)y = 0, \tag{7}$$

whereas in [1, chap. 10], j_n is defined as solutions of

$$x^2 y'' + 2xy' + (x^2 - n(n+1))y = 0. \tag{8}$$

These are connected by $j_n(x) = \sqrt{\frac{\pi}{2x}} J_{n+\frac{1}{2}}(x)$. Can such a relationship be deduced automatically? If we know that $j_n(x)$ should be of the form[6] $J_{n+\frac{1}{2}}(x)/f(x)$, the fact that $f(x)$ is of the form $c\sqrt{x}$ can be deduced relatively easily. If we assume merely that $j_n(x)$ is of the form $J_k(x)/f(x)$, the correct solution can still be deduced. Similarly, if we are faced with

$$x^2 y'' + 2xy' + 4(x^4 - n(n+1))y = 0, \tag{9}$$

and if we suspect that the solution is of the form $J_n(f(x))$, deducing $f(x) = x^2$ is not too hard. However, given

$$4\left(y''(x)\right)x^2 + \left(16\,x^4 - 16\,n^2 + 1\right)y(x) \tag{10}$$

and the suspicion that y is of the form $J_n(f(x))g(x)$, the author knows of no way of recovering the true answer — $J_n(x^2)\sqrt{x}$. It is possible that the results of [21] might help us to know which equation was related to which other equation, but, despite the call in [24], little seems to have been done in this direction.

Properties of the Function. The differential equation and suitable initial conditions will, in general, allow numerical evaluation, and thence plotting[3]. Formal integration of the differential equation will lead to a corresponding equation for the integral, so the question of integration reduces to the "is it related" question.

Even/odd simplification rules can be deduced from the differential equation, where appropriate. More general rules, and special values, are even more intractable than they are for integrals.

---

[6] The author is not sure whom the factor $\sqrt{\frac{\pi}{2}}$ is meant to please.

3.5 By Functional Equations

The simplest functional equation is the polynomial one: y such that $P(x,y) = 0$. If this is soluble by radicals, then we can import the branch cut for logarithm (though the result may be messy, and we need to worry about false solutions, as in Cardan's formula for the cubic [25]).. If it is not soluble by radicals, then there appears to be no "natural" placement for the branch cuts.

About the simplest non-algebraic functional equation is $ye^y = x$, whose solution is the Lambert W function [10]. This is not elementary or Liouvillian [6], but can also be defined by a non-linear differential equation: $W'(x) = \frac{W(x)}{(1+W(x))x}$. Just as log has infinitely many variants, separated by $2k\pi i$, which can be chosen to have a common branch cut, conventionally[7] the negative real axis, with the cut itself adhering [4] to the upper half-plane, so W has infinitely many branches, but the description is somewhat more complex [10,17].

The analysis of W was very much *ad hoc*, and the author knows of no systematic approach to such equations, unless they can be reduced to differential equations, as in the next section.

3.6 By Non-linear Differential Equations

The Lambert W function (see above) is one such. The question posed above "is this function definable in terms of that one", becomes even more relevant in this setting, and there are some surprising results: [14] gives the solution[8] to

$$(4y + 2x + 3)y' - 2y - x - 1 = 0 \tag{11}$$

as

$$4096\exp\left(-W\left(32768\exp(8x + c)\right) + 8x + c\right) - \frac{x}{2} - \frac{5}{8}. \tag{12}$$

The author knows of no way of deducing this in any sensible manner. The constant $5/8$ is easy enough to determine, as is the $1/2$, but the overall structure of the integral, necessary for any 'method of undetermined coefficients' to succeed, is not obvious.

3.7 By Definite Integration

The classic example of this is the Γ function, defined by

$$\Gamma(z) = \int_0^\infty t^{z-1}e^{-t}dt. \tag{13}$$

This is continuous over the whole of the complex plane, except for $z=0,-1,-2\ldots$.. It cannot be defined by a differential equation [16]. As far as the author was[9]

---

[7] There is nothing special about this choice: see [9].

[8] A reviewer pointed out that the solution can also be found by Maple as $W(c\exp(8x)) - \frac{x}{2} - \frac{5}{8}$, but this does not fully answer the question: does equation (11) have a solution in the form of (12) with fully undetermined coefficients?

[9] The reviewer pointed out [15], which has some techniques for negative results.

aware, there have been no attempts to systematise this analysis. There are heuristics in some packages (e.g. Maple), which sometimes produce differential equations. Hence it seems that, at the current time, there is nothing that a system can do *in general* except say "OK: you seem to have defined a function, which I can (generally) evaluate numerically".

3.8 Interrelations Between Methods

As we have seen, W can be defined either by a functional equation or by a (nonlinear) differential equation. In this case, going from the functional equation to the differential equation is fairly straight-forward, and mechanised in Maple's `PDEtools`, but the author knows no general way of reversing the process, or of knowing whether it is reversible.

4 Branch Cuts

These are inevitable for certain functions defined by integration or other analytic processes. Just to remind ourselves, let us look again at the branch cut for log. $\oint_C \frac{1}{x} dx = 2\pi i$, where C is the unit circle (traversed counter-clockwise). Hence any *continuous* definition of $\log z = \int_1^z \frac{1}{x} dx$ is bound to be multi-valued by multiples of $2\pi i$. Hence the minimum[10] branch cut necessary is a cut from 0 to (complex) infinity, with the value of log decreasing by $2\pi i$ as one crosses the branch cut in the direction of C (and increasing if one crosses it the other way). This poses two questions.

– What shape and where should the cut be?
– What happens *on* the cut?

In answer to the first, Occam's razor suggests that the cut might as well be a straight line from the origin to complex infinity. Note that this is not *mathematically necessary*, merely philosophically desirable[11]. Occam's razor again suggests that the cut might as well be along one of the axes. The current favourite [1] seems to be along the negative real axis, though the positive real axis has also been used.

In answer to the second question, clearly any behaviour is possible. Adherence to one side or the other (i.e. the value on the cut is the limit as you approach the cut from a given direction) seems a reasonable stipulation, as does the fact that the decision be taken consistently on the cut (if t parameterises the cut, we could insist on upper continuity for rational t, and lower continuity for irrational t, but this seems perverse) where possible. We stipulate "where possible" because

---

[10] We could always add "unnecessary" and cancelling branch cuts by arbitrary (but cancelling) amounts, but we will assume that this is not done.

[11] We note that the International Date Line, which can be viewed as a branch cut of $\sqrt[7]{}$ is, for geopolitical reasons, not a straight line, but is piecewise straight, at least in its current incarnation [13, `international_date.html`].

branch cuts may bifurcate or merge: see [17, Figure 2] Beyond this, logic and Occam's razor make no suggestions. There are two common schools of thought, both of which can lay claim to being "consistent" in their own ways.

Independent. consistency. Here we have a rule for all branch cuts. The common one is "counter-clockwise consistency", advocated in [18], see also [26]. Here one defines continuity on the branch cut as continuity with the region from which one approaches the cut when circling the origin counter-clockwise.

Dependent. consistency. Here one stipulates that, if h can be derived from g, i.e. $h(x) = f_1(g(f_2(x)))$ where f_1 and f_2 have no, or "simpler" cuts, then the branch cuts and adherence of h are derived from those of g. This is largely the approach taken in [1]: one defines the branch cuts for log and the rest follow. Difficulties occur when there are alternative definitions for h, say $h = \hat{f}_1(g(\hat{f}_2(x)))$, which might induce different branch cuts or adherence. Hence this approach only makes sense when a *particular* definition of h in terms of g is fixed. [26].

5   Conclusion

There has been comparatively little systematic work in this area: an early attempt was [24], which urged the consideration of [21], but little has been done in this direction. Perhaps the most interesting is [23].

To define functions completely, one has to know the branch cuts and their behaviour, and nothing has been done about automating this — largely because there is no consistent philosophy here. Indeed, it would be a significant step forward to have a system capable of checking that a definition of, say, a proper Liouvillian function and its branch cuts *was* consistent.

Hence the answer to the question "how should a system understand a new function" at the moment seems to be "we don't know, in general".

References

1. Abramowitz, M., Stegun, I.: Handbook of Mathematical Functions with Formulas, Graphs, and Mathematical Tables. US Government Printing Office (1964)
2. Avitzur, R., Bachmann, O., Kajler, N.: From Honest to Intelligent Plotting. RIACA Technical Report 5 (1995)
3. Baddoura, J.: A conjecture on integration in finite terms with elementary functions and polylogarithms. In: Proceedings ISSAC 1994, pp. 158–162 (1994)
4. Beaumont, J.C., Bradford, R.J., Davenport, J.H., Phisanbut, N.: Adherence is Better Than Adjacency. In: Kauers, M. (ed.) Proceedings ISSAC 2005, pp. 37–44 (2005)
5. Bronstein, M.: Symbolic Integration I, 2nd edn. Springer-Verlag, Heidelberg (2005)
6. Bronstein, M., Corless, R., Davenport, J.H., Jeffrey, D.J.: Algebraic properties of the Lambert W function from a result of Rosenstein and Liouville. To appear in J. Integral Transforms and Special Functions (2007)
7. Cherry, G.W.: Integration in Finite Terms with Special Functions: the Error Function. J. Symbolic Comp. 1, 283–302 (1985)

8. Cherry, G.W.: Integration in Finite Terms with Special Functions: the Logarithmic Integral. SIAM J. Computing 15, 1–21 (1986)
9. Corless, R.M., Davenport, J.H., Jeffrey, D.J., Watt, S.M.: According to Abramowitz and Stegun. SIGSAM Bulletin 2, 34, 58–65 (2000)
10. Corless, R.M., Gonnet, G.H., Hare, D.E.G., Jeffrey, D.J., Knuth, D.E.: On the Lambert W Function. Advances in Computational Mathematics 5, 329–359 (1996)
11. Davenport, J.H.: y'+fy=g. In: Fitch, J. (ed.) EUROSAM 84. LNCS, vol. 174, pp. 341–350. Springer, Heidelberg (1984)
12. Davenport, J.H.: Equality in computer algebra and beyond. J. Symbolic Comp. 34, 259–270 (2002)
13. United States Naval Observatory Astronomy Application Department. Frequently Asked Questions (2004) http://aa.usno.navy.mil/faq/docs
14. Dubinova, I.D.: Exact solution of some nonlinear differential equations. Diff. Eq. 40, 1195–1196 (2004)
15. Flajolet, P., Gerhold, S., Salvy, B.: On the non-holonomic character of logarithms, powers, and the n-th prime function (2005) http://arxiv.org/abs/math.CO/0501379
16. Hölder, O.: Über die Eigenschaft der Gamma Funktion keineralgebraischen Differentialgleichungen zu genügen. Math. Ann. 28, 1–13 (1887)
17. Jeffrey, D.J., Hare, D.E.G., Corless, R.M.: Unwinding the branches of the Lambert W function. Mathematical Scientist 21, 1–7 (1996)
18. Kahan, W.: Branch Cuts for Complex Elementary Functions. The State of Art in Numerical Analysis, pp. 165–211 (1987)
19. Knowles, P.H.: Integration of a Class of Transcendental Liouvillian Functions with Error Functions Part I. J. Symbolic Comp. 13, 525–543 (1992)
20. Knowles, P.H.: Integration of a Class of Transcendental Liouvillian Functions with Error Functions Part II. J. Symbolic Comp. 16, 227–241 (1993)
21. Kolchin, E.: Algebraic groups and algebraic dependence. Amer. J. Math. 90, 1151–1164 (1968)
22. Kovacic, J.J.: An Algorithm for Solving Second Order Linear Homogeneous Differential Equations. J. Symbolic Comp. 2, 3–43 (1986)
23. Meunier, L., Salvy, B.: ESF: An Automatically Generated Encyclopedia of Special Functions. In: Sendra, J.R. (ed.) Proceedings ISSAC 2003, pp. 199–206 (2003)
24. Moses, J.: Towards a General Theory of Special Functions. Comm. ACM 15, 550–554 (1972)
25. Nickalls, R.W.D.: A New Approach to Solving the Cubic: Cardan's Solution Revealed. Math. Gazette 77, 354–359 (1993)
26. Rich, A.D., Jeffrey, D.J.: Function Evaluation on Branch Cuts. SIGSAM Bulletin 2, 30, 25–27 (1996)
27. Risch, R.H.: The Problem of Integration in Finite Terms. Trans. A.M.S. 139, 167–189 (1969)
28. Singer, M.F.: Solving Homogeneous Linear Differential Equations in Terms of Second Order Linear Differential Equations. Amer J. Math. 107, 663–696 (1985)
29. Singer, M.F.: Liouvillian solutions of linear differential equations with Liouvillian coefficients. J. Symbolic Comp. 11, 251–273 (1991)
30. Singer, M.F., Saunders, B.D., Caviness, B.F.: An Extension of Liouville's Theorem on Integration in Finite Terms. SIAM J. Comp. 14, 966–990 (1985)
31. Stoutemyer, D.R.: Should computer algebra programs use ln x or ln |x| as their antiderivation of 1/x. DERIVE Newsletter 26, pp. 3–6 (Jun 1997)

Biform Theories in Chiron

William M. Farmer

McMaster University
Hamilton, Ontario, Canada
wmfarmer@mcmaster.ca

Abstract. An *axiomatic theory* represents mathematical knowledge declaratively as a set of *axioms*. An *algorithmic theory* represents mathematical knowledge procedurally as a set of *algorithms*. A *biform theory* is simultaneously an axiomatic theory and an algorithmic theory. It represents mathematical knowledge both declaratively and procedurally. Since the algorithms of algorithmic theories manipulate the syntax of expressions, biform theories—as well as algorithmic theories—are difficult to formalize in a traditional logic without the means to reason about syntax. *Chiron* is a derivative of von-Neumann-Bernays-Gödel (NBG) set theory that is intended to be a practical, general-purpose logic for mechanizing mathematics. It includes elements of type theory, a scheme for handling undefinedness, and a facility for reasoning about the syntax of expressions. It is an exceptionally well-suited logic for formalizing biform theories. This paper defines the notion of a biform theory, gives an overview of Chiron, and illustrates how biform theories can be formalized in Chiron.

1 Introduction

The mission of *mechanized mathematics* is to develop software systems that support the process people use to create, explore, and apply mathematics. There are historically two major approaches to mechanized mathematics, *computer theorem proving* and *computer algebra*. Computer theorem proving emphasizes the conjecture proving aspect of the mathematics process and usually represents mathematical knowledge as "axiomatic theories". On the other hand, computer algebra focuses on the computational aspect of the mathematics process and usually represents mathematical knowledge as "algorithmic theories".

An *axiomatic theory* is a set of formulas in a language L called *axioms* that serve as the background assumptions of the theory. The axioms encode a set of mathematical truths, namely, the formulas of L that are the logical consequences of the axioms. There is thus a clear demarcation between what is assumed (the axioms) and what is derived (the logical consequences of the axioms). The deduction and computation rules for reasoning within the theory are usually expressed in the metalanguage of L, not in L itself. This is because deduction and computation rules cannot directly manipulate values such as numbers, functions, and sets; they can only manipulate the expressions that denote these values. Traditional logics do not usually provide a facility for formalizing the syntax of

M. Kauers et al. (Eds.): MKM/Calculemus 2007, LNAI 4573, pp. 66–79, 2007.

expressions. As a result, neither the specifications of deduction and computation rules nor the algorithms that implement them can be directly expressed in an axiomatic theory.

An *algorithmic theory* is a set of algorithms that manipulate expressions in a language L. The background assumptions of the theory and the specifications of the algorithms are usually not part of an algorithmic theory; they are instead part of the informal metatheory of the theory. An algorithmic theory can be used to manipulate expressions, but it cannot be used to understand what the results of the manipulations mean. Also, unlike an axiomatic theory, there is no clear demarcation between the algorithms that are primitive in the theory and those that are derived from the primitive algorithms.

A *biform theory* T is a set Ω of formulas and rules in a language L. A *rule* in L consists of an algorithm called a *transformer* that transforms a tuple of input expressions of L into an output expression of L and a *meaning formula* that specifies how the values of the input expressions are related to the value of the output expression. For each tuple I of input expressions, the meaning formula M reduces to a formula M_I that specifies the relationship between the values of the members of I and the value of the resulting output expression. M_I is said to be an *instance* of the rule.

The notion of a biform theory merges the notions of an axiomatic theory and an algorithmic theory. In fact, a biform theory is simultaneously both an axiomatic theory and an algorithmic theory. The axiomatic theory of T, written T_{axm}, is the set of formulas in Ω together with the set of the instances of all the rules in Ω, while the algorithmic theory of T, written T_{alg}, is the set of the transformers of all the rules in Ω.

The formulas and rules in Ω are called the *axioms* of T. They are implicit background assumptions of T, and the axioms of T_{axm} are the explicit background assumptions of T. A rule is a *logical consequence* of T if its instances are logical consequences of T_{axm}. Thus in a biform theory there is a clear demarcation between primitive formulas and rules whose correctness is assumed and derived formulas and rules whose correctness is a logical consequence of the primitive formulas and rules.

In summary, a biform theory includes both formulas and rules as primitive assumptions. A rule consists of an algorithm that manipulates expressions and a formula that specifies what the manipulations of the expressions mean semantically. A biform theory is simultaneously both an axiomatic theory and an algorithmic theory. The meaning of an algorithm of the algorithmic theory is understood in the context of the axiomatic theory. And there is a clear definition of what a derived formula or rule is in a biform theory.

The notion of a biform theory was first introduced as part of FFMM, a Formal Framework for Managing Mathematics [11] developed as part of the MathScheme project [15] at McMaster University. One of the principal goals of FFMM is to integrate and generalize computer theorem proving and computer algebra. Biform theories play a central role in FFMM by providing a formal context in which deduction and computation can be merged. In general, biform theories are useful for formalizing mathematics in which deduction and computation are intimately related. For applications of biform theories outside of FFMM, see [4,5,6].

A mechanized mathematics that utilizes biform theories to represent mathematical logic needs a logic in which biform theories can be expressed. At the very least, it must be possible to express in the logic the meaning formulas of rules. Otherwise, there is no formal basis for understanding what a transformer of a rule means. This is problematic because a meaning formula expresses statements both about the syntax of expressions and what the expressions mean. Traditional logics are usually not equipped with the means to express statements about syntax and to reason about syntax.

The transformer of a rule does not need to be expressed in the logic. As long as its corresponding meaning formula is expressed in the logic, it can treated as a black-box algorithm that is assumed to behave according to its meaning formula. In other words, the transformer's rule would be considered as an axiom of the biform theory. Hence an algorithm in the form of a program in a high-level programming language can be made into a perfectly legitimate rule if a meaning formula for it can be expressed in the logic.

Chiron [7,9] is a derivative of von-Neumann-Bernays-Gödel (NBG) set theory that is intended to be a practical, general-purpose logic for mechanizing mathematics. It includes elements of type theory, a scheme for handling undefinedness, and a facility for reasoning about the syntax of expressions. Chiron has a high level of both theoretical and practical expressivity [7]. It is an exceptionally well-suited logic for formalizing biform theories. In particular, the meaning formulas of rules can be directly expressed in Chiron.

This paper defines the notion of a biform theory, gives an overview of Chiron, and illustrates how biform theories can be formalized in Chiron. Section 2 defines the notions of a transformer, a rule, and a biform theory. Section 3 gives a quick introduction to Chiron and shows how rules are expressed in Chiron. Section 4 sketches the development in Chiron of a nontrivial example of a biform theory. The paper ends with a conclusion in Section 5 that discusses related and future work.

2   Biform Theories

We present here a formulation of a biform theory that is simpler than the formulation given in [11].

2.1   General Logics

A *general language* is a pair $L = (\mathcal{E}, \mathcal{F})$ where $\mathcal{E}$ is a set of syntactic entities called the *expressions* of L and $\mathcal{F} \subseteq \mathcal{E}$ is a set of expressions called the *formulas* of L. For example, if F is a first-order language, then $L_F = (\mathcal{S} \cup \mathcal{F}, \mathcal{F})$ is a general language where $\mathcal{S}$ and $\mathcal{F}$ are the sets of terms and formulas of F, respectively. In the rest of this paper, let $L = (\mathcal{E}, \mathcal{F})$ be a general language.

A *general logic* is a set of general languages with a notion of logical consequence. In the rest of this paper, let **K** be a general logic. L is a *language* of **K** if it is one of the general languages of $\mathbf{K}$. If L is a language of $\mathbf{K}$ and $\Sigma \cup \{A\}$ is a set of formulas of L, then $\Sigma \models_{\mathbf{K}} A$ means A is a logical consequence of Σ in $\mathbf{K}$. For example, let $\mathbf{FOL}$ be a general logic representation of first-order logic such that L is a language of $\mathbf{FOL}$ iff $L = L_F$ for some first order language F and $\Sigma \models_{\mathbf{FOL}} A$ means A is a logical consequence of Σ in first-order logic.

An *axiomatic theory* in $\mathbf{K}$ is a pair $T = (L, \Gamma)$ where $L = (\mathcal{E}, \mathcal{F})$ is a language of $\mathbf{K}$ and $\Gamma \subseteq \mathcal{F}$. L is the *language* of T, and Γ is the set of *axioms* of T. A formula A of L is a *logical consequence* of T if $\Gamma \models_{\mathbf{K}} A$.

2.2 Transformers

For $n \geq 0$, an n-ary *transformer in* L is a pair $\Pi = (\pi, \hat{\pi})$ where π is a symbol and $\hat{\pi}$ is an algorithm that implements a (possibly partial) function $f_{\hat{\pi}} : \mathcal{E}^n \to \mathcal{E}$. The symbol π serves as a name for the algorithm $\hat{\pi}$. There is no restriction on how the algorithm is presented. For example, it could be a lambda-expression of L or a program written in a high-level programming language like C or Java.

Let $\mathsf{dom}(\Pi)$ denote the domain of $\hat{\pi}$, i.e., the subset of $\mathcal{E}^n$ on which $f_{\hat{\pi}}$ is defined. Suppose $E_1, \ldots, E_n$ are expressions in $\mathcal{E}$. If $(E_1, \ldots, E_n) \in \mathsf{dom}(\Pi)$, the expression $\pi(E_1, \ldots, E_n)$ denotes the output of $\hat{\pi}$ when given $E_1, \ldots, E_n$ as input, i.e., it denotes $f_{\hat{\pi}}(E_1, \ldots, E_n) \in \mathcal{E}$ (and is thus *defined*). If $(E_1, \ldots, E_n) \notin \mathsf{dom}(\Pi)$, $\pi(E_1, \ldots, E_n)$ does not denote anything (and is thus *undefined*). The expression $\pi(E_1, \ldots, E_n)$ is not required to be in $\mathcal{E}$; it will usually be an expression of the metalanguage of L but not of L itself.

Example 1. Suppose $L_F = \{\mathcal{E}_F, \mathcal{F}_F\}$ is the general language corresponding to a first-order language F. Let $\Pi = (\pi, \hat{\pi})$ be a unary transformer in L_F such that:

1. $\pi(E)$ is defined iff $E \in \mathcal{F}_F$.
2. If $\pi(A)$ is defined, it denotes a formula $B \in \mathcal{F}_F$ that is in prenex normal form and is logically equivalent to A.

That is, the algorithm $\hat{\pi}$ transforms any formula of L_F into a logically equivalent formula in prenex normal form. The expression $\pi(E)$ cannot be an expression in $\mathcal{E}_F$ (without some mechanism, such as Gödel numbering, for formalizing the syntax of L_F in L_F itself). $\square$

Example 2. Suppose $L_F = \{\mathcal{E}_F, \mathcal{F}_F\}$ is again the general language corresponding to a first-order language F. Let $\Pi = (\pi, \hat{\pi})$ be a ternary transformer in L_F such that:

1. $\pi(E_1, E_2, E_3)$ is defined iff E_1 is a term of F, E_2 is a variable of F, and E_3 is a formula of F.
2. If $\pi(t, x, A)$ is defined, it denotes the result of simultaneously substituting t for each free occurrence of x in A.

That is, given t, x, A, the algorithm $\hat{\pi}$ transforms the formula A into the formula $A[x \mapsto t]$. Again the expression $\pi(E_1, E_2, E_3)$ cannot be an expression in $\mathcal{E}_F$. $\square$

Example 3. Let **STT** be a general logic representation of simple type theory [8]. Suppose $T = (L, \Gamma)$ is an axiomatic theory of a complete ordered field in **STT** and that we have defined in T a type real of real numbers and the basic concepts of calculus such as limits, continuity, derivatives, etc. Let $\Pi = (\pi, \hat{\pi})$ be a unary transformer in L such that:

1. $\pi(E)$ is defined iff E is an expression of L of type real $\rightarrow$ real.
2. If $\pi(E)$ is defined, it is an expression of L of type real $\rightarrow$ real that denotes the derivative of the function denoted by E.

That is, $\hat{\pi}$ is an algorithm that differentiates expressions that denote functions on the real numbers. □

An *algorithmic theory* is a pair $T = (L, \Delta)$ where L is a general language and Δ is a set of transformers in L. L is called the *language* of T, and Δ is the set of *algorithms* of T. For more on transformers, see [10,11].

2.3 Rules

A *rule* in L is a pair $R = (\Pi, M)$ where:

1. $\Pi = (\pi, \hat{\pi})$ is an n-ary transformer in L.
2. M is a formula that uses π to relate the values of the inputs to $\hat{\pi}$ to the value of the output of $\hat{\pi}$.

The *transformer* of R, written trans(R), is Π, and the *meaning formula* of R, written mean(R), is M. The meaning formula M, which specifies the semantic relationship between the tuple of inputs and the output of the algorithm $\hat{\pi}$, will usually be an expression of the metalanguage of L but not of L itself. For each n-tuple $I = (E_1, \ldots, E_n)$ of inputs to $\hat{\pi}$, we assume that M reduces to a formula M_I of L which is called the *instance* of M with respect to I. An instance of M specifies the relationship between the values of a given tuple of input expressions and the value of the resulting output expression. Let inst(R) be the set of instances of M. M can often be conveniently expressed as a formula schema.

Example 4. Let $R = (\Pi, M)$ where:

1. $\Pi = (\pi, \hat{\pi})$ is the transformer in L_F given in Example 1.
2. M is the formula schema

$$A \equiv \pi(A)$$

where A is a formula of L_F.

If A is the formula $p(c) \supset \forall x \,.\, q(x)$ (where c is a constant) and the result of applying $\hat{\pi}$ to (A) is $\forall x \,.\, p(c) \supset q(x)$, then

$$(p(c) \supset \forall x \,.\, q(x)) \equiv (\forall x \,.\, p(c) \supset q(x))$$

is the instance of M with respect to (A). □

Example 5. Let $R = (\Pi, M)$ where:

1. $\Pi = (\pi, \hat{\pi})$ is the transformer in L_F given in Example 2.
2. M is the formula schema

$$(x = t \wedge A) \supset \pi(t, x, A)$$

where t is a term, x is a variable, and A is a formula of L_F and t is free for x in A.

If t is a term, x is a variable, and A is $f(x, y) = g(x)$, then

$$(x = t \wedge f(x, y) = g(x)) \supset f(t, y) = g(t)$$

is the instance of M with respect to (t, x, A).   □

Example 6. Let $R = (\Pi, M)$ where:

1. $\Pi = (\pi, \hat{\pi})$ is the transformer in the language L of the theory T given in Example 3.
2. M is the formula schema

$$\mathsf{derivative}(E) = \pi(E)$$

where E is of type real $\rightarrow$ real. derivative is an expression of L of type

$$(\mathsf{real} \rightarrow \mathsf{real}) \rightarrow (\mathsf{real} \rightarrow \mathsf{real})$$

that maps a function to its derivative. M thus asserts that the derivative of the function denoted by E is the function denoted by $\pi(E)$.

If E is $\lambda x : \mathsf{real} . x^2$, then

$$\mathsf{derivative}(\lambda x : \mathsf{real} . x^2) = (\lambda x : \mathsf{real} . 2 \cdot x)$$

is the instance of M with respect to (E).   □

For the sake of convenience, we will view a formula A of L as a (transformer-less) rule in L and assume that $\mathsf{trans}(A)$ is undefined, $\mathsf{mean}(A) = A$, and $\mathsf{inst}(A) = \{A\}$.

2.4   Biform Theories

A *biform theory* in $\mathbf{K}$ is a pair $T = (L, \Omega)$ where L is a language of $\mathbf{K}$ and Ω is a set of rules in L. (Ω may include formulas of L viewed as transformer-less rules.) L is the *language* of T, and Ω is the set of *axioms* of T.

T can be viewed as simultaneously both an *axiomatic theory* and an *algorithmic theory*. The *axiomatic theory* of T is the axiomatic theory $T_{\mathrm{axm}} = (L, \Gamma)$ in $\mathbf{K}$ where

$$\Gamma = \bigcup_{R \in \Omega} \mathsf{inst}(R),$$

while the *algorithmic theory* of T is the algorithmic theory $T_{\mathrm{alg}} = (L, \Delta)$ where

$$\Delta = \{\mathsf{trans}(R) \mid R \in \Omega \text{ and } \mathsf{trans}(R) \text{ is defined}\}.$$

The axioms of T—which are formulas and rules—are the background assumptions of T in an implicit form. The axioms of T_{axm}—which are formulas alone—are the background assumptions of T in an explicit form. A rule R in L is a *logical consequence* of T if, for all formulas $A \in \mathsf{inst}(R)$, A is a logical consequence of T_{axm}. Thus, the axioms of T are trivially logical consequences of T. Notice also that, since we are assuming that the formulas of L are rules in L, every logical consequence of T_{axm} is also a logical consequence of T.

3   Chiron

A formal, complete presentation of the syntax and semantics of Chiron is given in [9], and a shorter, more informal presentation is given in [7].

3.1   Values

The semantics of Chiron is based on the notion of a *standard model* which is an elaboration of a model of NBG set theory. The basic values or elements in a model of NBG are classes (which include sets and proper classes).[1] A standard model M includes other values besides classes, but classes are the most important. M is derived from a structure, consisting of a nonempty domain D_{c} of classes and a membership relation $\in$ on D_{c}, that satisfies the axioms of NBG set theory as given, for example, in [13] or [16]. The *values* of M include sets, classes, superclasses, truth values, the undefined value, and operations.

A *class* of M is a member of D_{c}. A *set* of M is a member x of D_{c} such that $x \in y$ for some member y of D_{c}. That is, a set is a class that is itself a member of a class. A class is thus a collection of sets. A class is *proper* if it is not a set. A *superclass* of M is a collection of classes in D_{c}. We consider a class, as a collection of sets, to be a superclass itself. Let D_{v} be the domain of sets of M and D_{s} be the domain of superclasses of M. The following inclusions hold: $D_{\mathrm{v}} \subset D_{\mathrm{c}} \subset D_{\mathrm{s}}$. D_{v} is the universal class (the class of all sets), and D_{c} is the universal superclass (the superclass of all classes).

T, F, and $\perp$ are distinct values of M not in D_{s}. T and F represent the truth values *true* and *false*, respectively. $\perp$ is the *undefined value* which serves as the value of undefined terms. For $n \geq 0$, an *n-ary operation* of M is a total mapping from $D_1 \times \cdots \times D_n$ to D_{n+1} where D_i is D_{s}, $D_{\mathrm{c}} \cup \{\perp\}$, or $\{\mathrm{T}, \mathrm{F}\}$ for each i with $1 \leq i \leq n+1$. Let D_{o} be the domain of operations of M. $D_{\mathrm{s}} \cup \{\mathrm{T}, \mathrm{F}, \perp\}$ and D_{o} are assumed to be disjoint.

3.2   Expressions

Let S be a fixed infinite set of symbols that includes the 30 *key words* in Table 1. The key words are used to classify expressions, identify different categories of expressions, and name the built-in operators (see below).

---

[1] Recall that values of a model of Zermelo-Fraenkel (ZF) set theory includes only sets, not proper classes.

Table 1. The Key Words of Chiron

op	type	formula	op-app	var
type-app	dep-fun-type	fun-app	fun-abs	if
exist	def-des	indef-des	quote	eval
true	false	set	class	expr
expr-op	expr-type	expr-term	expr-formula	in
type-equal	term-equal	formula-equal	not	or

An *expression* of Chiron is defined inductively by:

1. Each symbol $s \in \mathcal{S}$ is an expression.
2. If $e_1, \ldots, e_n$ are expressions where $n \geq 0$, then $(e_1, \ldots, e_n)$ is an expression.

Hence, an expression is an S-expression (with commas in place of spaces) that exhibits the structure of a tree whose leaves are symbols in $\mathcal{S}$. Let $\mathcal{E}$ be the set of expressions of Chiron.

There are four special sorts of expressions: *operators, types, terms,* and *formulas*. An expression is *proper* if it is one of these special sorts of expressions, and an expression is *improper* if it is not proper. Proper expressions denote values of M, while improper expressions are nondenoting (i.e., they do not denote anything). Operators are used to construct expressions. They denote operations. Types are used to restrict the values of operators and variables and to classify terms by their values. They denote superclasses. Terms are used to describe classes. They denote classes or the undefined value $\perp$. Formulas are used to make assertions. They denote truth values. A *kind* is the key word type, a type, or the key word formula.

A term is *defined* if it denotes a class and is *undefined* if it denotes $\perp$. Every term is assigned a type. Suppose a term a is assigned a type α. Then a is said to be a *term of type* α. Suppose further α denotes a superclass Σ_α. If a is defined, i.e., a denotes a class x, then x is in Σ_α. The value of a nondenoting term is the undefined value $\perp$, but the value of a nondenoting type or formula is D_c (the universal superclass) or F (false), respectively. That is, the values for nondenoting types, terms, and formulas are D_c, $\perp$, and F, respectively.

There are 13 proper expression categories. They are shown in Table 2 in both a compact notation in the middle and the official S-expression-like notation on the right. $O, P, Q, \ldots$ denote operators, $\alpha, \beta, \gamma, \ldots$ denote types, $a, b, c, \ldots$ denote terms, $A, B, C, \ldots$ denote formulas, $s, t, u, \ldots$ denote symbols, $e, e', \ldots$ denote expressions, and $k, k', \ldots$ denote kinds.

Table 3 defines additional compact notation for the built-in operators and the universal quantifier. The compact notation also includes some customary abbreviation rules (see [9]).

3.3   Quotation and Evaluation

If e is any expression, proper or improper, then

$$(\mathsf{quote}, e)$$

Table 2. Compact Notation

Expression Category	Compact Notation	Official Notation
Operator	$(s :: k_1, \ldots, k_{n+1})$	$(\text{op}, s, k_1, \ldots, k_{n+1})$
Operator application	$(s :: k_1, \ldots, k_{n+1})$ $(e_1, \ldots, e_n)$	$(\text{op-app}, (\text{op}, s, k_1, \ldots, k_{n+1}),$ $e_1, \ldots, e_n)$
Variable	$(x : \alpha)$	(var, x, α)
Type application	$\alpha(a)$	$(\text{type-app}, \alpha, a)$
Dependent Function Type	$(\Lambda x : \alpha . \beta)$	$(\text{dep-fun-type}, (\text{var}, x, \alpha), \beta)$
Function application	$f(a)$	$(\text{fun-app}, f, a)$
Function abstraction	$(\lambda x : \alpha . b)$	$(\text{fun-abs}, (\text{var}, x, \alpha), b)$
Conditional term	$\text{if}(A, b, c)$	(if, A, b, c)
Existential quantification	$(\exists x : \alpha . B)$	$(\text{exist}, (\text{var}, x, \alpha), B)$
Definite description	$(\iota x : \alpha . B)$	$(\text{def-des}, (\text{var}, x, \alpha), B)$
Indefinite description	$(\epsilon x : \alpha . B)$	$(\text{indef-des}, (\text{var}, x, \alpha), B)$
Quotation	$\lceil e \rceil$	(quote, e)
Evaluation	$\llbracket a \rrbracket_{\text{ty}}$	$(\text{eval}, a, \text{type})$
	$\llbracket a \rrbracket_{\alpha}$	(eval, a, α)
	$\llbracket a \rrbracket_{\text{fo}}$	$(\text{eval}, a, \text{formula})$

Table 3. Additional Compact Notation

Compact Notation	Defining Expression
T	$(\text{true} :: \text{formula})(\,)$
F	$(\text{false} :: \text{formula})(\,)$
V	$(\text{set} :: \text{type})(\,)$
C	$(\text{class} :: \text{type})(\,)$
E	$(\text{expr} :: \text{type})(\,)$
E_{op}	$(\text{expr-op} :: \text{type})(\,)$
E_{ty}	$(\text{expr-type} :: \text{type})(\,)$
E_{te}	$(\text{expr-term} :: \text{type})(\,)$
E_{fo}	$(\text{expr-formula} :: \text{type})(\,)$
$(a \in b)$	$(\text{in} :: \text{V}, \text{C}, \text{formula})(a, b)$
$(\alpha =_{\text{ty}} \beta)$	$(\text{type-equal} :: \text{type}, \text{type}, \text{formula})(\alpha, \beta)$
$(a =_{\alpha} b)$	$(\text{term-equal} :: \text{C}, \text{C}, \text{type}, \text{formula})(a, b, \alpha)$
$(a = b)$	$(a =_{\text{C}} b)$
$(A \equiv B)$	$(\text{formula-equal} :: \text{formula}, \text{formula}, \text{formula})(A, B)$
$(\neg A)$	$(\text{not} :: \text{formula}, \text{formula})(A)$
$(a \notin b)$	$(\neg(a \in b))$
$(a \neq b)$	$(\neg(a = b))$
$(A \vee B)$	$(\text{or} :: \text{formula}, \text{formula}, \text{formula})(A, B)$
$(\forall x : \alpha . A)$	$(\neg(\exists x : \alpha . (\neg A)))$

is a term of type E called a *quotation*. The value of the quotation is a set, called the *construction* of e, that represents the syntactic structure of the expression e. Thus a proper expression e has two different meanings:

1. The *semantic meaning* of e is the value denoted by e itself.
2. The *syntactic meaning* of e is the construction denoted by (quote, e).

If a is a term and k is a kind, then

$$(\mathsf{eval}, a, k)$$

is an expression called an *evaluation* that is a type if $k = \mathsf{type}$, a term of type k if k is a type, and a formula if $k = \mathsf{formula}$. Roughly speaking, if a denotes a construction that represents an expression e, then the evaluation denotes the value of e. If a denotes a construction that represents an expression in which the symbol eval occurs, then the evaluation is undefined. This provision is needed to block the liar paradox and similar semantically ungrounded expressions (see [9]).

3.4 Biform Theories in Chiron

Let L be a language of Chiron. An *n-ary transformer in L* is an n-ary transformer $\Pi = (\pi, \hat{\pi})$ where π is an n-ary operator $(s :: \mathsf{E}, \dots, \mathsf{E})$ in L (with E occurring $n + 1$ times). A *rule in L* is a rule $R = (\Pi, M)$ where $\Pi = (\pi, \hat{\pi})$ is an n-ary transformer in L and M is formula of Chiron having the form

$$\forall e_1 . \mathsf{E}_1, \dots, e_n : \mathsf{E}_n . M'$$

where E_i is E, E_{op}, E_{ty}, E_{te}, or E_{fo} for all i with $1 \leq i \leq n$. If $a_1, \dots, a_n$ are quotations (of type E), then the *instance* of M with respect to $(a_1, \dots, a_n)$ is the result of replacing each occurrence of $\pi(a_1, \dots, a_n)$ in

$$M[e_1 \mapsto a_1, \dots, e_n \mapsto a_1]$$

with $f_{\hat{\pi}}(a_1, \dots, a_n)$ if this is defined and with $\perp_{\mathsf{C}}$ (which denotes the undefined value) if this is undefined. A *biform theory in* Chiron is a pair $T = (L, \Omega)$ where L is a language of Chiron and Ω is a set of rules in L.

Example 7. Let R be the rule given in Example 4 expressed as a rule in a language of Chiron. Then M would be the formula

$$\forall e : \mathsf{E}_{\mathsf{fo}} . [\![e]\!]_{\mathsf{fo}} \equiv [\![\pi(e)]\!]_{\mathsf{fo}},$$

and the instance of M with respect to $(\lceil p(0) \supset \forall x . q(x) \rceil)$ would be

$$[\![\lceil p(0) \supset \forall x . q(x) \rceil]\!]_{\mathsf{fo}} \equiv [\![\lceil \forall x . p(0) \supset q(x) \rceil]\!]_{\mathsf{fo}},$$

which reduces to

$$(p(0) \supset \forall x . q(x)) \equiv (\forall x . p(0) \supset q(x))$$

as desired.   □

Example 8. Let R be the rule given in Example 5 expressed as a rule in a language of Chiron. Then M would be the formula

$$\forall e_1 : \mathsf{E}_{\mathsf{te}}, e_2 : \mathsf{E}_{\mathsf{te}}, e_3 : \mathsf{E}_{\mathsf{fo}} .$$

$$(\mathsf{is\text{-}var}(e_2) \wedge \mathsf{free\text{-}for}(e_1, e_2, e_3) \wedge [\![e_2]\!]_{\mathsf{te}} = [\![e_1]\!]_{\mathsf{te}} \wedge [\![e_3]\!]_{\mathsf{fo}}) \supset [\![\pi(e_1, e_2, e_3)]\!]_{\mathsf{fo}}$$

which says that, for all expressions E_1, E_2, E_3, if E_1 is a term t, E_2 is a variable x, and E_3 is a formula A such that t is free for x in A, then $x = t \wedge A$ implies the result of applying the algorithm $\hat{\pi}$ to (t, x, A). Notice that the syntactic side condition of the formula schema in Example 5 (that says t is free for x in A) has been directly incorporated into M.   □

Example 9. Let R be the rule given in Example 6 expressed as a rule in a language of Chiron. Assume that real → real is the type

$$(\Lambda x : \text{real . real})$$

and deriv is the operator

$$(\text{derivative} :: \text{real} \to \text{real}, \text{real} \to \text{real}).$$

Also let $(a \downarrow \alpha)$ mean that the term a is defined with a value in the denotation of the type α. Then M would be the formula

$$\forall e : \mathsf{E}_{\text{te}} . (\llbracket e \rrbracket_{\text{te}} \downarrow \text{real} \to \text{real}) \supset \text{deriv}(\llbracket e \rrbracket_{\text{te}}) = \llbracket \pi(e) \rrbracket_{\text{te}}$$

which says that, for all expressions E, if E is a term t that denotes a function f that maps real numbers to real numbers, then the result of applying the algorithm $\hat{\pi}$ to (t) is a term that denotes the derivative of f.   □

See [9] for further details, discussion, examples, and references concerning Chiron.

4   An Example

In this section we will sketch the development of a nontrivial biform theory. We will start with a theory $T = (L, \Omega)$ of (higher-order) Peano arithmetic where:

- L contains operators nat, $0, S$ that represent the type of natural numbers, zero, and the successor function, respectively.
- Γ contains three formulas that express that 0 does not succeed another natural number, that the successor function is injective, and the full induction principle over all sets of natural numbers.

The next step is to extend T to $T' = (L', \Omega')$ by introducing defined operators $1, +, *$ for one, the addition function, and the multiplication function, respectively. 1 is defined as the successor of 0. + and * are defined recursively.

The last step is to extend T' to $T'' = (L'', \Omega'')$ by introducing the machinery to add and multiply binary numerals. Define a *(binary) numeral* to be an expression $(a_1, \ldots, a_n)$ where $n \geq 1$ and a_i is 0 or 1 for each i with $1 \leq i \leq n$. As defined, a numeral is an improper expression, and thus it denotes nothing. However, if n is a numeral, then $\lceil n \rceil$ is a proper expression that denotes the construction of n. We can introduce defined operators num, num-val that represent the type of numerals and a function that maps the type of numerals onto the type of natural numbers.

We can then define a rule $R = (\Pi, M)$ for numeral addition where $\Pi = $ (add, add-alg) is a binary transformer and M is the formula

$$\forall\, m, n : \mathsf{num}\, .$$
$$\mathsf{num\text{-}val}(m) + \mathsf{num\text{-}val}(n) = \mathsf{num\text{-}val}((\mathsf{add} :: \mathsf{num}, \mathsf{num}, \mathsf{num})(m, n)).$$

This formula says that the sum of the values of two numerals equals the value of the output of the algorithm add-alg when given the two numerals as input. The formula also says implicitly that

$$(\mathsf{add} :: \mathsf{num}, \mathsf{num}, \mathsf{num})(m, n)$$

is defined iff m and n both denote numerals. The algorithm add-alg could be implemented, for example, as a lambda-expression of L'' or as a program in some convenient programming language. We can introduce a rule for numeral multiplication in a similar way.

5   Conclusion

The notion of a biform theory enables axiomatic mathematics and algorithmic mathematics to be expressed together in one theory. A biform theory consists of a set of axioms that includes both formulas and rules. A rule is an expression-manipulating algorithm called a transformer coupled with a meaning formula that defines its semantics. A biform theory can be viewed both as an axiomatic theory and as an algorithmic theory. The algorithmic theory provides the deduction and computation rules for reasoning within the theory, while the axiomatic theory provides the context in which to understand the reasoning that is done via these rules. The axioms of a biform theory are the implicit background assumptions of the theory that define what formulas and rules are logical consequences of the theory.

Since transformers are algorithms that manipulate expressions, the meaning formulas of biform theory rules can only be directly formalized in a logic with support for reasoning about the syntax of expressions. Traditional logics do not offer this kind of support. Chiron is a general-purpose logic with high theoretical and practical expressivity and a facility for reasoning about the syntax of expressions. As a result, it is exceptionally well-suited for formalizing biform theories. Meaning formulas—that would usually be expressed as formula schemas in tradition logics—can be directly expressed in Chiron.

Biform theories can also be formalized in other logics that provide a means to reason about syntax. Many approaches for formalizing the syntax of expressions have been proposed starting with K. Gödel's famous *arithmetization of syntax* via Gödel numbering [12]. Two good surveys of this research area are [14] and the extended version of [17].

A great deal of research has been directed to the problem of how to integrate computer theorem proving and computer algebra. Much of this research has been done in connection with the Calculemus Project [3] or has been presented at the Calculemus symposia that began in 1996. Two research initiatives that are closely related to biform theories and the MathScheme project are the Theorema project [2] at the RISC Research Institute for Symbolic Computation [18] and the work by H. Barendregt and F. Wiedijk on the foundations of computerized mathematics [1].

The development and application of Chiron is a long-range research project composed of the following four tasks:

1. The design of Chiron.
2. The design of a proof system for Chiron.
3. The development of an implementation of Chiron and its proof system.
4. The development of a series of applications to demonstrate Chiron's reach and level of effectiveness.

The first task is largely completed [7,9]. The last three tasks have hardly been started. This paper begins the fourth task.

Acknowledgments

The author is grateful to Marc Bender and Jacques Carette for many valuable discussions on the design and use of Chiron. Over the course of these discussions, Dr. Carette convinced the author that Chiron needs to include a powerful facility for reasoning about the syntax of expressions. The author would also like to thank the referees for their suggestions on how to improve the paper.

References

1. Barendregt, H., Wiedijk, F.: The challenge of computer mathematics. Philosophical Transactions of the Royal Society A: Mathematical, Physical and Engineering Sciences 363, 2351–2375 (2005)
2. Buchberger, B., Craciun, A., Jebelean, T., Kovacs, L., Kutsia, T., Nakagawa, K., Piroi, F., Popov, N., Robu, J., Rosenkranz, M., Windsteiger, W.: Theorema: Towards computer-aided mathematical theory exploration. Journal of Applied Logic 4, 470–504 (2006)
3. Calculemus Project: Systems for Integrated Computation and Deduction. Web site at http://www.calculemus.net/
4. Carette, J.: Understanding expression simplification. In: Gutierrez, J. (ed.) Proceedings of the 2004 International Symposium on Symbolic and Algebraic Computation (ISSAC 2004), pp. 72–79. ACM Press, New York (2004)
5. Carette, J., Farmer, W.M., Sorge, V.: A rational reconstruction of a system for experimental mathematics. In: Kauers, M., Windsteiger, W. (eds.) Towards Mechanized Mathematical Assistants. LNCS (LNAI), vol. 4573, Springer, Heidelberg (2007)

6. Carette, J., Farmer, W.M., Wajs, J.: Trustable communication between mathematics systems. In: Hardin, T., Rioboo, R. (eds.) Calculemus 2003, pp. 58–68, Rome, Italy, Aracne (2003)
7. Farmer, W.M.: Chiron: A multi-paradigm logic. Studies in Logic, Grammar and Rhetoric. Special issue: Matuszewski, R., Rudnicki, P., Zalewska, A. (eds.) From Insight to Proof, forthcoming
8. Farmer, W.M.: The seven virtues of simple type theory. SQRL Report No. 18, McMaster University, 2003. Revised (2006)
9. Farmer, W.M.: Chiron: A set theory with types, undefinedness, quotation, and evaluation. SQRL Report No. 38, McMaster University (2007)
10. Farmer, W.M., von Mohrenschildt, M.: Transformers for symbolic computation and formal deduction. In: Colton, S., Martin, U., Sorge, V. (eds.) Automated Deduction - CADE-17. LNCS, vol. 1831, pp. 36–45. Springer, Heidelberg (2000)
11. Farmer, W.M., von Mohrenschildt, M.: An overview of a formal framework for managing mathematics. Annals of Mathematics and Artificial Intelligence 38, 165–191 (2003)
12. Gödel, K.: Über formal unentscheidbare Sätze der Principia Mthematica und verwandter Systeme I. Monatshefte für Mathematik und Physik 38, 173–198 (1931)
13. Gödel, K.: The Consistency of the Axiom of Choice and the Generalized Continuum Hypothesis with the Axioms of Set Theory. In: Annals of Mathematical Studies, vol. 3, Princeton University Press, Princeton (1940)
14. Harrison, J.: Metatheory and reflection in theorem proving: A survey and critique. Technical Report CRC-053, SRI Cambridge, Millers Yard, Cambridge, UK (1995) Available at http://www.cl.cam.ac.uk/jrh13/papers/reflect.ps.gz
15. MathScheme: An Integrated Framework for Computer Algebra and Computer Theorem Proving. Web site at http://imps.mcmaster.ca/mathscheme/
16. Mendelson, E.: Introduction to Mathematical Logic, vol. 4. Chapman & Hall/CRC, Sydney (1997)
17. Nogin, A., Kopylov, A., Yu, X., Hickey, J.: A computational approach to reflective meta-reasoning about languages with bindings. In: Momigliano, A., Pollack, R. (eds.) MERLIN'05: Proceedings of the Third ACM SIGPLAN Workshop on Mechanized Reasoning about Languages with Variable Binding. An extended version is available as a California Institute of Technology technical report, CaltechCSTR:2005.003, pp. 2–12. ACM Press, New York (2005)
18. RISC Research Institute for Symbolic Computation. Web site at http://www.risc.uni-linz.ac.at//

Automatic Synthesis of Decision Procedures: A Case Study of Ground and Linear Arithmetic[*]

Predrag Janičić[1] and Alan Bundy[2]

[1] Faculty of Mathematics, University of Belgrade
Studentski trg 16, 11000 Belgrade, Serbia
janicic@matf.bg.ac.yu
[2] School of Informatics, University of Edinburgh
Appleton Tower, Crichton St, Edinburgh EH8 9LE, UK
A.Bundy@ed.ac.uk

Abstract. We address the problem of automatic synthesis of decision procedures. Our synthesis mechanism consists of several stages and sub-mechanisms and is well-suited to the proof-planning paradigm. The system (ADEPTUS), that we present in this paper, synthesised a decision procedure for ground arithmetic completely automatically and it used some specific method generators in generating a decision procedure for linear arithmetic, in only a few seconds of CPU time. We believe that this approach can lead to automated assistance in constructing decision procedures and to more reliable implementations of decision procedures.

1 Introduction

Decision procedures are often vital in theorem proving [2,7]. In order to have decision procedures usable in a theorem prover, it is necessary to have them implemented not only efficiently, but also flexibly. It is often very important to have decision procedures for new, user-defined theories. Also, the implementation of decision procedures should be such that it can be verified in some formal way. For all these reasons, it would be fruitful if the process (or, at least, all its routine steps) of implementing decision procedures can be automated. It would help in avoiding human mistakes in implementing decision procedures.

In this paper we follow ideas from the programme on proof plans for normalisations and for automatic generation of decision procedures from [4]. As discussed there, many steps of many decision procedures can be described via sets of rewrite rules (so, object level proofs could also be relatively easily derived). Following and extending the ideas from [4], we have developed a system ADEPTUS (coming from *Assembly of DEcision Procedures via TransmUtation and Synthesis*) capable of automatically synthesising normalisation procedures and decision procedures.[1] All the methods that ADEPTUS generates are built in

---

[*] First author supported by EPSRC grant GR/R52954/01 and Serbian Ministry of Science grant 144030. Second author supported in part by EPSRC grant GR/S01771.
[1] Adeptus (Lat.) is also "one with the alchemical knowledge to turn base metals into gold". ADEPTUS is implemented in PROLOG as a stand-alone system. The code and the longer version of this paper are available from www.matf.bg.ac.yu/~janicic.

M. Kauers et al. (Eds.): MKM/Calculemus 2007, LNAI 4573, pp. 80–93, 2007.
© Springer-Verlag Berlin Heidelberg 2007

the spirit of the proof planning paradigm (and are implemented in PROLOG). For some theories, the approach gives not only automatically generated decision procedures, but also — by generating structured procedures consisting of simple methods — a higher-level understanding of syntactical transformations within the theory. Also, thanks to their modular architecture, generated procedures can be easily modified to slightly changed circumstances. We believe that this approach can be helpful in providing an easier and more reliable implementation of decision procedures. In this paper we evaluate our techniques on ground arithmetic and linear arithmetic (over rationals). ADEPTUS synthesised the decision procedures for ground arithmetic in around 3 seconds, and a decision procedure for (quantified) linear arithmetic in around 5 seconds of CPU time.

2 Preliminaries

Decision procedure. A theory $\mathcal{T}$ is decidable if there is an algorithm, which we
call a *decision procedure*, such that for an input $\mathcal{T}$-sentence f, it returns *yes*
if and only if $\mathcal{T} \vdash f$ (i.e., if f is a theorem of $\mathcal{T}$), and returns *no* otherwise).

Ground and linear arithmetic. Ground arithmetic is a fragment of arithmetic
that does not involve variables. Linear arithmetic is a fragment of arithmetic
that involves only addition (nx is treated as $x + \cdots + x$, where x appears
n times). For both these theories, we assume that variables can range over
rational numbers. The Fourier/Motzkin procedure [9] is one of the decision
procedures for linear arithmetic.

Backus-Naur form. For describing syntactical classes, we use Backus-Naur form
— BNF (equivalent to context-free grammars). We assume that each BNF
specification has attached its *top nonterminal*. The language of a BNF is
a set of all expressions that can be derived from the top nonterminal. For
representing some infinite syntactical classes, for convenience, we use some
meta-level conditions. We define the relation *ec* (*element of class*) as follows:
$ec(b, e, c)$ holds iff e can be derived from c w.r.t. the BNF specification b.

Rewrite rules. Unconditional rewrite rules are of the form: $RuleName : l \longrightarrow r$.
Conditional rewrite rules are of the form: $RuleName : l \longrightarrow r$ if $p_1, p_2, \ldots$,
p_n, where $p_1, p_2, \ldots, p_n$ are literals. These rewrite rules may be used modulo
the underlying theory $\mathcal{T}$ (e.g., the rule $n_1 x + n_2 x \longrightarrow nx$ if $n = n_1 + n_2$
may be used modulo linear arithmetic). For a rule $RuleName : l \longrightarrow$
r if $p_1, p_2, \ldots, p_n$, we say that it is *sound* w.r.t. $\mathcal{T}$ if for arbitrary $\mathcal{T}$-formula
Φ and arbitrary substitution φ it holds that $\mathcal{T} \vdash \Phi$ if $\mathcal{T}, p_1\varphi, p_2\varphi, \ldots$,
$p_n\varphi \vdash \Phi[l\varphi \mapsto r\varphi]$, and we say that it is *complete* w.r.t. $\mathcal{T}$ if for arbitrary $\mathcal{T}$-formula and arbitrary substitution φ it holds that $\mathcal{T} \vdash \Phi$ only if
$\mathcal{T}, p_1\varphi, p_2\varphi, \ldots, p_n\varphi \vdash \Phi[l\varphi \mapsto r\varphi]$.[2]

Proof planning and methods. Proof-planning is a technique for guiding the
search for a proof in automated theorem proving. To prove a conjecture,
within a proof-planning system, a method constructs the proof plan and

---

[2] Note that $\mathcal{T}$ does not necessarily contain the theory of equality, so we define soundness and completeness of the rules this way.

this plan is then used to guide the construction of the proof itself [3]. These plans are made up of tactics, which represent common patterns of reasoning. A method is a specification of a tactic. A method has several slots: a name, input, preconditions, transformation, output, postconditions, and the name of the attached tactic. A method cannot be applied if its preconditions are not met. Also, with the transformation performed and the output computed, the postconditions are checked and the method application fails if they fail.[3]

3   Proposed Programme

Our programme (slightly modified from the first version [4]) for automated synthesis of normalisation methods and decision procedures has several parts:

- Given a syntactical class, a set of rewrite rules, and a kind of transformation, select (if it is possible) a subset of rewrite rules that is sufficient to transform any member of the input syntactical class in the required way. The output class should also be generated automatically. We call a *method generator* an algorithm capable of generating a method that transforms members of the input class to members of the output class.
- There are different kinds of methods, e.g., one for removing some function symbol, one for stratification, one for thinning etc. (see further text and [4] for explanation of these terms); for each of them, there is a method generator.
- Given several generated methods, it should be possible to combine them (automatically) into a compound method or, sometimes, into a decision procedure for some theory;
- Methods (and compound methods) should be designed in such a way that their soundness, completeness, and termination can be easily proved;
- Since some transformations (required for some procedures) are very complex, building methods may require human interaction and assistance.

Example 1. From any formula derivable from f w.r.t. the following BNF:
$$f ::= af|\neg f|f \vee f|f \wedge f|f \Rightarrow f|f \Leftrightarrow f|(\exists var : sort)f|(\forall var : sort)f$$
(where af is another nonterminal, describing atomic formulae) the symbol $\Leftrightarrow$ can be removed by exhaustively using the rewrite rule $f_1 \Leftrightarrow f_2 \longrightarrow (f_1 \Rightarrow f_2) \wedge (f_2 \Rightarrow f_1)$ and the resulting formula can be derived from f w.r.t. the following BNF:
$$f ::= af|\neg f|f \vee f|f \wedge f|f \Rightarrow f|(\exists var : sort)f|(\forall var : sort)f .$$

Following the above programme, we implemented our system ADEPTUS capable of generating code for real-world decision procedures. We have implemented several method generators. They take a given BNF, transform it into another one, and build a method that can transform any formula that belongs to the first BNF into a formula that belongs to the second BNF. On the set of all these generators, we can perform a (heuristically guided) search for a sequence of methods which

---

[3] Alteratively, instead of (active) postconditions, methods can have (passive) effects — conditions that are *guaranteed* to be true when the method succeeds.

goes from a starting BNF to a trivial BNF (consisting of only $\top$ and $\bot$). If the final syntactical class is equal to $\{\bot, \top\}$, then the whole of the sequence yields a decision procedure for the underlying theory (under some assumptions about available rewrite rules). If such a method can be built, soundness, termination, and completeness can be easily proved. Apart from these method generators, we also use special-purpose method generators. For simplicity, in the rest of the paper we assume that, in formulae being transformed, variables are standardised apart, i.e., there are no two quantifiers with the same variable symbol.

4   Method Generators and Generated Methods

Normalisation Method Generators. Normalisation methods are methods based on exhaustive application of rewrite rules. Each normalisation method has the following general form:

> **name:** *methodname*;
> **input:** f;
> **preconditions:** $ec(b, f, top\ nonterminal)$ (where b is the input BNF);
> **transformation:** transforms f to f' by exhaustive application of the set of rewrite rules (applying to positions that correspond to the attached language constructs);
> **output:** f';
> **postconditions:** $ec(b', f', top\ nonterminal)$ (where b' is the output BNF).

We have implemented generators for several kinds of methods:

Remove is a normalisation method used to eliminate a certain function symbol, predicate symbol, logical connective, or a quantifier from a formula. The method uses sets of appropriate rewrite rules and applies them exhaustively to the current formula until no occurrences of the specific symbol remain. For instance, as shown in Example 1, the given BNF specification can be transformed to the corresponding BNF specification without the symbol $\Leftrightarrow$.

Stratify is a normalisation method used to stratify one syntactical class into two syntactical classes containing some predicate or function symbols, logical connectives or quantifiers. For instance, a stratify method for moving disjunctions beneath conjunctions can be constructed if the following rewrite rules are available: st_conj_disj1: $f_1 \wedge (f_2 \vee f_3) \longrightarrow (f_1 \wedge f_2) \vee (f_1 \wedge f_3)$, st_conj_disj2: $(f_2 \vee f_3) \wedge f_1 \longrightarrow (f_2 \wedge f_1) \vee (f_3 \wedge f_1)$.

Thin is a normalisation method that eliminates multiple occurrences of a unary logical connective or a unary function symbol. For instance, we can use the rule $\neg\neg f \longrightarrow f$ in order to transform each formula derivable from $f ::= af \mid \neg f$ to a formula derivable from $f ::= af \mid \neg af$.

Absorb is a normalisation method that can eliminate some recursion rules. For instance, we can use the rule rm_mult: $c_1 \cdot c_2 \longrightarrow c_3$ if $c_3 = c_1 \cdot c_2$ in order to transform each term derivable from $t ::= t \cdot rc \mid rc$ (where the nonterminal rc denotes rational constants) to a term derivable from $t ::= rc$.

Left-assoc is one of the normalisation methods for reorganising within a class. If a syntactical class contains only one function symbol or a connective and if that symbol is both binary and associative, then members of this class can be put into left associative form. For instance, we will need the left association of addition and the left association of conjunction.

A *normalisation method generator* is a procedure with the following input: *(i)* a BNF b for the input expressions; *(ii)* a set of rewrite rules R; *(iii)* a kind of the required method (e.g., *remove*). It generates a method $\mathcal{M}$ and a BNF b' (for the output expressions).[4] By applying the rules from R, $\mathcal{M}$ transforms any expression derivable from b to an expression derivable from b'.

Example 2. Consider the BNF: $f ::= h(a)|h(b)|g_1(a)|g_2(b)$ where a and b are non-terminals, and the following rewrite rules: $R_1 : h(x) \longrightarrow g_1(x)$, $R_2 : h(x) \longrightarrow g_2(x)$. These rules are sufficient for eliminating the symbol h and for transforming the above BNF into: $f ::= g_1(a)|g_2(b)$. However, it cannot be reached by arbitrary use of exhaustive applications of the given rewrite rules: R_1 should be applied only to $h(a)$, and R_2 only to $h(b)$. The lesson is that we have to take care about which rule we use for specific language constructs. This sort of information has to be built into the method we want to construct.

A *normalisation method generator* works, basically, as follows: first it tries to eliminate non-recursive nonterminals in the input BNF, then searches for "problematic" BNF rules and generates the output BNF set, then, a generic algorithm for searching over available rewrite rules is invoked and it checks if all "problematic" language constructs can be rewritten in such a way that any input formula, when rewritten, is derivable from the top nonterminal of the output BNF. Also, this search mechanism attaches rewrite rules to particular language constructs. If there are no required rewrite rules, a method generator reports it, so the user could try to provide missing rules (in a planned, advanced version, which is not part of the work presented in this paper, the method generator would speculate the remaining necessary rules and/or try to redefine/relax the output class).

Example 3. The remove method generator can generate the method for removing the symbol $\neg$ from formula derivable from f w.r.t. the following BNF:
$$f ::= f \vee f|f \wedge f|\neg\bot|\neg\top|\neg t < t|\neg t = t|\bot|\top|t < t|t = t|$$
with the following rewrite rules attached to particular language constructs:

rm_bottom:	$\neg\bot \longrightarrow \top$	attached to $\neg\bot$
rm_top:,	$\neg\top \longrightarrow \bot$	attached to $\neg\top$
rm_neg_less:	$\neg(t_1 < t_2) \longrightarrow (t_2 < t_1) \vee (t_1 = t_2)$	attached to $\neg t < t$
rm_neg_eq:	$\neg(t_1 = t_2) \longrightarrow (t_1 < t_2) \vee (t_2 < t_1)$	attached to $\neg t = t$

The output BNF is: $f ::= f \vee f|f \wedge f|\bot|\top|t < t|t = t|$ and, by the generated remove method, the formula $\neg(3 < 2) \wedge \neg(1 = 2)$ will be transformed to $(2 < 3 \vee 3 = 2) \wedge (1 < 2 \vee 2 < 1)$.

Special-Purpose Method Generators. The first one of the following special-purpose generators can be used for a quantifier elimination procedure for any theory, while the remaining three are specific for linear arithmetic. Note, however,

---

[4] In our system, the tactics are not implemented yet. So, our procedures produce meta-level proof plans, not the object level proofs.

that it is essential to have these generators (although they are theory-specific): they can be used in an automatic search process and generate the required methods with the given preconditions (which are not known in advance).

Method Generator for Adjusting the Innermost Quantifier. It generates a method that transforms a formula in prenex normal form in the following way: if its innermost quantifier is existential, then keep it unchanged; if its innermost quantifier is universal, then rewrite the formula $(Qx_1)(Qx_2)$ $\ldots(Qx_n)(\forall x)f$ to $(Qx_1)(Qx_2)\ldots(Qx_n)\neg(\exists x)\neg f$ by using the following rewrite rule: rm_univ: $(\forall x)f \longrightarrow \neg(\exists x)\neg f$. The motive of this method is to deal only with elimination of existential quantifiers.

One-side Method Generator. It generates a method that transforms all literals in such a way that each of them has 0 as its second argument. For instance, for symbols $<, >, \leq, \neq, \geq, =$ as parameters, after applying the generated *one-side* method, each literal will have one of the following forms: $t < 0$, $t > 0$, $t \leq 0$, $t \neq 0$, $t \geq 0$, $t = 0$.

Method Generator for Isolating a Variable. It generates a method that isolates a distinguished variable x in all literals. After applying that method, each of the literals either does not involve x or has one of the forms: $\alpha x = \beta$, $x = \beta$, $\alpha x < \beta$, $x < \beta$ (where α and β have no occurrences of x).

Method Generator for Removing a Variable. The *cross multiply and add* step is the essential step of the Fourier/Motzkin's procedure [9]. It is applied for elimination of x from $\exists x F(x)$, where F is in disjunctive normal form and each of its literals either does not involve x or has one of the forms: $\alpha x = \beta$, $x = \beta$, $\alpha x < \beta$, $x < \beta$ (where α and β have no occurrences of x). After performing this step, x does not occur in the current formula and so the corresponding quantifier can be deleted. It is important to have this generator (instead of a single method) — it generates required methods with concrete specific preconditions and postconditions, which is vital for combining with other concrete methods, and for automatic search process.

Properties of Generated Methods. A normalisation method links two sets of formulae. From the syntactical point of view, each formula f_1 derivable from the top nonterminal of the input BNF should be transformed (in a finite number of steps) into a formula f_2 derivable from the top nonterminal of the output BNF. From the deductive point of view, it should hold that $\mathcal{T} \vdash f_1$ if (and only if) $\mathcal{T} \vdash f_2$. If the "if" condition holds, then the method is sound, and if the "only if" condition holds then the method is complete (w.r.t. $\mathcal{T}$).

Termination. For each generated method it must be shown that it is terminating (by considering properties of the rewrite rules used[5]). For some sorts of methods, their termination is guaranteed by the way they are generated.

Soundness. We distinguish soundness of a method w.r.t. syntactical restrictions and soundness of a method w.r.t. the underlying theory $\mathcal{T}$:

---

[5] Note that these sets of rewrite rules are not always confluent. Moreover, for certain tasks, such as, for instance, transforming a formula into disjunctive normal form, there is no confluent and terminating rewrite system [10].

- If a method transforms one formula into another one, then it is ensured by the method's postconditions that the second one does meet the required syntactical restrictions (given by the method specification), so the method is sound w.r.t. syntactical restrictions.[6]
- All available rewrite rules (all of them correspond to the underlying theory T) are assumed to be sound. Thus, since a method is (usually) based on exhaustive application of some (normally sound) rewrite rules, it is trivially sound w.r.t. T.

Completeness. We distinguish completeness of a method w.r.t. syntactical restrictions and w.r.t. the underlying theory T:

- It is not *a priori* guaranteed that a method can transform any input formula (which meets the preconditions) into some other formula (that belongs to the output class), i.e., it is not guaranteed that the method is complete w.r.t. syntactical restrictions. Namely, a method maybe uses some conditional rewrite rules (which cannot be applied to all input formulae). If a method uses only unconditional rewrite rules or conditional rewrite rules which cover all possible cases, then it can transform any input formula into a formula belonging to the output class.
- Completeness of a method w.r.t. T relies on the completeness of the rewrite rules used. If a method can transform any input formula into a formula belonging to the output class and if all the rewrite rules it uses are complete, then the method is complete w.r.t. T.

5   Search Engine for Synthesising Compound Methods

Given method generators, a BNF description of a theory T, and a set of available rewrite rules, a user can try to combine different generated methods and transform the initial BNF step by step, searching for some goal BNF. Also, an automatic search for compound methods or a decision procedure for T can be performed. The goal of this process is to generate a sequence of methods such that: *(i)* the output BNF of a non-last method is the input BNF of the next method in the sequence; *(ii)* the output BNF of the last method in the sequence is a goal BNF, for instance, a trivial BNF — consisting only of rules with $\top$ and $\bot$ for the top nonterminal. Of course, this sequence can have more methods that are different instances of the same kind of methods, or even the very same method more than once. In each step, our search procedure tries all available method generators,

---

[6] Conditional rules are the reason for using active postconditions in methods (instead of passive *effects*). For instance, for BNF $f ::= f \wedge f | n = n | n < n | \top | \bot$, the rewrite rules rm_ls1: $n_1 < n_2 \longrightarrow \top$ if $number(n_1)$, $number(n_2)$, $n_1 < n_2$ and rm_ls2: $n_1 < n_2 \longrightarrow \bot$ if $number(n_1)$, $number(n_2)$, $n_1 > n_2$ eliminate the symbol $<$. The method generator would take both these rules for building a remove method for $<$, but (since it works only in syntactical manner) it would not check if the conditions for rm_ls1 and rm_ls2 cover all cases, i.e., if the generated method can transform *any* input formula. That is why the methods have (active) postconditions that check if the input formula is really rewritten so the result belongs to the output class.

with all possible parameters (based on the underlying language). In order to ensure termination, the search procedure tries to find a sequence of methods that consists of subsequences, such that each of them is of length less than or equal to a fixed value M, and such that the last BNFs of the subsequences are of strictly decreasing size. So, in any generated procedure there might be some BNF size increasing steps (for instance, with introducing new symbols in the current BNF), but the whole of the generated procedure will be size decreasing. The size of BNF specification is a heuristic measure and we define it to be the sum of sizes of all its rules; the size of a rule $c ::= c'$ is equal to $100 \cdot n_c(c') + 10 \cdot n_1(c') + n_2(c')$, where n_c denotes the number of occurrences of c in c', n_1 the number of occurrences of all other nonterminals in c', and n_2 the number of all other symbols in c'. Defined this way, the measure forces the engine to try to get rid of recursive nonterminals and then of the nonterminals whose specifications involve some other nonterminals. The trivial, goal BNF (consisting of only $f ::= \top | \bot$) has the size 2. If the current sequence cannot be continued, the engine backtracks and tries to find alternatives.

Example 4. The size of the following BNF $f ::= af | \neg f$, $af ::= \top | \bot$ is 113 (10 for $f ::= af$, 1+100 for $f ::= \neg f$, 1 for $af ::= \top$, 1 for $af ::= \bot$).

The size of the following BNF specification (for ground arithmetic):
$$f ::= af | \neg f | f \vee f | f \wedge f | f \Rightarrow f | f \Leftrightarrow f$$
$$af ::= \top | \bot | t = t | t < t | t > t | t \leq t | t \geq t | t \neq t$$
$$t ::= rc | - t | t \cdot t | t + t$$
is 1556 (af denotes atomic formulae, t denotes terms, and rc denotes rational constants). The size of BNF for the full linear arithmetic is 2233.

Given a finite number of method generators and a finite number of rewrite rules, at each step a finite number of methods can be generated (there is also a finite number of possible parameters). Thus, since the algorithm produces subsequences (of maximal length M) of decreasing sizes (that are natural numbers) of corresponding BNF specifications, the given algorithm terminates. If method generators can generate all methods necessary for building the required compound method, then (thanks to backtracking) the given algorithm can build one such compound method (for M large enough). If we iterate the given algorithm (for $M = 1, 2, 3, \ldots$), then it will eventually build the required compound method, so this iterated algorithm is complete. However, we can also use it only with particular values for M (then the procedure is not complete, but it gives better results if it used only for an appropriate value for M).

The ordering of method generators is not relevant for termination and correctness of the search algorithm, but it is important for its efficiency. We used the following ordering (based on empirical tests, simpler than a potential theoretical analysis, specific for each case): `remove`, `thin`, `absorb`, `stratify`, `left_assoc`.

When normalisation methods themselves cannot build a decision procedure, we use special-purpose method generators and the basic search engine in a more complex way. The search for a decision procedure based on quantifier elimination is performed in three stages, by the following *compound search engine*:

- the first stage is reaching a BNF for which the method for adjusting the innermost quantifier is applicable;
- the second stage produces a sequence of methods (that will form a loop) for variable elimination; the output BNF of this sequence of methods has to be a subset of its input BNF;
- the third stage is for final simplifications, it starts with the output BNF of the first stage, but with all rules involving variables and quantifiers deleted; its goal BNF specification is the trivial one (i.e., it consists only of $\top$ and $\bot$).

For each of these stages we use the basic search engine and we use all method generators with higher priority given to the special-purpose method generators.

Properties of Compound Methods. A set of generated methods for some underlying theory T can be combined (by a human, or automatically) into a compound method (for that theory). Compound methods (in this context) can use primitive methods in a sequence or in a loop (but not conditional branching). The preconditions of a compound method are the preconditions of the first method, and the postconditions are the postconditions of the last method used.[7]

Termination. If a compound method is a sequence of terminating methods, then it is (trivially) terminating. If it has a loop, a deeper argument is required.

Soundness. Since it relies on the soundness of the used primitive methods, every compound method is also sound (both w.r.t. syntactical restrictions and w.r.t. the underlying theory T). Meeting the syntactical restrictions of the compound method is also ensured by its postconditions.

Completeness. If all the used methods are complete and if the compound method is terminating, then it is (trivially) complete. More precisely, if a compound method (i) is terminating; (ii) uses only (primitive) methods which never fail (i.e., the methods which transform any input formula to a formula belonging to the output class) and which use only complete rewrite rules, then that compound method is complete (w.r.t. T).

Based on the above considerations, we can make a crucial observation: if a compound method for some theory T has an input BNF corresponding to the whole of T, a trivial output BNF consisting only of $\top$ and $\bot$, and if it is terminating, sound, and complete (w.r.t. T)[8], then it is a decision procedure for T. This way, we can, in some cases, trivially get a proof that some (automatically generated) compound method is a decision procedure for some theory.

6   Evaluation

We ran the basic search engine, on the BNF specification for ground arithmetic given in Example 4, with $M = 3$, with the described method generators, and

---

[7] This way of constructing the preconditions and postconditions of a compound method is not adequate in general but suffices for the examples we were working on (recall that in compound methods that our system generates, the output BNF of a method is always the input BNF of the next method in the sequence).

[8] Soundness and completeness properties rely on properties of the rewrite rules used.

with 59 relevant rewrite rules available. We set the goal BNF specification to be the trivial one ($f ::= \top | \bot$), thus aiming at synthesising a decision procedures for ground arithmetic. The search algorithm took 2.91 seconds of *cpu* time[9], during the search there were 48 methods successfully generated and there are 22 of them in the final sequence. The search algorithm produced the sequence of methods DP_GA with the following "overview" (in bracket the sizes of the output BNFs are given): *remove* $\Leftrightarrow$ *(1345)*, *remove* $\Rightarrow$ *(1144)*, *remove* $\leq$ *(1123)*, *remove* $\geq$ *(1102)*, *remove* $\neq$ *(1081)*, *remove* $>$ *(1060)*, *remove* $-$ *(959)*, *stratify* $[\wedge, \vee]$ *(969)*, *thin* $\neg$ *(906)*, *remove* $\neg$ *(858)*, *stratify* $[\vee]$ *(868)*, *stratify* $[+]$ *(878)*, *left_assoc* $\vee$ *(788)*, *left_assoc* $+$ *(698)*, *left_assoc* $*$ *(608)*, *absorb* $*$ *(487)*, *absorb* $+$ *(366)*, *remove* $<$ *(345)*, *remove* $=$ *(324)*, *left_assoc* $\wedge$ *(327)*, *remove* $\wedge$ *(206)*, *remove* $\vee$ *(2)*.

Example 5. The method stratify $[+]$ from the above list was generated for the following input BNF:
$$f ::= f_1 | f \vee f$$
$$f_1 ::= f_1 \wedge f_1 | \bot | \top | t < t | t = t |$$
$$t ::= t \cdot t | t + t | rc$$
with the following rewrite rules attached to particular language constructs (derivable from t):
st_mult_plus1: $t_1 \cdot (t_2 + t_3) \longrightarrow (t_1 \cdot t_2) + (t_1 \cdot t_3)$ attached to $t \cdot (t + t)$
st_mult_plus2: $(t_2 + t_3) \cdot t_1 \longrightarrow (t_2 \cdot t_1) + (t_3 \cdot t_1)$ attached to $(t + t) \cdot t$
The output BNF is:
$$f ::= f_1 | f \vee f$$
$$f_1 ::= f_1 \wedge f_1 | \bot | \top | t < t | t = t |$$
$$t ::= t_1 | t + t$$
$$t_1 ::= t_1 \cdot t_1 | rc$$
By this method, the formula $2 \cdot (1 + 3) < 3$ will be transformed to $2 \cdot 1 + 2 \cdot 3 < 3$.

Theorem 1. *The procedure* DP_GA *for ground arithmetic is terminating, sound and complete, i.e., it is a decision procedure for ground arithmetic.*

Proof sketch. The procedure DP_GA is sound and terminating, as all generated methods are sound and terminating and there is no loop. We still don't claim that it is complete as there are some conditional rewrite rules used. For instance, in the step *absorb* $+$ of DP_GA, the conditional rule reduce_plus: $t_1 + t_2 \Rightarrow t_3$, if $t_3 = t_1 + t_2$ is used, but it is still not shown that its condition covers all possible cases. The user can show this by proving: $(\forall c_1 : rational)(\forall c_2 : rational)(\exists c_3 : rational)(c_3 = c_1 + c_2)$. It is easy to prove that such conjectures are theorems of arithmetic. Moreover, some of them can be proved by the decision procedure DP_LA for linear arithmetic (which we also automatically generated, see the subsequent text). All this leads us to conclude that the procedure DP_GA is correct.

Example 6. This example shows the formulae produced by the 22 subsequent methods of the procedure DP_GA applied to the formula $\neg(7 \leq 5) \Rightarrow \neg(2 \cdot (1 + 3) \geq 3)$ (it is assumed that $\wedge$ has higher priority than $\vee$).

---

[9] The system is implemented in SWI Prolog and tested on a 512Mb PC Celeron 2.4Ghz.

remove $\Leftrightarrow$	$\neg(7 \le 5) \Rightarrow \neg(2 \cdot (1 + 3) \ge 3)$
remove $\Rightarrow$	$\neg(\neg(7 \le 5)) \vee \neg(2 \cdot (1 + 3) \ge 3)$
remove $\le$	$\neg(\neg(7 < 5 \vee 7 = 5)) \vee \neg(2 \cdot (1 + 3) \ge 3)$
remove $\ge$	$\neg(\neg(7 < 5 \vee 7 = 5)) \vee \neg(3 < 2 \cdot (1 + 3) \vee 2 \cdot (1 + 3) = 3)$
remove $\neq$	$\neg(\neg(7 < 5 \vee 7 = 5)) \vee \neg(3 < 2 \cdot (1 + 3) \vee 2 \cdot (1 + 3) = 3)$
remove $>$	$\neg(\neg(7 < 5 \vee 7 = 5)) \vee \neg(3 < 2 \cdot (1 + 3) \vee 2 \cdot (1 + 3) = 3)$
remove -,	$\neg(\neg(7 < 5 \vee 7 = 5)) \vee \neg(3 < 2 \cdot (1 + 3) \vee 2 \cdot (1 + 3) = 3)$
stratify $[\wedge, \vee]$	$(\neg(\neg 7 < 5) \vee \neg(\neg 7 = 5)) \vee \neg 3 < 2 \cdot (1 + 3) \wedge \neg 2 \cdot (1 + 3) = 3$
thin $\neg$	$(7 < 5 \vee 7 = 5) \vee \neg 3 < 2 \cdot (1 + 3) \wedge \neg 2 \cdot (1 + 3) = 3$
remove $\neg$	$(7 < 5 \vee 7 = 5) \vee (2 \cdot (1 + 3) < 3 \vee 3 = 2 \cdot (1 + 3)) \wedge 2 \cdot (1 + 3) < 3 \vee 3 < 2 \cdot (1 + 3)$
stratify $[\vee]$	$(7 < 5 \vee 7 = 5) \vee ((2 \cdot (1 + 3) < 3 \wedge 2 \cdot (1 + 3) < 3) \vee 3 = 2 \cdot (1 + 3) \wedge 2 \cdot (1 + 3) < 3) \vee$
	$(2 \cdot (1 + 3) < 3 \wedge 3 < 2 \cdot (1 + 3)) \vee 3 = 2 \cdot (1 + 3) \wedge 3 < 2 \cdot (1 + 3)$
stratify $[+]$	$(7 < 5 \vee 7 = 5) \vee ((2 \cdot 1 + 2 \cdot 3 < 3 \wedge 2 \cdot 1 + 2 \cdot 3 < 3) \vee 3 = 2 \cdot 1 + 2 \cdot 3 \wedge 2 \cdot 1 + 2 \cdot 3 < 3) \vee$
	$(2 \cdot 1 + 2 \cdot 3 < 3 \wedge 3 < 2 \cdot 1 + 2 \cdot 3) \vee 3 = 2 \cdot 1 + 2 \cdot 3 \wedge 3 < 2 \cdot 1 + 2 \cdot 3$
left_assoc $\vee$	$((((7 < 5 \vee 7 = 5) \vee 2 \cdot 1 + 2 \cdot 3 < 3 \wedge 2 \cdot 1 + 2 \cdot 3 < 3) \vee 3 = 2 \cdot 1 + 2 \cdot 3 \wedge 2 \cdot 1 + 2 \cdot 3 < 3) \vee$
	$2 \cdot 1 + 2 \cdot 3 < 3 \wedge 3 < 2 \cdot 1 + 2 \cdot 3) \vee 3 = 2 \cdot 1 + 2 \cdot 3 \wedge 3 < 2 \cdot 1 + 2 \cdot 3$
left_assoc $+$	$((((7 < 5 \vee 7 = 5) \vee 2 \cdot 1 + 2 \cdot 3 < 3 \wedge 2 \cdot 1 + 2 \cdot 3 < 3) \vee 3 = 2 \cdot 1 + 2 \cdot 3 \wedge 2 \cdot 1 + 2 \cdot 3 < 3) \vee$
	$2 \cdot 1 + 2 \cdot 3 < 3 \wedge 3 < 2 \cdot 1 + 2 \cdot 3) \vee 3 = 2 \cdot 1 + 2 \cdot 3 \wedge 3 < 2 \cdot 1 + 2 \cdot 3$
left_assoc $\cdot$	$((((7 < 5 \vee 7 = 5) \vee 2 \cdot 1 + 2 \cdot 3 < 3 \wedge 2 \cdot 1 + 2 \cdot 3 < 3) \vee 3 = 2 \cdot 1 + 2 \cdot 3 \wedge 2 \cdot 1 + 2 \cdot 3 < 3) \vee$
	$2 \cdot 1 + 2 \cdot 3 < 3 \wedge 3 < 2 \cdot 1 + 2 \cdot 3) \vee 3 = 2 \cdot 1 + 2 \cdot 3 \wedge 3 < 2 \cdot 1 + 2 \cdot 3$
absorb $\cdot$	$((((7 < 5 \vee 7 = 5) \vee 2 + 6 < 3 \wedge 2 + 6 < 3) \vee 3 = 2 + 6 \wedge 2 + 6 < 3) \vee$
	$2 + 6 < 3 \wedge 3 < 2 + 6) \vee 3 = 2 + 6 \wedge 3 < 2 + 6$
absorb $+$	$((((7 < 5 \vee 7 = 5) \vee 8 < 3 \wedge 8 < 3) \vee 3 = 8 \wedge 8 < 3) \vee 8 < 3 \wedge 3 < 8) \vee 3 = 8 \wedge 3 < 8$
remove $<$	$((((\bot \vee 7 = 5) \vee \bot \wedge \bot) \vee 3 = 8 \wedge \bot) \vee \bot \wedge \top) \vee 3 = 8 \wedge \top$
remove $=$	$((((\bot \vee \bot) \vee \bot \wedge \bot) \vee \bot \wedge \bot) \vee \bot \wedge \top) \vee \bot \wedge \top$
left_assoc $\wedge$	$((((\bot \vee \bot) \vee \bot \wedge \bot) \vee \bot \wedge \bot) \vee \bot \wedge \top) \vee \bot \wedge \top$
remove $\wedge$	$((((\bot \vee \bot) \vee \bot) \vee \bot) \vee \bot) \vee \bot$
remove $\vee$	$\bot$

We applied the compound search engine on the BNF description of the full linear arithmetic, with $M = 3$ for the first and the third stage, with $M = 5$ for the second stage[10], with all the described method generators, and with 71 relevant rewrite rules available. The search algorithm took 4.80 seconds of *cpu* time and during the search there were 89 methods successfully generated, while there are 51 of them in the final sequence, yielding a decision procedure DP_LA with:[11]

- 9 methods in the first stage: *remove* $\Leftrightarrow$, *remove* $\Rightarrow$, *remove* $\le$, *remove* $\ge$, *remove* $\neq$, *remove* $>$, *remove* -, *remove* -, *stratify* $[\forall, \exists]$,
- 22 methods in the quantifier elimination loop: *adjust_innermost* x_0, *stratify* $[\wedge, \vee]$, *thin* $\neg$, *remove* $\neg$, *one_side* $[0, [<, >, \le, \neq, \ge, =]]$, *stratify* $[\vee]$, *stratify* $[+]$, *stratify* $[+]$, *left_assoc* $\vee$, *left_assoc* $\wedge$, *stratify* $[<, =]$, *remove* $\wedge$, *left_assoc* $+$, *left_assoc* $+$, *left_assoc* $\cdot$, *absorb* $\cdot$, *absorb* $+$, *absorb* $\cdot$, *absorb* $+$, *isolate* $[[x_0, rc \cdot x_0], [<, >, \le, \neq, \ge, =]]$, *eliminate* $[[x_0, rc \cdot x_0], [<, >, \le, \neq, \ge, =]]$,
- 20 methods for final simplifications: *stratify* $[\wedge, \vee]$, *thin* $\neg$, *remove* $\neg$, *stratify* $[\vee]$, *stratify* $[+]$, *left_assoc* $\vee$, *stratify* $[+]$, *left_assoc* $+$, *left_assoc* $+$, *left_assoc* $\cdot$, *absorb* $\cdot$ *absorb* $+$, *absorb* $\cdot$, *absorb* $+$, *remove* $<$, *remove* $=$, *left_assoc* $\wedge$, *remove* $\wedge$, *remove* $\vee$, *remove* $\neg$

Theorem 2. *The procedure* DP_LA *for linear arithmetic is terminating, sound and complete, i.e., it is a decision procedure for linear arithmetic.*

Proof sketch. Each of individual methods used in the generated procedure DP_LA is terminating. Since each loop eliminates one variable and since there are a finite number of variables in the input formula, the loop terminates. Hence,

---

[10] For lower values of M the system failed to generate the required procedure.

[11] Same methods (e.g., left_assoc $+$) are applied to different language constructs.

the procedure DP_LA is terminating. Since all methods in DP_LA use only sound rewrite rules, all of them are sound, and hence, the procedure is sound. The completeness relies not only on the completeness of the rewrite rules used, but also on the coverage property for the methods that use conditional rewrite rules. It can be shown (similarly as for DP_GA) that all required coverage properties are fulfilled (moreover, some of the coverage properties can be proved by the generated procedure itself, which is acceptable, as we know that the procedure is sound). Therefore, in each method, either unconditional rules are used or conditional rules that cover all possible cases. Hence, all methods always succeed and are complete, and the procedure DP_LA is complete. All in all, the procedure DP_LA terminates, it transforms an arbitrary input (linear arithmetic) formula Φ into $\top$ or $\bot$, while the output is $\top$ iff Φ is a theorem of linear arithmetic.

We don't claim that the generated procedure DP_LA is the shortest or the most efficient one. However, we doubt that a decision procedure for linear arithmetic can be described in a much shorter way (see, for instance, the description from [5]). This suggests that it is non-trivial for a human programmer to implement this procedure without flaws and bugs, even when provided with the code for the key step (*cross multiply and add*), because the most probable flaws are rather in correctly combining all the remaining steps.

7   Related Work

Our approach is based on ideas from [4] and apart from that strong link, as we are aware of, it can be considered basically original.

The work presented here is related to the Knuth-Bendix completion procedure [8] and its variants in a sense that it performs automatic construction of decision procedures. However, there are significant differences. While the completion procedure generates a confluent and terminating set of rewrite rules, and hence a way how to reach a normal form, it does not give a description of the normal form. In contrast, our system does not necessarily produce a decision procedure (or a normalisation procedure) whenever the completion procedure, but when it does, it also provides a finite description of the output (normalised) language. The completion procedure generates procedures that are based on exhaustive applications of rewrite rules, while our system produces procedures that use subsets of rewrite rules in stages and give structured proofs (easily understandable to a human). For instance, our system can generate a procedure for constructing conjunctive normal form, which cannot be done by the completion procedure and by a single rule set (because, as said, there is no confluent and terminating rewrite system for transforming a formula into disjunctive normal form [10]). We believe that it would be worthwhile to combine our work with the Knuth-Bendix completion procedure in the following way: the completion procedure can be used to find a confluent and terminating set of rules and then ADEPTUS can be used over them.

Our work is also related to work aimed at deriving decision procedures using superposition-based inference system for clausal equational logic [1]. That approach is an alternative to the congruence closure algorithm and to the

Knuth-Bendix completion procedure. It does not use subsets of rewrite rules in stages, and it cannot handle some transformations required for decision procedures for fragments of arithmetic.

The presented approach is also related to work that performs automatic learning of proof methods [6]. The system LEARNΩMATIC learns proof methods (including decision procedures) from proof traces obtained by brute force application of available primitive methods. This approach (unlike ours) does not give opportunities for simple proofs of termination or completeness of learnt methods.

8   Realm of the Approach and Further Automation

In the presented method generators, we take a method kind, input BNF, and a set of rewrite rules, and use them to generate a required method (with some output BNF). However, it would be fruitful if we could start with an input BNF and look at BNFs and methods that can be obtained by subsets of the available rewrite rules. It is interesting to consider if, for a given BNF and a set of (terminating) rewrite rules, we can compute the output BNF. The answer for the general case is negative, since the resulting set of expressions is not necessarily definable by a BNF. Even if there is an algorithm that (given a BNF and a terminating set of rewrite rules) constructs an output BNF *whenever it is possible* (this is subject of our current research[12]), it would still not ensure further automation of our programme in general case. Namely, if we want to synthesise a decision procedure, we would generate a sequence of BNFs looking for a trivial one and we would have to check if two BNFs give the same language, but that problem is undecidable. Therefore, it is likely that we cannot have a complete such procedure for synthesising decision procedures. On the other hand, we believe that the presented system can work well in many practical situations. It is heuristic and its realm is determined by the set of method generators available (so it is difficult to make a formal characterisation of the realm). Basically, it can be used for producing linear procedures, possibly with loops, but with no branching. In addition to linear arithmetic, it can be also used for producing decision procedures for other fragments of arithmetic (e.g., Presburger arithmetic) or for some normalisation procedures for some inductively defined data structures. Procedures for fragments of arithmetic are the most illustrative examples for the approach that we have found so far. We are looking for additional such illustrative theories.

The problem of combining decision procedures is not addressed by our approach: a decision procedure for a combination theory can be synthesised only if it as a whole can be described in terms of normalisation methods.

For future work we are planning the following lines of research: we will be looking for other challenging domains (for instance, it would be interesting to use our system in the context of SMT (satisfiability modulo theory) solving, for

---

[12] The algorithms for some special cases of this problem were presented by the authors and Alan Smaill at the workshops CIAO 2003, CIAO 2004, and at *Deduction and Applications* meeting at Dagstuhl, 2005., without publications. The work described in this paper has not been presented or published before.

producing modules for checking unsatisfiability for underlying theory); we will try to extend the set of our method generators and search engines and will try to further improve their efficiency; we will implement generators not only for methods, but also for the corresponding tactics; we will try to automate the process of checking if conditions in the rewrite rules used cover all possible cases (we will try to do it whenever possible by using the "self-reflection" principle, as discussed in the proofs of theorems 1 and 2); we will try to combine our system with Knuth-Bendix completion procedure.

9   Conclusions

We presented a system (ADEPTUS) for synthesising decision procedures, based on ideas from [4]. ADEPTUS consists of several method generators and mechanisms for searching over them and combining them. We have implemented the system and used it for automatically generating decision procedures (in PROLOG) for ground arithmetic and for linear arithmetic. These implementations are correct (and the system makes easier proving correctness, completeness and termination), which is not quite easy for a human programmer to achieve. The approach generates procedures that are structured and easy to understand, and also very modular, making it easy to adapt them to slightly changed circumstances (e.g., with new rules or terms introduced). We believe that our approach can be used in other domains as well and can lead to automation of some routine steps in different types of programming tasks.

References

1. Armando, A., Ranise, S., Rusinowitch, M.: Uniform Derivation of Decision Procedures by Superposition. In: Fribourg, L. (ed.) CSL 2001 and EACSL 2001. LNCS, vol. 2142, Springer, Heidelberg (2001)
2. Boyer, R.S., Moore, J.S.: Integrating Decision Procedures into Heuristic Theorem Provers: A Case Study of Linear Arithmetic. Machine Intelligence 11 (1988)
3. Bundy, A.: The Use of Explicit Plans to Guide Inductive Proofs. In: Lusk, R., Overbeek, R. (eds.) 9th Conference on Automated Deduction (1988)
4. Bundy, A.: The Use of Proof Plans for Normalization. In: Boyer, R.S. (ed.) Essays in Honor of Woody Bledsoe (1991)
5. Hodes, L.: Solving Problems by Formula Manipulation in Logic and Linear Inequalities. In: ProcIJCAI-71 (1971)
6. Jamnik, M., Kerber, M., Pollet, M., Benzmuller, C.: Automatic Learning of Proof Methods in Proof Planning. CSRP-02-5, University of Birmingham (2002)
7. Janičić, P., Bundy, A.: A General Setting for the Flexible Combining and Augmenting Decision Procedures. Journal of Automated Reasoning 28(3) (2002)
8. Knuth, D.E., Bendix, P.B.: Simple word problems in universal algebra. In: Leech, J. (ed.) Computational problems in abstract algebra, Pergamon Press, New York (1970)
9. Lassez, J.-L., Maher, M.: On Fourier's algorithm for linear arithmetic constraints. Journal of Automated Reasoning 9, 373–379 (1992)
10. Socher-Ambosius, R.: Boolean algebra admits no convergent rewriting system. In: Book, R.V. (ed.) Rewriting Techniques and Applications. LNCS, vol. 488, Springer, Heidelberg (1991)

Certified Computer Algebra on Top of an Interactive Theorem Prover

Cezary Kaliszyk and Freek Wiedijk

Institute for Computing and Information Sciences,
Radboud University Nijmegen, The Netherlands
{cek,freek}@cs.ru.nl

Abstract. We present a prototype of a computer algebra system that is built on top of a proof assistant, HOL Light. This architecture guarantees that one can be certain that the system will make no mistakes. All expressions in the system will have precise semantics, and the proof assistant will check the correctness of all simplifications according to this semantics. The system actually *proves* each simplification performed by the computer algebra system.

Although our system is built on top of a proof assistant, we designed the user interface to be very close in spirit to the interface of systems like Maple and Mathematica. The system, therefore, allows the user to easily probe the underlying automation of the proof assistant for strengths and weaknesses with respect to the automation of mainstream computer algebra systems. The system that we present is a prototype, but can be straightforwardly scaled up to a practical computer algebra system.

1 Introduction

Computer algebra systems do not always give correct answers. This happens because those systems do not certify the operations performed. There can be various reasons for errors in a CAS: assumptions can be lost, types of expressions can be forgotten [2], the system might get confused between branches of 'multi-valued' functions, and of course the algorithms of the system themselves may contain implementation errors [23].

As an example of the kind of error that we are talking about here, consider the following MAPLE [11] session that tries to compute $\int_0^\infty \frac{e^{-(x-1)^2}}{\sqrt{x}}\,dx$ numerically in two different ways:

```
> int(exp(-(x-t)^2)/sqrt(x), x=0..infinity);
```

$$\frac{1}{2}\frac{e^{-t^2}\left(-\frac{3(t^2)^{\frac{1}{4}}\pi^{\frac{1}{2}}2^{\frac{1}{2}}e^{\frac{t^2}{2}}K_{\frac{3}{4}}(\frac{t^2}{2})}{t^2}+(t^2)^{\frac{1}{4}}\pi^{\frac{1}{2}}2^{\frac{1}{2}}e^{\frac{t^2}{2}}K_{\frac{7}{4}}(\frac{t^2}{2})\right)}{\pi^{\frac{1}{2}}}$$

```
> subs(t=1,%);
```

M. Kauers et al. (Eds.): MKM/Calculemus 2007, LNAI 4573, pp. 94–105, 2007.
© Springer-Verlag Berlin Heidelberg 2007

$$\frac{1}{2}\frac{e^{-1}\left(-3\pi^{\frac{1}{2}}2^{\frac{1}{2}}e^{\frac{1}{2}}K_{\frac{3}{4}}(\frac{1}{2})+\pi^{\frac{1}{2}}2^{\frac{1}{2}}e^{\frac{1}{2}}K_{\frac{7}{4}}(\frac{1}{2})\right)}{\pi^{\frac{1}{2}}}$$

> evalf(%);

0.4118623312

> evalf(int(exp(-(x-1)^2)/sqrt(x), x=0..infinity));

1.973732150

(We are showing MAPLE here, but all major computer algebra systems make errors like this.)

To be sure that results are correct, one may use a proof assistant instead of a CAS. But in that case even calculating simple things, like adding fractions or calculating a derivative of a polynomial becomes a non-trivial activity, which requires significant experience with the system.

Our approach is to implement a computer algebra system on top of a proof assistant. For our prototype we chose the LCF-style theorem prover HOL LIGHT [16]. Thanks to this, we obtain a CAS system where the user can be sure of the correctness of the results. Such a system has strong semantics, that is all variables have types, all functions have precise definitions in the logic of the prover and for every simplification there is a theorem that ensures the correctness of this simplification.[1] The interface of our computer algebra system resembles most CAS systems. It has a simple read-evaluate-print loop. The language of the formulas typed into the system is as close as possible to the language in which formulas are generally entered in CAS and to the language in which mathematics is done on paper. Interaction with the system currently looks like this[2]:

```
In1 := (3 + 4 DIV 2) EXP 3 * 5 MOD 3
Out1 := 250
In2 := vector [&2; &2] - vector [&1; &0] + vec 1
Out2 := vector [&2; &3]
In3 := diff (diff (λx. &3 * sin (&2 * x) + &7 + exp (exp x)))
Out3 := λx. exp x pow 2 * exp (exp x) + exp x * exp (exp x) +
        -- &12 * sin (&2 * x)
In4 := N (exp (&1)) 10
Out4 := #2.7182818284 + ... (exp (&1)) 10 F
In5 := 3 divides 6 ∧ EVEN 12
Out5 := T
In6 := Re ((Cx (&3) + Cx (&2) * ii) / (Cx (-- &2) + Cx (&7) * ii))
Out6 := &8 / &53
```

---

[1] In HOL LIGHT *simplification* is implemented through what in the LCF world is called *conversions*. A conversion is a function that takes a term and returns an equational theorem. The theorem has the given term on its left side and a simplified version of the term on the right side.

In this paper 'simplification' should *not* be taken to be a fixed reduction hardwired into the logic of the proof assistant, the way it is in type theoretical systems like CoQ [12].

[2] The '&', 'Cx' and '#' are coercions to real, complex and floating point numbers.

```
In7  := x + &1 - x / &1 + &7 * (y + x) pow 2
Out7 := &7 * x pow 2 + &14 * x * y + &7 * y pow 2 + &1
In8  := sum (0,5) (λx. &x * &x)
Out8 := &30
```

One can distinguish three categories of systems that try to fill the gap between computer algebra and proof assistants:

- Theorem provers inside computer algebra systems:
  - ANALYTICA [6],
  - THEOREMA [8],
  - REDLOG [13],
  - logical extension of AXIOM [20].
- Frameworks for mathematical information exchange between systems:
  - MATHML [10],
  - OPENMATH [15],
  - OMSCS [7],
  - MATHSCHEME [9],
  - LOGIC BROKER [1].
- Bridges between theorem provers and computer algebra systems, also referred to as ad-hoc information exchange solutions:
  - PVS and MAPLE [14],
  - HOL and MAPLE [17],
  - ISABELLE and MAPLE [4],
  - NUPRL and WEYL [18],
  - OMEGA and MAPLE/GAP [21],
  - ISABELLE and SUMMIT [3].

An important distinction that one can make within the category of bridges is the *degree of trust* between the prover and the CAS. In some of these solutions the prover uses the simplification of the CAS as an axiom, i.e., without checking its correctness. But in other solutions the prover takes the CAS output and then builds a verified theorem out of it. In this case there are again two possibilities: either the result is verified independently of how the CAS obtained it, or the system takes a trace of the rules that the CAS applied, and then uses that as a suggestion for what theorems should be used to construct a proof of the result.

In the work that we referred to here either the proof assistant is built inside the CAS, or the proof assistant and the CAS are next to each other. In our work however, we have the CAS inside the proof assistant.

Of course in many proof assistants there already is CAS-like functionality, in particular many proof assistants have arithmetic procedures or even powerful decision procedures. However, we do not just provide the functionality, but also build a *system* that can be used in a similar way as most other computer algebra systems are used.

In our system it is the first time that anyone pursued the combination of a CAS *inside* a proof assistant (in which all simplifications are validated), with an interface that has the customary CAS look and feel.

Our way of combining theorem proving and computer algebra has advantages over the ones presented above. All calculations done by our system are certified by the architecture of our system. All formulas defined inside it have types assigned, all defined operators have explicit semantics and all simplifications performed have theorems associated with them. No translation of formulas or semantics is needed, as the CAS shares the internal data structures of the proof assistant. There is no need to worry about mistakes in the implementation of the CAS, since all conversions are certified using the logic of the underlying prover. There is no verification required after the result is obtained, thanks to the creation of theorems alongside with the results. All simplifications performed by our architecture are completely certified, that is if a certificate for a particular simplification does not exist [5] it can not be performed. All variables used in HOL LIGHT conversions have to be typed, so working in a proof assistant might seem less flexible than a traditional CAS implementation, but the abundance of decision procedures for HOL show that this probably is not a strong limitation.

The paper is organized as follows: in Section 2 we present the architecture of the system. In Section 3 we talk about the knowledge base. Finally in Section 4 we present a conclusion.

2   Architecture

We present a general architecture for a certified computer algebra system, and we will describe an implementation prototype. The source for the prototype is available from http://www.cs.ru.nl/~cek/holcas/. For the implementation we chose the proof assistant HOL LIGHT [16]. The factors that influenced our choice were: the possibility to manipulate terms to create the conversions, prove theorems and implement the system in the same language[3], as well as a good library of analysis and algebra. The system created is rather a proof of concept than a real product, which is why the efficiency of the underlying prover was not a decisive factor. In particular we perform all computations inside the proof assistant's logic, sometimes with the help of decision procedures.

Our system is divided in three independent parts (Fig. 1): the user interface (input-response loop), the abstract algorithm of dealing with a formula (we will call this *the CAS conversion*), and the knowledge that is specific to the CAS system. That architecture allows the user both to use it as a computer algebra system, as well as making it usable in the context of theorem proving[4].

2.1   Input-Response Loop

The system displays a prompt, where one can write expressions to be simplified and commands. It is necessary to distinguish expressions to be computed or simplified from commands that represent actions that do not evaluate anything, like listing theorems or modifying and printing assumptions.

---

[3] HOL LIGHT is written in OCAML and is provided as an extension of it.
[4] *The CAS conversion* can be applied to a goal to be proved using CONV_TAC.

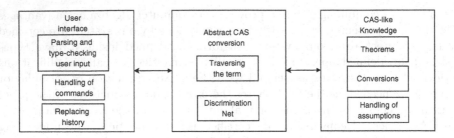

Fig. 1. Architecture of a CAS inside a TP system with responsibilities of the parts of our implementation marked. The prover is not marked on the figure, since all parts make use of it, by using it's type of terms and theorems, as well as tactics and conversions to build them.

Every expression that is not recognized as a command is passed to *the CAS conversion*, which will try to compute or simplify the expression. The theorem given back by *the CAS conversion* is the certificate that the output is correct. If *the CAS conversion* is not able to simplify the term, it returns an instance of reflexivity, and the result is then the same as the input.

In most CASs variables can be used without declaring them, but for certain algebraic operations one can define a variable to be of a particular type (necessary for example in MAGMA). Our system can handle expressions in both ways. The free variables are typed using HOL LIGHT type inference, but one can also require a specific type with the `assumetype` command (described in section 3.4).

Most computer algebra systems allow one to reuse previously typed in expressions and calculated outputs. One may calculate `In1 + Out2`. The loop has to have access to all expressions entered, theorems proved and outputs. In our framework every expression entered is stored with its type, so when it is reused, parsing the same expression, even in a different context, gives the same type.

2.2   Abstract CAS Conversion

To be able to benefit from the CAS simplification in theorem proving, it is useful to have the CAS functionality available as a single conversion (that we call here *the CAS conversion*). Since the underlying prover can be further developed and new theorems can be proved later, it is useful to separate *the CAS conversion* from the knowledge that it uses. For this reason *the CAS conversion* is parametrized. The general idea behind *the CAS conversion* is to try to apply all sub-conversions from the knowledge base at all positions in the term until it is saturated (Fig. 2). Applying the same conversions to a modified term is necessary, since some conversions return terms, parts of which can be again simplified. Particular implementations may include mechanisms to increase efficiency or to provide termination of simplification.

We are not particularly aiming at completeness for the algorithms in *the CAS conversion*, since completeness in practice can only be realized for rather basic theories. However any mathematically correct algorithm that exists for existing computer algebra systems can be implemented as a HOL LIGHT conversion too, that does the calculation while building the correctness proof in parallel. Examples include conversions that perform algorithms for integration, conversions that perform splitting and joining for branching calculations, or conversions that simplify terms involving higher order operations (like summation).

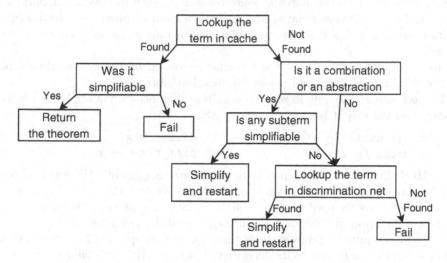

Fig. 2. Our implementation of *the CAS conversion* first tries to look up the term in a built-in cache (for efficiency). If the term is an application or an abstraction, then it tries to simplify subterms recursively (not performed if the term is known not to be expandable or is a suggestion that should not be expanded, for example NUMERAL or assuming). Finally it tries to apply all conversions from the knowledge base to the term.

3 CAS-Like Knowledge

The knowledge base is a separate part of the system. The knowledge is kept in a discrimination net (a structure that allows matching a term to a number of patterns efficiently). There is an interface on the theorem prover level that allows introducing knowledge to the knowledge base in the following three forms:

- Rewrite rules, for example:
  |- ∀z. abs (norm z) = norm z
- Conditional rewrite rules, for example:
  |- ∀x. &0 <= x ==> abs x = x
- Conversions meant to be used with an argument that matches a certain pattern and return ad-hoc rewrite rules. An ad-hoc rewrite rule is a theorem that is generated to be used for rewriting the formula, but it is not added to the knowledge base (although our implementation keeps all rewritten theorems in cache, implemented as a hash-table, for efficiency reasons). For example the HOL LIGHT conversion DIVIDES_CONV takes terms that match

the pattern n divides m and then returns ad-hoc rewrite rules for the given data like |- 33 divides 123453 <=> T.

The CAS conversion has to check whether the given term matches one of the rewrite rules and ad-hoc rewrite rules in the knowledge base. For efficiency it keeps all theorems and conversions included in the knowledge base in a discrimination net. To allow matching conversions with even less overhead, optional patterns for matching associated with conversions can be provided. The discrimination net is not changed, the particular used instances are only added to the cache.

To resemble a CAS system, the formulas processed by the system should be in the "evaluation" form and not in "verification" form.

Let us compare the ways in which one writes differentiation in the HOL LIGHT library and the way it is written in our CAS:

```
∀x. (f diffl (g x)) x        →        diff f = g
    (f diffl (g x)) x        →        diff f x = g x
```

In HOL LIGHT the diffl predicate takes three arguments: the function (on the left of the predicate), the value of its derivative and the point. To write a general derivative we need to generalize the point and replace the value with the derivative function in this point. Even then it is still a binary predicate.

In most computer algebra systems there exists a simple diff operator, that takes a function and returns its derivative. Using the Hilbert's choice operator, we created a such function, defined: diff f = λx. εv. (f diffl v) x. We also created a conversion that is able to calculate the derivative of a function, if HOL LIGHT's DIFF_CONV can.

Just like we defined a functional form of differentiation, we also defined a functional integration operator. Using these we can then compute the following expressions in the system:

```
 In9  := dint (&1,&2) sin
Out9  := -- &1 * cos (&2) + cos (&1)
 In10 := dint (&1,&2) (λx. x pow n)
Out10 := &1 / &(n + 1) * &2 pow (n + 1) +
             -- &1 * &1 / &(n + 1) * &1 pow (n + 1)
 In11 := diff (diff (λx. &3 * sin (&2 * x) + &7 + exp (exp x))) (&2)
Out11 := exp (exp (&2)) * exp (&2) pow 2 + exp (exp (&2)) * exp (&2) +
             -- &12 * sin (&4)
 In12 := diff (λx. dint (x,x + &2) (λx. x pow 3))
Out12 := λx. &6 * x pow 2 + &12 * x + &8
```

Our differentiation and integration definitions do not work well with partial functions. An approach for defining them in such a way that partial functions are handled better will be described in Section 4.

3.1   Numerical Approximations

In complex calculations computer algebra systems provide users with numerical approximations. They are usually implemented with an approximation algorithm, which keeps an error bound with every calculation. In a proof assistant a numerical approximation must have its semantics completely defined, and the algorithm has to respect the approximation definition and theorems that specify its properties.

The two main ways of rounding a real number are down to an integer and towards the nearest integer. Both these operations do not give rise to a computable function (see for example [19]). In [22] it is shown that if one computes non-deterministically either one of those values then one does get a computable function. We will use a conversion that calculates the value rounded both down and to nearest value, that terminates when one of those calculations terminate.

We propose to define the numerical approximation of a given number x to a precision p as identical to the number itself: N x p = x. It is only a hint for the system that the number has to be simplified to a decimal fraction plus a rest. It is the rest, that determines in which form is the number given: rounded down or rounded to the nearest. For rest defined in this way we provide a theorem, that states that the approximation can be different from the exact value only on the last digit, and the difference is less than one.

In the following HOL LIGHT definitions, N is the numerical approximation of a number to a precision (following the convention of MATHEMATICA) and ... is the rest of a number to the given precision with an additional argument that specifies the form of the rest. T stands for rounding to nearest and F stands for rounding down.

```
... x p F = x - floor (&10 pow p * x) / &10 pow p
... x p T = x - floor (&10 pow p * x + &1 / &2) / &10 pow p
|- abs(... x p v - x) < &1 / &10 pow p
```

The system is able to compute some numerical approximations with this scheme:

```
In13  := N (&1 / &3) 8
Out13 := #0.33333333 + ... (&1 / &3) 8 F
In14  := N (sqrt #5.123456789) 8
Out14 := #2.26350542 + ... (sqrt #5.123456789) 8 F
In15  := N (dint (#0.1,#0.4) exp) 7
Out15 := #0.3866537 + ... (-- &1 * exp #0.1 + exp (&2 / &5)) 7 F
```

3.2   Assumptions

In most CASs there is a possibility to make type assumptions or binary assumptions about variables. Examples include assuming a variable to be greater than zero, greater than another variable, natural or real. There are various methods of introducing assumptions in computer algebra systems:

- Assumptions associated with a simplification
  in MATHEMATICA: Simplify[Sin[n Pi]], Element[n,Integers]]
- Global list of assumptions
  in MAPLE: assume(x>0); sqrt(x*x);

– Asking the user for conditions on variables (e.g. MAXIMA)
– Adding assumptions automatically and silently to the prover environment (e.g. MATHXPERT)

In our system we keep a global list of assumptions, which are Boolean properties that may be later used to instantiate assumptions of rewrite rules and ad-hoc rewrite rules. In a big CAS the number of rules that can be used is so big that asking the user seems not to be a good choice. Also automated assuming will probably not behave too well in such a situation.

An assumption can be introduced by the user either using `assume`, which takes a Boolean, or `assumetype` which takes a typed variable. An assumption associated with a single simplification of a sub-term may be also introduced using `assuming`. The latter method temporarily changes the assumptions list to simplifying the sub-expression. The assumptions will be added to the assumptions of the theorem generated by *the CAS conversion*, which is why changing the assumptions list is only useful at the top-level of the expression to simplify.

The global list of assumptions is used by the conversions from the knowledge base, therefore we consider is a part of the latter. To ensure the usage of variables with correct types, type checking has to have access to this list. When an expression is typed in the system it is type-checked in a particular context. This context includes types already assigned to all free variables from the assumptions list, as well as all variables for which types have been assumed with `assumetype`. To do this, the latter are kept in another global list.

For example, $\sqrt{x^2}$ cannot be simplified to x, since we don't know whether x is positive or not. Also $\frac{x}{x}$ cannot be simplified to 1, since it is possible that $x = 0$.

```
In16  := sqrt (x * x)
Out16 := abs x
In17  := x / x
Out17 := x * inv x
```

When an assumption about x is introduced, stating that it is greater then 1, numeric things about x can be proved, and both of the above formulas can be simplified:

```
In18  := assume (x > &1)
Out18 := T
In19  := x > #0.5
Out19 := T
In20  := sqrt (x * x)
Out20 := x
In21  := x / x
Out21 := &1
```

There are two ways in which assumptions are used: direct and indirect. The first way is to use an assumption directly in the derivation in unchanged form. It can be used to a prove a reflexive theorem or to fill the requirement of a certain conditional rewrite rule (or a conditional ad-hoc rewrite rule). An assumption may be used as an indirect step in the derivation, for example simplifying $abs(x)$ to x requires $x \geq 0$, and the assumption $x > 1$ can be used for this.

3.3 Manipulating Assumptions

A CAS has to provide a mechanism for adding assumptions and listing defined assumptions. In our prototype we added the `assumptions` and `about` commands, which resemble their MAPLE equivalents.

```
Command: about Argument: x
  'x > &1'
```

An issue that is hard to handle in any approach are errors that may be caused by incorrect parsing and printing. We try to be as close as possible to the original HOL LIGHT's parsing and printing mechanism. In fact, the system currently uses HOL's term printing (with special output for errors) but, when parsing, the system has to add typing information and distinguish commands from terms. Special output is added, so that the user always knows when a given string has been interpreted as a command.

To further lower the risk of parsing and printing problems, we add the `theorems` command. It allows printing all theorems defined in a session. The standard HOL LIGHT theorem printing function is used for this. It is especially useful for conversions that use assumptions, because in that it case the assumptions that have been actually used will be shown. Below are the first five theorems proved by the examples from this document:

```
Command: theorems
  |- (3 + 4 DIV 2) EXP 3 * 5 MOD 3 = 250
  |- vector [&2; &2] - vector [&1; &0] + vec 1 = vector [&2; &3]
  |- diff (diff (λx. &3 * sin (&2 * x) + &7 + exp (exp x))) =
       (λx. exp x pow 2 * exp (exp x) + exp x * exp (exp x) +
       -- &12 * sin (&2 * x))
  |- N (exp (&1)) 10 = #2.7182818284 + ... (exp (&1)) 10 F
  |- 3 divides 6 ∧ EVEN 12 <=> T
```

4 Conclusion

Our work integrates computer algebra and proof assistant technology. We will now look at how our architecture compares with what one gets by just having a CAS or a proof assistant.

Developing a system according to our architecture (i.e., where the algorithms not only generate the results, but also generate certificates of the correctness of those results) will be slower than the development of traditional CAS systems (because that only has to generate the results). As far as the performance of the system is concerned, our architecture will be somewhat slower than a traditional CAS as well. This is mostly because generating the certificates for all simplifications also takes time. However, we expect this slow-down over traditional CAS to only multiply the running time by a constant factor. Our expectation is not experiment based, but based on the architecture, we trace what a traditional CAS does and provide proofs for every step.

When we compare our architecture to the way that one normally does CAS-like manipulations in an interactive theorem prover, the main difference is the interaction model. Our CAS system does not interactively work on propositions that are to be proved, but instead takes an expression and automatically simplifies it.

Our primary focus is to extend the knowledge base with a formalization of multivalued functions, to be able to handle more complicated expressions, like the MAPLE example of a complex function with multiple branches given in the introduction.

Another important feature that we plan to investigate are the coercions that many proof assistants use, like the embedding of the integers in the real numbers or the real numbers in the complex numbers. Currently a user of our prototype needs to use the '&' and 'Cx' symbols for this (as is customary in the HOL LIGHT library). A small improvement to the current situation might be to overload the '&' operator, but we would rather not make the user write these functions at all.

An issue that our approach does not cover is completeness of the conversions. In the case of rewrite rules the completeness is clear. But in the case of arbitrary algorithms, it is not guaranteed by our architecture that a given conversion will always terminate and never fail.

We believe that both computer algebra systems and proof assistants currently have a problem. In computer algebra the lack of explicit semantics and the lack of verification of the results inside the system makes the systems less reliable than one would like them to be. In proof assistants the powerful symbolic manipulations that are taken for granted in computer algebra often are missing and, even when present, it takes work and expertise to make use of it.

We claim that the architecture that we present here can solve both problems simultaneously. The computer algebra systems will get explicit semantics and certification. And the proof assistants will get CAS-like functionality that will make them more powerful and easier to use than they are today.

References

1. Armando, A., Zini, D.: Towards interoperable mechanized reasoning systems: the logic broker architecture. In: Corradi, A., Omicini, A., Poggi, A. (eds.) WOA, Pitagora Editrice Bologna, pp. 70–75 (2000)
2. Aslaksen, H.: Multiple-valued complex functions and computer algebra. SIGSAM Bulletin (ACM Special Interest Group on Symbolic and Algebraic Manipulation) 30(2), 12–20 (1996)
3. Ballarin, C., Paulson, L.C.: A pragmatic approach to extending provers by computer algebra - with applications to coding theory. Fundam. Inf. 39(1-2), 1–20 (1999)
4. Ballarin, C., Homann, K., Calmet, J.: Theorems and algorithms: an interface between Isabelle and Maple. In: ISSAC '95: Proceedings of the 1995 international symposium on Symbolic and algebraic computation, pp. 150–157. ACM Press, New York (1995)

5. Barendregt, H., Cohen, A.M.: Electronic communication of mathematics and the interaction of computer algebra systems and proof assistants. J. Symb. Comput. 32(1/2), 3–22 (2001)
6. Bauer, A., Clarke, E.M., Zhao, X.: Analytica - an experiment in combining theorem proving and symbolic computation. Journal of Automated Reasoning 21(3), 295–325 (1998)
7. Bertoli, P., Calmet, J., Giunchiglia, F., Homann, K.: Specification and integration of theorem provers and computer algebra systems. Fundam. Inform. 39(1-2), 39–57 (1999)
8. Buchberger, B., et al.: The Theorema Project: A Progress Report. In: Kerber, M., Kohlhase, M. (eds.) Symbolic Computation and Automated Reasoning (Proceedings of CALCULEMUS, Natick, Massachusetts, A.K. Peters (2000)
9. Carette, J., Farmer, W., Wajs, J.: Trustable communication between mathematics systems. In: CALCULEMUS 2003, Rome, Italy, Aracne, pp. 55–68 (2003)
10. Carlisle, D., Ion, P., Miner, R., Poppelier, N.: Mathematical Markup Language (MathML) Version 2.0, 2nd edn. (2003)
11. Char, B.W., Geddes, K.O., Gentleman, W.M., Gonnet, G.H.: The design of Maple: A compact, portable and powerful computer algebra system. Springer, London (1983)
12. Coq Development Team. The Coq Proof Assistant Reference Manual Version 8.0. INRIA-Rocquencourt (January 2005)
13. Dolzmann, A., Sturm, T.: Redlog: Computer algebra meets computer logic. ACM SIGSAM Bulletin 31(2), 2–9 (1997)
14. Adams, A., et al.: Computer algebra meets automated theorem proving: Integrating Maple and PVS. In: Boulton, R.J., Jackson, P.B. (eds.) TPHOLs 2001. LNCS, vol. 2152, pp. 27–42. Springer, Heidelberg (2001)
15. Buswell, S., et al.: The OpenMath Standard, version 2.0 (2002)
16. Harrison, J.: HOL light: A tutorial introduction. In: Srivas, M., Camilleri, A. (eds.) FMCAD 1996. LNCS, vol. 1166, pp. 265–269. Springer, Heidelberg (1996)
17. Harrison, J., Théry, L.: A skeptic's approach to combining HOL and Maple. Journal of Automated Reasoning 21, 279–294 (1998)
18. Jackson, P.B.: Enhancing the Nuprl Proof Development System and Applying it to Computational Abstract Algebra. PhD thesis, Cornell University, Ithaca, NY, USA (January 1995)
19. Lester, D.R.: Effective continued fractions. In: Proceedings 15th IEEE Symposium on Computer Arithmetic, pp. 163–170. IEEE Computer Society Press, Washington (2001)
20. Poll, E., Thompson, S.: Adding the axioms to Axiom: Towards a system of automated reasoning in Aldor. In: Calculemus and Types '98 (July 1998)
21. Sorge, V.: Non-trivial symbolic computations in proof planning. In: Kirchner, H. (ed.) Frontiers of Combining Systems. LNCS, vol. 1794, pp. 121–135. Springer, Heidelberg (2000)
22. Vuillemin, J.: Exact real computer arithmetic with continued fractions. In: LFP '88: Proceedings of the 1988 ACM conference on LISP and functional programming, pp. 14–27. ACM Press, New York (1988)
23. Wester, M.J. (ed.): Contents of Computer Algebra Systems: A Practical Guide, chapter A Critique of the Mathematical Abilities of CA Systems. John Wiley & Sons, Chichester, United Kingdom (1999)

Quantifier Elimination for Approximate Factorization of Linear Partial Differential Operators

Elena Kartashova[1] and Scott McCallum[2]

[1] RISC, J.Kepler University, Linz, Austria
lena@risc.uni-linz.ac.at
[2] Macquarie University, Sydney, Australia
scott@ics.mq.edu.au

Abstract. This paper looks at the feasibility of applying the quantifier elimination program QEPCAD-B to obtain quantifier-free conditions for the approximate factorization of a simple hyperbolic linear partial differential operator (LPDO) of order 2 over some given bounded rectangular domain in the plane. A condition for approximate factorization of such an operator to within some given tolerance over some given bounded rectangular domain is first stated as a quantified formula of elementary real algebra. Then QEPCAD-B is applied to try to eliminate the quantifiers from the formula. While QEPCAD-B required too much space and time to finish its task, it was able to find a partial solution to the problem. That is, it was able to find a nontrivial quantifier-free sufficient condition for the original quantified formula.

1 Introduction

Let $\mathbb{Q}$ denote the field of all rational numbers, and let R denote the polynomial ring $\mathbb{Q}[x, y]$ in the variables x and y over $\mathbb{Q}$. A *linear partial differential operator (LPDO)* in x and y over $\mathbb{Q}$ is an element of the noncommutative ring $R[\partial_x, \partial_y]$, where ∂_x and ∂_y denote the usual derivation operators on R. An LPDO is of *order n* if the highest order derivations occurring in it are of the n-th order.

While factorization of linear ordinary differential operators is well studied and has a well developed algorithmic theory, the theory of factorization of LP-DOs is much more difficult. One constructive factorization method for LPDOs – Beals-Kartashova (BK) factorization – was introduced in [1]. The method could roughly be described as a straightforward search for first order factors of a given LPDO from the *left*, so to speak. One simply expresses a given LPDO of order n as a symbolic product of a first order factor on the left and an $(n-1)$st order factor on the right. One writes down a system of equations for the symbolic coefficients of the factors. Then one tries to solve these equations. Usually one or more *factorization conditions* are thereby derived, necessary and sufficient for the existence of such factors. In case the factorization conditions are fulfilled, the factors can be obtained, and the method applied recursively to the factor on

M. Kauers et al. (Eds.): MKM/Calculemus 2007, LNAI 4573, pp. 106–115, 2007.

the right. A simple LPDO of order 2, together with its factorization condition, and its factorization when the condition is satisfied, is presented in Section 2.

The idea to use BK-factorization for the *approximate* factorization of an LPDO over some bounded domain is discussed in [2]. It is motivated by the important application area of numerical simulations. The processing time for such numerical simulations could be substantially reduced if instead of computation with one LPDO of order n we could proceed with n LPDOs all of order 1. In numerical simulations the coefficients of the given operator are given within some tolerance. It is thus not necessary to fulfil the factorization conditions exactly, but instead within some given tolerance, and over some bounded domain. This leads to the idea of *approximate factorization conditions* for an LPDO over some bounded domain. Approximate factorization conditions would be expected to be formulated using quantifiers over the real numbers.

The idea of the present paper is to look into the feasibility of obtaining quantifier-free approximate factorization conditions using *quantifier elimination by cylindrical algebraic decomposition (QE by CAD)* [3]. This idea was suggested to us by Bruno Buchberger [4]. In Section 2 of this paper we formulate an approximate factorization condition for a so-called hyperbolic LPDO of order 2 with simple polynomial coefficients. We use the language of Tarski algebra to do this. In Section 3 we provide a brief synopsis of QE by CAD. In Section 4 we report the results of applying a computer program for carrying out QE by CAD to the approximate factorization condition obtained in Section 2 in an attempt to find a quantifier-free version of the condition. We find that, while our program could not solve the problem given using a reasonable amount of time and space, it was able to find a partial solution to the problem. More specifically, it was able to find a nontrivial quantifier-free sufficient condition for the given quantified formula.

2   Factorization Conditions for a Hyperbolic LPDO of Order 2

Let us consider a hyperbolic LPDO of order 2 in canonical form:

$$H_2 = \partial_x^2 - \partial_y^2 + p\partial_x + q\partial_y + r, \tag{1}$$

where the coefficients p, q, r are arbitrary elements of R (that is, arbitrary polynomials in x and y over $\mathbb{Q}$). We say that H_2 is *factorizable* if H_2 can be expressed in the form

$$H_2 = (\partial_x \pm \partial_y + s)(\partial_x \mp \partial_y + t),$$

for some elements s and t of R. With $\omega = \pm 1$, put

$$\mathcal{R}_\omega = (\partial_x - \omega\partial_y) \left(\frac{p - \omega q}{2} \right) + \frac{p^2 - q^2}{4}.$$

It follows from equation system 2 in [1], specialised for H_2, that H_2 is factorizable if and only if

$$[r = \mathcal{R}_{-1}] \vee [r = \mathcal{R}_1]. \tag{2}$$

If the former disjunct is valid, then

$$H_2 = (\partial_x + \partial_y + \frac{p-q}{2})(\partial_x - \partial_y + \frac{p+q}{2}),$$

and if the latter disjunct is valid then

$$H_2 = (\partial_x - \partial_y + \frac{p+q}{2})(\partial_x + \partial_y + \frac{p-q}{2}).$$

The reader is reminded that multiplication of such LPDOs is in general non-commutative. Hence the former factorization need not imply the latter, and vice versa.

Suppose now that the polynomial coefficients of the operator H_2 are of the first degree: say $p(x,y) = p_3 x + p_2 y + p_1$, $q(x,y) = q_3 x + q_2 y + q_1$, $r(x,y) = r_3 x + r_2 y + r_1$. Then – by expanding the $\mathcal{R}_\omega$ in terms of x and y and equating the coefficients of r and the $\mathcal{R}_\omega$ – the factorization condition (2) can be expressed as a disjunction of two systems of equations in the nine variables p_i, q_j and r_k. The solution of this disjunction of equation systems yields all *exactly* factorizable hyperbolic LPDOs of this type.

For the remainder of this paper we will address the problem of trying to determine and simplify a condition for *approximate* factorization of hyperbolic operators of this type. We use the standard formal language of elementary real algebra, that is, Tarski algebra [3], to formulate a condition for approximate factorization of hyperbolic operators of this type as a quantifier elimination (QE) problem. In addition to the coefficients of the given operator H_2, we assume that we are also given:

(1) a constant ε;

(2) constants M and N, which define a bounded rectangular region in the plane: $-M < x < M$, $-N < y < N$.

With all this given, and with $\omega = \pm 1$, let us consider the quantified formula of elementary real algebra $\phi^* = \phi^*(p_i, q_j, r_k)$ which asserts that "for all x and y in the bounded region $-M < x < M$, $-N < y < N$, we have $-\varepsilon < r(x,y) - \mathcal{R}_\omega(x,y) < \varepsilon$." We wish to eliminate the quantifiers from $\phi^*(p_i, q_j, r_k)$. More precisely, we wish to find a formula of elementary real algebra $\phi' = \phi'(p_i, q_j, r_k)$, free of quantifiers, such that if $\phi'(p_i, q_j, r_k)$ is true then $\phi^*(p_i, q_j, r_k)$ is true. That is, we wish to find conditions on the coefficients of the polynomials $p(x,y)$, $q(x,y)$ and $r(x,y)$ which imply that the function $\mathcal{R}_\omega(x,y)$ differs not too much from one these polynomials, namely $r(x,y)$, throughout the bounded region $-M < x < M$, $-N < y < N$.

3   Synopsis of QE by CAD

Let A be a set of integral polynomials in $x_1, x_2 \ldots, x_r$, where $r \geq 1$. An A-*invariant cylindrical algebraic decomposition (CAD)* of $\mathbf{R}^r$, r-dimensional real space, is a decomposition D of $\mathbf{R}^r$ into nonempty connected subsets called cells such that

1. the cells of D are cylindrically arranged with respect to the variables $x_1, x_2, \ldots, x_r$;

2. every cell of D is a semialgebraic set (that is, a set defined by means of boolean combinations of polynomial equations and inequalities); and

3. every polynomial in A is sign-invariant throughout each cell of D.

The CAD algorithm as originally conceived [3,5] has inputs and outputs as follows. Given such a set A of r-variate polynomials and a nonnegative integer f with $f < r$, the algorithm produces as its output a description of an A-invariant CAD D of $\mathbf{R}^r$, in which explicit semialgebraic defining formulas are provided only for the cells of the CAD D_f of $\mathbf{R}^f$ induced (that is, implicitly determined) by D. The description of D comprises lists of indices and sample points for the cells of D. (Every cell is assigned an *index* which indicates its position within the cylindrical structure of D.)

The working of the original CAD algorithm can be summarized as follows. If $r = 1$, an A-invariant CAD of $\mathbf{R}^1$ is constructed directly, using polynomial real root isolation. If $r > 1$, then the algorithm computes a *projection set* P of $(r-1)$-variate polynomials (in $x_1, \ldots, x_{r-1}$) such that any P-invariant CAD D' of $\mathbf{R}^{r-1}$ can be extended to a CAD D of $\mathbf{R}^r$. If $f = r$ we set $f' \leftarrow f - 1$ and otherwise set $f' \leftarrow f$. Then the algorithm calls itself recursively on P and f' to get such a D'. Finally D' is extended to D. In order to produce semialgebraic defining formulas for the cells of D_f the algorithm must be used in a mode called *augmented projection*.

Thus for $r > 1$, if we trace the algorithm, we see that it computes a first projection set P, eliminating x_r, then computes the projection of P, eliminating x_{r-1}, and so on, until the $(r-1)$-st projection set has been obtained, which is a set of polynomials in the variable x_1 only. This is called the *projection phase* of the algorithm. The construction of a CAD of $\mathbf{R}^1$ invariant with respect to the $(r-1)$-st projection set is called the *base phase*. The successive extensions of the CAD of $\mathbf{R}^1$ to a CAD of $\mathbf{R}^2$, the CAD of $\mathbf{R}^2$ to a CAD of $\mathbf{R}^3$, and so on, until an A-invariant cad of $\mathbf{R}^r$ is obtained, constitute the *extension phase* of the algorithm.

Now we consider the *quantifier elimination (QE) problem* for the elementary theory of the reals: given a quantified formula (known as a *QE problem instance*) of elementary real algebra

$$\phi^* = (Q_{f+1} x_{f+1}) \ldots (Q_r x_r) \phi(x_1, \ldots, x_r)$$

where ϕ is a formula involving the variables $x_1, x_2, \ldots, x_r$ which is free of quantifiers, find a formula $\phi'(x_1, \ldots, x_f)$, free of quantifiers, such that ϕ' is equivalent to ϕ^*. The QE problem can be solved by constructing a certain CAD of $\mathbf{R}^r$. The method is described as follows.

1. Extract from ϕ the list A of distinct non-zero r-variate polynomials occurring in ϕ.

2. Construct lists S and I of sample points and cell indices, respectively, for an A-invariant CAD D of $\mathbf{R}^r$, together with a list F of semialgebraic defining formulas for the cells of the CAD D_f of $\mathbf{R}^f$ induced by D.

3. Using S, evaluate the truth value of ϕ^* in each cell of D_f. (By construction of D, the truth value of ϕ^* is constant throughout each cell c of D_f, hence can be determined by evaluating ϕ^* at the sample point of c.)

4. Construct $\phi'(x_1, \ldots, x_f)$ as the disjunction of the semialgebraic defining formulas of those cells of D_f for which the value of ϕ^* has been determined to be true.

The above algorithm solves any given particular instance of the QE problem in principle. However the computing time of the algorithm grows steeply as the number r of variables occurring in the input formula ϕ increases.

Collins and Hong [8] introduced the method of *partial CAD construction* for QE. This method, named with the acronym QEPCAD, is based upon the simple observation that we can often solve a QE problem by means of a partially built CAD. The QEPCAD algorithm was originally implemented by Hong. A recent implementation, denoted by QEPCAD-B, contains improvements by Brown, Collins, McCallum, and others – see [6]. QEPCAD-B has solved a range of reasonably interesting problems for which the original QE algorithm takes too much time. Nevertheless the worst case computing time of QEPCAD-B remains large (that is, it depends doubly-exponentially on r).

4   Application of QEPCAD to BK-Factorization

We consider only the first simple case of approximate factorization described in Section 2. Using the notation of Section 2, we suppose that ε, M and N have been given specific constant values, say $\varepsilon = M = N = 1$, and we put $\omega = -1$. We consider the formula $\phi^*(p_i, q_j, r_k)$ which asserts that

$$(\forall x)(\forall y)[(|x| < 1 \wedge |y| < 1) \Rightarrow |r(x,y) - \mathcal{R}_{-1}(x,y)| < 1]. \tag{3}$$

We wish to find a formula $\phi'(p_i, q_j, r_k)$, free of quantifiers, such that $\phi'(p_i, q_j, r_k)$ implies $\phi^*(p_i, q_j, r_k)$.

Remark 1. It would be of greatest interest to find the most general such $\phi'(p_i, q_j, r_k)$ – that is, to find quantifier-free $\phi'(p_i, q_j, r_k)$ *equivalent* to $\phi^*(p_i, q_j, r_k)$. But as we'll see it seems that the time and space resources needed to do this are prohibitive. We'll also see that it is not as time consuming, yet hopefully still of interest, to find quantifier-free conditions merely *sufficient* for ϕ^* to be true.

As a first step we rewrote the quantified formula 3 so that the variables p_i, q_j, r_k appear explicitly, and the denominator 4 is cleared from the right hand side of the implication. Expanding in terms of x and y formula 3 thus has the form:

$$(\forall x)(\forall y)[(|x| < 1 \wedge |y| < 1) \Rightarrow |ax^2 + bxy + cy^2 + dx + ey + f| < 4], \tag{4}$$

where a, b, c, d, e, f are integral polynomials in the p_i, q_j, r_k. (For example, $a = q_3^2 - p_3^2$, $b = 2q_2q_3 - 2p_2p_3$, and $c = q_2^2 - p_2^2$.) In fact it is computationally advantageous to use the general form of quantified formula 4, in which a, b, c, d, e, f

occur as distinct *indeterminates*, rather than as polynomial expressions in the p_i, q_j, r_k, because then the total number of variables in the formula is reduced from 11 to 8.

We attempted to find a solution to the above QE problem instance by running the program QEPCAD-B with the quantified formula 4 (in its general form) as its input. The variable ordering used was (a, b, c, d, e, f, x, y). The computer used for this and subsequent experiments was a Sun server having a 292 MHz ultraSPARC risc processor. Four megabytes of memory were made available for list processing. However the program ran out of memory after a few minutes. The program was executing the projection phase of the algorithm when it stopped. The first three projection steps – that is, successive elimination of y, x and f – were almost complete.

Increasing the amount of memory to eighty megabytes did not help – the program still ran out of memory during the fourth projection step (that is, during elimination of e).

Of course a very special, but completely trivial, quantifier-free sufficient condition for our QE problem instance is the formula

$$\phi'(p_i, q_j, r_k) := [\text{all } p_i = 0 \wedge \text{all } q_j = 0 \wedge \text{all } r_k = 0].$$

It could be of some interest to look for partial solutions to (that is, quantifier-free sufficient conditions for) our QE problem instance in which some but not all of the variables p_i, q_j, r_k are equal to zero. For example, suppose that we put $p_2 = q_2 = r_2 = 0$ in (4). We obtain:

$$(\forall x)(\forall y)[(|x| < 1 \wedge |y| < 1) \Rightarrow |a'x^2 + d'x + f'| < 4],$$

where a', d', f' are polynomials in $p_1, p_3, q_1, q_3, r_1, r_3$. (In fact we have $a' = a$ and $d' = d$.) Clearly this formula is equivalent to:

$$(\forall x)[(|x| < 1) \Rightarrow |a'x^2 + d'x + f'| < 4], \tag{5}$$

which we shall denote by $\psi^*(p_i, q_j, r_k)$.

The following theorem shows that a partial solution to the special QE problem instance $\psi^*(p_i, q_j, r_k)$ (that is, a quantifier-free sufficient condition for ψ^*) leads to a partial solution to the QE problem instance ϕ^* (that is, a quantifier-free sufficient condition for ϕ^*).

Theorem 1. *Suppose that $\psi'(p_i, q_j, r_k)$ is a quantifier-free formula, involving only $p_1, p_3, q_1, q_3, r_1, r_3$, which implies $\psi^*(p_i, q_j, r_k)$. Then the quantifier-free formula $\psi'(p_i, q_j, r_k) \wedge p_2 = 0 \wedge q_2 = 0 \wedge r_2 = 0$ implies $\phi^*(p_i, q_j, r_k)$.*

▶ Let p_i, q_j, r_k be real numbers and let (with slight abuse of notation) a, b, c, d, e, f denote the values of the polynomials a, b, c, d, e, f at the particular p_i, q_j, r_k. Assume $\psi'(p_i, q_j, r_k) \wedge p_2 = 0 \wedge q_2 = 0 \wedge r_2 = 0$. Then $\psi^*(p_i, q_j, r_k) \wedge p_2 = 0 \wedge q_2 = 0 \wedge r_2 = 0$ is true, by hypothesis. Take real numbers x and y, with $|x| < 1$ and $|y| < 1$, and (with slight abuse of notation) let a', d', f' denote the values of the polynomials a', d', f' at the particular p_i, q_j, r_k. Then

$$|a'x^2 + d'x + f'| < 4,$$

by virtue of (5) (since $|x| < 1$). Hence (4) is true (since $a = a'$, $d = d'$, $f = f'$, $b = c = e = 0$). ∎

The above discussion suggests that it would be worthwhile to try to find a solution to the simplified, special QE problem instance ψ^* using the program QEPCAD-B. We use the more general form of (5), in which a', d', f' occur as indeterminates, and hence reduce by 3 the number of variables in the formula. For simplicity of the notation hereafter we use the variables a, b, c in place of a', d', f', and thus treat the formula:

$$(\forall x)[(|x| < 1) \Rightarrow |ax^2 + bx + c| < 4]. \tag{6}$$

We ran program QEPCAD-B with (6) as its input. Eighty megabytes of memory were made available for list processing. After 191 seconds the program produced the following quantifier-free formula equivalent to (6):

```
c - b + a + 4 >= 0 /\ c - b + a - 4 <= 0 /\
c + b + a + 4 >= 0 /\ c + b + a - 4 <= 0 /\
[ 4 a c - b^2 + 16 a > 0 \/ 4 a c - b^2 - 16 a > 0 \/
[ b^2 - 16 a = 0 /\ b^2 + 16 a > 0 ] \/
[ b^2 - 16 a < 0 /\ b - 2 a >= 0 ] \/
[ b^2 - 16 a < 0 /\ b + 2 a <= 0 ] \/
[ b^2 - 16 a > 0 /\ b + 2 a >= 0 ] \/
[ b^2 - 16 a > 0 /\ b - 2 a <= 0 ] \/
[ b^2 - 16 a = 0 /\ c - b + a + 4 > 0 /\ c - b + a - 4 < 0 ] ] ].
```

We could transform this formula into a partial solution $\psi'(p_i, q_j, r_k)$ to ψ^* by setting $a = q_3^2 - p_3^2$, $b = 4r_3 - (p_3 - q_3)(p_1 + q_1) - (p_1 - q_1)(p_3 + q_3)$, and $c = 4r_1 - 2(p_3 + q_3) - (p_1 - q_1)(p_1 + q_1)$.

It is possible to induce the program to produce an arguably even simpler solution formula using less computing time by making three separate runs of QEPCAD-B. The first run uses the command

```
assume [a < 0].
```

After just 1.9 seconds the program produced the following quantifier-free formula equivalent to (6) under the assumption $a < 0$:

```
c - b + a + 4 >= 0 /\ c - b + a - 4 <= 0 /\
c + b + a + 4 >= 0 /\ c + b + a - 4 <= 0 /\
[ b - 2 a <= 0 \/ b + 2 a >= 0 \/ 4 a c - b^2 - 16 a > 0 ].    (7)
```

The above formula is perhaps more elegant and understandable. For it is a slight improvement of (that is, slightly more compact than) a formula seen to be equivalent to it (under assumption $a < 0$) which is quite straightforward to derive by hand from (6) using elementary properties of the parabola $y = ax^2 + bx + c$ on the interval $(-1, +1)$:

```
[ 2 a - b >= 0 /\ a + b + c + 4 >= 0 /\ a - b + c - 4 <= 0] \/
[2 a + b >= 0 /\ a - b + c + 4 >= 0 /\ a + b + c -4 <= 0] \/
[2 a - b < 0 /\ 2 a + b < 0 /\ 4 a c - b^2 - 16 a > 0 /\
   a - b + c + 4 >= 0 /\ a + b + c + 4 >= 0].                    (8)
```

Remark 2. To derive by hand (8) from (6) under the assumption $a < 0$, one has to notice that function $f(x) = ax^2 + bx + c$ has its maximum value for $f'(x) = 2ax + b = 0$, that is, for $x = -b/(2a)$, and consider three cases separately: (1) $-b/(2a) \leq -1$, (2) $-b/(2a) \geq +1$, and (3) $-1 < -b/(2a) < +1$. For each of the above three cases one can then write down necessary and sufficient conditions for (6) to be true. For example, in Case 1, (6) is clearly equivalent to $-4 \leq a + b + c \wedge a - b + c \leq 4$. After treating each of the above cases, we obtain (8) by forming the disjunction of the formulas corresponding to the cases.

To obtain a complete solution to the QE problem instance (6) we need to run QEPCAD two more times, for the cases $a > 0$ and $a = 0$, respectively. For the second run we use the command

```
assume [a > 0].
```

and obtain after 1.9 seconds the following quantifier-free formula equivalent to (6) under the assumption $a > 0$:

```
c - b + a + 4 >= 0 /\ c - b + a - 4 <= 0 /\
c + b + a + 4 >= 0 /\ c + b + a - 4 <= 0 /\
[ b + 2 a <= 0 \/ b - 2 a >= 0 \/ 4 a c - b^2 + 16 a > 0 ].       (9)
```

For the third run we put $a = 0$ in (6) and use the command

```
assume [b /= 0].
```

After 60 milliseconds the program produced the following formula equivalent to (6) with $a = 0$ under assumption $b \neq 0$:

```
c - b + 4 >= 0 /\ c - b - 4 <= 0 /\
c + b + 4 >= 0 /\ c + b - 4 <= 0                                 (10)
```

This is immediately seen to be correct! Finally we could obtain a complete solution to (6) by combining (7) for $a < 0$, (9) for $a > 0$, (10) for $b \neq a = 0$ and the formula $c - 4 < 0 \wedge c + 4 > 0$ (for $a = b = 0$). In fact a simple and elegant way to achieve such a combination is as follows:

```
c - b + a + 4 >= 0 /\ c - b + a - 4 <= 0 /\
c + b + a + 4 >= 0 /\ c + b + a - 4 <= 0 /\
[ b - 2 a <= 0 \/ b + 2 a >= 0 \/
4 a c - b^2 - 16 a > 0 \/ a >= 0] /\
[ b + 2 a <= 0 \/ b - 2 a >= 0 \/
4 a c - b^2 + 16 a > 0 \/ a <= 0].                               (11)
```

5 Discussion

As we remarked in Section 3 the worst case computing time of QEPCAD-B grows steeply as the number of variables in the given QE problem instance increases. Indeed, as is suggested by the results reported in Section 4, a complete solution of the QE problem instance (4) by QEPCAD-B using a reasonable amount of time and space seems to be unlikely for the foreseeable future.

Nevertheless the results of Section 4 also suggest that QEPCAD-B could be of help in searching for certain kinds of sufficient conditions for (3), especially those which involve setting some of the variables to zero.

We briefly mention here another kind of approach which a person could use to derive another kind of sufficient condition for (4) by hand. We simply notice that a sufficient condition for (4) is:

$$|a| < \frac{4}{6} \wedge |b| < \frac{4}{6} \wedge |c| < \frac{4}{6} \wedge |d| < \frac{4}{6} \wedge |e| < \frac{4}{6} \wedge |f| < \frac{4}{6}.$$

The above sufficient condition is unlikely to be obtained in a reasonable amount of time and space using QEPCAD-B applied to (4), even if one issues `assume` commands. The number of variables involved is probably too big. However a version of QEPCAD-B which is planned for the future, which will have the capability to determine adjacency relationships amongst the cells of the partial CAD, could be of some use in analyzing certain topological properties of the truth set in nine-dimensional space of the corresponding quantifier-free formula in p_i, q_j, r_k obtained from the above.

Acknowledgements

This paper has its origin in discussions between the authors at RISC-Linz during the second of S.M. to RISC-Linz in 2005. S.M. would like to thank Professors Franz Winkler and Bruno Buchberger, and all their colleagues and staff at RISC, for their hospitality during his stay. S.M. would also like to acknowledge helpful discussions and communications with Professors Daniel Lazard and Chris Brown. E.K. acknowledges the support of the Austrian Science Foundation (FWF) under projects SFB F013/F1304.

References

1. Beals, R., Kartashova, E.: Constructively factoring linear partial differential operators in two variables. TMPh 145(2), 1510–1523 (2005)
2. Kartashova, E., Rudenko, O.: Invariant Form of BK-factorization and its Applications. In: Calmet, J., Seiler, W.M., Tucker, R.W. (eds.) Proc. GIFT-2006, Universitätsverlag Karlsruhe, pp. 225–241 (2006)
3. Collins, G.E.: Quantifier elimination for real closed fields by cylindrical algebraic decomposition. In: Brakhage, H. (ed.) Automata Theory and Formal Languages. LNCS, vol. 33, pp. 134–183. Springer, Heidelberg (1975)

4. Personal communication with Bruno Buchberger (2005)
5. Arnon, D., Collins, G., McCallum, S.: Cylindrical algebraic decomposition I: the basic algorithm. JSC 13(4), 878–889 (1984)
6. Brown, C.: QEPCAD B: a program for computing with semi-algebraic sets using CADs. ACM SIGSAM Bulletin 37(4), 97–108 (2003)
7. Caviness, B.F., Johnson, J.R. (eds.): Quantifier Elimination and Cylindrical Algebraic Decomposition. Springer, Berlin (1998)
8. Collins, G.E., Hong, H.: Partial cylindrical algebraic decomposition for quantifier elimination. JSC 12(3), 299–328 (1991)

Rule-Based Simplification in Vector-Product Spaces

Songxin Liang and David J. Jeffrey

Department of Applied Mathematics, University of Western Ontario,
London, Ontario, Canada

Abstract. A vector-product space is a component-free representation of
the common three-dimensional Cartesian vector space. The components
of the vectors are invisible and formally inaccessible, although expressions for the components could be constructed. Expressions that have
been built from the scalar and vector products can be simplified using a rule-based system. In order to develop and specify the system,
an axiomatic system for a vector-product space is given. In addition, a
brief description is given of an implementation in Aldor. The present
work provides simplification functionality which overcomes difficulties
encountered in earlier packages.

1 Introduction

We start with a definition from J. W. Gibbs [7], which appeared in a privately
printed pamphlet that was dated 1881.

> An algebra or analytical method in which a single letter or other expres-
> sion is used to specify a vector may be called a vector algebra or vector
> analysis.

Vector analysis has found many applications throughout engineering and science.
It often simplifies the derivation of mathematical theorems and the statements
of physical laws, while vector notation can often clearly convey geometric or
physical interpretations that greatly facilitate understanding.

Almost all well known computer algebra systems provide basic vector datatypes and vector operations. The data types, however, consist of lists of components, and there is no provision for a letter or other symbol of the system to
have vector attributes such as was intended by Gibbs. This is a great pity, as a
perusal of any advanced textbook in physics will reveal vector expressions being
formed and worked with in a component-free way. In addition to the operations
that are built-in, meaning that they are distributed with the base system, the
major systems offer vector-analysis packages, for example, the *VectorAnalysis*
package for MATHEMATICA [15], the *Vector33* package for REDUCE [8] and the
VectorCalculus package in MAPLE. There are a number of vector analysis packages build on the well known systems, for example, the *VecCalc* package using
MAPLE [2], the *Vect* package using MACSYMA [14] and the *OrthoVec* package using REDUCE [5]. All these packages, however, rely on component representation.
The following shows some features of the *VectorCalculus* package in MAPLE.

M. Kauers et al. (Eds.): MKM/Calculemus 2007, LNAI 4573, pp. 116–127, 2007.
© Springer-Verlag Berlin Heidelberg 2007

```
> with(VectorCalculus)
> CrossProduct(A, B)
      Error, (in VectorCalculus:-CrossProduct) the first argument
      must either be the differential operator Del, or a three
      dimensional VectorField, got A
> DotProduct(A, B)
      A . B
```

It is curious that MAPLE checks the arguments of *CrossProduct* and requires an explicit set of components, but *DotProduct* does not. In any event, it is clear that the package does not expect abstract vectors. Once explicit components are given, the package is able to perform common tasks.

```
> A := <a, b, c>
      A := a e_x  + b e_y  + c e_z
> B := <d, e, f>
      B := d e_x  + e e_y  + f e_z
> CrossProduct(A, B)
      (b f - c e) e_x  + (c d - a f) e_y  + (a e - b d) e_z
> DotProduct(A, B)
      a d + b e + c f
```

Like the *VectorCalculus* package, almost all existing packages can only perform component-dependent operations. They cannot perform component-free operations. This means that before one can do any vector analysis, one must set the components for all vectors involved. This is inconvenient when one wants to deal with problems containing many vectors and only wants to know the relationship among them, because the expansion of vector expressions into specific components is usually tedious and error prone. With a component-free system, one is free from the distracting details of individual components and can concentrate on the meaningful results of the problems.

For more detailed commentary, we introduce some notation. Vectors are denoted by bold-face letters; for the vector product of $\mathbf{a}$ and $\mathbf{b}$, we use the notation of Chapman [3], namely $\mathbf{a} \wedge \mathbf{b}$. The scalar product is $\mathbf{a} \bullet \mathbf{b}$, and $(\mathbf{abc})$ or $(\mathbf{a}, \mathbf{b}, \mathbf{c})$ stands for the scalar triple product $(\mathbf{a} \wedge \mathbf{b}) \bullet \mathbf{c}$ of three vectors $\mathbf{a}$, $\mathbf{b}$ and $\mathbf{c}$.

Compared with component-dependent systems, which have a systematic method for deriving mathematical statements, component-free systems are more difficult and challenging because vector algebra has a strange and intriguing structure [12].

- Vectors are not closed under the scalar product operation. If $\mathbf{p}$, $\mathbf{q}$ are vectors, then $\mathbf{p} \bullet \mathbf{q}$ is a scalar.
- The scalar product is commutative while the vector product is anticommutative. If $\mathbf{p}$, $\mathbf{q}$ are vectors, then $\mathbf{p} \bullet \mathbf{q} = \mathbf{q} \bullet \mathbf{p}$ while $\mathbf{p} \wedge \mathbf{q} = -\mathbf{q} \wedge \mathbf{p}$.
- Neither is associative. If $\mathbf{p}$, $\mathbf{q}$, $\mathbf{r}$ are vectors, then $\mathbf{p} \wedge (\mathbf{q} \wedge \mathbf{r}) \neq (\mathbf{p} \wedge \mathbf{q}) \wedge \mathbf{r}$, whereas $\mathbf{p} \bullet (\mathbf{q} \bullet \mathbf{r})$ and $(\mathbf{p} \bullet \mathbf{q}) \bullet \mathbf{r}$ are invalid.
- Neither has a multiplicative unit. There does not exist a fixed vector $\mathbf{u}$ such that for any vector $\mathbf{p}$, $\mathbf{u} \wedge \mathbf{p} = \mathbf{p}$ or $\mathbf{p} \wedge \mathbf{u} = \mathbf{p}$ or $\mathbf{u} \bullet \mathbf{p} = \mathbf{p}$ or $\mathbf{p} \bullet \mathbf{u} = \mathbf{p}$.

- Both admit zero divisors. For any vector $\mathbf{p}$, $\mathbf{p} \wedge \mathbf{p} = \mathbf{0}$; if $\mathbf{q}$ is a vector perpendicular to $\mathbf{p}$, then $\mathbf{p} \bullet \mathbf{q} = 0$.
- The two operations are connected through the operation of scalar-vector multiplication $*$ by the strange side relation $\mathbf{p} \wedge (\mathbf{q} \wedge \mathbf{r}) = (\mathbf{p} \bullet \mathbf{r}) * \mathbf{q} - (\mathbf{p} \bullet \mathbf{q}) * \mathbf{r}$.

In contrast to the number of component-dependent vector analysis packages, to our knowledge, only the packages [6], [10] and [12] address component-free vector operations. However, the emphasis of these packages is still on component-dependent operations, and only the package by Stoutemyer [12] provides non-trivial simplification examples. Even in Stoutemyer's package, however, simplification problems remained unsolved. For example, when he tried to simplify the vector expression

$$(\mathbf{a} \wedge \mathbf{b}) \wedge (\mathbf{b} \wedge \mathbf{c}) \bullet (\mathbf{c} \wedge \mathbf{a}) - (\mathbf{a} \bullet (\mathbf{b} \wedge \mathbf{c}))^2$$

which should be simplified to zero, he only got

$$-\mathbf{a} \bullet \mathbf{c} \wedge (\mathbf{a} \bullet \mathbf{b} \wedge \mathbf{c} * \mathbf{b} - \mathbf{a} \bullet \mathbf{b} \wedge (\mathbf{b} \wedge \mathbf{c})) - (\mathbf{a} \bullet \mathbf{b} \wedge \mathbf{c})^2.$$

When he returned this expression to his simplification engine, instead of getting the desired result 0, he could only get

$$\mathbf{a} \bullet (\mathbf{a} \bullet \mathbf{b} \wedge \mathbf{c} * \mathbf{b}) \wedge \mathbf{c} - (\mathbf{a} \bullet \mathbf{b} \wedge \mathbf{c})^2.$$

He explained that the scalar factor $\mathbf{a} \bullet \mathbf{b} \wedge \mathbf{c}$ could be factored out, clearly revealing that the expression is zero, but the built-in scalar-factoring-out mechanism could not recognize that $\mathbf{a} \bullet \mathbf{b} \wedge \mathbf{c}$ is a scalar. Therefore, it is necessary and meaningful to develop a new component-free vector analysis package.

As a final comment on the background, we note that the vector product has not been generalized beyond 3 dimensions except in a limited way. A generalization of the vector product in 3 dimensional space to 7 dimensional case has been proposed, but the important property $\mathbf{p} \wedge (\mathbf{q} \wedge \mathbf{r}) = (\mathbf{p} \bullet \mathbf{r}) * \mathbf{q} - (\mathbf{p} \bullet \mathbf{q}) * \mathbf{r}$ is no longer valid [11]. So we confine ourselves here to the 3 dimensional case.

2   Axiomatic Theory and Transformation Rules

In order to provide a unified picture of component-free vector algebra and component-dependent vector algebra, we define an axiomatic theory $\mathcal{T}$ for a vector-product space. The language of $\mathcal{T}$ is the set of vector operations $\{+, *, \bullet, \wedge\}$, where $+, *,$ and $\bullet$ correspond to the addition, scalar multiplication and scalar product of $\mathbb{R}^3$ considered as an inner product space of component-free vectors, and $\wedge$ corresponds to the usual vector product of $\mathbb{R}^3$.

The axiomatic set of $\mathcal{T}$ consists of the axioms of a real inner-product vector space, as usually defined on $(\mathbb{R}^3, +, *, \bullet)$ (including the fact of three-dimensionality), together with the addition of axioms for vector product.

A0: $(a + b) \wedge c = a \wedge c + b \wedge c$.

A1: $a \wedge b = -b \wedge a$.

A2: $a \wedge (b \wedge c) = (a \bullet c) * b - (a \bullet b) * c$.

A3: $(abc) = (bca)$, where (abc) denotes $(a \wedge b) \bullet c$.

A4: $(abc) = 0$ implies that a, b, c are linearly dependent.

Based on the axioms of $\mathcal{T}$, we can now prove the following theorems in $\mathcal{T}$:

T1: $a \wedge a = 0$.

T2: $(abc) = -(acb)$.

T3: $(aab) = (abb) = (aba) = 0$.

T4: $(a \wedge b) \wedge (c \wedge d) = (abd) * c - (abc) * d$.

T5: $(a \wedge b) \bullet (c \wedge d) = (a \bullet c)(b \bullet d) - (a \bullet d)(b \bullet c)$.

T6: $(abc) * d = (d \bullet a) * (b \wedge c) + (d \bullet b) * (c \wedge a) + (d \bullet c) * (a \wedge b)$.

T7: $(abc) * d = (dbc) * a + (adc) * b + (abd) * c$.

T8: $(d \bullet h)(abc) = (d \bullet a)(hbc) + (d \bullet b)(ahc) + (d \bullet c)(abh)$.

T9: $(abc) * (d \wedge h) = (dha) * (b \wedge c) + (dhb) * (c \wedge a) + (dhc) * (a \wedge b)$.

Proof. For theorem T1, by axiom A1, we have $a \wedge a = -a \wedge a$. Then $2 * (a \wedge a) = 0$, and $(a \wedge a) = 0$. For theorem T2, by axioms A1 and A3, we have $(abc) = (a \wedge b) \bullet c = -(b \wedge a) \bullet c = -(bac) = -(acb)$. Theorem T3 follows immediately from theorem T1. Theorem T4 follows immediately from axiom A2 by viewing $a \wedge b$ as a single vector.

For theorem T5, by axioms A3 and A2, we have

$$(a \wedge b) \bullet (c \wedge d) = (a, b, c \wedge d) = (b, c \wedge d, a)$$
$$= (b \wedge (c \wedge d)) \bullet a = a \bullet (b \wedge (c \wedge d))$$
$$= a \bullet [(b \bullet d) * c - (b \bullet c) * d]$$
$$= (a \bullet c)(b \bullet d) - (a \bullet d)(b \bullet c).$$

Next we prove theorem T6. First, we claim that $(abc) = 0$ if and only if a, b, c are linearly dependent. By axiom A4, we only need to prove the "if". Without loss of generality, we can express a as $a = x * b + y * c$. Then by axiom A0 and theorem T3, $(abc) = (a \wedge b) \bullet c = [(x * b + y * c) \wedge b] \bullet c = x(bbc) + y(cbc) = 0$.

There are two cases for T6. If a, b and c are linearly independent, then $(b \wedge c, c \wedge a, a \wedge b) = (a, b, c)^2 \neq 0$. So $a \wedge b$, $b \wedge c$, $c \wedge a$ are also linearly independent, so we can express d as

$$d = x * (b \wedge c) + y * (c \wedge a) + z * (a \wedge b). \tag{1}$$

Taking scalar product of both sides of (1) with a, we get $d \bullet a = (abc)x$, so $x = (d \bullet a)/(abc)$. Similarly, we can get $y = (d \bullet b)/(abc)$ and $z = (d \bullet c)/(abc)$. Therefore, by (1), the desired result follows.

If a, b and c are linearly dependent then $(abc) = 0$. Without loss of generality, we can express c as $c = x * a + y * b$. Then, by axioms A0, A1 and theorem T1, the right hand side of T6 is

$$x(d \bullet a) * (b \wedge a) + y(d \bullet b) * (b \wedge a) - [x(d \bullet a) + y(d \bullet b)] * (b \wedge a) = 0 ,$$

the left hand side of T6.

For theorem T7, we can use a similar method as above so we omit the details. For theorem T8, taking scalar product of both sides of T6 with $\mathbf{h}$ and using axiom A3, the desired result follows immediately. For theorem T9, we can use a similar method as theorem T6, so we omit the details again. Consequently, all theorems are proved.

Now we define another axiomatic theory $\mathcal{T}'$ for component-dependent vectors in $\mathbb{R}^3$. The definitions of the vector operations in $\mathcal{T}'$ are just the standard ones. For $\mathbf{a}, \mathbf{b} \in \mathbb{R}^3$ with $\mathbf{a} = (a_1, a_2, a_3)$ and $\mathbf{b} = (b_1, b_2, b_3)$, and $k \in \mathbb{R}$, define

$$k * \mathbf{a} = (ka_1, ka_2, ka_3) \ .$$
$$\mathbf{a} + \mathbf{b} = (a_1 + b_1, a_2 + b_2, a_3 + b_3) \ .$$
$$\mathbf{a} \bullet \mathbf{b} = a_1 b_1 + a_2 b_2 + a_3 b_3 \ .$$
$$\mathbf{a} \wedge \mathbf{b} = (a_2 b_3 - a_3 b_2, a_3 b_1 - a_1 b_3, a_1 b_2 - a_2 b_1) \ .$$

It is routine to check that the vector operations defined above satisfy all the axioms of $\mathcal{T}$. Therefore, we have showed that $\mathcal{T}'$ is a model of $\mathcal{T}$. In this way, we provide a unified picture of the axiomatic specification of vector operations, the computation of abstract (component-free) vectors, and the computation of concrete vectors (component-dependent).

Now we turn to transformation rules. In order to achieve a normal form for an input vector expression, or to simplify a vector expression to its shortest normal form, we need transformation rules. The transformation rules consist of the 4 axioms (A0)-(A3) and 9 theorems (T1)-(T9) of $\mathcal{T}$, together with the *inverses* of theorems (T6)-(T9). By the *inverse* of a formula (theorem) $A = B$, we mean $B = A$.

Roughly speaking, the transformation rules can be divided into two types: expansion type and combination type. An *expansion type rule* expands a single term into a sum of different terms, or transfers a single term to another single term. For example, $(\mathbf{abc}) * \mathbf{d} = (\mathbf{d} \bullet \mathbf{a}) * (\mathbf{b} \wedge \mathbf{c}) + (\mathbf{d} \bullet \mathbf{b}) * (\mathbf{c} \wedge \mathbf{a}) + (\mathbf{d} \bullet \mathbf{c}) * (\mathbf{a} \wedge \mathbf{b})$ is an expansion type rule. A *combination type rule* combines a sum of different terms into a single term. For example, $(\mathbf{d} \bullet \mathbf{a})(\mathbf{hbc}) + (\mathbf{d} \bullet \mathbf{b})(\mathbf{ahc}) + (\mathbf{d} \bullet \mathbf{c})(\mathbf{abh}) = (\mathbf{d} \bullet \mathbf{h})(\mathbf{abc})$ is a combination type rule.

3   Structure and Description of the Package

Our package is developed using Aldor. Aldor [1] is a programming language with a two-level object model of categories and domains. Types and functions are first class entities allowing them to be constructed and manipulated within Aldor programs just like any other values. Aldor is an ideal tool for symbolic mathematical computations.

For the package, we first define a vector space category *VectorSpcCategory* as follows.

```
define VectorSpcCategory(R:Join(ArithmeticType, ExpressionType),
n:MI==3): Category==with {
  *: (R,%)->%;
  *: (%,R)->%;
  +: (%,%)->%;
  -: (%,%)->%;
  -:    %->%;
  =: (%,%)->Boolean;
  default
  {
    import from R;
    (x:%)-(y:%):%==x+(-1)*y;
    (x:%)*(r:R):%==r*x;
    -(x:%):%==(-1)*x;
  }
},
```

where *ArithmeticType* and *ExpressionType* are two categories defined in the algebra library of Aldor and n is the dimension of the space (the default dimension is 3). Those within the braces after *with* are the operations for the space.

Then, based on VectorSpcCategory, we define a vector algebra category *VectorAlgCategory* as follows.

```
define VectorAlgCategory(R:Join(ArithmeticType,
ExpressionType)):Category == VectorSpcCategory(R) with {
  vector:     Symbol->%;
  zero:       ()->%;
  Simplify:   %->%;
  s3p:        (%,%,%)->%;
  *:          (%,%)->%;
  apply:      (%,%)->%;
  /\:         (%,%)->%;
  <<:         (TextWriter,%)->TextWriter;
  default
  {
    s3p(x:%,y:%,z:%):%==apply(x/\y,z);
  }
},
```

where *with* means that besides the exports of VectorSpcCategory, VectorAlgCategory has additional exports (within the braces) which are normally related to component-free vector operations. For example, *apply* is the scalar product, *s3p* is the scalar triple product, and ∧ is the vector product.

Finally, we come to the point of implementing the vector algebra domain *VectorAlg* in detail.

```
VectorAlg(R:Join(ArithmeticType,ExpressionType)):
Join(ExpressionType,VectorAlgCategory R)== add
{
  ...
},
```

where the part within the braces after *add* is the implementation details of the domain which composes the rest of this paper.

3.1 Normal Form

A common strategy for simplifying expressions in computer algebra is to define a normal form. We define a normal form for vector expressions as follows.

Definition 1 (Normal Form). *A normal form for a vector expression is a sum of terms. Each term may consist of a real coefficient, a scalar part and a vector part. The vector part consists of either a single letter or a vector product, or there is no vector part if the term is not really a vector. If there is a scalar part for a term, then the factors of the scalar part may consist of single letters, scalar products and scalar triple products. The terms are in ascending order which is determined by the* termOrder *defined in section 3.2.*

In our vector algebra package, all input vector expressions will be transferred automatically into their normal forms. A vector expression can be presented in different normal forms. For example, according to theorem T6 in section 2, we can present the same expression in two different normal forms: $(\mathbf{abc}) * \mathbf{d}$ and $(\mathbf{c} \bullet \mathbf{d}) * (\mathbf{a} \wedge \mathbf{b}) - (\mathbf{b} \bullet \mathbf{d}) * (\mathbf{a} \wedge \mathbf{c}) + (\mathbf{a} \bullet \mathbf{d}) * (\mathbf{b} \wedge \mathbf{c})$. However, by using the simplification functionality (see section 4.2), we can get the shortest normal form for a given vector expression.

3.2 Data Representations and Term Order

There are two components for a normal form of a vector expression: data representation and term order. In our package, a vector expression is presented internally as:

Rep==List Term, and
Term==Record (coe:R, sca:List List String, vec:List String),
where *coe, sca* and *vec* are respectively the real coefficient, the scalar part and the vector part of a term. For example, the term $-2(\mathbf{a} \bullet \mathbf{b})(\mathbf{abc}) * (\mathbf{c} \wedge \mathbf{d})$ can be expressed as $[-2, [[\text{``}a\text{''}, \text{``}b\text{''}], [\text{``}a\text{''}, \text{``}b\text{''}, \text{``}c\text{''}]], [\text{``}c\text{''}, \text{``}d\text{''}]]$. Inspired by Stoutemyer's unsolved problem, we adopt such data representation because we want to separate the scalar part of a term from the vector part of the term.

In the scalar part *sca*, there are three different kinds of lists of strings. List of length 1 stands for a scalar of single letter, list of length 2 stands for a scalar product, and list of length 3 stands for a scalar triple product.

Similarly, for the vector part vec, list of length 0 means that the term is not really a vector, list of length 1 stands for a vector of single letter, and list of length 2 stands for a vector product.

On the other hand, in order to implement the transformation rules in section 2, we also need another minor data structure.

$mixTerm == Record(scal:scaList, vec2:List String)$, where
$scaList == List scaTerm$, and
$scaTerm == Record(coe2:R, sca2:List List String)$.

This representation is necessary when one wants to simplify a vector expression which contains terms with the same vector parts. For example, for vector expression $(\mathbf{d} \bullet \mathbf{a})(\mathbf{hbc}) * \mathbf{c} + (\mathbf{d} \bullet \mathbf{b})(\mathbf{ahc}) * \mathbf{c} + (\mathbf{d} \bullet \mathbf{c})(\mathbf{abh}) * \mathbf{c}$, we can combine the scalar parts of the three terms because they have same vector parts $\mathbf{c}$. Then, $[(\mathbf{d} \bullet \mathbf{a})(\mathbf{hbc}) + (\mathbf{d} \bullet \mathbf{b})(\mathbf{ahc}) + (\mathbf{d} \bullet \mathbf{c})(\mathbf{abh})] * \mathbf{c}$ is a $mixTerm$, the $scal$ part is $(\mathbf{d} \bullet \mathbf{a})(\mathbf{hbc}) + (\mathbf{d} \bullet \mathbf{b})(\mathbf{ahc}) + (\mathbf{d} \bullet \mathbf{c})(\mathbf{abh})$, the $vec2$ part is $\mathbf{c}$, and there are three $scaTerms$: $(\mathbf{d} \bullet \mathbf{a})(\mathbf{hbc})$, $(\mathbf{d} \bullet \mathbf{b})(\mathbf{ahc})$ and $(\mathbf{d} \bullet \mathbf{c})(\mathbf{abh})$. Using the transformation rules from section 2, we can simplify the expression to $(\mathbf{d} \bullet \mathbf{h})(\mathbf{abc}) * \mathbf{c}$.

As mentioned above, the second component for a normal form is term order. Unlike polynomials which have a natural way to define term order, we have to choose a term order for our purpose. The order $termOrder$ is defined step by step as follows.

Firstly, we use the natural order for strings.

Secondly, we define the order for lists of strings. Given two lists of strings L_1 and L_2, then $L_1 < L_2 \iff \#L_1 < \#L_2$, or $\#L_1 = \#L_2$ and there exists a natural number i such that for all $0 < j < i$, $L_1(j) = L_2(j)$ and $L_1(i) < L_2(i)$, where $\#L$ denotes the length of list L.

Thirdly, we define the order for lists of lists of strings in the same way as above for lists of strings.

Finally, we define the order of terms for vector expressions. Given two terms $T_1 = [coe_1, sca_1, vec_1]$ and $T_2 = [coe_2, sca_2, vec_2]$, then $T_1 < T_2 \iff vec_1 < vec_2$, or $vec_1 = vec_2$ and $sca_1 < sca_2$, or $vec_1 = vec_2$ and $sca_1 = sca_2$ and $coe_1 < coe_2$ (if R is an ordered arithmetic type).

For example, given a vector expression $5(\mathbf{abc})(\mathbf{a} \bullet \mathbf{d}) * (\mathbf{b} \wedge \mathbf{c}) + 2(\mathbf{b} \bullet \mathbf{c}) * (\mathbf{a} \wedge \mathbf{c}) + (\mathbf{acd})(\mathbf{a} \bullet \mathbf{b}) * \mathbf{c}$, after being sorted by $termOrder$, the expression will become $(\mathbf{a} \bullet \mathbf{b})(\mathbf{acd}) * \mathbf{c} + 2(\mathbf{b} \bullet \mathbf{c}) * (\mathbf{a} \wedge \mathbf{c}) + 5(\mathbf{a} \bullet \mathbf{d})(\mathbf{abc}) * (\mathbf{b} \wedge \mathbf{c})$.

4 Implementation of the Package

This section addresses the implementation details of the package. These include the implementation of transformation rules, the implementation of the simplification function and the implementation of vector algebra operations.

4.1 Implementations of Transformation Rules

Now we turn to the implementations of transformation rules. The main skill we use is pattern matching [13]. Since the implementations are detail-involved, we only give brief descriptions of the algorithms.

The implementations of expansion type rules are not complicated. For example, the following algorithm is to implement the transformation rule $(\mathbf{abc}) * \mathbf{d} = (\mathbf{d} \bullet \mathbf{a}) * (\mathbf{b} \wedge \mathbf{c}) + (\mathbf{d} \bullet \mathbf{b}) * (\mathbf{c} \wedge \mathbf{a}) + (\mathbf{d} \bullet \mathbf{c}) * (\mathbf{a} \wedge \mathbf{b})$:

Algorithm Rule01

Input: a Rep xx.

Output: a Rep yy.

- Let tx range over the terms of xx.
- If there are lists in tx which match the pattern $(\mathbf{abc}) * \mathbf{d}$, replace them with the lists representing $(\mathbf{d} \bullet \mathbf{a}) * (\mathbf{b} \wedge \mathbf{c}) + (\mathbf{d} \bullet \mathbf{b}) * (\mathbf{c} \wedge \mathbf{a}) + (\mathbf{d} \bullet \mathbf{c}) * (\mathbf{a} \wedge \mathbf{b})$.
- Output the resulting Rep.

Compared with the implementations of expansion type rules, the implementations of combination type rules are much more complicated. For example, in order to implement the transformation rule $(\mathbf{d} \bullet \mathbf{a})(\mathbf{hbc}) + (\mathbf{d} \bullet \mathbf{b})(\mathbf{ahc}) + (\mathbf{d} \bullet \mathbf{c})(\mathbf{abh}) = (\mathbf{d} \bullet \mathbf{h})(\mathbf{abc})$, first we have to implement a subalgorithm *Match?* to check if any given three scaTerms match the pattern of this rule, then we have to implement another subalgorithm *Comb* which combines the matching scaTerms into a single Term using this rule.

Then the algorithm for implementing the transformation rule $(\mathbf{d} \bullet \mathbf{a})(\mathbf{hbc}) + (\mathbf{d} \bullet \mathbf{b})(\mathbf{ahc}) + (\mathbf{d} \bullet \mathbf{c})(\mathbf{abh}) = (\mathbf{d} \bullet \mathbf{h})(\mathbf{abc})$ can be described as follows.

Algorithm Rule02

Input: a Rep xx.

Output: a Rep yy.

- Change the data representation of xx from Rep to List mixTerm.
- Let tx range over xx, and let L be the *scal* part of tx.
- If $\#L < 3$, then leave it for output. Otherwise, again let L denote the collection of scaTerms in L in which there are lists of length 2 and length 3 in their *sca2* part, and leave other scaTerms for output.
- In a while-loop, check if any given three scaTerms match the rule using the subalgorithm *Match?*.
- If there is no such three scaTerms or $\#L < 3$, then break the loop. Otherwise, combine the three scaTerms using the subalgorithm *Comb* and update L.
- output the remaining L.

4.2   Simplification

There are two levels of simplification. *Simplify0* is the basic level which, as part of the definitions of vector algebra operations, transfers an input vector expression into its normal form. *Simplify* is the advanced level which, based on *Simplify0*, transfers a vector expression into its shortest normal form.

The basic level can be described as follows.

Algorithm Simplify0

Input: an element x of VectorAlg.

Output: an element of VectorAlg.

- Let tx range over the data representation xx of x.
- Make the scalar part and vector part of tx in ascending order, and use the transformation rules to check if tx is zero.
- Collect non-zero terms tx, and denote is yy.
- Make yy in ascending order and combine like terms.
- Output the domain element y of yy.

The advanced simplification can be described as follows.

Algorithm Simplify
Input: an element x of VectorAlg.
Output: an element of VectorAlg.

- Perform basic simplification on x.
- Apply transformation rules to the resulting expression one by one within a while-loop and then perform basic simplification.
- If the resulting expression is shorter than the original one then we replace the original one with the resulting expression.
- If at any time the number of terms < 3, or the expression is not changed after applying all transformation rules, break the loop.
- Return after the loop.

5   Examples

We use Aldor Interpreter to test our package. After compiling our source file into a platform-independent object file *vectorAlg.ao*, we open the Aldor Interpreter using command: *aldor -gloop*. Then we use the following commands to do initialization:

```
%1 >> #include "algebra"
%2 >> #include "aldorinterp"
%3 >> #library aaa "vectorAlg.ao"
%4 >> macro MI==MachineInteger
%5 >> import from aaa, MI, String, Symbol, VectorAlg MI
```

Now we declare vectors **a**, **b**, **c**, **d**, **e**, **f**, **g** and **h** as follows.

```
%6  >> a:=vector(-"a")
%7  >> b:=vector(-"b")
%8  >> c:=vector(-"c")
%9  >> d:=vector(-"d")
%10 >> e:=vector(-"e")
%11 >> f:=vector(-"f")
%12 >> g:=vector(-"g")
%13 >> h:=vector(-"h")
```

At this point, we are ready to test some examples using our package. All computations have been performed on a Pentium IV PC with 3.2 GHz CPU and 1 GB RAM.

EXAMPLE 1.
We first test Stoutemyer's unsolved problem [12].

```
%14 >> ((a^b)^(b^c)).(c^a)-(a.(b^c))*(a.(b^c))
0 @ VectorAlg(MachineInteger)
                         Comp: 0 msec, Interp: 40 msec
```

From the example above, we can see that our package is quite efficient. It only takes 40 milliseconds to get the result.

EXAMPLE 2. Now, we test some more examples also from Stoutemyer:

- $(a-d) \wedge (b-c) + (b-d) \wedge (c-a) + (c-d) \wedge (a-b) - 2*(a \wedge b + b \wedge c + c \wedge a) = 0.$
- $(b \wedge c) \wedge (a \wedge d) + (c \wedge a) \wedge (b \wedge d) + (a \wedge b) \wedge (c \wedge d) + 2(abc)*d = 0.$

```
%15 >> (a-d)^(b-c)+(b-d)^(c-a)+(c-d)^(a-b)-2*(a^b+b^c+c^a)
0 @ VectorAlg(MachineInteger)
                         Comp: 0 msec, Interp: 40 msec
%16 >> v1:=(b^c)^(a^d)+(c^a)^(b^d)+(a^b)^(c^d)+2*s3p(a,b,c)*d
(bcd)*a-(acd)*b+(abd)*c-(abc)*d @ VectorAlg(MachineInteger)
                         Comp: 10 msec, Interp: 30 msec
%17 >> Simplify(v1)
0 @ VectorAlg(MachineInteger)
                         Comp: 10 msec, Interp: 30 msec
```

EXAMPLE 3. The following examples come from Cunningham [4].

- $a \wedge (b \wedge c) + b \wedge (c \wedge a) + c \wedge (a \wedge b) = 0.$
- $a \wedge (b \wedge (c \wedge d)) + b \wedge (c \wedge (d \wedge a)) + c \wedge (d \wedge (a \wedge b)) + d \wedge (a \wedge (b \wedge c)) = (a \wedge c) \wedge (b \wedge d).$

```
%18 >> a^(b^c)+b^(c^a)+c^(a^b)
0 @ VectorAlg(MachineInteger)
                         Comp: 0 msec, Interp: 20 msec
%19 >> a^(b^(c^d))+b^(c^(d^a))+c^(d^(a^b))+d^(a^(b^c))=(a^c)^(b^d)
T @ Boolean
                         Comp: 0 msec, Interp: 60 msec
```

EXAMPLE 4. The following examples come from Patterson [9].

- $(b \wedge c) \bullet (a \wedge d) + (c \wedge a) \bullet (b \wedge d) + (a \wedge b) \bullet (c \wedge d) = 0.$
- $((a \wedge b) \wedge c) \wedge d + ((b \wedge a) \wedge d) \wedge c + ((c \wedge d) \wedge a) \wedge b + ((d \wedge c) \wedge b) \wedge a = 0.$

```
%20 >> (b^c).(a^d)+(c^a).(b^d)+(a^b).(c^d)
0 @ VectorAlg(MachineInteger)
                         Comp: 0 msec, Interp: 30 msec
%21 >> ((a^b)^c)^d+((b^a)^d)^c+((c^d)^a)^b+((d^c)^b)^a
0 @ VectorAlg(MachineInteger)
                         Comp: 10 msec, Interp: 30 msec
```

6   Summary

In this paper, we have presented a rule-based component-free vector algebra package developed using the computer programming language Aldor. All the examples we encountered in the literature are simplified to their shortest normal forms. The key idea is to choose an appropriate data structure and a suitable set of transformation rules. In the future, we will add more functions to the package, for example, solving vector equations and systems of vector equations.

Acknowledgement. The authors would like to thank Dr. Marc Moreno Maza for his helpful suggestions and the referees for their valuable comments on improving this paper.

References

1. Aldor.org, Aldor User Guide (2002) http://www.aldor.org/docs/aldorug.pdf/
2. Belmonte, A., Yasskin, P.B.: A vector calculus package for Maple (2003) http://calclab.tamu.edu/maple/veccalc/
3. Chapman, S., Cowling, T.G.: The Mathematical Theory of Non-uniform Gases. Cambridge University Press, Cambridge (1939)
4. Cunningham, J.: Vectors. Heinemann Educational Books Ltd, London (1969)
5. Eastwood, J.W.: OrthoVec: version 2 of the Reduce program for 3-d vector analysis in orthogonal curvilinear coordinates. Comput. Phys. Commun. 64, 121–122 (1991)
6. Fiedler, B.: Vectan 1.1. Manual Math. Inst., Univ. Leipzig (1997)
7. Gibbs, J.W.: Elements of vector analysis. In: The Scientific Papers of J. Willard Gibbs, Dover (1961)
8. Harper, D.: Vector33: A Reduce program for vector algebra and calculus in orthogonal curvilinear coordinates. Comput. Phys. Commun. 54, 295–305 (1989)
9. Patterson, E.M.: Solving Problems in Vector Algebra. Oliver & Boyd Ltd, Edinburgh-London (1968)
10. Qin, H., Tang, W.M., Rewoldt, G.: Symbolic vector analysis in plasma physics. Comput. Phys. Commun. 116, 107–120 (1999)
11. Silagadze, Z.K.: Multi-dimensional vector product. arXiv: math.ra, 0204357. (2002)
12. Stoutemyer, D.R.: Symbolic computer vector analysis. Computers & Mathematics with Applications 5, 1–9 (1979)
13. Tanimoto, S.L.: The Elements of Artificial Intelligence Using Common Lisp. Computer Science Press, New York (1990)
14. The Mathlab Group, Macsyma Reference Manual, vol. II, MIT (1983)
15. Wolfram, S.: The Mathematica Book. 3rd edn. Wolfram Media/Cambridge University Press (1996)

Mathematics and Scientific Markup

Peter Murray Rust

Unilever Centre for Molecular Science Informatics
Department of Chemistry, University of Cambridge, UK
http://www.ch.cam.ac.uk/staff/pm.html

Abstract. The development of e-Science (cyberScience, Grid, etc.) is starting to become a reality with formalised data resources, services on demand, domain-specific search engines, digital repositories, etc. Increasingly STM[1] information will be contained in compound XML documents, representing scientific communication (articles, theses, repository entries, etc.). In physical sciences such as chemistry, materials science, engineering, physics, earth sciences, these "datuments" [1] normally contain hypertext, graphics, tables, graphs and numerical data, mathematical objects and relationships. In addition they may also contain domain-specific content such as chemical formula and reactions, thermodynamic and mechanical properties, electric, magnetic and optical properties.

Among the domain-specific languages, CML (Chemical Markup Language) is the oldest and broadest, and is now being actively used for publishing by the Royal Society of Chemistry (Project Prospect [2]) which gives an idea of what chemistry in datuments can look like. CML has had to develop the domain-specific objects (molecules, atoms, bonds, spectra, crystallography, etc.) and the relationships between them. However, due to the text-based nature of early XML, it has also had to design an implement domain-independent infrastructure which can support much of physical science. Originally called STMML [3] it supports data types (float, integer, complex, etc.), data structures (arrays, lists, matrices, etc.), geometrical concepts (points, planes, lines, etc.) and scientific units of measurement. In addition CML bases much of its flexibility one user-created dictionaries (ontologies) which are hyperlinked from objects in the datuments.

It is now clear that the domain-independent parts of CML (and by extension some other markup languages in physical science) are loosely isomorphic with approaches in MathML and OMDOC. If a synthesis can be found, then CML can happily forget about the "non-chemistry" knowing that the mathematical and physical science community has a general way forward. In easiest-first order, the following are suggested:

- Mathematical variables and equations in chemical documents. An obvious challenge is that the variables represent types, often physical quantities (but also chemical objects such as atomTypes). This would be one of the first areas to explore with publishers.
- Graphs and tables. A high proportion of graphs are functions of one of more dependent variables against one or more independent variables, currently supported by <table>.

---

[1] STM: Scientific, Technical (Engineering) Medical.

M. Kauers et al. (Eds.): MKM/Calculemus 2007, LNAI 4573, pp. 128–129, 2007.

- Dictionaries. The CML dictionaries and OMDOC content dictionaries seem fairly similar in approach.
- Mathematical relationships. A large area of physical science is based on theoretically and experimentally validated relationships which have been proved over many years (e.g. Maxwell's equations in thermodynamics). Often a quantity can be most easily determined by measuring different ones and transforming them. However most transformations are currently hidden in procedural non-portable code and it would be an exciting challenge to create a self-consistent declarative model of parts of physical science. It would be very exciting to have a discovery engine which could, on demand, decide which quantities were deducible from which (with similarity to theorem proving).

A major challenge for distributed mathematics and science is discovery through search engines. These currently work on "free text" and are optimised to recognise strings. In a few cases domain-specific canonicalisations can be used (e.g. our Google Inchi [4] transforms a molecular graph into a string which is recognised by search engines). However most cases require mathematical operations (arithmetic, transformations, subgraph-matching, etc.). How – and where – can these be performed? A new generation of domain-independent and domain-specific indexing and searching tools needs to be developed.

Recently CML has had to evolve a grammar to support fuzzy concepts representing sets of molecules. These have a distinguished mathematical history (see, e.g. enumeration of alkanes [5] and references therein). Polymers and chemicals in patents ("Markush") are often expressed in text when a grammar would be more precise. Chemical searches are also often expressed in a grammar and evaluation or comparison of representations is a common activity.

The presentation will give a number of interactive demonstrations. No chemical knowledge is required!

References

1. http://jodi.tamu.edu/Articles/v05/i01/Murray-Rust/
2. http://www.rsc.org/Publishing/Journals/ProjectProspect/
3. http://www.ch.ic.ac.uk/rzepa/codata2/
4. http://wwmm-svc.ch.cam.ac.uk/wwmm/html/googleinchiserver.html
5. http://www.cs.uwaterloo.ca/journals/JIS/cayley.html

The On-Line Encyclopedia of Integer Sequences

N.J.A. Sloane

AT&T Shannon Labs, Florham Park, NJ, USA
http://www.research.att.com/~njas/

Abstract. The **On-Line Encyclopedia of Integer Sequences** (or **OEIS**) is a database of some 130000 number sequences. It is freely available on the Web (http://www.research.att.com/~njas/sequences/) and is widely used.

There are several ways in which it benefits research:

1. It serves as a dictionary, to tell the user what is known about a particular sequence. There are hundreds of papers which thank the OEIS for assistance in this way.
2. The associated Sequence Fans mailing list is a worldwide network which has evolved into a powerful machine for tackling new problems.
3. As a direct source of new theorems, when a sequence arises in two different contexts.
4. As a source of new research, when one sees a sequence in the OEIS that cries out to be analyzed.

The 40-year history of the OEIS recapitulates the story of modern computing, from punched cards to the internet.

The talk will be illustrated with numerous examples, emphasizing new sequences that have arrived in the past few months. Many open problems will be mentioned.

Because of the profusion of books and journals, volunteers play an important role in maintaining the database. If you come across an interesting number sequence in a book, journal or web site, please send it and the reference to the OEIS. (You do not need to be the author of the sequence to do this.) There is a web site for sending in "Comments" or "New sequences".

Several new features have been added to the OEIS in the past year. Thanks to the work of Russ Cox, searches are now performed at high speed, and thanks to the work of Debby Swayne, there is a button which displays plots of each sequence. Finally, a "listen" button enables one to hear the sequence played on a musical instrument (try Recamán's sequence A005132!).

M. Kauers et al. (Eds.): MKM/Calculemus 2007, LNAI 4573, p. 130, 2007.
© Springer-Verlag Berlin Heidelberg 2007

First Steps on Using OpenMath to Add Proving Capabilities to Standard Dynamic Geometry Systems

Miguel A. Abánades, Jesús Escribano, and Francisco Botana

Ingeniería Técnica en Informática de Sistemas, CES Felipe II (UCM), 28300
Aranjuez, Spain
mabanades@cesfelipesegundo.com
Departamento de Sistemas Informáticos y Computación, Universidad Complutense de
Madrid, 28040 Madrid, Spain
scribano@sip.ucm.es
Departamento de Matemática Aplicada I, Universidad de Vigo, Campus A
Xunqueira, 36005 Pontevedra, Spain
fbotana@uvigo.es

Abstract. A prototype for a web application designed to symbolically process locus, proof and discovery tasks on geometric diagrams created with the commercial dynamic geometry systems *Cabri, The Geometer's Sketchpad* and *Cinderella* is presented. The application, named LAD (acronym for *Locus-Assertion-Discovery*) and thought of as a remote *add-on* for the considered DGS, follows the Groebner basis method relying on CoCoA and a Mathematica kernel for the involved symbolic computations. From the DGS internal textual representation of a geometric diagram, an OpenMath (i.e. semantic based) description of the requested task is created using the elements in the *plangeo* OpenMath *content dictionaries*. A review of the elements included in these CDs is given and two new elements proposed, namely *locus* and *discovery*. Everything is finally thoroughly illustrated with examples. LAD is freely accessible at http://nash.sip.ucm.es/LAD/LAD.html.

1 Introduction

The name of *Dynamic Geometry Systems* (DGS) is given to the family of computer applications that allow exact on-screen drawing of (generally) planar geometric diagrams and, their main characteristic, the manipulation of these diagrams by mouse dragging certain elements, making all other elements to automatically self adjust to the changes. This is also known as *Interactive* geometry. Since the appearance of the French *Cabri Géomètre* ([15]) and the American *The Geometer's Sketchpad* ([11]) in the late 80's, many others have been created with slightly different functionalities (*C.a.R. Euklides, Dr. Genius, Dr. Geo, Gambol, GeoGebra, Geometrix, Geonext, The Geometric Supposer, GeoProof, GEUP, GRACE, Kig, Kgeo, KSEG,...*). Special mention deserve applications oriented to the automatic proving of (Euclidean) geometric theorems where an ad-hoc

M. Kauers et al. (Eds.): MKM/Calculemus 2007, LNAI 4573, pp. 131–145, 2007.

DGS has been implemented as frontend for a symbolic *prover* ([9,24,10]). Being more an input device for the prover than a final tool, these DGS have not really left the academic circles. Among the commercial DGS it is worth mentioning the German *Cinderella* ([21]) whose main technical property is the *complex tracing* by which certain singularities are avoided when moving a point along a path.

From the beginning, DGS have been the paradigm of new technologies applied to Math education, area where they have found their most applications. Their convenience in the classroom is almost unanimous among education experts. However, questions have been raised on the influence of the use of DGS on the development of the concept of proof in school curricula ([14]). Being able to produce a great number of examples of a configuration is usually taken as a substitute for a formal proof in what has come to be known as a *visual proof*. This is a symptom of the incompleteness of general DGS relative to extra manipulation of configurations. In particular, out of the three DGS considered, only Cabri comes equipped with some property checker, and even in that case its numeric nature does not really provide a sound substitute for a formal proof.

1.1 Computational Abilities of DGS

To compensate the computational limitations of DGS, it seems clear that some symbolic capabilities have to be added to DGS. There are two different approaches to implement this idea. Some systems incorporate their own code to perform symbolic computations (e.g. [9,10]), while other systems choose to reuse existing *Computer Algebra Systems* (CAS) (e.g. [1,22,24]). Both strategies have been partially successful on addressing some of the three main issues in dynamic geometry, namely the continuity problem, the implementation of proof and discovery, and the determination of loci (see [3] for details). However, none of the main commercial DGS has tried to add ample symbolic capabilities in its latest versions. This is due in part to the very different nature of DGS and CAS. While DGS are based on immediate interactive manipulations, CAS are based on question/answer queries. One exception is provided by the recent *Geometry Expressions* ([13]), a commercial DGS based on constraints with an integrated symbolic kernel, that is, an interactive *symbolic* geometry system.

Besides the reasons mentioned above, the main problem to connect existing DGS and CAS is the fact that different applications speak different languages. Commercial interests, frequently at the heart of developing considerations, do not seem to be compatible with the idea of intercommunication (see [16]).

Every Math computer application has its own way to describe and internally process information also due to the lack of a *lingua franca* for computer Mathematics. Not having a universally accepted way to write, store and communicate Mathematics is also making the whole field lay behind in the IT revolution. How much information is unavailable in practice because it is impossible to efficiently perform a search for a Math term or expression on the Internet?

Being true that no language has been able to establish itself as *the* Math language for the computer era, there are two main different, and somehow complementary, candidates. On one hand there is *MathML*, a W3C Recommendation oriented to Math presentation on the Web ([17]). It is supported by most Web browsers and hence it is the most extended Math encoding. On the other hand there is *OpenMath*, a standard for representing mathematical objects together with their semantics ([18]). The emphasis on semantics is absolutely crucial in Math representation and makes OpenMath objects more suitable to be exchanged between computer programs or stored in databases. However, the amount of implemented OpenMath applications is very limited (see *Software and Tools* section on [18]). As any new idea or trend, OpenMath has not been free of questioning ([8]) but it is, in our opinion, a promising approach to establishing a common Math language.

In this note we present an example of DGS-CAS communication for three of the most relevant commercial systems, namely Cabri, The Geometer's Sketchpad and Cinderella with OpenMath as the communicating language. The tool is a web application programmed to symbolically process locus, proof and discovery tasks on geometric diagrams constructed with the considered DGS. Named LAD (acronym for *Locus-Assertion-Discovery*) it has been thought of as a remote *add-on* for the three DGS.

2   Descriptions of Geometric Diagrams

When creating a diagram in a DGS, the user draws a *static picture* whose elements are sequentially defined in terms of relations of dependence with respect to previous elements. Initially, all the elements have a concrete position on the canvas determined by concrete numeric values (coordinates of free and semi-free points, radius length of circles given by a radius, slope of lines determined by a slope, etc...). But a DGS construction is a *dynamic* picture, it is given by an infinite family of pictures parameterized by the variables associated to the starting numeric values. If we define two static pictures in a DGS to be equivalent when one can be dynamically modified into the other, a DGS construction can then be identified with an equivalent class of pictures. This mathematical formalism, clear to a qualified DGS user, can be confusing for a student. Let us consider for instance a diagram given by the circle $c1$ with center at the free point $A1$ passing through the free point $B1$ and the circle $c2$ with center at the free point $A2$ passing through the free point $B2$. Clearly there are four possible relative positions for these two circles depending on the number of (real) intersection points (see Fig. 1).

Although examples of all four possibilities are instances of the same diagram according to the definition above, one could rightly argue that they are completely different constructions. Moreover, we could take into considerations other aspects, such as the topological distribution of the ovals determined by the circles and obtain even more *different* pictures. This shows how the concept of a DGS diagram is elusive. The definition given by a high school student would probably differ considerably from that of an expert in Automated Theorem Proving (ATP). The same way that Differential and Algebraic Geometry study geometric

134   M.A. Abánades, J. Escribano, and F. Botana

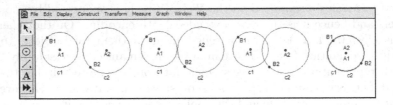

Fig. 1. Four equal constructions?

objects up to diffeomorphism and algebraic equivalence respectively, Dynamic Geometry needs its own definition, which is out of the scope of this short note.

Once pointed out the ambiguity in the concept of a diagram generated by a DGS, a brief description is given on the way the three chosen DGS describe and store their own diagrams, and how it can be done using OpenMath.

2.1   Descriptions of Geometric Diagrams by Standard DGS

Different DGS use different ways to describe their internal data. We will focus on the particular cases of the three systems considered, namely Cabri, The Geometer's Sketchpad (GSP) and Cinderella, which already show enough peculiarities to be representative of the existing variety.

When creating a diagram, all three applications generate a file in the local directory after saving the construction. The format of this file and its accessibility are the first differences among these DGS. Cabri files (with extension .fig) are readable as text files by just opening them with a text editor. Cinderella files (with extension .cdy) are compressed files and they have to be uncompressed to obtain a file readable as a text file. GSP files (with extension .gsp) are coded and their contents not accessible. However, one can save a .gsp file as an .html file (to be used with JAVA Sketchpad [12]) readable as a text file. It has to be noted that the JAVA Sketchpad description of a construction does not include all the information from the construction. For instance, labels assigned to elements other than points are not included.

As expected, in all three cases the text describing the diagram follows the basic ideas mentioned before. It gives an ordered list of elements described with basic data that include the type, numeric values and parents of an element. Auxiliary data are also specified for each element when needed. This auxiliary data include all graphic and internal DGS codifications. The following are textual descriptions of a line passing through two points as coded by Cabri, GSP and Cinderella respectively

```
6: Line, Const: 4, 3,
{3} Line(2,1)[color(0,0,255)];
("c"):=Join("E","C");
```

where in the case of Cinderella, 15 additional lines with auxiliary data have been removed.

Note that Cabri and GSP assign a nonnegative integer to each element. This number, besides ordering all the elements of the construction according to their relative position in the time sequence of the construction, also serves as the actual name of the element in all the references to it. The label that the user can attach in Cabri and GSP to any element is only used as a decorative item, it is never used by the applications as an identifier.

While Cabri and GSP descriptions of diagrams are based on a strict constructive order, a different strategy is followed by Cinderella. No numbers are associated to the different elements in the construction and all the references are made with (modifiable) alphanumeric labels automatically assigned by the application. In fact, the description does not even follow the order in which the elements where created by the user. If the two intersection points of a line and a circle are included in a diagram at, say steps 13 and 21, the description of both of them will be given in the same line, exactly as if they had been introduced in the diagram at the same time. The elements keep, of course, the natural parent-offspring ordering.

These few comments are enough to illustrate the variety that is found in the ways of describing a geometric construction as we consider different DGS. At the heart of this discussion is the ultimate goal of intercommunicating different DGS. The situation, as we see it, makes this goal unattainable at this moment. In this direction, it is worth mentioning the use by the recent Cabri 3D of XML files, technology specifically oriented to the standardization in communication.

A related but slightly less ambitious general project is that of providing a way to share repositories of constructions among DGS. This is clearly calling for a neutral *language* to which all DGS could translate their diagram descriptions. As mentioned in Sect. 1, OpenMath has all the characteristics to be that language.

2.2   OpenMath Description of Geometric Diagrams

The OpenMath representation of a mathematical structure is referred to as an *OpenMath object*. Although several encodings are available (functional, SGML, binary) the current preferred notation when describing OpenMath objects is XML. Formally, an OpenMath object is a labelled tree whose leaves are the basic OpenMath objects. Among these are the *symbols* which ultimately represent the Math symbols. They consist of a name and a reference to an explicit definition in an external document called a *content dictionary* (CD). OpenMath objects are then built up recursively (see [18]). The following is the XML encoding of an OpenMath object representing the line l determined by the points A and B.

```
<OMOBJ xmlns="http://www.openmath.org/OpenMath" version="2.0">
  <OMA>
    <OMS cd="plangeo1" name="line"/>
    <OMV name="l"/>
    <OMA>
      <OMS cd="plangeo1" name="incident"/>
      <OMV name="A"/>
```

```
        <OMV name="l"/>
      </OMA>
      <OMA>
        <OMS cd="plangeo1" name="incident"/>
        <OMV name="B"/>
        <OMV name="l"/>
      </OMA>
    </OMA>
  </OMOBJ>
```

presentation that can be schematically represented by

line(l,incident(A,l),incident(B,l)).

Note that this does not make a sensible diagram description since the referred points are not described.

As illustrated in Sect. 2 the description of a diagram by a DGS contains basic (i.e. geometric) as well as auxiliary data. The translation of the description of a diagram given by Cabri/GSP/Cinderella to an OpenMath element has been made to preserve only the basic data, namely the type of an element (point, line, segment,...), its label and its defining properties. The OpenMath descriptions given by LAD have been made, following the basic OpenMath philosophy, application independent. Hence, although LAD has been programmed as a symbolic prover and does not work with specific coordinates, these have been included in the OpenMath description of free and semi-free points.

We tried to use only existing OpenMath elements when possible. The basic elements considered came from the plangeo1 to plangeo6 content dictionaries. In particular, the general approach of representing a diagram by an instance plus its constraints has been followed. This structure is found in the configuration element from the CD plangeo1.

The absence of certain natural OpenMath elements, such as the length of a segment, has made some descriptions look a bit awkward. That is the case, for instance, of the description of a circle defined by a point (as center) an a segment (as radius). Given the length of the OpenMath descriptions, the reader is referred to http://nash.sip.ucm.es/LAD/LAD.html for complete examples.

The following shows the schematic description of distance element used to encode the circle with center C and radius given by the length of the segment S with endpoints A and B.

distance(point(A,in(A,endpoints(S))),point(B,in(B,endpoints(S))))

3   On the *plangeo* Content Dictionaries

Content Dictionaries are the most important aspect of OpenMath for they define the meaning of the objects being coded. A CD is a collection of related symbols and their definitions encoded in an XML format. A symbol definition in an OpenMath CD consists of a symbol name, a symbol description in plain text and optionally of some properties and examples of use. Collections of related CDs are usually grouped together as CD Groups. The set of CDs managed by the *OpenMath Society* and all the details about how they are used can be found at the official OpenMath web site ([18]).

The list of CD groups grows constantly from contributors in different areas. In particular, the `plangeo` CDs are part of the `riaca_algebra` CD group developed by the RIACA at *Technische Universiteit Eindhoven* with the DGS Cinderella in mind ([19]). Besides a previous work by the authors presented at ADG 2006 ([7]), as far as we know, these CDs have basically been used only by Roozemond to implement an automatic prover for Cinderella using GAP ([23]).

Despite the intention of using only the elements defined in the `plangeo` CDs to formalize the OpenMath representations needed in LAD, we found necessary a slight modification in one definition and the inclusion of two more elements, namely `locus` and `discovery`.

We modified the OpenMath description of the `segment` element in the plangeo2. Given only in terms of the two defining points, it did not include a name for the segment itself. Being the cross reference a basic ingredient in a description of a DGS configuration, we considered a slightly modified definition including a name for the encoded segment. The following schematic descriptions of Open-Math objects correspond to the *old* an *new* representations of the segment S defined by points A and B:

> segment(A,B)   segment(S,A,B)

The absence of a denomination for the defined element does not happen only in the case of the segment. While the OpenMath representations of line, point, circle or conic include a name for the encoded element, such a name is missing in the representations of, for instance, angle, midpoint or halfline.

3.1 New Elements: Locus, Discovery

The elements in the plangeo CDs are well suited to represent a geometric diagram (a `configuration` in OpenMath terms) and the element `assertion` from `plangeo1` can be used to codify the request to prove a condition on a diagram. However the plangeo CDs do not include elements to codify locus or discovery tasks.

With respect to the locus task, two different classes of loci have been distinguished, namely the *parametric* and the *implicit*. We have restricted to the locus set determined by one point in a configuration. This is clearly generalizable to sets with several points or even higher dimensional objects.

In a geometric configuration, given a point T, dependent of a point M, which is a point on the object a (line, circle,...), the *parametric locus* defined by T is the set of points *traced* by T as M moves along a. The points T and M are generally referred to as the tracer and the mover respectively. Standard DGS (such as the three considered by LAD) come equipped with a locus function that allows the user to visualize the plotting of a parametric locus.

In a geometric configuration, given a point T (dependent or independent), an *implicit locus* defined by T and a set C of conditions in the elements of the configuration, is the set of points *traced* by T when the conditions in C are satisfied. The point T is still referred to as the tracer in LAD although most implicit loci are not constructible. Standard DGS (in particular the three considered by LAD) do not come equipped with a locus function to visualize implicit loci (see example in Sect. 5.1).

A parametric locus is an *a priori* object. It is determined by the relations given *before* its definition. It is hence a natural object for a DGS. A sample of instances of the mover and the plotting of the corresponding instances of the tracer are enough to give a (numerical) representation of the locus. Let us mention that Cabri uses these sample of coordinates to numerically produce equations of some parametric loci.

On the other hand, implicit loci are *a posteriori* objects. They are determined by extra relations imposed on the configuration elements. This does not fit the logical ordering in standard DGS. Moreover, determining implicit loci require symbolic computations.

Although computationally different, both loci are identical conceptually. A locus is the set determined by the different instances of the tracer when considered all possible instances of the configuration satisfying all required conditions. A locus, by definition, is an element in a configuration. Hence, its OpenMath representation has been defined with a name and a reference to the tracer with the possibility of including some extra conditions (i.e. besides those coming from the relations among the different previous elements in the configuration). The following are the schematic representation of the OpenMath representation of the parametric locus a determined by the tracer point P and the implicit locus b determined by the tracer point Q with the condition that point $P3$ is incident to line $L8$.

```
locus(a,P)              locus(a,P,incident(P3,L8))
```

A new `plangeoX` has been used as CD reference in the XML codification of these elements. The authors will submit a geometry OpenMath CD after an experimental period.

A Discovery task in ATP enquires about the necessary conditions for some properties to hold with respect to the elements in a diagram. Being similar to a *proving* task, the assertion OpenMath element has been taken as model for the `discovery` element. The following is an example of the (schematic) representation of the discovery of necessary conditions on the n elements in the configuration for the lines $L11$ and $L12$ to be parallel and points A, B and C to be collinear.

```
discovery(
    configuration(element_1,...,element_n),
    parallel(L11,L12),
    are_on_line(A,B,C)
)
```

4   LAD: Locus, Assertion Discovery

The three basic guidelines in the developing process of the tool have been the use of CAS from standard DGS, the use of OpenMath as communication language and the final implementation of a web accessible tool. In particular, LAD is a web application that symbolically processes locus, proof and discovery tasks over geometric diagrams constructed with either Cabri, The Geometer's Sketchpad or Cinderella. It acts as a *remote plug-in* for these three DGS adding symbolic computational capabilities. From the DGS own textual representation of a geometric diagram, an OpenMath description of the requested task is created using the elements in the plangeo CDs as well as the new locus and discovery elements. This OpenMath representation of the task is then translated into a new description only in terms of points and relations among these points. As an example, the incidence of the point P on the line l determined by the points A and B, is translated as `aligned(P,A,B)`. Although the geometric diagram is described in terms of specific numeric data, symbolic coordinates are assigned to the points. These symbolic coordinates (variables) and the equations given by the relations among the points provide the algebraic setting of the problem in terms of polynomial ideals in those variables. Depending on the task requested a different parameter elimination algorithm is applied. The answer is obtained then in terms of equations in the appropriate variables with symbolic parameters.

Observe that this CAS-DGS communication is currently only one directional. Making the original DGS *understand* the answer given by LAD is currently work in progress.

4.1   User Interface and Architecture

Considering its natural applications in education, simplicity and ease of use were considered basic in the design of LAD. Consequently, its interface was designed to look like a simple web page (see Fig. 2) where the user is required to make use of an applet and a text editor to create a text file with the description of the task in the local directory. This file is finally uploaded to the server and a new browser window displays the answer produced by the application.

To obtain the final description of the task from the DGS description of the diagram, a double translating process takes place. First the original DGS diagram description is processed, producing an XML codification of the OpenMath description of the task. The task type as well as the different conditions for the task have to be input using the applet menu. The OpenMath description of the task is then translated into code for *webDiscovery*, application developed by Botana (see [2]) whose kernel has been appropriately modified to be integrated in LAD as the final computational tool.

LAD is based on webMathematica [25], a Java servlet technology allowing remote access to the symbolic capabilities of Mathematica. Once the user has created and uploaded the appropriate text file, a Mathematica Server Page is launched, reading the file and initializing variables. An initialization file for CoCoA [5] containing the ideal generated by the appropriate defining polynomials

Fig. 2. LAD user interface as a web page

is also written out, and CoCoA, launched by Mathematica, computes a Groebner basis for this ideal. Each generator is factored (a task also done by CoCoA), and a process of logical expansion is performed on the conjunction of the generators in order to remove repeated factors.

The JAVA application that processes the translation from the OpenMath to the webDiscovery description of the tasks was programmed using the RIACA OpenMath Library 2.0 ([20]).

It is important to emphasize that all the computations done by LAD are symbolic (i.e. *number-free*) so the answers to the tasks are completely general, result of the study of deep algebraic relations using sound algorithms ([4]).

However, a diagram with many elements could result in a task with unmanageable computations or in an answer with long symbolic expressions. An ad-hoc assignment of numerical coordinates to free points can solve the problem (see example 5.1). This is a standard procedure in ATP (cf. [6]).

As mentioned in Sect. 1.1, the approach of reusing existing CAS to add computational abilities to DGS has been followed before by other authors. In particular, the implementations in [24] and [22] lack, in our opinion, a sufficient degree of automatism.

5   Examples

LAD has been designed with a clear scenario in mind, namely that of a student in a Math lab working geometric experiments lead by a qualified instructor.

Besides adding exactness and rigorous proofs to ideas and conjectures, LAD can help illustrate subtle concepts such as symbolic equation, generic position, free variable or semantic description. The possibility of directly manipulating the final encoding of the task makes LAD also useful for advance college students and ATP researchers.

The following are three examples covering all tasks managed by LAD. We have chosen simple exercises related to well-known facts for the sake of clarity. These examples can be easily generalized.

5.1   Locus

When experimenting with different instances of a construction, the most natural question (*when is this property true?*) leads many times naturally to an implicit locus. Consider the triangle ABT and the orthogonal projection Af of A onto the opposite side (see the Cabri construction in Fig. 3, left). When is Af the midpoint of that side? Or in other words: what is the locus set of points T such that Af is the midpoint of BT?

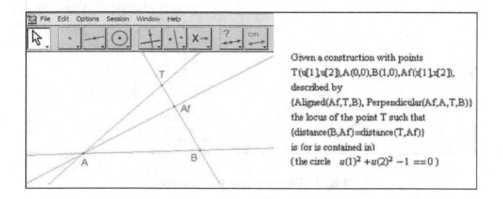

Fig. 3. When is Af the midpoint of BT?

After pasting the Cabri textual description of the construction in the applet, we select **Find a LOCUS** from the applet menu and add the condition **Equidistance** for the points Af, T and B. Finally we also have to specify T as the tracer point. Pressing the **To OM** and the **To wD** buttons generates the webDiscovery description of the task with symbolic variables assigned to the free points in the construction. If we upload the .txt file with that description, the answer given by LAD says that Af is the midpoint of the segment BT only if T lies in a conic whose exact symbolic equation is the following:

$$u(1)^2 - 2u(3)u(1) + 2u(2)^2 - u(5)^2 - u(6)^2 - 2u(2)u(4) + 2u(3)u(5) + 2u(4)u(6) = 0$$

This quadratic equation in $u(1)$ and $u(2)$ is in fact the general equation for the locus set described by T considering two general points A and B. However this kind of equation is in general too much information for a student in a Math lab. To simplify things we can assign numeric coordinates to the free points in the construction as explained in Sect. 4 to obtain the answer to a convenient particularization of the question. In particular if we assign numeric coordinates so the vertices A and B become the points $(0,0)$ and $(1,0)$ respectively and upload the modified file to the application, we get that T has to be in the circle with center B and radius equal to the distance AB (see Fig. 3, right).

5.2 Assertion

The fact that the three altitudes of a triangle meet in one point is one of the first general results that students are capable of guessing when using a DGS. To prove this result with LAD we can construct the triangle ABC and the intersection point P of two altitudes. We just have to construct the third altitude and ask the tool to prove that the point P lies in it (see the GSP construction in Fig. 4, left).

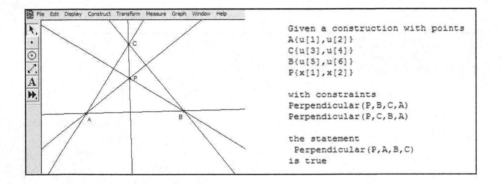

Fig. 4. Altitudes meet in one point

After pasting the GSP textual description of the construction from the .html file in the upper left text area in the applet, we select PROVE from the applet menu and add the condition Incidence for the point P and the line $L10$. Let us recall that GSP does not keep the line labels in its .html file. The assigned name of the last constructed altitude is $L10$ because it appears in the line numbered as 10 in the .html file. The name can also be looked up in an OpenMath description of the construction (obtained by pressing the To OM button). Once the condition to prove has been added, one just have to press the To OM and the To wD buttons to generate the webDiscovery description of the task. Uploading the .txt file to the application, the answer is obtained (see Fig. 4, right).

5.3 Discovery

The ideal situation when learning in an practice setting such as a Math lab is that after careful study of a situation with several examples one is able to make a good guess on the condition for some property to be true. But this is not always the case. LAD implements the idea of automatic discovery which can lead the user through the investigation. As an example, one could be interested in knowing the conditions for the circumcenter to lie in one side of the triangle. To *discover* the necessary conditions we have to construct the triangle ABC and its circuncenter as the intersection point O of the perpendicular bisectors of two sides (see the Cinderella construction in Fig. 5, left).

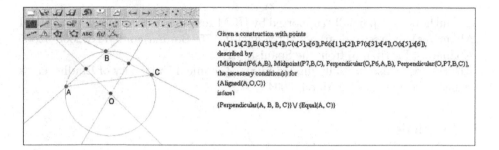

Fig. 5. The circumcenter is incident to one side only if it is a right triangle

After pasting the Cinderella textual description of the construction (from the file obtained after decompressing the .cdy file) in the applet, we select DISCOVER from the applet menu and add the condition Collinearity for the points A, O and C. Once the condition to discover has been added, one just have to press the To OM and the To wD buttons to generate the webDiscovery description of the task. Uploading the .txt file to the application, the answer is obtained (see Fig. 4, right). As expected, the condition is that we have right triangle. Note that LAD also gives the degeneracy condition $A = C$.

6   Conclusions and Future Work

LAD shows the possibilities of the CAS-DGS communication by successfully interconnecting Cabri, The Geometer's Sketchpad and Cinderella to Mathematica and CoCoA. The use of a standard semantic representation of mathematical objects is, as we see it, one of the main challenges in the computer math community today. LAD shows that OpenMath can be efficiently used to describe geometric configurations. Moreover, by choosing a ready-to-use web presentation, LAD becomes a real available tool to any user of the three considered GDS with an internet connection.

When processing a task, LAD makes all the details of the different translations available to the user, unlike black-box type applications. It is our opinion that being able to manipulate the insides of *the box* multiply its applications in education as well as in ATP research.

As future work, two main aspects of LAD are under study for generalization. First its two dimensional nature. An extension to 3D Dynamic Geometry Systems is being undertaken. This, however, requires first the definition of the three dimensional analogues of the plangeo OpenMath CDs. The second aspect refers to the unidirectionality of LAD. Incorporating the information provided by the CAS in the DGS would certainly improve enormously its applications.

Acknowledgements

The authors were partially supported by UCM research groups ACEIA and DOSI (Abánades) and research grants MTM2004-03175 (Botana) and MTM2005-02865 (Abánades, Escribano) from the Spanish MEC.

Botana was also partially funded by the Technical University of Eindhoven as a visiting scholar during March 2004.

References

1. Botana, F., Valcarce, J.L.: A dynamic-symbolic interface for geometric theorem discovery. Computers and Education 38(1-3), 21–35 (2002)
2. Botana, F.: A Web-based intelligent system for geometric discovery. In: Sloot, P.M.A., Abramson, D., Bogdanov, A.V., Gorbachev, Y.E., Dongarra, J.J., Zomaya, A.Y. (eds.) ICCS 2003. LNCS, vol. 2657, pp. 801–810. Springer, Heidelberg (2003)
3. Botana, F., Recio, T.: Towards solving the dynamic geometry bottleneck via a symbolic approach. In: Hong, H., Wang, D. (eds.) ADG 2004. LNCS (LNAI), vol. 3763, pp. 92–110. Springer, Heidelberg (2006)
4. Buchberger, B.: Groebner Bases: An algorithmic method in polynomial ideal theory. Multidimensional Systems Theory, D. Reidel Publishing Company, pp. 184–231 (1985)
5. Capani, A., Niesi, G., Robbiano, L.: CoCoA, a system for doing Computations in Commutative Algebra. Available via anonymous ftp from: cocoa.dima.unige.it
6. Chou, S.: Mechanical Geometry Theorem Proving. Reidel, Dordrecht Boston (1988)
7. Escribano, J., Abánades, M., Valcarce, J., Botana, F.: On Using OpenMath for Representing Geometric Constructions. In: Proc. 6th Int. Workshop on Automated Deduction in Geometry (ADG 2006), pp. 26–30 (2006)
8. Fateman, N.: A Critique of OpenMath and Thoughts on Encoding Mathematics (January 2001) http://www.cs.berkeley.edu/~fateman
9. Gao, X.S., Zhang, J.Z., Chou, S.C.: Geometry Expert. Nine Chapters, Taiwan (1998)
10. http://www.geogebra.at
11. Jackiw, N.: The Geometer's Sketchpad v 4. Key Curriculum Press, Berkeley (2002)
12. http://www.dynamicgeometry.com/javasketchpad
13. Todd, P.: Geometry Expressions: A Constraint Based Interactive Symbolic Geometry System. Computeralgebra-Rundbrief, vol. 39 (2006)
14. Hoyles, C., Jones, K.: Proof in Dynamic Geometry Contexts. In: Perspectives on the teaching of Geometry for the 21st Century, pp. 121–128. Kluwer, Dordrecht (1998)
15. Laborde, J.M., Bellemain, F.: Cabri Geometry II. Texas Instruments, Dallas (1998)

16. http://mathforum.org/kb/message.jspa?messageID=1095184\&tstart=450
17. http://www.w3.org/Math//
18. http://www.openmath.org/
19. http://www.win.tue.nl/~amc/oz/om/cds/
20. http://www.riaca.win.tue.nl/products/openmath/lib/index.html
21. Richter–Gebert, J., Kortenkamp, U.: The Interactive Geometry Software Cinderella. Springer, Berlin (1999)
22. Roanes–Lozano, E., Roanes–Macías, E., Villar, M.: A bridge between dynamic geometry and computer algebra. Mathematical and Computer Modelling 37(9–10), 1005–1028 (2003)
23. Roozemond, D.A.: Automated proofs using bracket algebra with Cinderella and OpenMath. In: Proc. 9th Rhine Workshop on Computer Algebra (RWCA 2004) (2004)
24. Wang, D.: GEOTHER: A geometry theorem prover. In: McRobbie, M.A., Slaney, J.K. (eds.) Automated Deduction - Cade-13. LNCS, vol. 1104, Springer, Heidelberg (1996)
25. http://www.wolfram.com/products/webmathematica/index.html

Higher Order Proof Reconstruction from Paramodulation-Based Refutations: The Unit Equality Case

Andrea Asperti and Enrico Tassi

Department of Computer Science, University of Bologna
Mura Anteo Zamboni, 7 — 40127 Bologna, Italy
{asperti,tassi}@cs.unibo.it

Abstract. In this paper we address the problem of reconstructing a higher order, checkable proof object starting from a proof trace left by a first order automatic proof searching procedure, in a restricted equational framework. The automatic procedure is based on superposition rules for the unit equality case. Proof transformation techniques aimed to improve the readability of the final proof are discussed.

1 Introduction

The integration of technologies developed by the automatic theorem proving (ATP) community with modern interactive theorem provers seems a fruitful research objective. ATP technologies showed their effectiveness in many occasions [7,17] and the lack of comfortable automation is one of the most commonly issues reported by users of interactive theorem provers. This challenge gets even more interesting when the target interactive theorem prover follows the independent verification principle, building proof objects that can be validated by third party checkers. Providing a valuable proof trace is not the main goal of ATP systems, and even when they do, the information can be too ambiguous to be checked by a different prover (see [5]).

Among the activities of interactive proving that one would like to be supported by powerful automation techniques a major one is rewriting. In this paper we describe our approach to this problem in relation with the interactive theorem prover Matita[1]. In particular we integrated Matita with a first order, paramodulation[11] based solver (currently restricted to the unit equality case). The solver is able to return a trace informative enough to be read back into a proof object of Matita, that is a term of the Calculus of Inductive Constructions[13,21] (CIC). In this paper we focus on the information that must be embedded in traces, on the reconstruction of typable proof objects, and finally on the refinement of the resulting proofs to enhance readability. In particular we prove that any equational proof based on rewriting can be transformed into a transitivity chain, where each step is justified by a simple side argument (an axiom, or an already proved lemma). This format is really close to the standard mathematical display of this kind of proofs.

M. Kauers et al. (Eds.): MKM/Calculemus 2007, LNAI 4573, pp. 146–160, 2007.

The paper starts introducing the interactive theorem prover Matita and giving an overview of the automatic procedure we implemented (Sec. 2). In particular Sec. 2.1 describes how the notion of equality is encoded in CIC, while Sec. 2.2 and 2.3 describe the variant of the paramodulation calculus implemented, the proof searching algorithm and the lightweight representation of proofs adopted during proof search. A proof reconstruction procedure is then presented in Sec. 3 and its result is refined with some transformations that are detailed in Sec. 4.

We will introduce notational conventions when needed, but as a general rule we will use a different syntax for functions living in the proof language CIC or living in the meta level and manipulating CIC terms. Proofs will essentially be applicative lambda terms written using the notation (f a b c), while we will write $\theta(a, b, c)$ for functions at the meta level.

2   Automatic Proof Search Procedure Implementation

Matita is an interactive theorem prover under development at the university of Bologna (see [1] for a description of the innovative features of the system).

Matita is based on the Curry-Howard isomorphism, adopting the Calculus of Inductive Constructions as its logical framework.

The automatic proof search procedure is a component of Matita, but is essentially orthogonal to the rest of the system. It has been extensively tested with unit equality problems of the TPTP[18] library. The results obtained by the procedure can be browsed on TPTP website[1] (we solve 512 problems out of 700 in the standard TPTP time limit of 10 minutes).

CIC terms are translated into first order terms by a forgetful procedure that simply erases all type information, and transforms into opaque constants all terms not belonging to the first order framework (fixpoints, pattern matching terms, etc.).

The inverse transformation takes advantage by the so called *refiner*, that is a type inference procedure typical of higher order interactive provers.

An overview of the rules used by the solver is given in Section 2.2. These rules are decorated with proofs; the next section gives the few notions needed to understand the proof terms.

2.1   Rewriting in the Calculus of Inductive Constructions

In the calculus of inductive constructions, equality is not a primitive notion, but it is defined as the smallest predicate containing (induced by) the reflexivity principle.

$$\text{Inductive eq } (A : Type) \ (x : A) : A \to Prop \overset{\text{def}}{=} \text{refl\_eq} : eq \ A \ x \ x.$$

For the sake of readability we will use the notation $a_1 =_A a_2$ for (eq A a_1 a_2).

---

[1] http://www.cs.miami.edu/~tptp/

As a consequence of this inductive definition, and similarly to all inductive types, it comes equipped with an elimination principle named eq_ind that, for any type A, any elements a_1, a_2 of A, any property P over A, given a proof h of $(P\ a_1)$ and a proof k that $a_1 =_A a_2$ gives back a proof of $(P\ a_2)$.

$$\frac{h : P\ a_1 \qquad k : a_1 =_A a_2}{(\text{eq\_ind}\ A\ a_1\ P\ h\ a_2\ k) : P\ a_2}$$

Similarly, we may define a higher order elimination principle eq_ind_r such that

$$\frac{h : P\ a_2 \qquad k : a_1 =_A a_2}{(\text{eq\_ind\_r}\ A\ a_2\ P\ h\ a_1\ k) : P\ a_1}$$

These are the building blocks of the proofs we will generate. With this definition of equality standard properties like reflexivity, symmetry and transitivity can be easily proved and are part of the standard library of lemmas available in Matita.

2.2 Superposition Rules

Paramodulation is precisely the management of equality by means of rewriting: given a formula (clause) $P(s)$, and an equality $s = t$, we may conclude $P(t)$. What makes paramodulation a really effective tool is the possibility of suitably constraining rewriting in order to avoid redundant inferences without loosing completeness. This is done by requiring that rewriting always replace *big* terms by *smaller* ones, with respect to a special ordering relation $\succ$ among terms, that satisfies certain properties, called the *reduction ordering*. This restriction of the paramodulation rule is called *superposition*.

Equations are traditionally split in two groups: facts (positive literals) and goals (negative literals). We have two basic rules: superposition right and superposition left. Superposition right combines facts to generate new facts: it corresponds to a forward reasoning step. Superposition left combines a fact and a goal, generating a new goal: logically, it is a backward reasoning step, reducing a goal G to a new one G'. The fragment of proof that can be associated to this new goal G' is thus not a proof of G' , but a proof of G *depending* on proof of G' (i.e. a proof of $G' \vdash G$).

We shall use the following notation: an equational fact will have the shape $\vdash M : e$, meaning that M is a proof of e; an equational goal will have the shape $\alpha : e \vdash M : C$, meaning that in the proof M of C the goal e is still open, i.e. M may depend on α.

Given a term t we write $t|_p$ to denote the subterm of t at position p, and $t[r]_p$ for the term obtained from t replacing the subterm $t|_p$ with r. Given a substitution σ we write $t\sigma$ for the application of the substitution to the term, with the usual meaning.

The logical rules, decorated with proofs, are the following:

Superposition left

$$\frac{\vdash h : l =_A r \qquad \alpha : t =_B s \vdash M : C}{\beta : t[r]_p\sigma =_B s\sigma \vdash M\sigma[R/\alpha\sigma] : C\sigma}$$

if $\sigma = mgu(l, t|_p)$, $t|_p$ is not a variable, $l\sigma \succ r\sigma$ and $t\sigma \succ s\sigma$; and
$R = (\text{eq\_ind\_r } A \ r\sigma \ (\lambda x : A.t[x]_p =_B s)\sigma \ \beta \ l\sigma \ h\sigma) : t\sigma =_B s\sigma$

Superposition right

$$\frac{\vdash h : l =_A r \qquad \vdash k : t =_B s}{\vdash R : t[r]_p\sigma =_B s\sigma}$$

if $\sigma = mgu(l, t|_p)$, $t|_p$ is not a variable, $l\sigma \succ r\sigma$ and $t\sigma \succ s\sigma$; and
$R = (\text{eq\_ind } A \ l\sigma \ (\lambda x : A.t[x]_p =_B s)\sigma \ k\sigma \ r\sigma \ h\sigma) : t[r]_p\sigma =_B s\sigma$

Equality resolution

$$\frac{\alpha : t =_A s \vdash M : C}{\vdash M[\text{refl\_eq } A \ t\sigma/\alpha] : C}$$

if there exists $\sigma = mgu(t, s)$; (notice $refl\_eq \ A \ t : t =_A t$, being refl_eq the constructor of the equality).

The main theorem is that, given a set of facts S, and a goal e, an instance e' of e is a logical consequence of S if and only if, starting from the trivial axiom $\alpha : e \vdash \alpha : e$ we may prove $\vdash M : e'$ (and in this case M is a correct proof term).

Simplification rules such as tautology elimination, subsumption and especially demodulation can be added to the systems, but they do not introduce major conceptual problems, and hence they will not be considered here.

2.3   Proof Search and Its Representation

Given the three superposition rules above, proof search is performed using the "given clause" algorithm (see [14,15]). The algorithm keeps all known facts and goals split in two sets: active, and passive. At each iteration, the algorithm carefully chooses an equation (given clause) from the passive set; if it is a goal (and not an identity), then it is combined via superposition left with all active facts; if it is a fact, superposition right is used instead. The selected equation is added to the (suitable) active set, while all newly generated equations are added to the passive set, and the cycle is repeated.

As the reader may imagine a huge number of equations is generated during the proof search process, but only few of them will be actually used to prove the goal. Even if demodulation and subsumption are effective tools to discard equations without loosing completeness, all automatic theorem provers adopt clever techniques to strike down the space consumption of each equation. This usually leads to an extensive use of sharing in the data structures, and to drop the idea of carrying a complete proof representation in favor of recording a minimal and lightweight proof trace. The latter choice is usually not a big concern for ATP systems, since proofs are mainly used for debugging purposes, but for an interactive theorem prover that follows the independent verification principle like Matita, proof objects are essential and thus it must be possible to reconstruct a complete proof object in CIC from the proof trace.

In our implementation the proof trace is composed by two slightly different kind of objects, corresponding to the two superposition steps. Superposition right steps are encoded with the following tuple:

$$\text{type } rstep \overset{\text{def}}{=} ident * ident * direction * substitution * predicate$$

The two identifiers are unambiguous names for the equations involved (h and k in the former presentation of the superposition rule), $direction$ can be either Left or Right, depending if h has been used left to right or right to left (i.e. if a symmetry step has to be kept into account). The $substitution$ and the $predicate$ are respectively the σ (i.e. the most general unifier between l and $t|_p$) and the predicate used to build the proof R (i.e. the third element applied to eq_ind), that is essentially a representation of the position $|_p$ identifying the subterm of t that has been rewritten with r once l and $t|_p$ were unified via σ.

This representation of the predicate is not optimal in terms of space consumption; we have chosen this representation mainly for simplicity, and left the implementation of a more compact coding as a future optimization.

The representation of a superposition left step is essentially the same, but the second equation identifier has been removed, since it implicitly refers to the goal. We will call the type of these steps $lstep$.

A map $\Sigma : ident \rightarrow (pos\_literal * rstep)$ from identifiers to pairs of positive literal (i.e. something of the form $\vdash a =_A b$) and proof step represents all the forward reasoning performed during proof search, while a list Λ of $lstep$ together with the initial goal (a negative literal) represent all backward reasoning steps.

3   Proof Reconstruction

The functions defined in Fig. 1 build a CIC proof term given the initial goal g, Σ and Λ. We use the syntax "let ($\vdash l =_A r$, π_h) $= \Sigma(h)$ in" for the irrefutable pattern matching construct "match $\Sigma(h)$ with ($\vdash$ eq $A\ l\ r$), $\pi_h \Rightarrow$".

The function ϕ produces proofs corresponding to application of the superposition right rule, with the exception that if h is used right to left and eq_ind_r is used to represent the hidden symmetry step. ψ builds proofs associated with the application of the superposition left rule, and fires ϕ to build the proof of the positive literal h involved.

Unfortunately this simple structurally recursive approach has the terrible behavior of inlining the proofs of positive literals even if they are used non linearly. This may (and in practice does) trigger an exponential factor in the size of proof objects. The obtained proof object is thus of a poor value, because type checking it would require an unacceptable amount of time.

As an empirical demonstration of that fact we report in Fig. 2 a graphical representation of the proof of problem GRP001-4 available in the TPTP[18] library version 3.1.1. Axioms are represented in squares, while positive literals have a circular shape. The goal is an hexagon.

Every positive literal points to the two used as hypothesis in the corresponding application of the superposition right rule. In this example a, b, c and e are

$\phi(\Sigma, (h, k, dir, \sigma, P)) =$
 let $(\vdash l =_A r, \pi_h) = \Sigma(h)$ and $(\vdash t =_B s, \pi_k) = \Sigma(k)$ in
 match dir with
 $|$ Left $\Rightarrow$ eq_ind A $l\sigma$ $P\sigma$ $\phi(\Sigma, \pi_k)\sigma$ $r\sigma$ $\phi(\Sigma, \pi_h)\sigma$
 $|$ Right $\Rightarrow$ eq_ind_r A $r\sigma$ $P\sigma$ $\phi(\Sigma, \pi_k)\sigma$ $l\sigma$ $\phi(\Sigma, \pi_h)\sigma$

$\psi'(\Sigma, (h, dir, \sigma, P), (t =_B s, \pi_g)) =$
 let $(\vdash l =_A r, \pi_h) = \Sigma(h)$ in
 match dir with
 $|$ Left $\Rightarrow$ $(P\ r)\sigma$, eq_ind A $l\sigma$ $P\sigma$ $\pi_g\sigma$ $r\sigma$ $\phi(\Sigma, \pi_h)\sigma$
 $|$ Right $\Rightarrow$ $(P\ l)\sigma$, eq_ind_r A $r\sigma$ $P\sigma$ $\pi_g\sigma$ $l\sigma$ $\phi(\Sigma, \pi_h)\sigma$

$\psi(g, \Lambda, \Sigma) =$
 let $(t =_B s) \vdash \_ = g$ in
 $snd(\text{fold\_right}(\lambda x.\lambda y.\psi'(\Sigma, x, y), (t =_B s, \text{refl\_eq } A\ s), \Lambda))$

$\tau \stackrel{\text{def}}{=} term$
$\phi : (ident \rightarrow (pos\_literal * rstep)) * rstep \rightarrow \tau$
$\psi' : (ident \rightarrow (pos\_literal * rstep)) * lstep * (\tau * \tau) \rightarrow \tau$
$\psi : neg\_literal * lstep$ list $* (ident \rightarrow (pos\_literal * rstep)) \rightarrow \tau$
fold_right $: (lstep * (\tau * \tau) \rightarrow (\tau * \tau)) * (\tau * \tau) * lstep$ list $\rightarrow (\tau * \tau)$

Fig. 1. Proof reconstruction

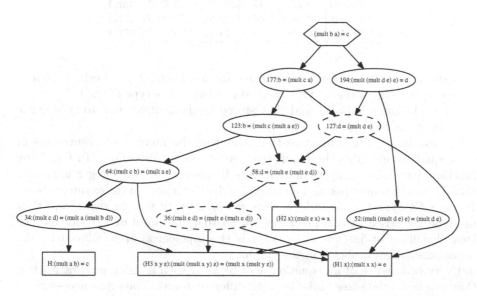

Fig. 2. Proof representation (shared nodes)

constants, the latter has the identity properties (axiom H2). The thesis is that
a group (axioms H3, H2) in which the square of each element is equal to the
unit (axiom H1) is abelian (compose H with the goal to obtain the standard

formulation of the abelian predicate). Equation 127 is used twice, 58 is used three times (two times by 127 and one by 123), consequently also 36 is not used linearly. In this scenario, the simple proof reconstruction algorithm inflates the proof term, replicating the literals marked with a dashed line.

The benchmarks reported in Tab. 1show that this exponential behavior makes proof objects practically untractable. The first column reports the time the automatic procedure spent in searching the proof, and the second one the number of iterations of the given clause algorithm needed to find a proof. The amount of time necessary to typecheck a non optimized proof is dramatically bigger then the time that is needed to find the proof. With the optimization we describe in the following paragraph typechecking is as fast as proof search for easy problems like the ones shown in Tab. 1.As one would expect, when problems are more challenging, the time needed for typechecking the proof is negligible compared to the time needed to find the proof.

Table 1. Timing (in seconds) and proof size

Problem	Search	Steps	Typing		Proof size	
			raw	opt	raw	opt
BOO069-1	2.15	27	79.50	0.23	3.1M	29K
BOO071-1	2.23	27	203.03	0.22	5.4M	28K
GRP118-1	0.11	17	7.66	0.13	546K	21K
GRP485-1	0.17	47	323.35	0.23	5.1M	33K
LAT008-1	0.48	40	22.56	0.12	933K	19K
LCL115-2	0.81	52	24.42	0.29	1.1M	37K

Fortunately CIC provides a construct for local definitions LetIn : *ident* ∗ *term* ∗ *term* → *term* that is type checked efficiently: the type of the body of the definition is computed once and then stored in the context used to type check the rest of the term.

We can thus write a function that, counting the number of occurrences of each equation, identifies the proofs that have to be factorised out. In Fig. 3 the function γ returns a map from identifiers to integers. If this integer is greater than 1, then the corresponding equation will be factorised. In the example above, 127 and 58 should be factorised, since γ evaluates to two on them, and they must be factorised in this precise order, so that the proof of 127 can use the local definition of 58. The right order is the topological one, induced by the dependency relation shown in the graph.

Every occurrence of an equation may be used with a different substitution, that can instantiate free variables with different terms. Thus it is necessary to factorise closed proofs obtained λ-abstracting their free variables, and applying them to the same free variables where they occur before applying the local substitution. For example, given a proof π whose free variables are $x_1 \ldots x_n$ respectively of type $T_1 \ldots T_n$ we generate the following let in:

$$\text{LetIn } h \overset{\text{def}}{=} (\lambda x_1 : T_1, \ldots \lambda x_n : T_n, \pi) \text{ in}$$

$\delta'(\Sigma, h, f) =$
    let $g = (\lambda x.\text{if } x = h \text{ then } 1 + f(x) \text{ else } f(x))$ in
    if $f(h) = 0$ then
        let $(\_, \pi_h) = \Sigma(h)$ in
        let $(k_1, k_2, \_, \_, \_) = \pi_h$ in
        $\delta'(\Sigma, k_1, \delta'(\Sigma, k_2, g))$
    else g

$\delta(\Sigma, (h, \_, \_, \_), f) = \delta'(\Sigma, h, f)$

$\gamma(\Lambda, \Sigma) = \text{fold\_right}(\lambda x.\lambda y.\delta(\Sigma, x, y), \lambda x.0, \Lambda)$

$\delta' : (ident \rightarrow (pos\_literal * rstep)) * ident * (ident \rightarrow int) \rightarrow (ident \rightarrow int)$
$\delta : (ident \rightarrow (pos\_literal * rstep)) * lstep * (ident \rightarrow int) \rightarrow (ident \rightarrow int)$
$\gamma : lstep \text{ list} * (ident \rightarrow (pos\_literal * rstep)) \rightarrow (ident \rightarrow int)$

Fig. 3. Occurrence counting

and the occurrences of π will look like $(h\ x_1 \ldots x_n)\sigma$ where σ will eventually differ.

3.1 Digression on Dependent Types

ATP systems usually operate in a first order setting, where all variables have the same type. CIC provides dependent types, meaning that in the previous example the type T_n can potentially depend on the variables $x_1 \ldots x_{n-1}$, thus the order in which free variables are abstracted is important and must be computed keeping dependencies into account.

Consider the case, really common in formalisations of algebraic structures, where a type, functions over that type and properties of these operations are packed together in a structure. For example, defining a group, one will probably end up having the following constants:

carr : $Group \rightarrow Type$ inv : $\forall g : Group, \text{carr } g \rightarrow \text{carr } g$
e : $\forall g : Group, \text{carr } g$ mul : $\forall g : Group, \text{carr } g \rightarrow \text{carr } g \rightarrow \text{carr } g$
                     id_l : $\forall g : Group, \forall x : \text{carr } g, \text{mul } g \ (e\ g)\ x = x$

Saturation rules work with non abstracted (binder free) equations, thus the id_l axiom is treated as $(mul\ x\ (e\ x)\ y = y)$ where x and y are free. If these free variables are blindly abstracted, an almost ill typed term can be obtained:

$$\lambda y :?_1, \lambda x :?_2, mul\ x\ (e\ x)\ y = y$$

where there is no term for $?_1$ such that $?_1 = (\text{carr } x)$ as required by the dependency in the type of mul: the second and third arguments must have type carr of the first argument. In the case above, the variable y has a type that depends on x, thus abstracting y first, makes it syntactically impossible for its type to depend on x. In other words $?_1$ misses x in its context.

When we decided to integrate automatic rewriting techniques like superposition in Matita, we were attracted by their effectiveness and not in studying a generalisation of these techniques to a much more complex framework like CIC. The main, extremely practical, reason is that the portion of mathematical problems that can be tackled using first order techniques is non negligible and for some problems introduced by dependent types, like the one explained above, the solution is reasonably simple. Exploiting the explicit polymorphism of CIC, and the rigid structure of the proofs we build (i.e. nested application of eq_ind) it is possible to collect free variables that are used as types, inspecting the first arguments of eq_ind and eq: these variable are abstracted first. Even if this simple approach works pretty well in practice and covers the probably most frequent case of type dependency, it is not meant to scale up to the general case of dependent types, in which we are not interested.

4   Proof Refinement

Proofs produced by paramodulation based techniques are very difficult to understand for a human. Although the single steps are logically trivial, the overall design of the proof is extremely difficult to grasp. This need is also perceived by the ATP community; for instance, in order to improve readability, the TPTP[18] library, provides a functionality to display proofs in a graphical form (called YuTV), pretty similar to the one in Fig. 2.

In the case of purely equational reasoning, mathematicians traditionally organize the proof as a chain of rewriting steps, each one justified by a simple side argument (an axiom, or an already proved lemma). Technically speaking, such a chain amounts to a composition of transitivity steps, where as proof leaves we only admit axioms (or their symmetric variants), possibly contextualized. Formally, the basic components we need are provided by the following terms:

$$\text{trans} : \forall A : Type. \forall x, y, z : A. x =_A y \rightarrow y =_A z \rightarrow x =_A z$$
$$\text{sym} : \forall A : Type. \forall x, y : A. x =_A y \rightarrow y =_A x$$
$$\text{eq\_f} : \forall A, B : Type. \forall f : A \rightarrow B. \forall x, y : A. x =_A y \rightarrow (f\ x) =_B (f\ y)$$

The last term (function law) allows to contextualize the equation $x =_A y$ in an arbitrary context f.

The normal form for equational proofs we are interested in is described by the following grammar:

Definition 1 (Proof normal form)

$$\begin{aligned} \pi = {} & \text{eq\_f}\ B\ C\ \Delta\ a\ b\ axiom \\ & |\ \text{eq\_f}\ B\ C\ \Delta\ a\ b\ (\text{sym}\ B\ b\ a\ axiom) \\ & |\ \text{trans}\ A\ a\ b\ c\ \pi\ \pi \end{aligned}$$

We now prove that any proof build by means of eq_ind and eq_ind_r may be transformed in the normal form of definition 1. The transformation is defined

Fig. 4. Natural language rendering of the (refined) proof object of GRP001-4

in two phases. In the first phase we replace all rewriting steps by means of applications of transitivity, symmetry and function law. In the second phase we propagate symmetries towards the leaves.

In Figure 4 we show an example of the kind of rendering obtained after the transformation, relative to the proof of GRP001-4.

4.1 Phase 1: Transitivity Chain

The first phase of the transformation is defined by the ρ function of Fig. 5. We use Δ and Γ for contexts (i.e. unary functions). We write $\Gamma[a]$ for the application of Γ to a, that puts a in the context Γ, and $(\Delta \circ \Gamma)$ for the composition of

$\rho(\pi) \rightsquigarrow \rho'(\lambda x{:}C.x,\ \pi)$ when $\pi : a =_C b$

$\rho'(\Delta,\ \text{eq\_ind } A\ a\ (\lambda x.\Gamma[x] =_B m)\ \pi_1\ b\ \pi_2) \rightsquigarrow$
    $\text{trans } C\ (\Delta \circ \Gamma)[b]\ (\Delta \circ \Gamma)[a]\ \Delta[m]$
        $(\text{sym } C\ (\Delta \circ \Gamma)[a]\ (\Delta \circ \Gamma)[b]\ \rho'(\Delta \circ \Gamma,\ \pi_2))\ \rho'(\Delta,\ \pi_1)$

$\rho'(\Delta,\ \text{eq\_ind\_r } A\ a\ (\lambda x.\Gamma[x] =_B m)\ \pi_1\ b\ \pi_2) \rightsquigarrow$
    $\text{trans } C\ (\Delta \circ \Gamma)[b]\ (\Delta \circ \Gamma)[a]\ \Delta[m]\ \rho'(\Delta \circ \Gamma,\ \pi_2)\ \rho'(\Delta,\ \pi_1)$

$\rho'(\Delta,\ \text{eq\_ind } A\ a\ (\lambda x.m =_B \Gamma[x])\ \pi_2\ b\ \pi_1) \rightsquigarrow$
    $\text{trans } C\ \Delta[m]\ (\Delta \circ \Gamma)[a]\ (\Delta \circ \Gamma)[b]\ \rho'(\Delta,\ \pi_2)\ \rho'(\Delta \circ \Gamma,\ \pi_1)$

$\rho'(\Delta,\ \text{eq\_ind\_r } A\ a\ (\lambda x.m =_B \Gamma[x])\ \pi_1\ b\ \pi_2) \rightsquigarrow$
    $\text{trans } C\ \Delta[m]\ (\Delta \circ \Gamma)[a]\ (\Delta \circ \Gamma)[b]$
        $\rho'(\Delta,\ \pi_1)\ (\text{sym } C\ (\Delta \circ \Gamma)[b]\ (\Delta \circ \Gamma)[a]\ \rho'(\Delta \circ \Gamma,\ \pi_2))$

$\rho'(\Delta,\ \pi) \rightsquigarrow \text{eq\_f } B\ C\ \Delta\ a\ b\ \pi$ when $\pi : a =_B b$ and $\Delta : B \to C$

Fig. 5. Transitivity chain construction

contexts, so we have $(\Delta \circ \Gamma)[a] = \Delta[\Gamma[a]]$. The auxiliary function ρ' takes a context $\Delta : B \to C$, a proof of $(c =_B d)$ and returns a proof of $(\Delta[c] =_C \Delta[d])$. In order to prove that ρ is type preserving, we proceed by induction on the size of the proof term, stating that if Δ is a context of type $B \to C$ and π is a term of type $a =_B b$, then $\rho'(\Delta,\ \pi) : \Delta[a] =_C \Delta[b]$.

Theorem 1 (ρ' injects). *For all B and C types, for all a and b of type B, if $\Delta : B \to C$ and $\pi : a =_B b$, then $\rho'(\Delta,\ \pi) : \Delta[a] =_C \Delta[b]$*

Proof. We proceed by induction on the size of the proof term.

Base case. By hypothesis we know $\Delta : B \to C$, and $\pi : a =_B b$, thus a and b have type B and $(\text{eq\_f } B\ C\ \Delta\ a\ b\ \pi)$ is well typed, and proves $\Delta[a] =_C \Delta[b]$

Inductive case. (We analyse only the first case, the others are similar) By hypothesis we know $\Delta : B \to C$, and

$$\pi = (\text{eq\_ind } A\ a\ (\lambda x.\Gamma[x] =_B m)\ \pi_1\ b\ \pi_2) : \Gamma[b] =_B m$$

From the type of eq_ind we can easily infer that $\pi_1 : \Gamma[a] =_B m$, $\pi_2 : a =_A b$, $\Gamma : A \to B$, $m : B$ and both a and b have type A. Since $\Delta : B \to C$, $\Delta \circ \Gamma$ is a context of type $A \to C$. Since π_2 is a subterm of π, by inductive hypothesis we have

$$\rho'(\Delta \circ \Gamma,\ \pi_2) : (\Delta \circ \Gamma)[a] =_C (\Delta \circ \Gamma)[b]$$

Since $(\Delta \circ \Gamma) : A \to C$ and a and b have type A, both $(\Delta \circ \Gamma)[a]$ and $(\Delta \circ \Gamma)[b]$ live in C. We can thus type the following application.

$$\pi_2' \stackrel{\text{def}}{=} (\text{sym } C\ (\Delta \circ \Gamma)[a]\ (\Delta \circ \Gamma)[b]\ \rho'(\Delta \circ \Gamma,\ \pi_2)) : (\Delta \circ \Gamma)[b] =_C (\Delta \circ \Gamma)[a]$$

We can apply the induction hypothesis also on $\pi_1' \stackrel{\text{def}}{=} (\rho' \Delta \pi_1)$ obtaining that is has type $(\Delta \circ \Gamma)[a] =_C \Delta[m]$. Since $\Delta[m] : C$, we can conclude that

$$\pi_3 \stackrel{\text{def}}{=} (\text{trans } C \ (\Delta \circ \Gamma)[b] \ (\Delta \circ \Gamma)[a] \ \Delta[m] \ \pi_2' \ \pi_1') : (\Delta \circ \Gamma)[b] =_C \Delta[m]$$

Expanding $\circ$ we obtain $\pi_3 : \Delta[\Gamma[b]] =_C \Delta[m]$

□

Corollary 1 (ρ is type preserving)

Proof Trivial, since the initial context is the identity.   □

4.2   Phase 2: Symmetry Step Propagation

The second phase of the transformation is performed by the θ function in Fig.6.

$\theta(\text{sym } A \ b \ a \ (\text{trans } A \ b \ c \ a \ \pi_1 \ \pi_2)) \rightsquigarrow$
    $\text{trans } A \ a \ c \ b \ \theta(\text{sym } A \ c \ a \ \pi_2) \ \theta(\text{sym } A \ b \ c \ \pi_1)$
$\theta(\text{sym } A \ b \ a \ (\text{sym } A \ a \ b \ \pi)) \rightsquigarrow \theta(\pi)$
$\theta(\text{trans } A \ a \ b \ b \ \pi_1 \ \pi_2) \rightsquigarrow \theta(\pi_1)$
$\theta(\text{trans } A \ a \ a \ b \ \pi_1 \ \pi_2) \rightsquigarrow \theta(\pi_2)$
$\theta(\text{trans } A \ a \ c \ b \ \pi_1 \ \pi_2) \rightsquigarrow$
    $\text{trans } A \ a \ c \ b \ \theta(\pi_1) \ \theta(\pi_2)$
$\theta(\text{sym } B \ \Delta[a] \ \Delta[b] \ (\text{eq\_f } A \ B \ \Delta \ a \ b \ \pi)) \rightsquigarrow$
    $\text{eq\_f } A \ B \ \Delta \ b \ a \ (\text{sym } A \ a \ b \ \pi)$
$\theta(\pi) \rightsquigarrow \pi$

Fig. 6. Canonical form construction

The third and fourth case of the definition of θ are merely used to drop a redundant reflexivity step introduced by the equality resolution rule.

Theorem 2 (θ is type preserving). *For all A type, for all a and b of type A, if $\pi : a =_A b$, then $\theta(\pi) : a =_A b$*

Proof. We proceed by induction on the size of the proof term analysing the cases defining θ. By construction, the proof is made of nested applications of sym and trans; leaves are built with eq_f. The base case is the last one, where θ behaves as the identity and thus is type preserving. The following cases are part of the inductive step, thus we know by induction hypothesis that θ is type preserving on smaller terms.

First case. By hypothesis we know that

$$(\text{sym } A \ b \ a \ (\text{trans } A \ b \ c \ a \ \pi_1 \ \pi_2)) : a =_A b$$

thus $\pi_1 : b =_A c$ and $\pi_2 : c =_A a$. Consequently $(\text{sym } A\ c\ a\ \pi_2) : a =_A c$ and $(\text{sym } A\ b\ c\ \pi_1) : c =_A b$ and the induction hypothesis can be applied to them, obtaining $\theta(\text{sym } A\ c\ a\ \pi_2) : a =_A c$ and $\theta(\text{sym } A\ b\ c\ \pi_1) : c =_A b$. From that we obtain

$$(\text{trans } A\ a\ c\ b\ \theta(\text{sym } A\ c\ a\ \pi_2)\ \theta(\text{sym } A\ b\ c\ \pi_1)) : a =_A b$$

Second case. We know that $(\text{sym } A\ b\ a\ (\text{sym } A\ a\ b\ \pi)) : a =_A b$, thus $(\text{sym } A\ a\ b\ \pi) : b =_A a$ and $\pi : a =_A b$. Induction hypothesis suffices to prove $\theta(\pi) : a =_A b$

Third case. Since $(\text{trans } A\ a\ b\ b\ \pi_1\ \pi_2) : a =_A b$ we have $\pi_1 : a =_A b$. Again, the induction hypothesis suffices to prove $\theta(\pi_1) : a =_A b$

Fourth case. Analogous to the third case

Fifth case. By hypothesis we know that

$$(\text{sym } B\ \Delta[a]\ \Delta[b]\ (\text{eq\_f } A\ B\ \Delta\ a\ b\ \pi)) : \Delta[b] =_B \Delta[a]$$

Thus $\pi : a =_A b$ and $(\text{eq\_f } A\ B\ \Delta\ a\ b\ \pi) : \Delta[a] =_B \Delta[b]$. Hence $(\text{sym } A\ a\ b\ \pi) : b =_A a$ and

$$(\text{eq\_f } A\ B\ \Delta\ b\ a\ (\text{sym } A\ a\ b\ \pi)) : \Delta[b] =_B \Delta[a]$$

Sixth case. Follows directly from the inductive hypothesis   □

5   Conclusion and Related Works

In this paper we have presented a procedure to transform a minimal proof trace left by an automatic proof searching procedure to a valuable proof term in the calculus of inductive constructions. We then refined this proof object with type preserving transformations, making it suitable for the natural language rendering engine of the Matita interactive theorem prover.

The problem of reconstructing a proof from some sort of trace left by an automatic prover is addressed by Hurd in [5] and by Kreitz and Schmitt in [6] while developing JProver[16]. In the former work, Hurd has to face the problem of reconstructing a proof from the ambiguous and incomplete output of the Gandalf[19] prover, and he solves it inferring the missing information with a prolog-style search. On the contrary, when we wrote the automatic procedure we had in mind that the output would have been a formal proof, thus we paid attention in not trading the proof trace completeness down for efficiency. The latter work describes several proof reconstruction methodologies in order to obtain natural deduction style or sequent style proofs from resolution and matrix based proof traces. Since we restricted our automatic procedure to the unit equality case, we do not have real clauses and we implement only a trivial subset of the resolution calculus with the equality resolution rule, thus these approaches do not fit well in our setting.

There is a wide literature on the integration of automated procedures with interactive provers, but they usually focus on slightly different aspects or drop some of the requirements we consider essential, anyway they give good suggestions on possible improvements of our work. Meng and Paulson were interested in integrating one of the best ATP systems, Vampire[14], with Isabelle[10] and studied a set of transformations[9,8] to encode (fragments of) the expressive HOL logic into the first order one implemented by Vampire. Some of these techniques could be applied in our case too, allowing us to treat a larger fragment of CIC with our automatic procedure. Ayache and Filliâtre have integrated many ATP systems, like haRVey[4] and CVC Lite[3] with the Coq[20] interactive theorem prover, encoding a fragment of the logic of Coq (CIC) into the intermediate polymorphic first order logic[2] (PFOL) logic, which is meant to be easily convertible to the logics understood by the ATP systems. While the translation to PFOL could be relevant for future improvements of our work, the rest of the paper drops the requirement of producing proof objects, trusting the essentially boolean answer of the ATP systems. Matita follows the De Bruijn principle, stating that proofs generated by the system should be verifiable with a small tool, and since in general an ATP system cannot be considered small, we consider the generation of a proof object that can be verified with a small kernel mandatory. Consider for example that haRVey counts nearly 50,000 lines of code and CVC Lite more then 70,000 while the kernel (type checker) of Matita only 10,000.

The main distinctive characteristic of our work is in the way we take care of the proofs we found; first encoding them in a formal calculus, then improving them both from a practical (space/type-checking efficiency) and an esthetical (natural language rendering) point of view. As suggested above, a natural continuation of this work would be to study how to treat a bigger fragment of CIC with the automatic procedure we implemented without dropping the fundamental requirement of being able to exhibit a valuable CIC proof term once a proof is automatically found.

References

1. Asperti, A., Coen, C.S., Tassi, E., Zacchiroli, S.: User Interaction with the Matita Proof Assistant. Journal of Automated Reasoning, Special Issue on User. Interfaces for Theorem Proving (To appear)
2. Ayache, N., Filliâtre, J.C.: Combining the Coq proof assistant with first-order decision procedures (Unpublished)
3. Barrett, C., Berezin, S.: CVC Lite: A New Implementation of the Cooperating Validity Checker. In: Alur, R., Peled, D.A. (eds.) CAV 2004. LNCS, vol. 3114, pp. 515–518. Springer, Heidelberg (2004)
4. Déharbe, D., Ranise, S., Fontaine, P.: haRVey, a cocktail of theories. http://harvey.loria.fr/
5. Hurd, J.: Integrating Gandalf and HOL. In: Proceedings of Theorem Proving in Higher Order Logics (TPHOL), pp. 311–321 (1999)

6. Kreitz, C., Schmitt, S.: A Uniform Procedure for Converting Matrix Proofs into Sequent-Style Systems. Information and Computation 162(1-2), 226–254 (2000)
7. Mc Cune, W.: Solution of the Robbins Problem. Journal of Automated Reasoning 19(3), 263–276 (1997)
8. Meng, J., Paulson, L.: Experiments On Supporting Interactive Proof Using Resolution. In: Basin, D., Rusinowitch, M. (eds.) IJCAR 2004. LNCS (LNAI), vol. 3097, Springer, Heidelberg (2004)
9. Meng, J., Paulson, L.: Translating Higher-Order Problems to First-Order Clauses. In: Sutcliffe, G., Schmidt, R., Schulz, S. (eds.) ESCoR: Empirically Successful Computerized Reasoning. CEUR Workshop Proceedings, vol. 192, pp. 70–80 (2006)
10. Meng, J., Quigley, C., Paulson, L.: Automation for Interactive Proof: First Prototype. Information and Computation 204(10), 1575–1596 (2006)
11. Nieuwenhuis, R., Rubio, A.: Paramodulation-Based Theorem Proving. Handbook of Automated Reasoning, pp. 371–443 (2001)
12. Obua, S., Skalberg, S.: Importing HOL into Isabelle/HOL. In: Furbach, U., Shankar, N. (eds.) IJCAR 2006. LNCS (LNAI), vol. 4130, pp. 298–302. Springer, Heidelberg (2006)
13. Paulin-Mohring, C.: Définitions Inductives en Théorie des Types d'Ordre Supŕieur. Habilitation à diriger les recherches, Université Claude Bernard Lyon I (1996)
14. Riazanov, A., Voronkov, A.: The design and implementation of VAMPIRE. AI Communications 15(2-3), 91–110 (2002)
15. Riazanov, A.: Implementing an Efficient Theorem Prover. PHD thesis, The University of Manchester (2003)
16. Schmitt, S., Lorigo, L., Kreitz, C., Nogin, A.: Jprover: Integrating connection-based theorem proving into interactive proof assistants. In: Goré, R.P., Leitsch, A., Nipkow, T. (eds.) IJCAR 2001. LNCS (LNAI), vol. 2083, pp. 421–426. Springer, Heidelberg (2001)
17. Slaney, J., Fujita, M., Stickel, M.: Automated Reasoning and Exhaustive Search: Quasigroup Existence Problems. Computers and Mathematics with Applications (1993)
18. Sutcliffe, G., Suttner, C.B.: The TPTP Problem Library: CNF Release v1.2.1. Journal of Automated Reasoning 21(2), 177–203 (1998)
19. Tammet, T.: A resolution theorem prover for intuitionistic logic. In: McRobbie, M.A., Slaney, J.K. (eds.) Automated Deduction - Cade-13. LNCS, vol. 1104, Springer, Heidelberg (1996)
20. The Coq Development Team: The Coq Proof Assistant Reference Manual (2006) http://coq.inria.fr/doc/main.html
21. Werner, B.: Une Théorie des Constructions Inductives. PHD thesis, Université Paris VII (1994)

A Framework for Interactive Proof

David Aspinall[1], Christoph Lüth[2], and Daniel Winterstein[1]

[1] LFCS, School of Informatics, The University of Edinburgh, U.K.
[2] Deutsches Forschungszentrum für künstliche Intelligenz (DFKI), Bremen, Germany

Abstract. This paper introduces *Proof General Kit*, a framework for software components tailored to interactive proof development. The goal of the framework is to enable flexible environments for managing formal proofs across their life-cycle: creation, maintenance and exploitation. The framework connects together different kinds of component, exchanging messages using a common communication infrastructure and protocol called *PGIP*. The main channel connects *provers* to *displays*. Provers are the back-end interactive proof engines and displays are components for interacting with the user, allowing browsing or editing of proofs. At the core of the framework is a *broker* middleware component which manages proof-in-progress and mediates between components.

1 Introducing Proof General Kit

The use of interactive machine proof is becoming more widespread, and larger and more complex formalisations are being undertaken in application areas such as hardware or software verification, and formalisation of mathematics, even up to formalising deep proofs of recently established results. Examples of interactive provers include general purpose provers such as Mizar, HOL, Isabelle, PVS, Coq, ACL2, or NuPrl, and domain-specific provers such as the Forte system [19]. Of course, this is to name just a few systems: Freek Wiedijk's database [26] currently lists almost 300 systems for doing mathematics on computer! Although many of these may be classed as small-scale experiments or obsolete, it is natural to expect researchers to continue investigating new logical foundations, and to build domain-specific provers for new application areas.

For interactive provers such as those mentioned, the record of instructions of how to create the proof, or a representation of the proof itself, is kept in a text file with a programming language style syntax. We call these files *proof scripts*. About 100 systems on Wiedijk's list are based on textual proof script input. Each system uses its own proof script language, and while there are similarities across languages, there are crucial differences as well, particularly concerning the underlying logic. For large proofs, the proof scripts are themselves large: by now there are individual developments and mathematical libraries which reach hundreds of thousands of lines of code and represent many person-years of work.

Yet, compared with the facilities available to the modern programmer, the facilities for developing and maintaining formal proofs are lamentably poor, in general.[1] Modern software development uses sophisticated Integrated Development

---

[1] We note a few exceptions in a related work section in the conclusions.

M. Kauers et al. (Eds.): MKM/Calculemus 2007, LNAI 4573, pp. 161–175, 2007.
© Springer-Verlag Berlin Heidelberg 2007

Environments (IDEs), which support features such as automatic documentation lookup, completion of identifiers, and integration with version control and the build process. Modern knowledge management facilities help further: context-aware search finds related definitions; content assistance mechanisms insert declarations and instantiations; advanced software engineering methods like refactoring help improve design, making large-scale structural changes easy.

One reason why these facilities have not yet been provided for theorem proving is the fragmentation of the community across so many different systems, which dilutes the effort available. We believe the community should invest in shared tools as much as possible, and keep only the underlying logical proof engines as separate, distinct implementations. Thus, we are arguing not just for exchanging and relating mathematical knowledge between systems, provided by formats such as OMDoc, but also for the component-based construction of proof management environments themselves, using a uniform protocol.[2]

In this paper we introduce the *Proof General Kit* (PG Kit for short). This is a framework for proof management, based on the *PGIP* protocol. We believe that PG Kit will provide sophisticated and useful development environments for a whole class of interactive provers, and also be a vehicle for research into the foundations of such environments.

Outline. Sect. 2 motivates the PG Kit framework, describing the contribution of the current Proof General system and the component architecture for the new framework. Sect. 3 introduces the PGIP protocol, Sect. 4 describes the central role of the broker component, and Sect. 5 describes several display components. In Sect. 6 we conclude, mentioning future and related work.

2   Proof General Kit Architecture

The claim that we can provide a uniform framework for interactive proof seems bold, especially considering that those provers do not just differ in their underlying languages, but also in their existing interaction mechanisms as well.

2.1   Proof General and Script Management

The *Proof General* project [2] provides evidence that at least some of our aims are feasible. Proof General is a successful generic interface for interactive proof assistants, where a proof script can be sent line-by-line to the prover with the prover responding at each step. It has been adapted to a variety of provers and is in common use for several, most notably Isabelle and Coq.

The central feature is an advanced version of *script management*. To interactively "run" a script like Fig. 1, we send each line to the prover; thus, each line corresponds to a prover state, and the prover's current state always corresponds to one particular line of the script called the prover's *focus*. Script management divides a proof script into three consecutive regions: a part which

---

[2] Very roughly: OMDoc is to PGIP as HTML is to HTTP.

```
lemma fn1: "(EX x. P (f x)) ⟶ (EX y. P y)"
proof
  assume "EX x. P (f x)"
  thus "EX y. P y"
  proof
    fix a
    assume "P (f a)"
    show ?thesis ..
  qed
qed
```

Fig. 1. An short example proof script in Isabelle/Isar

has been processed, a part which is currently being processed, and a part which has not yet been processed. Proof General displays this partitioning to the user by colouring processed text blue and busy (being-processed) text pink. Editing is prevented in the coloured region to ensure synchronisation with the prover. A toolbar provides buttons for navigating within the proof, moving the focus. The navigation buttons behave identically across numerous different systems, despite behind-the-scenes using rather different system-specific control commands.

Although successful, there are several drawbacks to the present Proof General. Users are required to learn Emacs and tolerate its idiosyncratic UI. Developers must contend with the Emacs Lisp API which is restrictive, often changing, and inconsistent across the many flavours of Emacs. For configuring provers, the instantiation mechanism has become fragile and too complex. This is because Proof General arose by successively generalising a common basis to a growing number of proof systems, with the design goal not to change the systems themselves. But this leaves the interface vulnerable to breakage by even small changes in the prover output format, and it does not itself offer a clear API, relying on regular expression matching of the prover output.

2.2   The Framework Architecture

Instead of trying to anticipate a range of slightly different behaviours, we propose a uniform protocol and model of proof development which captures behaviour reasonably common to all provers at an abstract level, and ask that each proof system implements that. We want to generalise away from Emacs and allow other front-ends, and possibly several at once, so that the proof progress can be displayed in different ways, or to other users. We also want to allow connecting to more than one prover at once, to allow easy switching between different developments and systems. We even want to allow connecting other components that provide assistance during the proof process (e.g., for recommendation [13] or proof planning [8]). In the end, what we need is exactly a software framework: a way of connecting together interacting components customised to the domain.

The PG Kit framework has three main component types: interactive *prover* engines, front-end *display* components, and a central *broker* component which

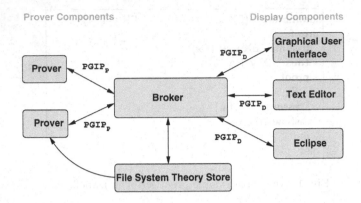

Fig. 2. PG Kit Framework architecture

orchestrates proofs-in-progress. The architecture is pictured in Fig. 2. The components communicate using messages in the PGIP protocol, described in the next section. The general control flow is that a user's action causes a command to be sent from the display to the broker, the broker sends commands to the prover, which sends responses back to the broker which relays them to the displays. The format of the messages is defined by an XML schema. Messages are sent over channels, typically sockets or Unix pipes.

3   A Protocol for Interactive Proof

The protocol for directing proof used by PG Kit is known as *PGIP*, for *Proof General Interaction Protocol* [4]. It arose by examining and clarifying the communications used in the existing Proof General system. As we developed prototype systems following the ideas outlined above, the protocol has been revised and extended to encompass graphical front-ends, a document model markup for proof scripts, and authoring extensions [3,4,5]. PGIP is an abstraction of the communication between provers and interfaces. It allows for prover-specific behaviour and syntax (e.g. in the proof scripts), but specifies an abstract model of behaviour which all provers have to follow.

The syntax of PGIP messages is defined by an XML schema written in RELAX NG [17]. Every message is wrapped in a <pgip> packet which uniquely identifies its origin and contains a sequence number and possibly a referent identifier and sequence number. PGIP comprises three sub-protocols, corresponding to the different types of components from Fig. 2:

– The *prover protocol* $PGIP_P$ defines messages exchanged between provers and the broker. This includes: commands sent to the prover, which correspond to the commands in a conventional proof script and may affect the internal (proof-relevant) state of the prover; messages from the prover in reaction to these commands such as <normalresponse>, <errorresponse> or

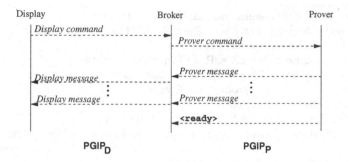

Fig. 3. Message exchange in the PGIP protocol

<ready>, which reflect the internal state; and configuration messages which describe some elements of its concrete syntax, preference settings available to the user, or which icons to use in a graphical interface.

- The *display protocol* PGIP$_D$ defines messages exchanged between displays and the broker. This includes: display commands sent from the display to the broker, corresponding to user interaction, such as starting a prover, loading a file <loadparsefile>, or editing <editcmd>; and display messages, which contain output directed to the user, either relayed from the prover, or generated from the broker.
- The *inter-broker protocol* PGIP$_I$ defines messages exchanged between different brokers, allowing running the prover on a remote machine (see Sec. 4).

The sub-protocols are not disjoint: some prover output (e.g., <normalresponse> or <errorresponse>) is relayed to the displays, so these messages are part of both PGIP$_D$ and PGIP$_P$. The broker analyses messages from the prover, and keeps an abstract view of the internal state of the prover which behaves according to a model described in Sect. 3.2. There is a secondary schema called *PGML*, for *Proof General Markup Language*, used to markup messages from the prover.[3]

Fig. 3 shows a schematic message exchange. The pattern of exchanges between the components is more permissive than in simple synchronous RPC mechanisms like XML RPC or most web services. This is necessary because interactive proof may diverge (e.g. during proof search); it is essential that feedback can be displayed eagerly so the user can take action as soon as possible. The message exchange between the display and the broker is asynchronous (single request, non-waiting multiple response): the display sends a command, and the broker may send several responses later. The message exchange between the broker and the prover can be asynchronous or synchronous (single request, waiting single response). In the default asynchronous message exchange between prover and broker (corresponding to a command that may cause a proof attempt), the prover will send several responses, eventually followed by a <ready> message, which signals availability of the prover to the broker.

---

[3] A standard markup language, e.g., MathML, could be used instead, but PGML is designed for easy support by existing systems by marking up concrete syntax.

```
<opengoal name="fn1">lemma fn1: &quot;(EX x. P (f x)) <sym
        name="longrightarrow">--&gt;</sym> (EX y. P y)&quot;</opengoal>
<openblock/><proofstep>proof</proofstep>
  <proofstep>assume &quot;EX x. P (f x)&quot;</proofstep>
  <opengoal>thus &quot;EX y. P y&quot;</opengoal>
  <openblock/><proofstep>proof</proofstep>
    <proofstep>fix a</proofstep>
    <proofstep>assume &quot;P (f a)&quot;</proofstep>
    <opengoal>show ?thesis</opengoal><openblock/><closegoal>..</closegoal>
                                                      <closeblock/>
  <closegoal>qed</closegoal><closeblock/>
<closegoal>qed</closegoal><closeblock/>
```

Fig. 4. A proof script in Isabelle/Isar, marked up in PGIP

On top of this exchange mechanism, interactive proof proceeds in an edit-parse-prove cycle. The user enters a command via the display; it gets parsed and inserted into the proof script as parsed commands; and eventually it is evaluated, giving a new prover state. Repeating this builds up a sequence of prover commands inside the broker interactively, which form a proof script.

3.1 Proof Scripts in PGIP

Proof scripts are the central artefact of the system. Provers usually just check proof scripts to guarantee their correctness, but do not construct them, relying on external tools (mostly, humans with text editors). The basic principle for representing them in PGIP is to use the prover's native language and *mark up* the content with PGIP commands which explain the proof script structure. Fig. 4 shows the PGIP representation of the example proof script from Fig. 1 with the structural markup, including a PGML <sym> symbol element (we omit other PGML symbols and markup such as <whitespace> for white spaces for brevity). Notice the named and unnamed <opengoal> elements, and the indentation structure introduced by <openblock> and <closeblock>.

Proof scripts consist of prover commands, but not all prover commands appear in a proof script; we distinguish between *proper* commands which can appear and *improper* commands which must not. Proper commands are sent to the prover in plain text, so the prover can interpret them as it would do ordinarily when reading a file. The broker does not know how to generate the prover-specific concrete syntax of proper commands; it is usually written directly by the user. However, the prover can offer a configuration of *prover types* and *prover operations* for building up commands which then enable interface features to help the user. The operations are defined by textual substitution. A trivial example for Isar is an operation taking an identifier *id* and a term string *tm*, and produces the command **lemma** *id* : "*tm*" which opens a goal. For textual interfaces, these operations allow a *template* mechanism; for graphical interfaces, they define operations which can be invoked when the user employs certain gestures.

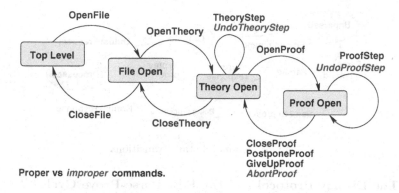

Fig. 5. Proof states during development

Improper commands are only used for controlling the prover's state, and do not appear in the proof script being developed; examples are the three italicised undo commands appearing in Fig. 5. Improper commands are not treated as markup, so the prover must interpret these directly.

3.2   The Prover Protocol: Modelling the Prover State

PG Kit has an abstract model of incremental interactive proof development, where we suppose there are four fundamental states occupied by the prover, with transitions between the states triggered by both proper and improper prover commands. Fig. 5 shows the states, and the commands to change between them. The four states illustrated are:

1. the *top level* state where nothing is open yet;
2. the *file open* state where a file is currently being processed;
3. the *theory open* state where a theory is being built;
4. the *proof open* state where a proof is currently in progress.

These fundamental states give rise to a hierarchy of named items: The top level may contain a number of files. A file contains a proof script, structured into theories. Theories in turn may contain theory items (declarations etc.) and proofs consisting of proofsteps. Within the fourth state, we allow arbitrary nesting (e.g., a proof that contains sub-lemmas).

The reason for distinguishing the states is that the undo behaviour is different in each state, and that different commands are available in each state. In the theory state, for example, we may issue *theory steps* which extend the theory, or we may undo the additions. In the proof state, we can issue *proof steps* and undo these steps, or finish the current proof attempt in a number of ways. After finishing a proof, the history is forgotten, and we can only undo the whole proof.

This model is based on abstracting the common behaviour of many interactive proof systems, acting as a clearly specified virtual layer that must be emulated in each prover to cooperate properly with the broker.

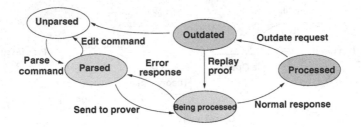

Fig. 6. Command state transitions

3.3 The Display Protocol and the Edit-Parse-Prove Cycle

The markup on a proof script makes the structure of the proof script explicit, and splits the source code into non-overlapping text spans each containing a prover command (see Fig. 4). Each text span has a status ranging over five main[4] possible values, shown in Fig. 6. A text starts off as *unparsed*, and after parsing becomes one (or more) freshly *parsed* prover commands. Actual proving consists of sending the command to the prover. While waiting for a response from the prover, the command is *being processed*. Once the prover has sent a positive answer, the command becomes *processed*; on the other hand, if the prover sends an error, the command reverts to *parsed*. To successfully process a command all commands it is depending on will have to been processed first. Similarly, when we *outdate* a command, all commands depending on it are outdated as well; the difference between outdated and parsed is that outdated regions have been successfully processed before. To edit a processed command, we have to outdate it first. Displays can either make the outdate step explicit, requiring the user first to outdate the text range manually, or they can perform the outdate tacitly.

The transitions between the commands refine the current script management in Proof General. By controlling the state of text spans independently, we can exploit a more fine-grained dependency analysis (if the prover reports the necessary dependency information): to process a command we only need to process those commands which are really needed. The broker handles all this dependency analysis behind the scenes. If the prover does not provide dependency information, the broker automatically assumes linear dependency, where every line potentially depends on all lines that come before.

To demonstrate the edit-parse-prove cycle in action, we consider the message exchange in a typical situation: the user requests a file to be loaded, then edits a part of the text, and finally runs the proof. Fig. 7 shows the resulting messages being sent between display, broker and prover. Note that the proof is "run" by requesting a command be processed (`<setcmdstatus>`), which causes a lot of other commands to be processed first. If an error occurs at some point in this scenario, the prover sends an `<errorresponse>` and the broker flushes all outstanding requests. If the error occurs during the parsing, it will insert the

---

[4] To be precise, there are other transient states besides *Being processed* but they are not distinguished to the user, so omit them from Fig. 6.

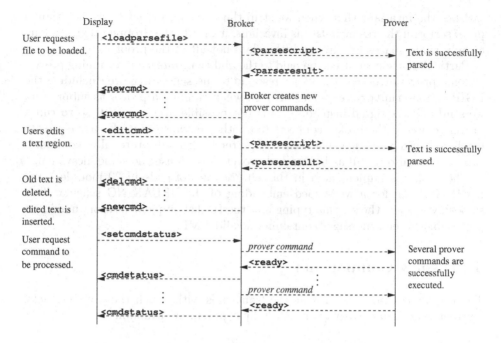

Fig. 7. The edit-parse-prove cycle in a typical situation

corresponding text as an unparsed element into the proof script, to allow the user to edit (and correct) it later.

4   Brokering Electronic Proof

The broker is the central middleware component of the PG Kit framework. It gathers input from the displays, sends prover commands to the provers, handles the responses and does the house-keeping, keeping track of the files and the commands, their respective status and the dependencies between them as provided by the prover. Using this dependency information, it can translate abstract display commands such as <setcmdstatus> into a series of prover commands.

Provers and displays are handled uniformly as *components*, but they differ in their communication pattern: prover commands are sent to one specific prover, whereas display messages are broadcast to all connected displays. For each prover the broker models its state according to the abstract state model from Sect. 3.2. It keeps a queue of all pending prover commands, sending the next one only once it has received a <ready> message from the prover. If the prover sends an <errorreponse>, the queue of pending messages is cleared, as it makes little sense to continue. On the other hand, displays have no internal notion of the prover state, but need to keep track of the displayed text and its state.

The broker sends parsing requests to the prover, and extracts the new commands from the answer, checking that the parsing result returned by the prover satisfies the invariant that when we strip the markup, we get back the original proof script; if the result fails this invariant, it inserts the dropped text. As long as only white spaces are dropped, this does not affect the proof.

Particular attention needs be paid to the ability to *interrupt* a running prover. When a prover diverges, it may not respond to messages anymore (including the PGIP <interruptprover> message), so when running a prover as subprocess, we send a Posix signal instead. This is not possible over a socket, so to run a prover remotely, the broker connects to another instance of itself on the remote machine called a *proxy*, using the PGIP$_I$ inter-broker sub-protocol to communicate. This is also useful as broker and prover have to use the same filesystem.

The broker is implemented in Haskell (7k lines of code in 20 modules), using HaXml [25] for a well-typed embedding of the RELAX NG schema. This smoothly extends the schema typing into the Haskell implementation, making it impossible to send messages containing invalid XML.

5   Display Components

The display components provide the front-ends with which the user interacts. Currently, an Emacs display and an Eclipse plugin are available.

5.1   Emacs Proof General Revisited

The Emacs display for PG Kit will eventually replace the present Proof General. By moving complex functionality into the broker, the Elisp in Emacs can be greatly simplified. The Emacs display may be somewhat limited in facilities, but it has the advantage of portability, including functioning in a plain terminal.

Emacs has a built-in notion of text region which can have special properties attached, called "spans". Spans are used to directly capture the commands described by the broker. Emacs keeps a record of which spans have been altered, and automatically sends requests to the broker to re-parse them, either when the file is saved, or during editor idle time. Additionally each span provides a context sensitive menu to adjust its state according to the diagram in Fig. 6. Spans which are in the "being processed" state cannot be edited, and there is customisable protection against editing those which are in the "processed" state. Compared with the present Emacs interface, this now allows non-sequential dependencies within proof scripts, under control of the broker. However, the same toolbar and navigation metaphor for processing the next step is still available.

5.2   Eclipse Proof General

Eclipse [20,22] is an open-source IDE and tool integration platform written in Java. Most prominently it provides a powerful and attractive IDE for Java, but its plugins and extension points mechanism allows great adaptability. Many plugins are now available, supporting different programming languages, profiling and testing tools, graphical modelling, etc.

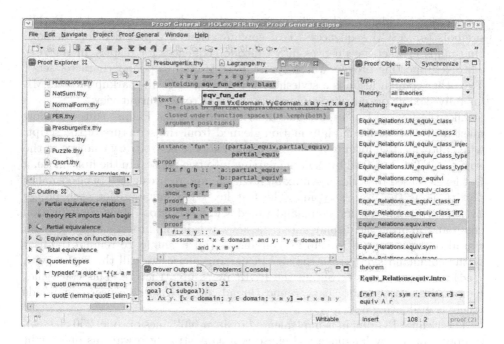

Fig. 8. Eclipse Proof General Display

Eclipse Proof General is a truly powerful IDE for formal proof which, we hope, will enable a dramatic improvement in usability and productivity for proof development. Graphical views are possible [23], but the primary mode of working remains the editing and scripting management of proof script files.

A screenshot in Fig. 8 shows it in action. The main editor window displays the proof script PER.thy; a tool-tip hover shows a definition under the mouse. The Prover Output view below shows the latest subgoal message. The Problems view (obscured) in the tab behind lists outstanding problems, such as syntax errors or unfinished proofs. To the left of the editor window is an Outline View of the proof script showing its structure; above that, the Proof Explorer shows proof scripts in the present folder and indicates their status in the prover with coloured decorators. The colouring metaphor (blue means completed, pink means busy) is used in both of these views as well as the editor window. Above the editor, the toolbar buttons trigger proof or undo steps by sending appropriate PGIP instructions. On the right hand panel, the Proof Objects view allows browsing the theories and theorems currently loaded in the running session. In the tab behind, the standard Synchronize view (obscured) allows synchronising the development with a version control system (e.g., CVS).

Further features include code folding to hide parts of the text (a sub-proof in the PER.thy file is folded in), integrated Javadoc-style help, and hyperlinked indexes for quick access to theorems, definitions and unfinished proofs. Completion is available for identifiers both found in proof script files and given in PGIP messages from the prover containing identifier tables. Completion also provides support for templates and mathematical symbols which are encoded by PGML symbols (or ASCII sequences). Two configurations are provided for symbols. One maps character sequences into Unicode sequences for display in the text editor (supporting provers whose syntax is restricted to poorer character sets). The other configuration is a stylesheet which maps PGML markup into HTML for fully-flexible output display used e.g. in the Prover Output view.

Like the revised Emacs interface described above, the Eclipse editor window must deal with managing information gleaned from the structure of the script, while allowing free form text edits — which can wreak arbitrary changes to the structure. This is solved by dividing parsing into two phases. In the first phase, a fast lexer is used to perform syntax highlighting and to break scripts into smaller partitions as the user is typing. The fast lexer is configured for each prover by a PGIP configuration command called <proverinfo>. This configuration command informs the display about the keywords in the prover's language, and can also provides tool-tip help for commands (for example, to remind the user of the command syntax). In the second parsing phase, we call the broker with <editcmd> messages to obtain the PGIP mark-up structure. This can either happen in a low-priority background thread, or with specific user commands (such as evaluating a script).

The Eclipse PG plugin is implemented in Java (40k lines of code, 250 classes). Support for a new language in Eclipse is not as straightforward as one might hope, as much of the advanced functionality is still specific to Java. But, paralleling our own development of PGIP, the Eclipse platform is rapidly evolving to migrate Java functionality to platform-level generic mechanisms.

5.3   Other Displays

A different kind of display is the lightweight "theorem proving desktop" providing a more abstract, less syntax-oriented interface based on direct manipulation and supported by the visual metaphor of a *notepad* [10]. All objects of interest, such as proofs, theorems, tactics, sets of rewriting rules, etc., are visualised by icons on the notepad, and manipulated using mouse gestures. The icon is given by the type of the object, which determines the available operations. PGIP supports this style of GUI with the <operationsconfig> specification, which describes prover types and operations as mentioned in Sect. 3.1, and can also include icons and hints for selecting operations. We have implemented a prototypical graphical display called PGWin for an earlier version of PGIP [3], where display commands and messages were not represented in XML. It is currently being adapted to the revised architecture, and made into a separate PGIP component.

Another display currently in development is a web-based display, which will allow users to connect to a running broker with a web-browser, and view the proof scripts as they are being developed. This is an example of a read-only display, which does not provide editing facilities.

6 Conclusions

The Proof General Kit is a framework for connecting interactive proof systems to interface tools and other components. This paper has provided an overview; elsewhere we provide full details including the RELAX NG schemas and protocol descriptions [4]. Ultimately, we hope that implementers of existing proof systems will have a compelling reason to add PGIP support to their systems to access powerful front-ends, and we hope that implementers of new systems will now have a clear model to follow to gain interface support with minimal effort.

At the time of writing, the broker component, the Emacs display and the Eclipse plugin are available as beta releases. These have been developed for the upcoming 2007 version of Isabelle, to which support for PGIP has been added by the first author. While straightforward in principle, supporting PGIP in Isabelle turned out to be harder than expected because of difficulties with parsing proof scripts independently of their execution: the Isabelle code uses functional combinators to build combined parse-execute functions that are hard to unravel. We expect that this will usually be easier to do in other systems.

PG Kit is unique in proposing a generic framework customised for interactive proof, although there is related work in different settings. Efforts to publish formalised mathematical content on the web include HELM [1] and MoWGLI [15]. The MathWeb project [12] provides a standardised interface using OMDoc [9] as an exchange language. OMDoc elaborates the semantical content of documents, which goes beyond the scope of PG Kit. Other systems such as MONET [14], the MathBroker [18] and MathServe [27] have an architecture similar to ours, but integrate fully automated provers (Otter, Spass etc.) wrapped up as web services, with a broker orchestrating proofs between different provers with little user interaction during the actual proof. In contrast, PG Kit is geared towards connecting interactive theorem provers to user interfaces.

Other frameworks in theorem proving include Prosper [7], which connects several automatic provers to an LCF core to ensure logical consistency. The Prosper Integration Interface (PII) is similar to the low-level aspects of PG Kit, in particular in the way in which interrupts to running components are routed.

Other interfaces similar in spirit to ours include Alcor [6], which extends the Mizar system [24] with knowledge management services such as searching and authoring assistance, and Plato [11], which uses TeXmacs as authoring tool for the Omega system [21]. The architecture is somewhat similar to ours, with a middleware component mediating between prover and interface. Unlike PG Kit, both Alcor and Plato are geared to a specific prover.

There are many possible lines for future development. Foremost among them, we want to use the framework to investigate foundations for *Proof Engineering*, exploring an analogy with software engineering to study useful ways to support the construction, maintenance and understanding of large proof developments. Analogues of code browsing, refactoring, and model driven development would all be intriguing to investigate. Because proofs (in practice) are quite different beasts from programs, and their development is a rather different process, this is a significant research programme.

Another promising direction lies in pushing the generic aspects of the framework, by providing extra language layers or enhancements which work for different systems. For example, we have already designed a generate literate style markup or a document-driven development methodology [5]. We can also use the broker to control proof construction and search: PGIP contains almost enough functionality to support a tactic language at a generic level.

We welcome contact from researchers interested in working with us on future directions or in connecting their systems to PG Kit. Please contact either of the first two authors directly, or visit the Proof General web pages [16] for more information and software downloads.

Acknowledgments. We would like to acknowledge contributions over the years made to the Proof General project by its many users and past developers; not just bug reports, but useful suggestions for improvements, some of which have influenced work described here. Contributors to the Eclipse front-end have included Graham Dutton, Ahsan Fayyaz and Alex Heneveld. The Isabelle developers, particularly Makarius Wenzel, have provided essential help with supporting PGIP. DA benefited from support of the TYPES project (Types for Proofs and Programs, EU IST-2004-510996) and EPSRC platform grant GR/S01771 (The Integration and Interaction of Multiple Mathematical Reasoning Processes). DW was supported by a 2004 Eclipse Innovation Grant awarded by IBM to work on Eclipse Proof General.

References

1. Asperti, A., Padovani, L., Coen, C.S., Schena, I.: HELM and the semantic math-web. In: Boulton, R.J., Jackson, P.B. (eds.) TPHOLs 2001. LNCS, vol. 2152, pp. 59–74. Springer, Heidelberg (2001)
2. Aspinall, D.: Proof General: A generic tool for proof development. In: Schwartzbach, M.I., Graf, S. (eds.) ETAPS 2000 and TACAS 2000. LNCS, vol. 1785, pp. 38–42. Springer, Heidelberg (2000)
3. Aspinall, D., Lüth, C.: Proof General meets IsaWin. In: Aspinall, D., Lüth,C. (eds.) User Interfaces for Theorem Provers UITP'03, Electronic Notes in Theoretical Computer Science, vol. 103 (2003)
4. Aspinall, D., Lüth, C.: Commentary on PGIP (2003-7) Available from http://proofgeneral.inf.ed.ac.uk/kit/
5. Aspinall, D., Lüth, C., Wolff, B.: Assisted proof document authoring. In: Kohlhase, M. (ed.) MKM 2005. LNCS (LNAI), vol. 3863, pp. 65–80. Springer, Heidelberg (2006)
6. Cairns, P., Gow, J.: Integrating searching and authoring in Mizar. To appear in Journal of Automated Reasoning
7. Dennis, L.A., Collins, G., Norrish, M., Boulton, R.J., Slind, K., Melham, T.F.: The PROSPER toolkit. International Journal on Software Tools for Technology Transfer 4(2), 189–210 (2003)
8. Dixon, L., Fleuriot, J.D.: Higher order rippling in IsaPlanner. In: Slind, K., Bunker, A., Gopalakrishnan, G.C. (eds.) TPHOLs 2004. LNCS, vol. 3223, pp. 83–98. Springer, Heidelberg (2004)

9. Kohlhase, M.: OMDoc – An Open Markup Format for Mathematical Documents [version 1.2]. LNCS (LNAI), vol. 4180. Springer, Heidelberg (2006)
10. Lüth, C., Wolff, B.: Functional design and implementation of graphical user interfaces for theorem provers. Journal of Functional Programming 9(2), 167–189 (1999)
11. Marc Wagner, C.B., Autexier, S.: PLATO: A mediator between text-editors and proof assistance systems. In: Benzmüller, C., Autexier, S. (eds.) 7th Workshop on User Interfaces for Theorem Provers (UITP'06) (August 2006)
12. Mathweb homepage. http://www.mathweb.org/
13. Mercer, A., Bundy, A., Duncan, H., Aspinall, D.: PG Tips, a recommender system for an interactive prover. Presented at MathUI workshop (2006)
14. MONET — mathematics on the web. Home page at http://monet.nag.co.uk/
15. MoWGLI.: mathematics on the web: Get it right by logics and interfaces. http://www.mowgli.cs.unibo.it/
16. Proof General Kit home page. http://proofgeneral.inf.ed.ac.uk/kit/
17. RELAX NG XML schema language (2003). Home page at http://www.relaxng.org/
18. Schreiner, W., Caprotti, O., Baraka, R.: The MathBroker project Johannes-Kepler-Universität Linz (2002) http://www.risc.uni-linz.ac.at/projects/basic/mathbroker/
19. Seger, C.-J.H., Jones, R.B., O'Leary, J.W., Melham, T., Aagaard, M.D., Barrett, C., Syme, D.: An industrially effective environment for formal hardware verification. IEEE Transactions on Computer-Aided Design of Integrated Circuits and Systems 24(9), 1381–1405 (2005)
20. Shavor, S., D'Anjou, J., Fairborther, S., Kehn, D., Kellerman, J., McCarthy, P.: The Java Developer's Guide to Eclipse. Addison-Wesley, New York (2003)
21. Siekmann, J., Benzmüller, C., Autexier, S.: Computer supported mathematics with OMEGA. Journal of Applied Logic, special issue on Mathematics Assistance Systems 4(4) (December 2006)
22. The Eclipse Foundation. Project web site. http://www.eclipse.org
23. Timiriassova, E.: Tracking and visualizing dependency information within theories. Master's thesis, School of Informatics, University of Edinburgh (2005)
24. Trybulec, A., et al.: The Mizar project University of Bialystok, Poland (1973) http://mizar.org
25. Wallace, M., Runciman, C.: Haskell and XML: Generic combinators or type-based translation? In: International Conference on Functional Programming ICFP'99, pp. 148–159. ACM Press, New York (1999)
26. Wiedijk, F.: Digital math: systems implementing mathematics in the computer, http://www.cs.ru.nl/~freek/digimath/
27. Zimmer, J., Autexier, S.: The MathServe system for semantic web reasoning services. In: Furbach, U., Shankar, N. (eds.) IJCAR 2006. LNCS (LNAI), vol. 4130, pp. 140–144. Springer, Heidelberg (2006)

Supporting User-Defined Notations When Integrating Scientific Text-Editors with Proof Assistance Systems

Serge Autexier[1,2], Armin Fiedler[2], Thomas Neumann[2], and Marc Wagner[2]

[1] German Research Center for Artificial Intelligence (DFKI GmbH), Saarbrücken, Germany,
autexier@dfki.de
[2] FR 6.2 Informatik, Saarland University, Saarbrücken, Germany
{autexier,fiedler,tneumann,wagner}@ags.uni-sb.de

Abstract. In order to foster the use of proof assistance systems, we integrated the proof assistance system ΩMEGA with the standard scientific text-editor $T_{E}X_{MACS}$. We aim at a document-centric approach to formalizing and verifying mathematics and software. Assisted by the proof assistance system, the author writes her document entirely inside the text-editor in a language she is used to, that is a mixture of natural language and formulas in LATEX style. We present a basic mechanism that allows the author to define her own notation inside a document in a natural way, and use it to parse the formulas written by the author as well as to render the formulas generated by the proof assistance system. To make this mechanism effectively usable in an interactive and dynamic authoring environment, we extend it to efficiently accommodate modifications of notations, to track dependencies to ensure the right order of notations and formulas, to use the hierarchical structure of theories to prevent ambiguities, and to reuse concepts together with their notation from other documents.

1 Introduction

The vision of a powerful mathematical assistance environment that provides computer-based support for most tasks of a mathematician has stimulated new projects and international research networks in recent years across disciplinary boundaries. Even though the functionalities and strengths of proof assistance systems are generally not sufficiently developed to attract mathematicians on the edge of research, their capabilities are often sufficient for applications in e-learning and engineering contexts. However, a mathematical assistance system that shall be of effective support has to be highly user oriented. We believe that such a system will only be widely accepted by users if the communication between human and machine satisfies their needs, in particular only if the extra time spent on the machine is by far compensated by the system support. One aspect of the user-friendliness is to integrate formal modeling and reasoning tools with software that users routinely employ for typical tasks in order to promote the use of formal logic based techniques.

One standard activity in mathematics and areas that are based on mathematics is the preparation of documents using some standard text preparation system like LATEX. $T_{E}X_{MACS}$ [10] is a scientific text-editor in the WYSIWYG paradigm that provides professional type-setting and supports authoring with powerful macro definition facilities

M. Kauers et al. (Eds.): MKM/Calculemus 2007, LNAI 4573, pp. 176–190, 2007.
© Springer-Verlag Berlin Heidelberg 2007

like those in LATEX. As a first step towards assisting the authoring of mathematical documents, we integrated the proof assistance system ΩMEGA into TEX_MACS using the generic mediator PLATΩ [12]. In this setting the formal content of a document must be amenable to machine processing, without imposing any restrictions on how the document is structured, on the language used in the document, or on the way the document can be changed. The PLATΩ system [11] transforms the representation of the formal content of a document into the representation used in a proof assistance system and maintains the consistency between both representations throughout the changes made on either side.

Such an integrated authoring environment should allow the user to write her mathematical documents in the language she is used to, that is a mixture of natural language and formulas in LATEX style with her own notation. To understand the meaning of the natural language parts in a mathematical document we currently rely on annotations for the document structure that must be provided manually by the user. Although it might still be acceptable for an author to indicate the macro-structures like theories, definitions and theorems, writing annotated formulas (e.g. "\F{in}{\V{x},\F{cup}{\V{A}, \V{B}}}" instead of "x \in A \cup B") is definitely not. Aiming at a document-centric approach to formalizing mathematics, we present a mechanism that allows authors to define their own notation and to use it when writing formulas within the same document. Furthermore, this mechanism enables the proof assistance system to access the formal content and use the same notation when presenting formulas to the author.

The paper is organized as follows: Section 2 presents the annotation language for documents of the PLATΩ system, in particular for formulas. Inspired by notational definitions in text-books, we then present the means the author should have to define notations. The goal consists of starting from such notations to obtain an *abstraction* parser that allows to read formulas using that notation and also a corresponding *rendering* parser to render formulas generated by the proof assistance system. Section 3 describes how the notational definitions can automatically be transformed into grammar rules defining the *abstraction* and *rendering* parsers, which are created by a parser generator that allows to integrate arbitrary disambiguators. Section 4 presents a basic mechanism how to accommodate efficiently modifications of the notations. In Section 5 we extend the basic framework to restrain ambiguities, allow for the redefinition of notations and use notations defined in other documents. We discuss related works in Section 6 before concluding in Section 7.

Presentational convention: The work presented in this paper has been realized in TEX_MACS. Although the TEX_MACS markup-language is analogous to LATEX-macros, one needs to get used to it: For instance a macro application like \frac{A}{B} in LATEX becomes <frac|A|B> in TEX_MACS-markup. Assuming that most readers are more familiar with LATEX than with TEX_MACS, we will use a LATEX-syntax for sake of readability.

2   Towards Dynamic Notation

The PLATΩ system supports users to interact with a proof assistance system from inside the text-editor TEX_MACS by offering service menus and by propagating changes of the document to the system and vice versa. Mediating between a text-editor and a

Table 1. Grammar of PLATΩ's *Formula Annotation Language*

Element	Arguments
\F	{*name*}{\B?, (\F |\V |\S)⋆ }
\B	{\V+ }
\V	{*name*, (\T |\TX |\TF)?}
\S	{*name*}
\T	{*name*}
\TX	{(\T |\TX |\TF), (\T |\TX |\TF) }
\TF	{(\T |\TX |\TF), (\T |\TX |\TF) }

proof assistance system requires to extract the formal content of a document, which is already a challenge in itself if one wants to allow the author to write in natural language without any restrictions. Therefore we currently use a semantic annotation language to semantically annotate different parts of a document. The annotations can be nested and subdivide the text into dependent theories that contain definitions, axioms, theorems and proofs, which themselves consist of proof steps like for instance subgoal introduction, assumption or case split. The annotations are a set of macros predefined in a T_EX_{MACS} style file and must be provided manually by the author (see [11] for details). We were particularly cautious that adding the annotations to a text does not impose any restrictions to the author about how to structure her text. Note that for the communication with the proof assistance system, also the formulas must be written in a fully annotated format whose grammar is shown in Tab. 1.

\F{name}{args} represents the application of the function name to the given arguments args. \B{vars} specifies the variables vars that are bound by a quantifier. A variable name is denoted by \V{name} and may be optionally typed by \V{name,type}. A type name is represented by \T{name}. Complex types are composed using the function type constructor → represented by \TF{type1,type2}, or the operator × represented by \TX{type1,type2} as syntactic sugar for argument types. Finally, a symbol name is denoted by \S{name}. For instance, the formula $x \in A \cap (B \cup C)$ is represented by the fully annotated form \F{in}{\V{x}, \F{cap}{\V{A},F{cup}{\V{B}, V{C}}}} and the quantified formula $\forall x.\ x = x$ as \F{forall}{\B{\V{x}}, \F{=} {\V{x}, \V{x}}}. In many cases type reconstruction allows to dertermine the type of a variable, and therefore typing variables is optional in our system.

Currently the macro-structures like theories, definitions, theorems, and proof steps must be annotated manually by the user. However, it is not acceptable to require to write formulas in fully annotated form. This motivates the need for an *abstraction* parser that converts formulas in LATEX syntax into their fully annotated form. Furthermore, we also need a *rendering* parser to convert fully annotated formulas obtained from the proof assistance system into LATEX-formulas and using the user-defined notation. In the future, we plan on the one hand to combine our approach with techniques in the tradition of using a *mathematical vernacular* (e.g. MATHLANG [4]) and on the other hand to use natural language analysis techniques for the semi-automatic annotation of the document structure, e.g. to automatically detect macro-structures.

The usual software engineering approach would be to write grammars for both directions and integrate the generated parsers into the system. Of course, this method is highly efficient but the major drawback is obvious: the user has to maintain the grammar files together with her documents. In our document-centric philosophy, the only source of knowledge for the mediator and the proof assistance system should be the document in the text-editor.

Therefore, instead of maintaining special grammar files for the parser, the idea of dynamic notation is to start from basic abstraction and rendering grammars for types and formulas, where only the base type *bool*, the complex type constructors $\rightarrow, \times$ and the logic operators $\forall, \exists, \lambda, \top, \bot, \wedge, \vee, \neg, \Rightarrow, \Leftrightarrow$ are predefined. Based on that initial grammar the definitions and notations occurring in the document are analyzed in order to extend incrementally both grammars for dealing with new symbols, types and operators. The scope of a notation should thereby respect the visibility of its defining symbol or type, i.e. the transitive closure of dependent theories. Finally, all formulas are parsed using their theory specific *abstraction* parser.

Notations defined by authors are typically not specified as grammar rules. Therefore, we first need a user friendly WYSIWYG method to define notations and to automatically generate grammar rules from it. Looking at standard mathematical textbooks, one observes sentences like *"Let x be an element and A be a set, then we write $x \in A$, x is element of A, x is in A or A contains x."*. Supporting this format requires the ability to locally introduce the variables x and A in order to generate grammar rules from a notation pattern like $x \in A$. Without using a linguistic database, patterns like *x is in A* are only supported as pseudo natural language. Beside that, the author should be able to declare a symbol to be right- or left-associative as well as precedences of symbols.

We introduce the following annotation format to define the operator $\in$ and to introduce multiple alternative notations for $\in$ as closely as possible to the textbook style.

```
\begin{definition}{Predicate $\in$}
  The predicate \concept{\in}{elem \times set \rightarrow bool}
  takes an individual and a set and tells whether that
  individual belongs to this set.
\end{definition}
```

A definition may introduce a new type by \type{name} or a new typed symbol by \concept{name}{type}. We allow to group symbols to simplify the definition of precedences and associativity. By writing \group{name} inside the definition of a symbol, this particular symbol is added to the group name which is automatically created if it does not exist. Any new concept is first introduced as a prefix symbol. This can be changed by declaring concept specific notations.

```
\begin{notation}{Predicate $\in$}
  Let \declare{x} be an individual and \declare{A} a set,
  then we write \denote{x \in A}, \denote{x is element of A},
  \denote{x is in A} or \denote{A contains x}.
\end{notation}
```

A notation may contain some variables declared by \declare{name} as well as the patterns written as \denote{pattern}. Furthermore, by writing \left{name} or \right{name} inside the notation one can specify a symbol or group of symbols to be left or right associative. Finally, precedences between symbols or groups are defined by \prec{name1,...,nameN}, which partially orders the precedence of these symbols

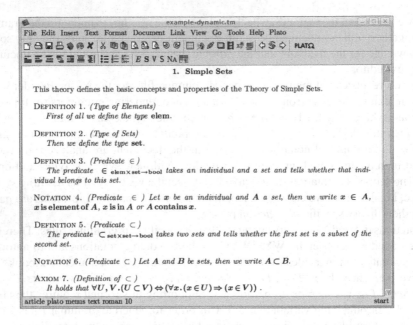

Fig. 1. A nnotated T_EX_{MACS} document with dynamic notation in text mode

and groups of symbols from low to high. Please note that a notation is related to a specific definition by refering its name, in our example `Predicate $\in$`.

Fig. 1 shows how the above example definition and notation appear in a T_EX_{MACS} document. Using a keyboard shortcut the author can easily switch into a so-called "box-mode" that visualizes the semantic annotations contained in the document (cf. Fig. 2, p.181). The author is free to develop the document in either view.

3   Creating Parsers from User-Defined Notations

We first present how the grammar rules for *abstraction* and *rendering* are obtained from a document and then briefly describe the parser generator that is a slight variant of standard implementations.

3.1   Obtaining the Grammar Rules

Starting the processing of a semantically annotated document, as for example the document shown in Fig. 1, all surrounding natural language parts in the document are removed and the *abstraction* and *rendering* parsers and scanners are initialized with the initial grammars for types and formulas. The grammar rules can be divided into rules for types, symbols and operator applications. The syntax of the rules is

```
NONTERMINAL ::= (TERMINAL|NONTERMINAL)+ --> PRODUCTION
```

and is best explained using an example rule:

```
APPLICATION ::= FORMULA.1 "\wedge" FORMULA.2
            --> \F{and}{FORMULA.1, FORMULA.2}
```

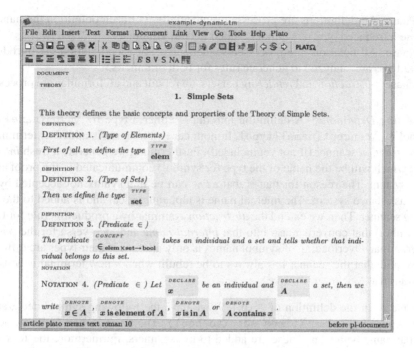

Fig. 2. Annotated T_EX_{MACS} document with dynamic notation in box mode

This is a rule for the non-terminal symbol APPLICATION used for any kind of application of an operator. If it could successfully recognize two chunks of text in the class of FORMULA and that are separated by "\wedge", it substitutes the obtained results for FORMULA.1 and FORMULA.2 in \F{and}{FORMULA.1, FORMULA.2} to create the parsing result.

The goal of processing the document is to produce a set of grammar rules for the respective grammars from the notational definitions given in the text. A top-down approach, that processes each definition or notation on its own, extending the grammars and recompiling the parsers before processing the next element, is far too inefficient for real time usage due to the expensive parser generation process[1]. The following procedure tries to minimize the amount of parser generations as much as possible.

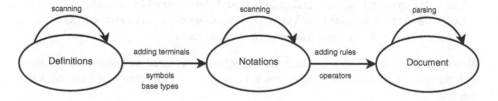

---

[1] The parser generator is currently implemented in Scheme to be part of T_EX_{MACS} and is not compiled. This causes, for instance, the parser generation for the example document to take $\approx 1 min$.

1. Phase: All definitions are processed sequentially. For each definition the name of the introduced type or symbol is added to both grammars.
2. Phase: All notations are processed sequentially. For each notation the introduced patterns are analyzed to generate rules for both grammars.
3. Phase: *Abstraction* and *rendering* parsers are rebuilt and all formulas are processed.

Processing Definitions: A definition introduces either a type by \type{name} or a symbol by \concept{name}{type}. In both cases, the name is added as terminal to the *abstraction* scanner (if not yet included). Furthermore, we generate a fresh *internal name* which will be the name of the type (or symbol) communicated to the proof assistance system. The reason for that is that name can be in a syntax not accepted by the proof assistance system. The internal name is alphanumerical and is added to the *rendering* scanner. Then we extend the *abstraction* grammar by a production rule for types (or symbol) that converts name into that *internal name* and vice-versa for the *rendering* grammar. Overloading of symbol names is only allowed if their types are different. Please note that the scanner has always to be rebuilt when a new terminal is added to the grammar.

Example 1. In the definition of ∈ (p.179) we have the following symbol declaration: \concept{\in}{elem \times set \to bool}. The name is \in and assume the internal name being in. These are added to the scanners. Furthermore the following rules are added to the *abstraction* and *rendering* grammars respectively:

- SYMBOL ::= "\in" --> "in" is added to the *abstraction* grammar
- SYMBOL ::= "in" --> "\in" is added to the *rendering* grammar

The type information in the symbol declaration is only processed in the third phase because we first have to collect all type declarations. This also complies with future extensions towards dependent types. Our system currently supports only simple types.

Processing Notations: A notation defines one or more alternative notations for some symbol. The author can introduce local variables by \declare{x1},...,\declare{xN} and use them in the patterns defining the different notations: \denote{pattern1},..., \denote{patternM}.

Example 2. As an example consider our running example from p. 179:

Let \declare{x} be an individual and \declare{A} a set, then we write \denote{x \in A}, \denote{x is element of A}, \denote{x is in A} or \denote{A contains x}.

We impose that the ordering in which the variables are declared by \declare complies with the domain of the associated operator, i.e. x is the first argument of \in and A the second.

First of all, the *abstraction* scanner is locally extended by the terminals for the local variables x1,..., xn. Then each notation pattern is tokenized by the scanner, that returns a list of terminals including new terminals for unrecognized chunks. For instance, in our example above the scanner knows the terminals for the local variables x and A when tokenizing "x is element of A". The unknown chunks are is, element and of, that are added on the fly. This behavior of the scanner is non-standard, but is an essential feature to efficiently accommodate new notations. More details about the scanner are presented at the end of this section.

A notation pattern is only accepted if all declared argument variables are recognized by the scanner, namely all x1, ..., xn occur in the pattern. For every notation pattern, the *abstraction* grammar is extended by a production rule for function application that converts the notation *pattern* into the semantic function application with respect to the argument ordering. To this end we must modify the pattern by replacing the occurrences of the local variables by the non-terminals FORMULA.1 ... FORMULA.N and add the *abstraction* grammar rule

```
APPLICATION ::= pattern1' --> \F{f}{FORMULA.1, ..., FORMULA.N}
```

where pattern1' is the modified pattern.

For instance, the above pattern [x "is" "element" "of" A] is transformed into [FORMULA.1 "is" "element" "of" FORMULA.2] and we obtain the following grammar rule:

```
APPLICATION ::= FORMULA.1 "is" "element" "of" FORMULA.2
                --> \F{f}{FORMULA.1, FORMULA.2}
```

For the *rendering* grammar we have the choice which pattern to use to render the terms. We currently just take the first possibility. The *rendering* grammar is then extended by the production rule

```
APPLICATION ::= \F{f}{FORMULA.1, FORMULA.2} --> pattern1'
```

For our running example we get the *rendering* grammar rule

```
APPLICATION ::= \F{f}{FORMULA.1, FORMULA.2}
                --> FORMULA.1 "is" "element" "of" FORMULA.2
```

Note that the notation pattern can permute the arguments of the symbol, as would be the case when using the pattern [A contains x]. All *abstraction* grammar rules of a symbol are grouped together in order to support dependency tracking.

Additionally the author can define the symbol name to be left- or right-associative by \left{name} or \right{name} as well as the precedence of operators by using \prec{name1},...{nameN} where namei is the symbol name given in the definitions. This declares the relative precedence of these symbols, where namei is lower than name(i+1).

Processing Formulas: The generated *abstraction* parser is finally used to parse the type information in symbol declarations and all formulas. The parser returns the fully annotated version of the types and formulas and also the list of grammar rules used. These rules are stored along with the type or formula and can serve to detect dependencies between definitions, notations and formulas. When the *abstraction* parsing process returns more than one possible reading, the author must advise which possibility is retained. How these situations can be reduced to a minimum is discussed in the next Section and in Section 5.

The *rendering* parser is used to convert fully annotated formulas generated by the proof assistance system into LATEX formulas for the text-editor. Again the grammar rules used are saved to allow for tracking the dependencies.

The Modified Scanner: The presented procedure to analyze the patterns in notational definitions makes extensive use of a modified scanner algorithm that returns all tokens including unknown chunks. The scanner has been implemented such that it guarantees the tokenizing of the longest possible prefixes. A hard wired scanner couldn't be used because the alphabet of the language is unknown. By using the standard scanner generation algorithm described in [14] a deterministic finite automaton is directly generated out of the given grammar without generating a non-deterministic finite automaton. Furthermore, the scanner can be built and used independently of the parser. Due to the small size of the automaton, the generation of a scanner is relatively fast with respect to grammar extensions.

3.2   Generating the Abstraction and Rendering Parsers

In this section we describe the parser generator used to create the *abstraction* and *rendering* parser from the collected grammar rules. The main features are that it returns all possible readings that are due to ambiguities in the grammar and returns for each reading the list of grammar rules used. In order to eliminate possible but incorrect readings as early as possible, an external function can be specified that is called at runtime to eliminate ambiguities.

The LALR parsers are generated using well-known algorithms [13,14] together with the usual action- and goto-tables. Since we don't restrict our input grammar too much, i.e. we allow also non-LALR grammars, we have to disambiguate the input grammar and provide a handling for non resolvable ambiguities. The disambiguation of the grammar is performed using standard methods from the BISON system[2]. In case there are still ambiguities remaining in the grammar, the parser allows to define an external callback function that is used at runtime to rule out possible readings that result from ambiguities. Thus it is possible to integrate a so-called "refiner" [3], which uses type reconstruction to filter the well-typed readings from all alternatives. Furthermore, since we are in an interactive setting, we could ask the user to resolve the remaining ambiguities in contrast to situations where there is no possibility of user feedback. However, that has still to be implemented in the PLATΩ system. If no external callback is defined, the parser splits itself by default into multiple subparsers, such that all possible readings are returned at the end. It has to be mentioned that the runtime of a splitted parser increases exponentially if the ambiguities are not completely removed. Since the formulas that need to be parsed in practice are usually not too large, we don't think this really poses a problem.

4   Management of Change for Notations

The *abstraction* parser constructed so far is for one version of the document. When the author continues to edit the document, it may be modified in arbitrary ways, including

---

[2] http://www.gnu.org/software/bison/

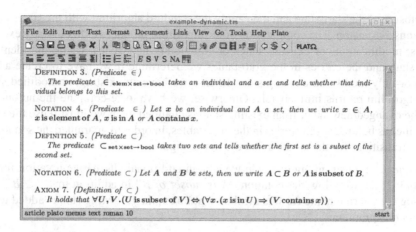

Fig. 3. Modified T_EX_{MACS} document with dynamic notation in text mode

the change of existing definitions and notations. Before the modified semantic content of the document is uploaded into the proof assistance system, we need to recompute the parsers and parse the formulas in the document. Always starting from scratch following the procedure described in the previous sections is not efficient and may jeopardize the acceptance by the author of the system if that process takes too long. Therefore there is a need for management of change for the notational parts of a document and those parts that depend on them. The management of change task has two aspects:

1. First, we must determine any modifications in the notational parts.
2. Second, we must adjust only those parts of the grammar that are affected by the determined modifications, adjust the parsers accordingly and then re-parse the formulas of the document.

Determining changes: Using the procedure from Section 3, we re-process all definitions and notations of the document and obtain a new set of grammar rules for each defined symbol. By caching these sets of rules for each symbols, we can determine how the grammar has changed using a differencing mechanism.

Adjusting the scanner and the parser: If the modification of the grammar is non-monotonic, i.e. some rules have been removed or changed, we currently have to re-compute the whole scanner and the parser from scratch using the procedure from Section 3.2. This is for instance the case if we change a notation for some symbol, e.g. if we replaced "$A \subset B$" by "$B \supset A$", but not if we add an additional alternative notation for a symbol, like allowing "A *is subset of* B" in addition to "$A \subset B$" as shown in Fig. 3.

If the grammar is simply extended, we can optimize the creation of the new parser (recreating the scanner is fast anyway and there is no need to optimize that). In this case we can reuse our previous parser and extend it using a variant of the standard parser generation algorithm (cf. [13], p. 138ff.): Aside from the action- and goto-tables on which the parser operates, we also have access to the states of the automaton. Storing these data is expensive but enables the extension of the automaton. We first compute the closure of the start-state and then apply the standard algorithm with the states of the previous automaton. The computation of the algorithm produces either already existing states, modifications of existing states or completely new states. Using the identifier of a state and the entries in the goto-tables we determine if we have to create a new state or only modify an existing state. In case we have an existing, unmodified state, the algorithm returns immediately. Otherwise, we have to re-compute the automaton for the changed/new state. If an existing state is changed, then we must reuse the same identifier as before for the entries in the goto-tables, in order to guarantee the soundness of the transitions for those states that did not change.

Example 3. In our running example we add an additional alternative notation for the symbol $\subset$, i.e. we allow the notation *"A is subset of B"* in addition to *"A $\subset$ B"*. Assuming the internal name for this symbol is `subset`, the following rule is added to the *abstraction* grammar:

```
APPLICATION ::= FORMULA.1 ''is'' ''subset'' ''of'' FORMULA.2
            --> \F{subset}{FORMULA.1, FORMULA.2}
```

Re-parsing and re-rendering of formulas: Once the parser has been adjusted we need to re-parse those formulas the author has written or changed manually, e.g. the formula $\forall U, V. (U$ *is subset of* $V) \Leftrightarrow (\forall x. (x$ *is in* $U) \Rightarrow (V$ *contains* $x))$ in the axiom of Fig. 3. Furthermore, we have to re-render those formulas that were generated by the proof assistance system. To this end we store the following information on formulas in the document: for each formula we have a flag indicating if it was generated by the proof assistance system, the corresponding fully annotated formula, and the set of grammar rules that were used for parsing or rendering that formula:

- If the formula was written by the author, the associated fully annotated formula and the grammar rules are the result of the *abstraction* parser.
- If a formula was generated by the proof assistance system, that formula is the result of *rendering* the fully annotated formula obtained from the proof assistance system. The stored grammar rules are those returned by the *rendering* parser.

Note that we do not prevent the author to edit a generated formula. As soon as the author edits such a formula, the flag attached to the formula is toggled to "user" and the cached fully annotated version and grammar rules are replaced during the next *abstraction* parsing of the formula.

The stored information is used to optimize the next parsing or rendering pass over the document: A formula is only parsed from scratch if at least one of the grammar rules used has been modified or deleted, or if either the user or the proof assistance system has changed the formula or the fully annotated formula.

This provides the basic mechanism that allows the author to define in a document her own notation that is used to extract the formal semantics, and that efficiently deals with modifications of the notation.

5   Ambiguities, Dependencies and Libraries

In order to enable a document-centric approach for formalizing mathematics and software, the added-values offered by the authoring environment must outweigh the additional burden imposed to the author compared to the amount of work for a non-assisted preparation of a document. In the following we present techniques to reduce the burden for the author by exploiting the theory structure contained in a document to reduce the ambiguities the author would have to deal with and also support the redefinition of notations. With respect to added-values, we provide checks if a notation is used before it has been introduced in a document and, most importantly, how we support an author to build on formalizations contained in other documents.

Ambiguities: The *abstraction* parser returns all possible readings and during the parsing process can try to make use of type-checking provided by the proof assistance system to eliminate possible readings that are not type-correct. It requires to provide type-information that depends on the context in which a formula is parsed, but even then it will require some amount of automated type-reconstruction. Therefore, there may always be situations in which we obtain more than one parsing result which requires the user to inspect the different possibilities and select the right one by menu interaction in the text-editor.

The logical context of a formula is determined by the theory it occurs in: The different parts of a document must be assigned to specific theories. New theories can be defined inside a document and build on top of other theories. The notion of theory is that of OMDOC [5] respectively development graphs [6]. One way to resolve an arising ambiguity is to provide the context of a formula to the *refiner*. The other possibility consists of having different parsers for different theories, hence avoiding some ambiguities that would arise when sticking to have a single parser.

Example 4. Consider a theory of the integers with multiplication with the notation "$x \times y$" and a completely unrelated theory about sets and Cartesian products with the same notation. This typically is a source of ambiguities that would require the use of type information to resolve the issue. Note that standard parser generators would not support the definition of two grammar rules that have the same pattern but different productions as would be necessary in this case. Using different parsers for different theories completely avoids that problem. However, when theories with overloaded notations are imported by another theory, e.g. a theory of Cartesian products of sets of integers, ambiguities may only be resolved using type information or user interaction.

From these observations we decided to not have a single parser for a whole document, but to exploit the theory structure contained in the document and allow for one parser for each theory. This entails that if we have a theory T that is included into two independent theories T_1 and T_2, then there is a parser for all three of them. Note that we could only have parsers for T_1 and T_2 and use either one when parsing a formula in the theory T. On the one hand, this would be more efficient for changes because we need only to maintain the parsers for T_1 and T_2, but on the other hand this would prevent the user to redefine in the context of say T_1 the notation for some symbol inherited from T, which is something quite common (and discussed in the next paragraph). Nevertheless, the drawback of our approach is that if some notation inside T is changed or added, all three dependent parsers must be adjusted rather than only two.

In that structured theory approach for parsers, the grammar of a parser for some theory consists of all rules obtained from notations for each symbol that is imported in that theory. The management of change mechanism from Section 4 is adapted in a straight-forward manner.

Redefining Notations: When importing a theory, we want to reuse the formal content, but possibly adapt the notation used to write formulas. This occurs less frequently inside a single document, but occurs very often when using a theory formalized in a different document. Since we linked the parsing (and hence the rendering) to the individual theories, we allow to redefine notations for symbols inherited from other theories. The grammar rules for a parser are determined by including for each imported symbol that notation that is closest in the import hierarchy of theories. If there are two such theories[3], a conflict is raised and the author aksed for advise which notation to use.

Dependencies: A parser and the associated renderer are attached to a theory and each position in the document belongs to a theory. Therefore, it is possible that within a specific theory, a formula uses the notation of some symbol although the definition of that notation only occurs afterwards in the document. We notice such situations by comparing the position in the document where grammar rules are defined and where they have been used to parse a formula. If we determine such a situation, we notify the author. The same problem can occur when rendering a formula: If the proof assistance system generates a formula with some concept c at a position that precedes the definition of the notation for c (but still in the same theory), the renderer uses the grammar rules before they are actually defined in the document. Sometimes the proof assistance system has no choice about where that formula is included: for instance if some proof steps are inserted into an existing partial proof that occurs before the definition of the notation. In other situations, for instance if the proof assistance system has used an automated theory exploration system[4] to derive new properties, we could try to determine an appropriate insertion position for these lemmas by inspecting the grammar rules used for *rendering*. However, our impression is that most authors would be upset if their document is rearranged automatically. Therefore we prefer to leave it to the user to move the text parts including surrounding descriptions into the appropriate places.

Note that another, much simpler dependency is that between the definition of a concept and the definition of its notation, that should not occur before the definition in the text. In that case we also simply notify the author.

Libraries: A library mechanism is the key prerequisite to support the development of large structured theories. We carry over that concept to the document-centric approach we are aiming at by extending the citation mechanism that is commonly used in documents. PLATΩ provides a macro to cite a document *semantically*, i.e. it will not only be included in the normal bibliography of the document, but the formalized content of the document is included. Currently, the document must be present in the filc system and is included when the macro is evaluated and the process is recursive. Aside from the extracted formalizations that are sent to the proof assistance system to setup the background for the current document, we extract the notations contained in that document. Importing a theory defined in a semantically cited document into a theory of our current

---

[3] The theories can form an acyclic graph which may lead to a Nixon diamond scenario when determining grammar rules.

[4] For instance, MATHSAID [9] is connected to ΩMEGA.

document, allows to determine the set of grammar rules using the same mechanism as described above for structured theories. Furthermore, since we allow the redefinition of notations, the author can redefine the notation for the concepts imported from cited documents, in order to adapt it to her preferences.

6 Related Work

Supporting specific mathematical notations is a major concern in all proof assistance systems. Wrt. to supporting the definition of new notations that are used for type-setting, the systems ISABELLE [7] and MATITA [2] are closest. ISABELLE comes with type-setting facilities of formulas and proofs for LATEX and supports the declaration of the notation for symbols as prefix, infix, postfix and mixfix. Furthermore, it allows the definition of *translations* which are close to our style of defining notations. The main differences are: the notations are not defined in the LATEX document but have to be provided in the input files of ISABELLE. Due to the batch processing paradigm of IS-ABELLE, there are no mechanisms to efficiently deal with modifications of the notation, which is crucial in our interactive authoring environment.

In the context of MATITA Padovani and Zacchiroli also proposed a mechanism of *abstraction* and *rendering* parsers [8] that are created from notational equations which are comparable to the grammar rules we generate from the notational definitions. Their mechanism is mainly devoted to obtain MathML representations [1] where a major concern also is to maintain links between the objects in MathML to the internal objects. Similar to ISABELLE, the notations must be provided in input files of MATITA that are separate from the actual document. Also, they do not consider the effect of changing the notations and to efficiently adjust the parsers.

7 Conclusion

In order to enable a document-centric approach for formalizing mathematics and software, the added-values offered in an assisted authoring environment must outweigh the additional burden imposed to the author compared to the amount of work for a non-assisted preparation of a document. One step in that direction is to give the freedom to define and use her own notation inside a document back to the author. In this paper we presented a mechanism that enables the author to define her own notation in a natural way in the text-editor TEX$_{MACS}$ while being able to get support from the proof assistance system, such as type checking, proof checking, interactive and automatic proving, and automatic theory exploration. The notations are used to parse formulas written by the user in the LATEX-style she is used to, as well as to render the formulas produced by the proof assistance system. Ambiguities are reduced by allowing one parser for each defined theory in the document and by integrating type-checking to resolve remaining ambiguities during the parsing process. The structure of theories also form the basis to include formalizations and notations defined in other documents. Finally, we developed maintenance techniques to accommodate the interactive and dynamic process of preparing a document by simultaneously reducing the amount of work for the user.

Future work will consist of supporting LaTeX-documents and using OMDoc to exchange the formalized content and notations contained in documents of different formats.

References

1. Mathematical Markup Language (MathML) Version 2.0. W3c recommendation 21 february 2001. Technical report (2003) http://www.w3.org/TR/MathML2
2. Asperti, A., Sacerdoti-Coen, C., Tassi, E., Zacchiroli, S.: User interaction with the matita proof assistant. Journal of Automated Reasoning, Special Issue on User Interfaces for Theorem Proving (2007)
3. Coen, C.S., Zacchiroli, S.: Efficient ambiguous parsing of mathematical formulae. In: Asperti, A., Bancerek, G., Trybulec, A. (eds.) MKM 2004. LNCS, vol. 3119, pp. 347–362. Springer, Heidelberg (2004)
4. Kamareddine, F., Maarek, M., Wells, J.B.: Toward an object-oriented structure for mathematical text. In: Kohlhase, M. (ed.) MKM 2005. LNCS (LNAI), vol. 3863, pp. 217–233. Springer, Heidelberg (2006)
5. Kohlhase, M.: OMDoc – An Open Markup Format for Mathematical Documents [version 1.2]. LNCS (LNAI), vol. 4180. Springer, Heidelberg (2006)
6. Mossakowski, T., Autexier, S., Hutter, D.: Development graphs - proof management for structured specifications. Journal of Logic and Algebraic Programming, special issue on Algebraic Specification and Development Techniques 67(1-2), 114–145 (2006)
7. Nipkow, T., Paulson, L.C., Wenzel, M.: Isabelle/HOL. LNCS, vol. 2283. Springer, Heidelberg (2002)
8. Padovani, L., Zacchiroli, S.: From notation to semantics: There and back again! In: Borwein, J.M., Farmer, W.M. (eds.) MKM 2006. LNCS (LNAI), vol. 4108, Springer, Heidelberg (2006)
9. Bundy, A., McCasland, R.: Mathsaid: A mathematical theorem discovery tool. In: Proceedings of SYNASC 2006 (2006)
10. van der Hoeven, J.: Gnu T_EX_{MACS}: A free, structured, wysiwyg and technical text editor. Number 39-40 in Cahiers GUTenberg (May 2001)
11. Wagner, M.: Mediation between text-editors and proof assistance systems. Diploma thesis, Saarland University, Saarbrücken, Germany (2006)
12. Wagner, M., Autexier, S., Benzmüller, C.: PlatΩ: A mediator between text-editors and proof assistance systems. In: Autexier, S., Benzmüller, C. (eds.) 7th Workshop on User Interfaces for Theorem Provers (UITP'06), ENTCS, Elsevier, North-Holland (2006)
13. Waite, W., Goos, G.: Compiler Construction. Springer, Heidelberg (1985) ISBN 0-387-90821-8
14. Wilhelm, R., Maurer, D.: Übersetzerbau - Theorie, Konstruktion, Generierung, 2. Auflage. Springer, Heidelberg (1997)

Mizar Course in Logic and Set Theory

Ewa Borak and Anna Zalewska

University of Bialystok,
Institute of Computer Science, Białystok, Poland
ewag@ii.uwb.edu.pl,
zalewska@uwb.edu.pl

Abstract. From the very beginning of the development of the MIZAR system experiments with using MIZAR as a tool for teaching mathematics have been conducted. Numerous organized courses were based on different versions of the system: starting from the first implementation of its processor, through MIZAR-MSE, MIZAR–4 and PC–MIZAR up till its present version. Now MIZAR with its mathematical library gives us quite new didactic possibilities.

The purpose of this paper is to present a certain course on logic and set theory offered by our Institute for freshman students. The course employs MIZAR as the main tool of instruction. In the paper we discuss the organization of this course and describe some examples of students' tasks. Finally, some conclusions and remarks are given.

1 Introduction

The MIZAR system[1] is a proof–assistant based on classical logic, i.e. it is a computer system for representing mathematical proofs in such a way that the computer checks their correctness. During the development of the system a lot of experiments concerning education were conducted. We would like to recall here only a few of them.

The first implementation of the MIZAR processor was used to teach propositional logic in Poland (1975–1976). Next some didactic experiments with MIZAR–MSE [9] were conducted. A correspondence course based on it was run for 10 months (September 1983 through June 1984) by a popular Polish science monthly *Delta* [2,5], a magazine aimed at secondary school students. MIZAR–MSE was also used to teach elementary logic in Bialystok and foundations of geometry at the Departament of Mathematics at Warsaw University [8,11]. In 1987–88 a richer version of MIZAR (called MIZAR–4) was applied in teaching "Introduction to mathematics" and "Lattice theory" at the Bialystok Branch of Warsaw University. PC–MIZAR (the early 90s) was used in teaching topology. Five scripts [1] with exercises and their solutions (written in the Mizar language and checked by the computer) were created and used in teaching topology the traditional way (the laboratory classes were not possible to conduct at that time).

---

[1] http://mizar.org

M. Kauers et al. (Eds.): MKM/Calculemus 2007, LNAI 4573, pp. 191–204, 2007.

MIZAR was also used for teaching introductory logic courses at many universities in other countries, e.g.: USA (by A. Trybulec), Canada (by P. Rudnicki), Japan (by Y. Nakamura), Belgium and has been used as a formal environment to prepare lots of diploma theses, from a bachelor degree to PhD.

Now MIZAR with its mathematical library (MML) provides us with new challenges and gives us quite new didactic possibilities.

During the spring semester of the academic year 2003–2004, MIZAR was applied for the first time to conduct a course in "Formalization of Mathematics" [6]. During the next two academic years two courses (each year) were conducted with the aid of the system: "Introduction to Logic and Set Theory" and "Formaliztion of Mathematics" both as an obligatory part of the curriculum for all first-year students of our Institute[2].

It seems that our teaching experience allowed to work out a certain method of conducting such courses. The method and its results are presented in the next sections of this paper concentrating on the Mizar course in logic for beginners. In the current academic year the course was conducted in the autumn semester. It consisted of 30 hours of lecture (given by A. Trybulec) and 30 hours of laboratory classes organized in 6 groups, each group led by a different teacher (A. Trybulec, A. Naumowicz, A. Korniłowicz, R. Milewski and the authors), with an average number of 10 students in each group.

2   Mizar Course for Beginners

2.1   Organization of the Course

Before conducting any Mizar course it is necessary to make some didactic decisions. One can present them in the form of questions as follows.

What are the main goals of the course? As far as the course in logic for beginners is concerned its main goals are:

- developing students' skills in deductive reasoning by selecting exercises, within the scope of the preselected areas of mathematics, requiring the usage of various proof techniques,
- teaching a selected mathematical theory presented during the lecture,
- teaching students how to justify their intuitive mathematical solutions in the formal way.

What kind of mathematical domains to choose? The theory chosen for the course should be rich enough to enable presenting all proof techniques and solving a considerable number of tasks within one field of mathematics. Following the experience of past years and having in mind the university curriculum two theories - set theory and the theory of binary relations - have been found as suitable for our aims.

---

[2] The courses were reported at the TYPES workshops in Nijmegen (2004) and Chambery (2005).

How to use MML? There is no doubt that from the educational point of view the Mizar repository is a large source of mathematical knowledge. MML provides us with new didactic possibilities but the main question is how students should work with it. There are three different ways of the work:

- students can work on the whole MML repository,
- they can work only on its parts,
- they can work on specially dedicated Mizar environments based on selected parts of MML.

It all depends on a given level of students' mathematical knowledge and their understanding of MIZAR. Following our previous experience it seems that for beginner students the third approach is the most proper and it was chosen to conduct the course described here. Consequently, an environment consisting of three parts built successively on top of previous ones was created: ENUMSET (for set theory), RELATION and RELAT_AB (for the theory of binary relations). The notions and definitions included in the environment are presented below.

Table 1. Environment

ENUMSET

functors	{} (empty set), {x} (singleton), {x,y} (unordered pair) \/ (union of two sets), /\ (intersection of two sets) \ (difference of sets), \+\ (symmetric difference of sets) **union, meet**
attributes	**empty** (set)
predicates	c= (inclusion of sets), = (equality of sets), <> (inequality of sets) **misses, meets** (as antonym for **misses**)

RELATION

modes	**Relation**
functors	{} (empty relation), ~ (converse of relation) \/ (union of relations), /\ (intersection of relations) \ (difference of relations), \+\ (symmetric difference of relations) * (composition of relations)
attributes	**symmetric, transitive, asymmetric, antisymmetric, irreflexive**
predicates	c= (inclusion of relations), = (equality of relations), <> (inequality of relations)

RELAT_AB

modes	**Relation of** A,B (relation on two sets)
functors	**dom** (domain of relation), **rng** (co–domain of relation) .: (image of relation), " (inverse image of relation) **id** (identity of relation), [: , :] (Cartesian product)
attributes	**reflexive, weakly-connected, connected**

Let us give some explanations of predicates and functors occurring in the ENUMSET part of environment.

1. The **misses** predicate is defined as follows:

$$A \text{ misses } B \quad \text{means} \quad A \cap B = \emptyset.$$

2. The **union** functor is union of arbitrary collection of sets and

$$x \in \text{union } A \quad \text{iff} \quad \exists_B \ (x \in B \land B \in A).$$

3. The meaning of the **meet** functor (intersection of arbitrary collection of sets) [3] is following:

$$\text{if } A \neq \emptyset \quad \text{then} \quad x \in \text{meet } A \quad \text{iff} \quad \forall_B \ (B \in A \ \rightarrow \ x \in B),$$
$$\text{otherwise } (A = \emptyset) \quad \text{meet } A = \emptyset.$$

What kind of software to choose for students? As far as software for students is concerned it is the same as in previous academic years. We would like only to recall here that:

- GNU Emacs was chosen as the best editor to write students' answers with (J. Urban's MizarMode for Emacs [10], with a web–accessible manual, offers the students a complete user–friendly interface to the MIZAR system),
- the version 7.8.01 of the MIZAR system was chosen to verify students results, (because of the steady development of the system this version was frozen for the course time),
- the software prepared by A. Naumowicz specially for Mizar courses was used (it allows students to download new part of the environment for subsequent classes and store the results of their work after each class to individual accounts on a server).

2.2   Students Work and Examples of Their Tasks

During the course students work proceeds within 15 computer sessions (2 hours per session): the first one is the introductory one, eleven of them are simply laboratory classes conducted by teachers and three of them are scored achievement tests (2 in-term exams and a final one).

The introductory session

The session is devoted to instruct students how to use the software prepared for them, i.e. the text editor Emacs, the MIZAR system and Naumowicz's special Internet application. The session can proceed as follows:

1. after short exercises within Emacs students try to check two very simple Mizar articles (for example, one of them can be the empty Mizar article only with two reserved words: **environ** and **begin**) in order to know how to use MizarMode for Emacs and how the MIZAR system reports errors,

2. next, students connect with the course's website (see Fig. 1) and sign-up for their individual web-accounts,

3. after downloading a trial part of the environment (PROPOSIT) they install it on their laboratory computers (the installation for ENUMSET, RELATION and RELAT_AB will be proceeded later in the same way; after the installation four folders are created: `abstr`, `dict`, `prel` and `text`; two of them: the first and the last one are important from the students' point of view; in the first one, except PROPOSIT, there are necessary files including the notions and definitions of a given part of the environment; in the last one there is the file `test.miz` including the first part of the Mizar article suitable for a given part of the environment),

4. next, students open in Emacs the file `test.miz` and try to write and check their first simple Mizar formulas of propositional calculus,

5. after finishing the work, each student sends `test.miz` (the file with the results) to his (her) individual web–account.

Below, the course's website is presented. There are two panels: the left one with our parts of the environment for downloading (the first four "download") and the right one with the MIZAR system and GNU Emacs (both for Windows) for downloading as well as some instructions how to install them.

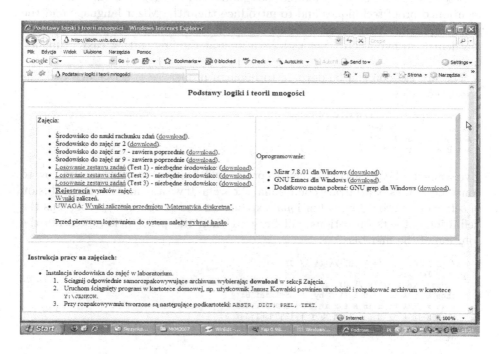

Fig. 1. The course's website: `http://alioth.uwb.edu.pl`

Laboratory classes

Laboratory classes can be divided into three parts. Each of them is composed of
3–4 computer sessions and is connected essentially with a given part of the environment. The students' task is to prove 5 theorems (as a rule) during each class
(depending on the level of complexity of these theorems and students' activity
from 3 to 5 tasks were usually solved). Each session lasts 90 minutes. Taking
into consideration the amount of students participating in the session, the teacher
leading the class can devote about 9 minutes per student for consultations.

Part I

During the first part of laboratory classes the ENUMSET part of the environment is used. Referring to students' knowledge of simple boolean properties of
sets (learning at the secondary school) we concentrate on teaching them how to
prove theorems in the formal way. By selecting exercises, within the scope of set
theory, students learn various proof techniques e.g.

- proofs by definitional expansion,
- conditional proofs,
- proofs by "reductio ad absurdum",
- proofs "per cases".

It seems that set theory is simple enough in order to teach students the above–
mentioned proof techniques and to introduce them the Mizar language and the
most important Mizar constructions that will be used later for justifying their
mathematical hypotheses. Let us have a look at the first, rather easy, students'
tasks.

The first session can be devoted to boolean properties of sets. Such tasks as
presented below

```
A \/ (B /\ C) c= (A \/ B) /\ (A \/ C);
(A \/ B) \ C = (A \ C) \/ (B \ C);
A c= B implies A c= A /\ B;
A \ B = {} iff A c= B;
```

give students the knowledge of how to prove by definitional expansion and how to
construct proofs of arbitrary conditional theorem. If we assume that the inclusion
of sets can be proved by definitional expansions and in the next proof steps the
definitions of sets operations will be used then the proof below presents the
solution of the first task.

```
A \/ (B /\ C)   c= (A \/ B) /\ (A \/ C)
  proof  let x;
    assume x in A \/ (B /\ C);
    then x in A or x in B /\ C by ENUMSET:def 6;
    then x in A or x in B & x in C by ENUMSET:def 7;
    then x in A \/ B & x in A \/ C by ENUMSET:def 6;
    hence x in (A \/ B) /\ (A \/ C) by ENUMSET:def 7;
  end;
```

Now, knowing how to prove the inclusion of sets, it is easy to prove their equality.
In the ENUMSET part of the environment equality of two sets X and Y is simply
defined as conjunction of two inclusions: X c= Y and Y c= X.

The conditional proof construction is used to prove the third task. Its solution can be as follows:

```
A c= B implies A c= A /\ B
proof
  assume E1: A c= B;
  thus A c= A /\ B
  proof let x;
    assume x in A;
      then x in A & x in B by E1, ENUMSET:def 10;
    hence x in A /\ B by ENUMSET:def 7;
    end;
  end;
```

In the above example the word **assume** indicates that the following sentence is an assumption of the proof. To justify its current thesis i.e. A c= A /\ B the technique of nested proofs has been applied. This technique is also used to proving equivalence of two sentences. Since each equivalence of two sentences is identified with conjunction of two implications, for the proof of the last task it suffices to prove the implications: A \ B = {} implies A c= B and A c= B implies A \ B = {}.

The next proof techniques ("per cases" and "reductio ad absurdum") and the Mizar construction **consider** can be introduced in solutions of the following tasks:

```
A\/B c= union {A,B};
A c= B implies union A c= union B;
A misses {};
A c=B & A misses C implies A c= B \ C;
A meets B or A meets C implies A meets B \/ C;
meet {A} c= union {A};
A <> {} & A c= B implies meet B c= meet A;
```

Of course, the first one can be solved in the other way, but we present below its "per cases" proof (for details of this proof technique see http://markun.cs.sh inshu-u.ac.jp/kiso/projects/proofchecker/mizar/skeletons/toc.html).

```
A\/B c= union {A,B}
  proof
  let x;
    assume x in A\/B;
    then E: x in A or x in B by ENUMSET:def 6;
    per cases by E;
    suppose x in A;
      then x in A & A in {A,B} by ENUMSET:def 4;
      hence x in union {A,B} by ENUMSET:def 13;
    end;
    suppose x in B;
      then x in B & B in {A,B} by ENUMSET:def 4;
      hence  x in union {A,B} by ENUMSET:def 13;
    end;
  end;
```

The best proof technique for solving the third task is the proof by "reductio ad absurdum". It is a special kind of conditional proofs because of a certain internal representation of the formula α in MIZAR[3]. The representation is given below:

---

[3] The notion "internal representation" is connected with the notion of "semantical correlates" (for details see [7]).

not α **implies contradiction**.

So instead of proving that **A misses {}** we assume the opposite, namely that
A meets {}. To end the proof it is enough to deduce **contradiction**.

```
A misses {}
 proof
   assume A meets {};
    then A /\ {} <> {} by ENUMSET:def 12;
    then consider x such that E: x in A /\ {} by ENUMSET:def 1;
    x in A & x in {} by E,ENUMSET:def 7;
    hence contradiction by ENUMSET:def 1;
 end;
```

Let us observe that in the above proof the Mizar construction **consider**, called
the choice construction, have been applied. In general, it allows to introduce a
new constant (in our case it is **x**) satisfying certain conditions (in our case it is
x in A /\ {}). What is important here (for students), it is necessary to justify
the conditions.

The remaining tasks allow students to train their deduction skills using the
Mizar constructions learned up till now.

Part II
It seems that most tasks of Part I can be recognized as some kind of exercises on
the level of symbol transformations: student starting from their assumptions try
to manipulate symbols in a certain way in order to obtain their conclusions. But
the next laboratory classes give them the possibility, during the solving of a given
task, to pass from "the context of discovery" to "the context of justification" [4].

The laboratory classes of Part II and Part III introduce students the theory
of binary relations. It is a rich enough theory in order to develop students'
deduction skills. From the didactic point of view one of its key features is very
important, namely it allows to visualize the students' "context of discovery" by
graphs that represent their reasonings. The Mizar proofs are simply the formal
justifications of students' intuitive solutions.

The computer sessions of Part II are connected essentially with the RELA-
TION part of the environment. Solving, for example, the following tasks

```
R*(S*T)=(R*S)*T;
R*(S /\ T) c= (R*S) /\ (R*T);
ex R,S st R*(S /\ T) <> (R*S) /\ (R*T);
ex R,S st R c= S & S is symmetric & not R is symmetric;
R is transitive & S is transitive & R*S=S*R implies R*S is transitive;
R is asymmetric iff R = R \ R~;
```

students learn basic notions and definitions connected with the theory of relations and practice several relation properties (e.g. symmetry, asymmetry, transitivity, etc.). Let us have a look at more complicated students' tasks.

The most creative tasks are those which require students to discover an example which is used to prove that a given theorem is false. We call them "tasks on
counterexamples". Usually counterexamples are not that easy to come up with.
We propose the below method of introducing such proofs:

– first students are asked to prove, for example, the following theorem

```
R*(S /\ T) c= (R*S) /\ (R*T);
```

– next they are asked to search for a proof of the inverse inclusion

```
(R*S) /\ (R*T) c= R*(S /\ T);
```

The beginning of their proof in MIZAR is presented below

```
let x,y;
  assume [x,y] in (R*S) /\ (R*T);
    then E1: [x,y] in (R*S) & [x,y] in (R*T) by RELATION:def 5;
    then consider a such that E2: [x,a] in R & [a,y] in S by RELATION:def 7;
    consider b such that E3: [x,b] in R & [b,y] in T by E1, RELATION:def 7;
```

Now this is the place in the proof when students realize that: a and b are not the same and it is impossible to conclude that some ordered pair with x as its first element can belongs to R. It is some weakness in their proof.
– then students are asked to use this weakness in their proof strategy to search for counterexamples visualizing this by a graph such as, for example, the following one:

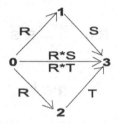

– finally students justify their reasoning in a formal way by the Mizar proof as presented below:

```
ex R, S, T st not (R*S)/\(R*T) c= R*(S/\T)
proof
  reconsider R={[0,1], [0,2]} as Relation by RELATION:9;
  reconsider S={[1,3]} as Relation by RELATION:8;
  reconsider T={[2,3]} as Relation by RELATION:8;
    take R,S,T;
    assume z1: (R*S)/\(R*T) c= R*(S/\T);
    [0,1] in R & [1,3] in S & [0,2] in R & [2,3] in T by ENUMSET:def 3,def 4;
    then [0,3] in R*S & [0,3] in R*T by RELATION:def 7;
    then z2: [0,3] in (R*S)/\(R*T) by RELATION:def 5;
    not [0,3] in R*(S/\T)
  proof
    assume [0,3] in R*(S/\T);
      then consider x such that z3: [0,x] in R & [x,3] in S/\T by RELATION:def 7;
      [x,3] in S & [x,3] in T by z3,RELATION:def 5;
      then [x,3] = [1,3] & [x,3] = [2,3] by ENUMSET:def 3;
      then x=1 & x=2 by ENUMSET:2;
    hence contradiction;
  end;
  hence contradiction  by z1,z2,RELATION:def 9;
end;
```

The next theorem to prove

```
R is transitive & S is transitive & R*S=S*R implies R*S is transitive;
```

is also interesting from the didactic point of view. Students, after the assumption that the relations R and S are both transitive and that their composition is associative, start to represent their reasoning on a graph. First, they draw two arrows (as successive assumption): one arrow from x to y and the other one – from y to z (see the graph below). The arrows above them represent their middle steps of reasoning that in the end should lead to draw the last arrow from x to z (the arrow number 8).

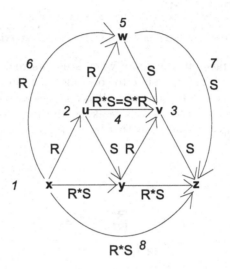

Now, students may formalize their reasoning in MIZAR. Below we give this formal proof with all marked steps.

```
R is transitive & S is transitive & R*S=S*R implies R*S is transitive
 proof
   assume    E1: R is transitive & S is transitive & R*S=S*R;
   thus R*S is transitive
    proof
     let x,y,z;
     assume E2: [x,y] in R*S & [y,z] in R*S;                              ::1
       then consider u such that E3: [x,u] in R & [u,y] in S by RELATION:def 7;   ::2
       consider v such that E4: [y,v] in R & [v,z] in S by E2,RELATION:def 7;     ::3
       [u,v] in S*R by E3,E4,RELATION:def 7;                             ::4
       then [u,v] in R*S by E1;                                         ::4
       then consider w such that E5: [u,w] in R & [w,v] in S by RELATION:def 7;   ::5
       E6: [x,w] in R by E1,E3,E5,RELATION:def 12;                      ::6
       [w,z] in S by E4,E5,E1,RELATION:def 12;                         ::7
     hence[x,z] in R*S by E6,RELATION:def 7;                           ::8
    end;
 end;
```

Part III
During the third part of laboratory classes students work on the RELAT_AB part of the environment connected (like the RELATION part of the environment)

with relations. But in fact students must still use notation, definitions and theorems from all previous parts of the environment. Therefore students prove properties of sets as well as properties of relations. Solving following tasks (for example)

```
rng (P*R) c= rng R;
ex  P,R st dom(P \ R) <> dom P \ dom R;
(dom R) /\ X c= (R~).:(R.:X);
(P*R)"X = P"(R"X);
ex A be set, Q be Relation of A,A st Q <> {} & Q is connected;
ex A,B be set, R be Relation of A,A, S be Relation of B,B
                    st A c=B & S is reflexive & not R is reflexive;
```

students perfect their deduction skills and formalization in the Mizar style. Some tasks are more complicated though essentially they do not step over knowledge learned up till now. One of that kind of tasks is

$$\exists_{A-\text{set},Q-\text{Relation of A,A}} (Q \neq \emptyset \ \wedge \ Q \text{ is connected})$$

and its Mizar proof is presented below.

```
ex A be set,Q be Relation of A,A st Q <> {} & Q is connected
 proof
   set A = {1};
   reconsider Q'={[1,1]} as Relation by RELATION:8;
   a: dom Q' c= A
     proof
       let x; assume x in dom Q';
         then consider y such that z: [x,y] in Q' by RELAT_AB:def 1;
         [x,y] = [1,1] by z,ENUMSET:def 3;
         then x=1 by ENUMSET:2;
       hence thesis by ENUMSET:def 3;
     end;
   rng Q' c= A
     proof
       let x; assume x in rng Q';
         then consider y such that z: [y,x] in Q' by RELAT_AB:def 2;
         [y,x] = [1,1] by z,ENUMSET:def 3;
         then x=1 by ENUMSET:2;
       hence thesis by ENUMSET:def 3;
     end;
   then reconsider Q=Q' as Relation of A,A by a,RELAT_AB:def 5;
   take A,Q;
   [1,1] in Q by ENUMSET:def 3;
   hence Q <> {} by RELATION:def 2;
   thus Q is connected
     proof
       let x,y; assume x in A & y in A;
         then x=1 & y=1 by ENUMSET:def 3;
         then [x,y] in Q by ENUMSET:def 3;
       hence [x,y] in Q or [y,x] in Q;
     end;
 end;
```

The material chosen for the exercises is not restricted and gives each student the possibility to construct his own examples of the objects (sets or relations) with given properties.

Tests

During the course, three tests were conducted. At the beginning of each test, students logged into their accounts on the server and downloaded the individual (generated by a special program) set of 5 tasks to solve (see Table 2). After downloading the test set, students solved tasks themselves with their computers. Results of finished work they sent back as usual by Internet onto the server.

Table 2. Examples of test tasks

Test 1
| ```
reserve X,Y,a,b,c,d,e for set, P,R,Q for Relation;

P is transitive & R is transitive implies P/\R is transitive;
dom (P\/R) = dom P \/ dom R;
{} is asymmetric;
(P \/ R)~ = P~ \/ R~;
R c= R~ iff R is symmetric;
``` |

Test 2
| ```
reserve X,Y,a,b,c,d,e for set, P,R,Q for Relation;

{} = {}.:X;
R"X c= dom R;
R.:(X/\Y) c= (R.:X) /\ (R.:Y);
ex P,R st P is symmetric & R is symmetric & not P*R is symmetric;
ex P,R st rng(P\R) <> rng P \ rng R;
``` |

Test 3
| ```
reserve X,Y,a,b,c,d,e for set, P,R,Q for Relation;

ex P,R st P is transitive & R is transitive & not P*R is transitive;
R.:X c= R.:dom R;
P c= R & R is irreflexive implies P is irreflexive;
rng (P\/R) = rng P \/ rng R;
(P /\ R)~ = P~ /\ R~;
``` |

All students' submissions were later checked by the teacher using the MIZAR system. A task was admitted as solved correctly when the Mizar verifier reported no errors. The following scoring scheme was used in the first and the second test: 0 points – any error appeared, 3 points – no errors, and in the third test: 0 points – any error appeared, 4 points – no errors.

The scores contributed to each student's final grade as 30 – 30 – 40 per cent. Our proposition of the scale of marks is given below in Table 3.

3   Conclusions and Remarks

Discussed here course has been conducted since the 2003–2004 academic year, in autumn semester each year. MIZAR as a tool for writing formal proofs was introduced gradually. At the beginning, it were only classes organized in traditional
```

Table 3. Statistics of marks

Points	Marks	ECTS	Number of students
50 - 44	5	A	6
43 - 35	4.5	B	9
34 - 27	4	C	14
26 - 20	3.5	D	8
19 - 14	3	E	9
0 - 13	2	F	10

way (30 hours). Next, in the 2003–2004 academic year the course consisted of 15 hours of common classes and 15 hours of laboratory classes (half to half) and for three years now – only 30 hours of laboratory classes. The university curriculum planned for the next year contains computer–aided subjects based on MIZAR as obligatory for all–year computer science students in our Institute. So, we treat this course as some kind of the long–term "students' investment in Mizar knowledge".

Students must devote some time for mastering all Mizar constructions needed for writing formal proofs in Mizar language. That is why we encourage them to solve some tasks at home. We noticed that students who worked at home reached much better results in tests than students who solved tasks only during the laboratory classes. For the very beginning of this course students were being pressed for:

- installing the text editor Emacs and the MIZAR system on their own computers at home,
- mastering Emacs (which far speeds–up writing proofs and checking their correctness),
- doing at home at least these task that were given to solve during the laboratory sessions,
- asking teacher for help in solving their problems.

We learned that the difference between good students and poorer ones is more apparent than usually. Weaker students need much more time to solve their tasks. The most problems of weaker students' work were caused by the following factors:

- they did not pay enough attention to the comments flagged by the system; some of students did not understand them at all or misinterpreted them (moreover, they did not report these problems to the teacher which slowed down teaching process in classes),
- they had problems with constructing the formal proofs as they did not have deduction skills developed enough; they lacked knowledge of what proof strategy can be used in a given context, what is the premiss, and what is the current thesis,

- they tried to solve their tasks only on the level of symbol transformations; starting from premiss(es) they try to manipulate symbols in such a way that MIZAR reports no errors – without understanding the proof,
- they did not try to solve tasks themselves at home,
- they manifest small activity during the classes.

In spite of some troubles, the didactic result of this course was positive. The average grade was 3,5 (3,5625 – on the 2 – 5 scale). This effect was caused by the following factors:

- individual students' work under control of the teacher,
- the ability of individual work at home (controlled by the MIZAR system) enabling more intensive training in the art of proving,
- the activity of good students mobilizes others to work more intensively (if at least one student solves a given task, then the others realize that the task is not so difficult after all and they start working more intensively in order to prove it themselves).

Acknowledgments

We would like to thank A. Trybulec and A. Naumowicz for their help in preparation of this paper.

References

1. Bajguz, W., Czuba, S.T.: Zbiór zadań z topologii w PC Mizar (A Collection of Exercises on Topology in PC Mizar), Zeszyt 1–5, Oddzial Regionalny Ogólnopolskiej Edukacji Komputerowej, Bialystok (1989-90)
2. Mostowski, M., Trybulec, Z.: A Certain Experimental Computer Aided Course of Logic in Poland. In: Proceedings of World Conference on Computer in Education, IFIP/AFIPS, Norfolk, North Holland (1985)
3. Padlewska, B.: Family of sets. Formalized Mathematics 1, 147–152 (1990), avaliable also from http://www.mizar.org/fm/1990-1/pdf1-1/setfam_1.pdf
4. Popper, K.R.: The Logic of Scientific Discovery, Hutchinson Education, London (1959)
5. Prażmowski, K., Rudnicki, P.: Kurs Logiki w Mizarze–MSE (Course in Logic in Mizar–MSE), Monthly DELTA, No 9 and next ones, Warsaw (1983–1984)
6. Retel, K., Zalewska, A.: Mizar as a Tool for Teaching Mathematics. Mechanized Mathematics and Its Applications 4, 25–33 (2005)
7. Suszko, R.: Ontology in the Tractatus of L. Wittgenstein. Notre Dame Journal of Formal Logic 9, 7–33 (1968)
8. Szczerba, L.W.: The Use of Mizar–MSE in a Course in Foundations of Geometry. In: Initiatives in Logic, Nijhoff Publishers, Dordrecht (1987)
9. Trybulec, A.: Logic Information Language MIZAR–MSE, Institute of Computer Science Polish Academy of Sciences, Warsaw, vol.465 (1982)
10. Urban J.: Mizar Mode for Emacs avaliable from http://alioth.uwb.edu.pl/twiki/bin/view/Mizar/MizarMode
11. Zalewska, A.: An Application of Mizar–MSE in a Course in Logic. In: Initiatives in Logic, Nijhoff Publishers, Dordrecht (1987)

Using Formal Concept Analysis
in Mathematical Discovery

Simon Colton and Daniel Wagner

Combined Reasoning Group
Department of Computing
Imperial College, London
{sgc,dwagner}@doc.ic.ac.uk
http://www.doc.ic.ac.uk/crg/

Abstract. Formal concept analysis (FCA) comprises a set of powerful algorithms which can be used for data analysis and manipulation, and a set of visualisation tools which enable the discovery of meaningful relationships between attributes of the data. We explore the potential of combining FCA and mathematical discovery tools in order to better facilitate discovery tasks. In particular, we propose a novel lookup method for the Encyclopedia of Integer Sequences, and we show how conjectures from the Graffiti discovery program can be better understood using FCA visualisation tools. We argue that, not only can FCA tools greatly enhance the management and visualisation of mathematical knowledge, but they can also be used to drive exploratory processes.

1 Introduction

Formal Concept Analysis (FCA) consists of a set of well established techniques for the analysis and manipulation of data. FCA has a strong theoretical underpinning, efficient implementations of fast algorithms, and useful visualisation tools. There are strong links between FCA and machine learning, and the connection of both fields is an active area of research [8,12,13]. We concentrate here on the combination of FCA tools with systems developed to aid mathematical discovery. In particular, we are interested in addressing (i) whether FCA algorithms can be used to enhance the discovery process and (ii) whether FCA visualisation tools can enable better understanding of the discoveries made.

To facilitate this study, we have implemented ways to integrate two FCA tools with (a) a system that uses HR [4] and Maple [18] to discover graph theory conjectures in a manner similar to the Graffiti program [5], and (b) the Online Encyclopedia of Integer Sequences, which is a very important mathematical database. In the first case, we show that the large number of conjectures which are produced can be efficiently organised and better visualised using a lattice structure afforded by representing the conjectures as a formal context. In the second case, we show that the data manipulation aspects of FCA enable a new way to mine information from the Encyclopedia. In particular, we describe a system we have implemented which can return sensible matches to query sequences

M. Kauers et al. (Eds.): MKM/Calculemus 2007, LNAI 4573, pp. 205–220, 2007.

that neither the Online Encyclopedia nor its more powerful sister program – the superseeker server – can explain.

This paper is organised as follows. In the next section, we briefly describe the theory, applications and implementations of Formal Concept Analysis. In section 3, we describe the Graffiti program, our HR/Maple simulation of it and the Encyclopedia of Integer Sequences. In section 4, we describe the methods we have developed in order to use FCA tools in conjunction with the HR/Maple simulation and the Encyclopedia. Following this, in section 5 we describe some experiments with both the HR/Maple/FCA combination and the Encyclopedia/FCA combination. In particular, we look at the sensitivity and selectivity of the lookup method afforded by the Encyclopedia/FCA combination, and we give some examples of it in use. We also provide an illustrative example of using FCA visualisation tools to better understand a set of graph theory conjectures produced by the HR/Maple system. We conclude in section 6 by suggesting that FCA visualisation and data analysis tools could be used not only to enhance mathematical discovery, but also to drive the discovery process.

2   Formal Concept Analysis

Formal concept analysis is a mathematical theory of conceptual hierarchies. A *formal context* is defined as a set of formal objects $\mathcal{O}$, a set of formal attributes $\mathcal{A}$ and a relation I on $\mathcal{O} \times \mathcal{A}$. The relation oIa for $o \in \mathcal{O}$ and $a \in \mathcal{A}$ can be read as "object o has attribute a" or "attribute a is true for object o". Such a binary context can be represented as a "cross table" (cf. Fig. 1). The set of common attributes of a set of objects $O \in \mathcal{O}$ is defined as $O' = \{a \in \mathcal{A}: oIa \ \forall o \in O\}$. Analogously, the set of common objects of a set of attributes $A \in \mathcal{A}$ is defined as $A' = \{o \in \mathcal{O}: oIa \ \forall a \in A\}$. A *formal concept* is a pair of sets (O, A) where $O' = A$ and $A' = O$. The set O is called the *extent* and A the *intent* of the formal concept $C = (O, A)$. Formal concepts correspond to maximally filled rectangles in the cross table. Together with the subconcept relation, which is the sub-set (resp. super-set) relation on intents (resp. extents), a formal context forms a complete lattice. For more details on the mathematical theory of FCA, see [10]. For details of how FCA has been formalised in PVS and Mizar, see [11] and [16].

FCA has been successfully applied in various domains [9], and FCA theory is a good foundation for efficient algorithms as well as for human understandable reasoning. Visualisation of the concept lattice in what are known as *Hasse diagrams* is a particularly useful tool in human centred knowledge discovery. If the context is too large, its visualisation may become confusingly complex, but there are various methods in FCA to handle this complexity, e.g., conceptual scaling, nested line diagrams, product of sublattices [7,10,15]. To further clarify the diagrams, a reduced labeling is often used in FCA. Reduced labelings can be summarised as follows: (i) each node represents a formal concept (ii) objects are annotated below the most specific concept which contains them in its extent (iii) analogous attributes are annotated above the most general concept which contains the attributes in its extent (iv) to retrieve the complete extent (resp. intent) of a concept, one simply collects all objects (resp. attributes) which can be reached from the concept through a path downwards (resp. upwards) in the lattice (v) the top concept contains all objects but often no other attribute besides "being an object of the domain", whereas the bottom node contains all attributes and often no object (vi) implications between intents can be read from the lattice by following the arcs upwards. See Fig. 1 for a simple example of a Hasse diagram with reduced labeling.

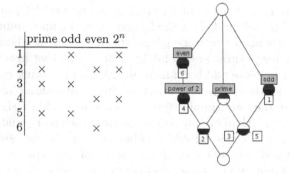

	prime	odd	even	2^n
1		×		×
2	×		×	×
3	×	×		
4			×	×
5	×	×		
6			×	

Fig. 1. A cross table and corresponding Hasse diagram produced by ConExp

There are several software tools available which implement FCA techniques. One of the first implementations of FCA algorithms was Peter Burmeister's ConImp[1]. More recent tools include ToscanaJ[2] and QuDA[3]. For the experiments described here, we have used the Galicia[4] and ConExp[5] [19] systems which are written in Java. Galicia can be used as a command line tool, e.g., within shell scripts, as a library, or as a stand-alone tool with a graphical user interface. ConExp offers a convenient user interface and implementations of most of the most powerful current algorithms, e.g., for the calculation of implication bases and for concept exploration.

3   Mathematical Discovery Systems

3.1   The Graffiti Program simulated by HR and Maple

The Graffiti program [5] by Siemion Fajtlowicz makes conjectures of a numerical nature in graph theory. Given a set of well known, interesting graph theory invariants, such as the diameter, independence number, rank, and chromatic number, Graffiti uses a database of graphs to empirically check whether one

---

[1] http://www.mathematik.tu-darmstadt.de/~burmeister/
[2] http://toscanaj.sourceforge.net/
[3] http://kirk.intellektik.informatik.tu-darmstadt.de/~quda/
[4] http://www.iro.umontreal.ca/~galicia/
[5] http://conexp.sourceforge.net/

sum of invariants is less than another sum of invariants. The empirical check is time consuming, so Graffiti employs two techniques, called the beagle and dalmation heuristics, to discard certain trivial or weak conjectures before the empirical test. If a conjecture passes the empirical test and Fajtlowicz cannot prove it easily, it is recorded in [6] and forwarded to interested graph theorists.

As an example, conjecture 18 in [6] states that, for any graph G:

$$cn(G) + r(G) \leq md(G) + fmd(G),$$

where $cn(G)$ is the chromatic number of G, $r(G)$ is the radius of G, $md(G)$ is the maximum degree of G and $fmd(G)$ is the frequency of the maximum degree of G. This conjecture was passed to some graph theorists, one of whom found a counterexample. The conjectures are useful because calculating invariants is often computationally expensive and bounds on invariants may bring computation time down. Moreover, these types of conjecture are of substantial interest to graph theorists, because they are simply stated, yet often provide a significant challenge to resolve – the mark of an important theorem such as Fermat's Last. In terms of adding to mathematics, Graffiti has been extremely successful. The conjectures it has produced have attracted the attention of scores of mathematicians, including many luminaries from the world of graph theory. There are over 60 graph theory papers published which investigate Graffiti's conjectures.

Unfortunately, Graffiti is not available for experimentation. However, in [14], we used a combination of the Maple computer algebra system and the HR automated theory formation system [4] to simulate Graffiti, and we produced very similar results. The way in which HR operates and the tools it has available for management of the mathematical information it produces have been described in [3]. A detailed description of how we used HR and Maple in graph theory is beyond the scope of this paper. However, we note that in order to further describe the conjectures produced, we calculated (i) a *tightness* measure which determined how close the inequality was to being equality (which is useful when using one summation of invariants to bound the calculation of another), and (ii) a slack constant for each conjecture which is the largest real number which can be added to the left hand side of the inequality without breaking it. For instance, the conjecture that for all graphs, G, $rank(G) \leq connectivity(G) + num\_vertices(G)$ has slack value 1, which means we can strengthen the conjecture to: $rank(G) + 1 \leq connectivity(G) + num\_vertices(G)$.

3.2   The Online Encyclopedia of Integer Sequences

The Online Encyclopedia of Integer Sequences,[6] contains more than 127,000 integer sequences, such as prime numbers, square numbers, the Fibonacci series, etc. They have been collected over 40 years by Neil Sloane, with contributions from hundreds (possibly thousands) of mathematicians. The Encyclopedia is very popular, receiving tens of thousands of queries every day. The first terms

---

[6] http://www.research.att.com/~njas/sequences

of each sequence are stored, and the user queries the database by providing the first terms of a sequence they wish to find a hit for. Sloane has recorded many times when using the Encyclopedia has led to a conjecture being made. For instance, in [17], he describes how a sequence that arose in connection with a quantization problem was linked via the Encyclopedia with a sequence that arose in the study of three-dimensional quasicrystals. Enabling such discoveries is one of the main purposes of the Online Encyclopedia. In addition to this web-service, there is also an email-service called the *superseeker*, which, as well as searching the Encyclopedia, performs extensive transformations on a given query sequence and on the Encyclopedia entries in order to find a match which explains the query.

4   Combining Mathematical Discovery Software and FCA

In the following sub-sections, we describe two applications where we have used FCA to (i) help visualise the results of a mathematical discovery system, and (ii) enhance the abilities of a mathematical discovery system.

4.1   Visualising Inequality Conjectures

A difficulty we had with the simulation of the Graffiti program described above was the sheer volume of the conjectures it produced. We ordered these in terms of the inequality tightness, but it was still difficult to pick out conjectures for certain graph invariants. Moreover, it was difficult to keep track of the chains of inequalities. For instance, it might look promising to investigate the conjecture that $I_1 \leq I_2 + I_3$, but this was made irrelevant by finding elsewhere in the list the stronger conjectures that, say, $I_1 \leq I_{17}$ and $I_{17} \leq I_2$.

In order to better manage and visualise the inequality conjectures produced, we used the ConExp FCA system. To do so, after a session with HR/Maple simulating Graffiti, we extracted the set of summations, S, of invariants in the theory, and formed a binary context with them (note that S also contained the initial set of invariants). We used S both as the set of formal objects and the set of formal attributes in the binary context. Then, we extracted the set of inequalities produced and used the binary relation that an object s_o (summation of invariants) has attribute s_a (also a summation of invariants) if the conjecture $x_o \leq s_a$ has been made. Deriving the formal context in this way enables us to read off inequality conjectures from the Hasse diagram: each node represents a summation of invariants or a set of summations, and if one node n_1 is joined to another node, n_2, which is higher in the lattice, then the set of summations represented by n_1 is conjectured to be less than or equal to the set of summations represented by n_2. We present a lattice for the first 120 conjectures produced by HR/Maple and we describe how the lattice size grows as the number of conjectures grows in section 5.1.

4.2 Integer Sequence Lookup

As discussed above, the Online Encyclopedia of Integer Sequences is an extremely powerful and popular mathematics tool. However, there is some room for improvement in the way it searches for conjunctions of sequences. For instance, searching for this query sequence: *1, 9, 36, 225* returns a single hit, namely sequence A036907, which are square refactorable numbers (see [1]). However, searching for the sequence *1, 36, 136, 276*, which is in fact the first four *triangular* refactorable numbers returns no results. In the first case, the hit was successful, because someone had entered the concept of square refactorables into the Encyclopedia, but as no-one had done likewise for triangular refactorables, no hit was made in the second case.

We have implemented a routine which is able to efficiently construct such missing conjunctions of pairs, triples, quadruples, etc., of number types from the Encyclopedia. We start with the database, D, which was embedded in the NumbersWithNames program [2]. This is a snapshot of a fragment of the Encyclopedia taken in 2001, and contains 990 integer sequences which are of sufficient importance to have been given names, such as prime numbers, even numbers, pernicious numbers, etc. D contains only strictly increasing sequences, and represents roughly one hundredth of the current Encyclopedia entries. Given a query sequence $Q = \{q_1, \ldots, q_n\}$, we first extract the set $S = \{s \in D : \forall q_i \in Q, q_i \in s\}$. This is effectively the set of database sequences which are supersequences of the query. We then set up a binary context with formal objects being the integers 1 to q_n, formal attributes being the set S, and the binary relation between objects and attributes expressed as the relationship of whether an integer (object) is part of a sequence (attribute). We then form a Hasse diagram with reduced labeling, and inspect the bottom node of the lattice. The intent of this node will be a conjunction of sequences in S, and is returned as the definition of a sequence which covers the query sequence. Note that the intent may include fewer terms than the set of supersequences of the query, due to the reduced labeling performed by FCA, yet the extent may contain more terms than the query sequence, as the intent may describe more than just the query terms.

As an illustrative example, take the sequence *88, 124, 216, 246*. Note that this returns no hit from the Online Encyclopedia, and even superseeker fails to find a way of explaining this sequence. Our system returns three hits from the NumbersWithNames database, and constructing the binary context as described above produces the lattice presented in Fig. 2. We see that, in this case, the intent covers perfectly the query sequence in its extent, i.e., our system has discovered that the query sequence can be described as the set of untouchable Erdos-Woods numbers which are palindromic when written in base 5. While this is certainly not a simple definition, it is considerably easier to understand than many returned from the superseeker server. As another example, if we start with the query sequence *12, 30, 42, 56*, our system informs us that the sequence of heteromecic, semi-perfect, balanced numbers has this sequence: *6, 12, 30, 42, 56*, which is a super-sequence of our query sequence.

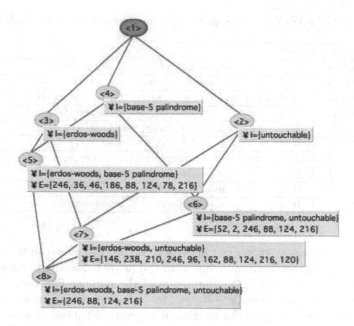

Fig. 2. Hasse diagram for sequence lookup (produced by Galicia)

Successful database lookup methods have to find a balance between sensitivity and selectivity. Sensitivity is the ability of the method to find genuine hits for a high proportion of queries it receives. Selectivity, on the other hand, is the ability of the method to avoid false positives, i.e., answers to queries which are not genuine hits. We have found that selectivity can be a problem for our method. For instance, if we start with the sequence *2, 18, 68, 102, 116*, then there are only two sequences from the database which contain all the query terms, namely A005843 (even numbers) and A045954 (even lucky numbers). In this case, therefore, our method will return the concept of integers which are both even and even lucky numbers as the hit for this sequence. The extent of this concept is, of course, the sequence of even lucky numbers: *2, 4, 6, 10, 12, 18, 20, 22, 26, 34, 36, 42, 44, 50, 52, 54, 58, 68, 70,...* Clearly this is too general, and hence not a genuine answer to the query. Hence, to keep the selectivity high, such answers should not be output by our method.

Another way in which an answer might be perceived as being not genuine is if it is too specialised. For instance, given the sequence *5, 20, 23, 29*, left unchecked, our method returns the answer that these integers are congruent, undulating, fibonacci-lucky, evil, cube-free, babylonian, noncube, nonsquare, unlucky, arc-cotangent irreducible/stormer, biquadratefree and weak numbers. This conjunction of 13 sequence definitions covers perfectly the integers 5, 20, 23, and 29, and no others between 1 and 29. However, the answer is hardly satisfying. We therefore need a way in which to constrain our method to output concepts which are not too specific in their intent, yet not too general in their extent.

Table 1. HR/Maple's tightest inequalities (from 120 conjectures in graph theory)

Tightness	Slack	Conjecture
0.998	0	$connectivity(G) \leq mindegree(G)$
0.918	0	$rank(G) \leq num\_vertices(G)$
0.842	0	$connectivity(G) + diameter(G) \leq num\_vertices(G)$
0.823	0	$av\_degree(G) \leq max\_degree(G)$
0.803	1	$max\_degree(G) \leq num\_vertices(G)$
0.799	0	$radius(G) \leq diameter(G)$
0.793	1	$rank(G) \leq connectivity(G) + num\_vertices(G)$
0.781	0	$chromatic\_number(G) \leq rank(G)$
0.769	0	$min\_degree(G) \leq av\_degree(G)$
0.767	0	$connectivity(G) \leq av\_degree(G)$

To improve the selectivity of our approach, we use two constraints to rule out certain sequences from D from inclusion in the answer. Firstly, sequences must have a density less than 0.5 over the region of the number line they occupy (a sequence $s_1, \ldots, s_n$ has density $\frac{n}{s_n - s_1}$). This rules out very general sequences such as square free integers. As roughly 6 in every 10 integers less than 100 are square-free, this concept is unlikely to form a property of a genuine hit. Secondly, we maintain a *black-list* of sequences which are found in answers too often (and hence are likely to be too general in their extent). We derived this list experimentally, by randomly generating 10 sequences, using our method to construct an answer, and then adding any sequence to the black-list if it appeared in three or more answers. We repeated this a number of times until no sequence appeared more than three times for a few turns.

The black-list currently contains the following sequences (with an asterix signifying a wild-card): nilpotent, nialpdrome*, smooth*, panconsummate, loeschian, odd, even, odd square free, even cototient, harshad/niven, equidigital, prime power, practical*, digitised partition, cyclic, amino acid, contracted, power-sum, flimsy, higgs' prime. Note that some black-listed sequences have density greater than 0.5, and are hence caught by both constraints. We have kept them in the list for the purpose of experimentation (see section 5.2). These two constraints very much reduce the set of database sequences which can be conjoined, and so the intent of an answer rarely contains more than a few sequences. However, we have a final filter on the output: if the intent contains a conjunction of more than five sequences, it is not shown to the user. We present some experiments with this lookup method in section 5.2.

5   Experiments and Results

5.1   Graph Theory Visualisation

For our experiments in graph theory, we used the HR/Maple simulation of Graffiti as described above to generate conjectures about inequalities of graph

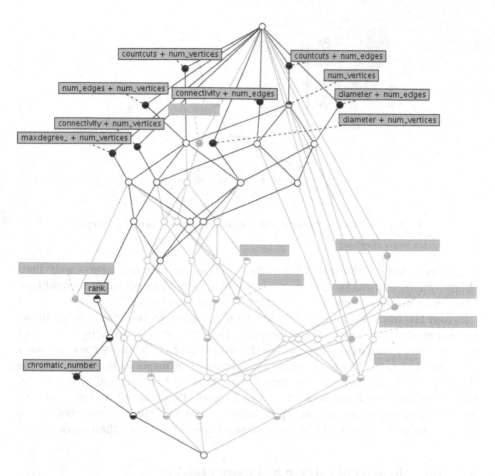

Fig. 3. Concept lattice for graph invariants, focused on chromatic numbers

invariants. Given the background definitions *number of vertices, number of edges, diameter, connectivity, countcuts, max-degree, min-degree, counttrees, rank, average degree, average temperature, radius* and *chromatic number*, we ran HR for 10000 steps which produced 66 new summations of the given 13 graph invariants, and 820 inequality conjectures. Using subsets of these conjectures, we calculated the formal context with the graph invariants as objects and attributes and the less than or equal to relation and then used ConExp to generate the Hasse diagram for inspection. For instance, in Fig. 3, we present the lattice derived from the first 120 conjectures that HR/Maple produced. We have used the ConExp interaction facilities to highlight the chromatic number node, which enables us to read off the conjectures involving that invariant. For comparison, in table 1, we show the kind of output that HR/Maple is able to produce, namely the top 10 conjectures from the 120 conjectures when they are ordered in decreasing tightness of the inequality in the conjecture. Compared to just a list of the inequality

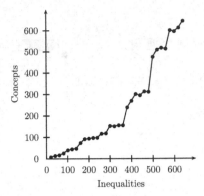

Fig. 4. Number of inequalities versus number of formal concepts

conjectures, we see that the FCA lattice enables us to much more easily read chains of inequalities. For instance, we can clearly read the chain of inequalities: $chromatic\_number \leq rank \leq max\_degree + num\_vertices$.

Given the highlighting abilities that ConExp has, we are able to mine information easily from more complex lattices. However, it is informative to look at how complex the lattices become as the number of conjectures they are used to represent grows. In Fig. 4, we plot the number of nodes in the lattice versus the number of inequality conjectures. We see that up to around 400 conjectures, the lattice contains fewer nodes than conjectures. However, after this point, the number of nodes rises quite sharply. Indeed, after 600 conjectures, there is no gain in clarity from using FCA because there are more nodes than conjectures.

5.2   Encyclopedia of Integer Sequences Lookup

Initial experiments with the integer sequence lookup approach described in section 4.2 showed much promise. To begin to determine whether the method could be used in practice (perhaps as part of the superseeker server), we need to test three aspects: (i) lookup speed (ii) sensitivity (iii) selectivity. The first of these is easy to deal with - the time taken to generate an answer to a query has been around 40 milliseconds, which would probably make it fast enough for a server.

Assessing the sensitivity of our method is somewhat problematic, as we need to assess the probability of a suitable answer being returned for any reasonable query. Such a measure of sensitivity will become clear as the lookup service is used for real queries. However, it is clear that, if the user submits a sequence which is a perfect conjunction of two to five number types in the Numbers-WithNames database, then the lookup method will return an answer. The only exceptions to this would be if the number types were too dense on the number line, or if one of the sequences was black-listed, and both these scenarios are unlikely for reasonable query sequences. A more tractable question involves the selectivity of our method – if it consistently returns spurious answers to queries,

Table 2. Sequence lookup for 1000 randomly generated sequences

experiment	Constraints			Random		Extent				Intent			Speed
	density	black-list	def-limit	random add	random tried	av. seq. length	av. hit length	perfect hits	contig. hits	av. super	av. defs	biggest def	av. time (ms)
1a	0.5	yes	yes	25	129845	4.117	8.157	73	168	2.434	2.323	5	20.665
2a	0.5	yes	no	25	131046	4.112	8.147	72	138	2.447	2.368	10	19.491
3a	0.5	no	yes	25	9943	4.396	16.161	8	40	2.812	2.707	5	15.656
4a	0.5	no	no	25	9048	4.4	15.878	26	53	3.216	3.034	13	14.933
5a	1.0	yes	yes	25	2281	5.477	38.186	1	1	4.863	3.477	5	33.973
6a	1.0	yes	no	25	1405	5.138	30.318	4	10	6.84	5.283	13	34.594
7a	1.0	no	yes	25	2603	5.641	37.519	0	0	4.801	3.46	5	24.949
8a	1.0	no	no	25	1341	5.19	27.56	13	29	7.091	5.512	18	32.153
9a	0.5	no	no	25	-	4.117	6.946	158	314	6.354	5.679	15	18.602
10a	1.0	yes	no	25	-	4.117	5.849	293	360	11.009	8.553	16	42.948
11a	1.0	no	no	25	-	4.117	5.285	440	512	15.12	11.512	23	83.639
1b	0.5	yes	yes	50	253540	4.09	12.566	25	47	2.274	2.234	5	29.786
9b	0.5	no	no	50	-	4.09	10.936	62	112	4.777	4.589	15	33.631
10b	1.0	yes	no	50	-	4.09	10.456	88	117	6.123	5.272	15	61.951
11b	1.0	no	no	50	-	4.09	9.067	159	213	8.786	7.406	21	50.171
1c	0.5	yes	yes	100	829628	4.065	17.851	7	20	2.377	2.368	5	45.432
9c	0.5	no	no	100	-	4.065	16.048	24	58	3.956	3.904	14	32.083
10c	1.0	yes	no	100	-	4.065	16.944	23	36	3.524	3.298	15	37.392
11c	1.0	no	no	100	-	4.065	15.242	58	88	5.182	4.782	21	37.305

then it will be de-valued and not used. To address selectivity, we ran three sets of 11 experiments, with the results presented in table 2.

In experiment 1a, we used the lookup method with the maximum density, black-list and maximum definition constraints all imposed. We generated random sequences in the following way: we generated a random integer between 1 and 25 as the first entry in the sequence, then added this to another random integer between 1 and 25 to get the second entry, and continued in this way until our sequence contained between 4 and 7 (inclusive) integers, with the length also determined randomly. This gives strictly increasing sequences with integers roughly between 1 and 100 on the number line, which is the kind of query sequence we may expect (with perhaps a bias towards smaller numbers than for usual queries). As the method with all three constraints is particularly selective, it only returned hits for 1000 out of 129845 randomly generated sequences.

We recorded a number of statistics about these 1000 sequences. Firstly, we compared the length of the query sequence with the length of the sequence which was returned (i.e., the extent of the conjunction of sequence definitions). A large difference in the two lengths means that the sequence returned is unlikely to be a genuine hit for the sequence. In experiment 1a, the average sequence length was 4.1, and the average hit length was 8.157, hence we could expect approximately double the integers in the hit than the query. However, for 73 of the 1000 sequences, a perfect hit was returned, i.e., the hit was exactly the query sequence. Also, 168 of the returned sequences contained the query sequence contiguously, i.e., the hit was either perfect or had some initial integers which were not part of the query sequence (our method precludes trailing integers). Note that this kind of hit is also returned by the Encyclopedia of Integer Sequences, as the initial terms of a query sequence are often omitted.

In addition to details of the extent of the concept returned, we also recorded details of the intent returned. In particular, we recorded the average number of sequence definitions which were conjoined in the definition of the hit. In experiment 1a, the average was 2.323. Note that the average number of supersequences of the query sequence was 2.434. Hence, we see that the FCA technique of reducing the intent to remove redundant definitions is effective, even when there are only a few definitions conjoined (this effect becomes greater in later experiments, e.g., experiment 11a, where the reduction is by between 3 and 4 terms on average). In summary, if our method does return an answer, then there is a roughly 1 in 10 chance that it will be high quality, i.e., perfect and/or contiguous. Moreover, the definition of the hit will be fairly understandable – on average a simple conjunction of either two or three sequence definitions.

It is difficult to know in advance whether this will be selective enough for users of the lookup method. However, we can show that our method could be much less selective. In experiments 2a to 8a, we varied the usage of the definition length, density and black-list constraints. As table 2 shows, with the exception of removing only the definition length constraint, the method is far less selective when we remove the constraints. For instance, in experiment 8a, there were no constraints used. The method returned an answer for most of the query sequences (the exceptions were queries able to be answered with only one sequence – which were still ruled out). However, the answers returned were very low quality. On average, the answer to a query contained 22 more terms than the query – effectively making them useless answers – and the method found only 13 perfect hits. Moreover, the definitions of the answers were more complicated, being a conjunction of on average 5.5 sequence definitions.

In experiments 9a, 10a and 11a, we tested how the unconstrained methods perform on the sequences which passed the selectivity test from experiment 1, i.e., the 1000 sequences which returned a hit from the constrained method. The results from these tests were very interesting: it appears that if a query sequence does have some semantic value with respect to the number types in the database (i.e., should have a description according to the selectivity criteria), then the unconstrained lookup methods can often synthesise a more complicated but more accurate answer. Experiment 9a is particularly interesting: here the returned solutions were on average conjunctions of only 6.354 sequence definitions, but they included 158 perfect hits, which was more than double than in experiment 1a. In experiment 11a, the accuracy of the hits was striking: in 440 cases, the

Table 3. Perfect hits from experiment 1a which are missing from the Encyclopedia

Sequence	Description
6,30,36,60	eban & evil & pseudo-perfect/semi-perfect & highly abundant
2,11,23,48	semi-fibonnaci & problime(3)
13,31,41,53,71	prime & primitive congruent & regular prime & short-period prime
13,37,41,53	prime & primitive congruent & gaussian prime & short-period prime
4,25,38,58,74	fibonacci-lucky & semi-prime/2-almost prime & twin fibonacci-lucky(1)
1,8,22,34	multiplicatively perfect & polyomino
8,21,32,50	arc-cotangent reducible/non-stormer & duffinian
3,18,20,30	evil & base-4 colombian/self & highly abundant
24,30,43,53	evil & strict egyptian(1)
3,17,21,38	base-4 palindrome & arc-cotangent reducible/non-stormer
6,15,23,30	evil & binary colombian/self & primitive congruent
2,5,25,38,59	fibonacci-lucky & base-4 palindrome & twin fibonacci-lucky(1)
15,23,48,51	rhombic & evil & binary colombian/self
4,6,30,46	eban & 2-knodel & binary colombian/self
6,12,27,36	evil & truncated triangular(1)
6,18,36,54	evil & pseudo-perfect/semi-perfect & base-8 palindrome
15,23,43,48	rhombic & evil & problime(3)
1,4,25,32	fibonacci-lucky & base-7 armstrong(2) & perfect power
7,19,23,32	rhombic & fibonacci-lucky & friendly/happy
2,7,13,37	prime & fibonacci-lucky & exceptional prime &
	class-1/pierpoint prime & absolute prime
1,10,20,26	semimorphic & base-3 palindrome & davenport-schinzel(1)

method returned a perfect hit. However, the cost for this was more complicated definitions of the hits: on average they were conjunctions of 11.512 sequences. This suggests a two-tier approach for the lookup method: (i) filter the sequence using the selectivity criteria, and return nothing if it fails the test (ii) for any sequence which passes the test, return both the answer from the constrained lookup and from the unconstrained lookup. The former answer will be easier to understand but perhaps less accurate, while the latter answer will be more difficult to understand, but is likely to be more accurate.

We repeated the set of 11 experiments twice. In the second set, we used 50 rather than 25 as the interval with which to generate random sequences, and in the third set we used 100. We report only the results from the first experiment in the set (with all the constraints imposed), and the final three experiments (with fewer constraints, applied to the sequences passing the selectivity criteria of the first experiment). We observe that finding solutions becomes increasingly difficult as the sequences spread out over the number line, which was a trend we expected. However, we didn't expect the trend that the number of definitions in the unconstrained answers reduces as the sequences spread out. For instance, in the first set of experiments, the unconstrained method produced around 6 times more perfect hits than the constrained method, but needed to conjoin around 4 times the number of definitions. In contrast, in the third set of experiments, the unconstrained lookup method found more than 8 times more perfect hits, but used only twice the number of definitions. While this phenomenom is explainable – as fewer sequences from the Encyclopedia will match sparse sequences – it adds weight to our proposal of using a two-tiered approach to sequence lookup.

In table 3, we list the 21 sequences for which our fully constrained method (experiment 1a) returned a perfect hit, whereas the Encyclopedia of Integer Sequences returned no hits. The average number of conjoined definitions for these sequences is 2.95, which is higher than the average for experiment 1a, suggesting that sequences have to be more complex to be missing from the Encyclopedia.

6   Conclusions and Future Work

Formal Concept Analysis is a well developed area with much to offer for data analysis in various application domains. We have investigated the usage of FCA for analysis, manipulation and visualisation of mathematical knowledge/data. In particular, we have addressed the question of whether FCA could be used to enhance systems used for automated or semi-automated mathematical discovery. In the application to database lookup from the Encyclopedia of Integer Sequences, we have shown that FCA tools are useful for manipulation of mathematical information, and our system was able to find sensible answers to numerous query sequences which the Encyclopedia (and in some cases superseeker) couldn't answer. In the application to visualising the graph theory conjectures produced by our HR/Maple system, we showed that FCA visualisation tools have much potential for better management of machine generated mathematics.

In a further set of experiments which there hasn't been space to describe here, we have used the implication generation tools within ConExp with the results from using HR in number theory. In a particular test, we wanted to see whether using HR followed by FCA concept exploration could help us find a particular conjecture quicker than using HR alone. Starting with just the ability to multiply two numbers, HR can discover the conjecture that odd refactorable[7] numbers are squares. With fairly strict restrictions on the search that HR can perform, it still takes 721 steps to produce 1883 conjectures, the last of which is the one we wanted. However, if we stop HR after only 64 steps, and then use ConExp to generate all implication conjectures possible from the 22 concepts that HR has produced, the conjecture that we want is output. We need to perform further tests, but it seems likely that a permanent combination of HR with an FCA system could dramatically reduce the combinatorial burden that HR has when making conjectures, and FCA could improve HR's discovery process as a whole.

This not only strengthens our claim that the combination of FCA tools with discovery systems has much potential to enhance discovery, but it also suggests a more fine-grained involvement of FCA in discovery tasks. In particular, in future work, we plan to build a hybrid FCA/machine learning system which is of benefit to both the machine learning and the FCA communities. Among other benefits, we intend the system to improve upon (a) FCA systems, by using concept formation abilities similar to those from machine learning, and

---

[7] Refactorable numbers, n, are such that the number of divisors of n is itself a divisor.

(b) the visualisation and user-interaction abilities of machine learning systems. The system will enable FCA tools to drive the exploration process using machine learning enhancements. We hope to show that the hybrid system enables users to make more interesting discoveries in mathematical domains than they would using FCA or machine learning tools alone. We also intend to apply the hybrid system to discovery tasks in other domains, such as bioinformatics.

Acknowledgements

We would like to thank the anonymous reviewers for their useful comments, and also Bernhard Ganter for very helpful discussions on Formal Concept Analysis.

References

1. Colton, S.: Refactorable numbers - a machine invention. Journal of Integer Sequences, 2 (1999)
2. Colton, S., Dennis, L.: The numberswithnames program. In: Proceedings of the Seventh AI and Maths Symposium (2002)
3. Colton, S., Torres, P., Cairns, P., Sorge, V.: Managing automatically formed mathematical theories. In: Proceedings of the 5th International Conference on Mathematical Knowledge Management (2006)
4. Colton, S.: Automated Theory Formation in Pure Mathematics. Springer, Heidelberg (2002)
5. Fajtlowicz, S.: On conjectures of Graffiti. Discrete Mathematics 72, 23, 113–118 (1988)
6. Fajtlowicz, S.: The writing on the wall. Unpublished preprint (1999) available from http://math.uh.edu/~clarson/
7. Ganter, B.: Formal Concept Analysis. Foundations and Applications. In: chapter Contextual Attribute Logic of Many-Valued Attributes, pp. 101–113. Springer, Heidelberg (2005)
8. Ganter, B., Kuznetsov, S.: Hypotheses and version spaces. In: Ganter, B., de Moor, A., Lex, W. (eds.) ICCS 2003. LNCS, vol. 2746, pp. 83–95. Springer, Heidelberg (2003)
9. Ganter, B., Stumme, G., Wille, R. (eds.): Formal Concept Analysis. Foundations and Applications. Springer, Heidelberg (2005)
10. Ganter, B., Wille, R.: Formal Concept Analysis: Mathematical Foundations. Springer, Heidelberg (1999)
11. Hidalgo, M., Martin-Mateos, F., Ruiz-Reina, J., Alonso, J.A., Borrego, J.: Verification of the Formal Concept Analysis. Rev. R. Acad. Cien. Serie A. Mat. 98(1), 3–16 (2004)
12. Kuznetsov, S.: Machine learning and formal concept analysis. In: Eklund, P.W. (ed.) ICFCA 2004. LNCS (LNAI), vol. 2961, pp. 287–312. Springer, Heidelberg (2004)
13. Liquiere, M., Sallantin, J.: Structural machine learning with galois lattice and graphs. In: International Conference on Machine Learning (1998)
14. Mohamadali, N.: A rational reconstruction of Graffiti. Master's thesis, Department of Computing, Imperial College, London (2003)

15. Scheich, P., Skorsky, M., Vogt, F., Wachter, C., Wille, R.: Information and Classification - Concepts, Methods and Applications. In: chapter Conceptual Data Systems, pp. 72–84. Springer, Heidelberg (1992)
16. Schwarzweller, C.: Mizar formalization of concept lattices. Mechanized Mathematics and its Application 1(1), 1–10 (2000)
17. Sloane, N.J.A.: My favorite integer sequences. In: Proceedings of the International Conference on Sequences and Applications (1998)
18. Waterloo Maple. Maple Manual at http://www.maplesoft.on.ca
19. Yevtushenko, S.: System of data analysis concept explorer. In: Proceedings of the 7th national conference on Artificial Intelligence KII, pp. 127–134 (2000)

Cooperative Repositories for Formal Proofs[*]
A Wiki-Based Solution

Pierre Corbineau and Cezary Kaliszyk

Institute for Computing and Information Science
Radboud University Nijmegen, Postbus 9010
6500GL Nijmegen, The Netherlands
{corbinea,cek}@cs.ru.nl

Abstract. We present a new framework for the online development of formalized mathematics. This framework allows wiki-style collaboration while providing users with a rendered and browsable version of their work. We describe a prototype based on Coq, its web interface as implemented by the second author, and a modified version of the MediaWiki code-base. We discuss open issues such as dependencies and repository consistency. We explain limitations of the current prototype and we give a perspective towards a more robust solution.

1 Introduction

1.1 Motivations

Proof assistants are software tools used for expressing properties and checking proofs of those properties, be it about mathematical concepts or models of computer systems or software. Nowadays, most proof assistants follow the *interactive paradigm*: the user enters the statement of a theorem; the system checks the well-formedness of the statement. The user then enters a proof commands and the systems responds by validating the command and giving the remaining facts to be proven. This process is then iterated until the proof is complete. Thus, the resulting sequence of commands, called *proof script*, has barely any meaning without the succession of *proof states* it yields. However, most formal developments only consist of the bare proof script, maybe with some comments.

Two solutions are available for people who want to understand the proofs better: HTML rendering and local execution. With web rendering, the proof scripts are processed by a documentation tool that turns the files into HTML documents and provides some facilities such as hyperlinks from symbols to their definition, indexes of symbols and searching. Some even provide pretty-printing of comments, rendering of mathematical formulae.

But to understand the proof script itself, one has to first locate and download the files containing the proofs, then install the proof assistant, and finally run

---

[*] This work was funded by NWO Bricks/Focus Project 642.000.501 (Advancing the Real use of Proof Assistants) and partially funded by NWO FEAR Project.

M. Kauers et al. (Eds.): MKM/Calculemus 2007, LNAI 4573, pp. 221–234, 2007.
© Springer-Verlag Berlin Heidelberg 2007

the proof assistant on the file to inspect the sequence of proof states. When doing this, one loses the ability to browse the code using hyperlinks, and it can sometimes be complicated to get the proof assistant to run on one's computer in the first place.

Recent work by Kaliszyk[1] shows that the *Asynchronous DOM Modification* web technology (sometimes referred to as *AJAX* or *Asynchronous Javascript and XML* [2]) can be used to build a web interface for interactive proof assistants: PROOFWEB. This means that users can use their favorite web browser to run proof assistants sessions, so they can perform themselves the checking of the formal proof. However this work still lacks essential features: it is not designed to support multi-file developments properly, no proper HTML rendering is implemented and there are no tools to store and retrieve multiple versions of files.

A popular web architecture supporting all those features is called wiki. The wiki concept actually covers many implementations, but all are aimed toward a *cooperative authoring* of knowledge repositories. The key feature of a wiki system is the ability to follow an 'edit' link and be able to immediately modify and publish a new version of the page being viewed.

The popularity of wiki based solutions made us think of integrating the web interface for proof assistants within a wiki repository: the web interface would be used as the viewing and editing window for files containing proof scripts. The main difference between our work and common wiki usage is that our framework handles formal content that requires a consistent environment (i.e. file dependencies) to run interactive sessions. Thus (semi-)automated maintaining of cross-file consistency is crucial.

1.2   Related Work

Most proof assistants already have a more or less user-friendly way of *rendering* formalisations as a set of interlinked web pages. Some provide a standalone tool that allows users to render their own files: this is the case for Isabelle[3] and Coq[4]. Isabelle also provides a way to navigate the dependency graph of multi-file developments.

The Mizar[5] system has a proof repository called the MML (Mizar Mathematical Library) [6]. This repository is modified by human editors: duplicates are eliminated, results are moved to appropriate sections, new sections are created. This gives the MML a monolithic and consistent look[7]: it is handled as an encyclopedia, where new content is added with many authors but one central editing comitee. However, the rendering tool is not available for the common user to work with his local development. The MML (and its associated journal JML) is the *de facto* standard way to publish a Mizar proof.

The Logiweb System[8] provides a way to submit and retrieve articles from a network of distributed repositories. It allows reliable cross-references to fixed versions of already published articles. However it still relies on a locally installed checker to verify articles before submitting them.

The HELM[9] (Hypertextual Electronic Library of Mathematics) and the Whelp[10] search engine give users a good rendering of distributed formal libraries along with a powerful search engine. The Matita[11] proof assistant offers native support for queries and browsing of these libraries.

The Logosphere[12] project aims at presenting developments from different proof assistants (Nuprl, PVS, ...) using a unified framework.

The Mizar and Coq proof assistants already have wiki web sites for their documentation. The Mizar wiki[13] is an official, general purpose web site whereas the Coq wiki, called Cocorico[14], is a community website more dedicated to the sharing of specific knowledge about Coq usage, hints and tips, dirty tricks ...

1.3 The Future of Proof Interfaces

The aim of this new web-based cooperative proof environment is to provide — as an IT marketing representatives would say — a *complete solution* for the development of formalized mathematics or software verification. It brings together the availability of a web-interface with the accessibility of a web-rendered archive.

The unique feature of this environment is that, beyond the separation between the raw editable and rendered read-only versions of the files (a characteristic of wiki environments), both of those versions can be processed by the proof assistant at the request of the user, giving him more information as to how the proof script works. Where standard online formal libraries tend to treat proof scripts as minor, here their contents can give the user insights on how the proof was made: the proof script is not write-only anymore.

Therefore, this environment provides a useful tool for specialists to communicate about proofs with a broader audience: non-specialists, general audience. It provides a simple way for article writers to give referees easy access to their formal development.

The repository is also a convenient way for proof authors to work from everywhere simply using a network access, and to learn from others' proof idiosyncrasies.

The repository can also be used for education about proof assistants and formal logics. A permissions system can allow students to cooperate on multi-files projects and their supervisor to provide guidance.

1.4 Document Contents

In the rest of the paper we present the technologies relevant for creating a wiki for proof assistants (Section 2), and the components that are used. Then we present the global architecture of our system and discuss our library consistency policy (Section 3). We describe our current prototype (Section 4) and discuss performance and security issues. Finally, we give our road map towards a more stable system and present our conclusion (Section 5).

2 Web Technologies

2.1 Asynchronous DOM Modification

With the growing usage of the Internet, more technologies are available for de-signers of web services. Recently asynchronous DOM modification technology has allowed the creation of interfaces that are completely available in a web browser, but have similar functionality and responsiveness to local ones.

The *asynchronous DOM modification* technology (sometimes referred to as *AJAX* [2] or *Web Application*) is a combination of three commonly available web technologies. JavaScript is a scripting programming language interpreted by the web browsers. *Document Object Model (DOM)* is a way of referring to sub elements of a web page, allowing to modify it on the fly to create dynamic elements. XmlHttp is an API available to client side scripts, that allows sending requests to the web server without reloading the page.

The asynchronous DOM modification technique consists in creating web pages that capture events and processes without reloading the page. Events that can be processed locally modify the web page in place. For actions that require interaction with the server, the data is sent using an asynchronous XmlHttp transaction and the page is modified by the script when receiving the answer, therefore making the interface as responsive as an application run locally. For a more detailed description see [1].

One application of this technology is an architecture for the creation of web interfaces for proof assistants, that are completely available in a web browser, but resemble and behave as local ones [1]. PROOFWEB, an implementation of this architecture, keeps a prover session for every user on the server (Fig. 1). It allows similar interaction as ProofGeneral [15] does, but using a web browser.

2.2 Wikis

Wikis are dynamic web sites that behave as static ones: they contain a number of fixed pages that can link to each other. The unique feature of wikis is that each of those pages includes an 'edit' link that displays the contents of the page in an editable textbox (or in a more fancy WYSIWYG box in advanced implementations). This page allows the user to actually modify the contents of the box and publish the new version on the web site, simply using a web browser.

This is what makes wikis dynamic: this online edition feature requires files to be served in a more fancy way than just static HTML files. Usually they are stored in a database system rather than in a filesystem. Unfortunately, current proof assistants do not include the functionality to access such databases.

The wiki technology is now very popular, especially for documentation of soft-ware tools: it allows to start with a very small, general (and somewhat imprecise) documentation which is then improved by visitors when finding inconsistencies, errors and missing items. The most famous wiki is obviously the Wikipedia web site, which aims at being an online encyclopedia where information is added by visitors. In each of the 14 most popular languages on Wikipedia, more than

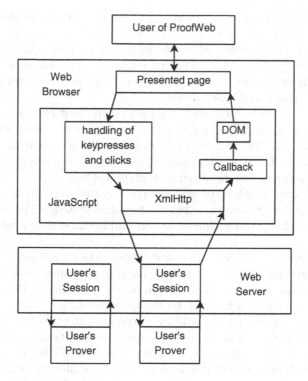

Fig. 1. PROOFWEB architecture

100,000 articles are available. This shows that the wiki architecture can support large amounts of data and heavy activity.

The file format used by wikis is usually a simplified markup and formatting language, tailored to make references to other pages simple to add. Wikis usually have a permissions system to forbid reading or writing for particular users. Most of them also allow modification by unregistered users. In that case the IP address of the client is used as an identifier.

Wikis also offer the possibility to explore the history of any article: what was modified, when. They allow renaming of pages, and provide indexes of available pages. They usually include tools that allow searching the page database.

Those features match our requirements for a content management system to be usable with the PROOFWEB framework.

3   Architecture

3.1   Main Components

The system we propose uses proof assistants with some of their companion utilities and some web serving utilities. The first element we need is the interactive toplevel of a prover. It is required on the server side to be able to verify the input of the user in an incremental manner and to go to particular positions in them.

For efficiency reasons some provers allow compilation of their input files. Such files can then be quickly reused, without verifying all proofs contained in them again. For such provers we want to use the compiler on the server to generate a compiled version of proofs that are saved.

Many provers include documentation generators, that process raw prover input files and generate rendered output. The output of a documentation generator is usually HTML or PDF format. Links between files are created, different conceptual elements of the prover input are colored in different color, and sometimes mathematical formulas are rendered in a graphical way.

We need to keep a history of versions for every prover file. Usually, collaborative developments are done using version control systems. The source files are kept in the repository and each user has to build compiled or rendered versions him/herself.

Not only would we like the source prover files to be stored for all versions, but also the rendered and the compiled files (for provers that include this concept). This way users can see a rendered version of older versions of files. Referring to older versions of compiled files will be discussed in section 3.3.

Wikis already include some kind of versioning of the files they contain. Generally file versions are numbered in sequence. The user that made every change is stored with the file, and viewing changes is possible. The history mechanism is more limited than the ones provided by file versioning systems, but the simplicity can also be an advantage: in particular wikis do not include branches, tags, etc and the casual user does may not have a good understanding of it.

A wiki infrastructure will be used for tracking changes done by users and allowing them to see the history of files and changes. It needs to store files in a way that is accessible by the prover toplevel. The wiki should allow generating indexes and searching for terms. Most wikis generate text indexes and allow searching for text only, whereas prover scripts are highly contextual.

Finally we need a web part that allows interactive edition of a proof script in a way that resembles local work, to allows efficient work. Additionally we would like to be able to step interactively over the proof regardless of whether we are in view or edit mode. The PROOFWEB framework can be modified to allow those two modes.

3.2   Global Design

Our architecture is composed of a web server running a modified version of a wiki that redirects some requests to a PROOFWEB server (Fig. 2).

Editors of most wikis are standard HTML textboxes, and the flat text includes special markup for marking links and elements that should be formatted in a special way in the read-only version. Recent wikis allow WYSIWYG editing in an editable IFrame. The HTML formatting introduced by user's browsers is combined with wiki links to create the read-only version. In our architecture we embed the PROOFWEB editor as the editor of the wiki. This way, the user can

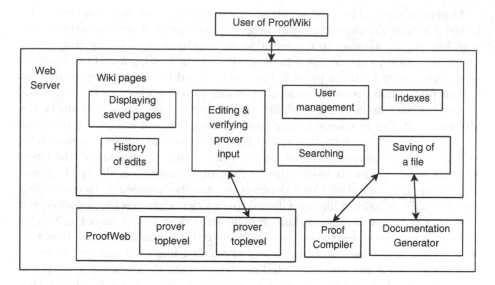

Fig. 2. Our architecture

edit the script in an interactive way seeing the output of the prover. Additionally we include a readonly version of the PROOFWEB interface for the read-only version to allow examining the prover state at any point in the buffer.

The next necessary change is the way the wiki stores users files. For every saved file the wiki tries to compile it and to make a rendered version of it. The rendered version should be linked with the original, and is therefore stored in the same way the wiki stores the original in its database. Whenever the user requests the file for viewing or editing one of those two versions is used. The compiled version is stored as a standard file in the filesystem in order to make it available to the compiler and to the toplevel used in prover sessions on files that refer to this compiled file.

This change in the wiki behavior should not prevent users from storing and editing standard wiki pages in the repository. Those would include textual description of the formal content, discussions, tutorials with hyper-links to formal content.

The documentation generators of provers have to be able to generate a wiki compatible output. The format that a wiki displays is usually very close to HTML which many prover documentation generators already support. The important difference with respect to HTML is that since we will process the rendered version of the script we need to be able to distinguish active parts of the file from comments.

3.3 Consistency Issues

In usual wikis links to nonexistent pages lead to a new editable page. This is perfectly acceptable for the usual informal content but not for a formal development referring to another: think of it as a program missing libraries.

Moreover, we need to make sure that dependencies are always consistent. Files in the database can depend on each other, sometimes in an indirect (transitive) way. First of all, we want to require all saved files to be valid (compilable); they can still contain incomplete proofs terminated with the Coq `Admitted` keyword or its equivalents for other provers. For a valid saved file we want to ensure that the current version remains valid after changes to the files it relies on. Some provers already include compatibility verification mechanisms. Coq stores the checksums of files to ensure binary compatibility between compiled proofs. To solve the problem, we have to consider the static and the dynamic approach.

The dynamic approach is the convention that a file always refers to the latest versions of other files. It means that saving any change to a file will induce a costly recompilation of all files it depends on. Another problem is that changing definitions deep inside a library will make many developments incompatible and thus correct files will stop working. Saving only valid files does not solve this problem since the objects they contain (their interface) might be modified too. This approach also makes older versions of existing files immediately obsolete.

The opposite approach is static linking, where a saved file always refers to the same version of other files. In other words, we never change a file, but rather add a new version of that file, with a fresh name. This means that the user will have to manually update the version number of files that are referred to if newer versions of those become available. The main advantage of this approach is that of integrity: provided you can safely assign new version numbers, you can enable concurrent access. Moreover, changing a file will never *break* another file. However, when changing a file deep in the library, one has to manually modify all the files in the dependency chain between that file and the files in which the changes should be reflected, which can sometimes be quite heavy.

3.4   Towards a Hybrid Approach

We believe that the static approach is a more adequate way to store older (historical) versions of a given file, whereas up-to-date files should use the dynamic approach towards dependency. This way, older versions of files still make sense by statically referring to older versions of files they depend on. The latest versions can remain up-to-date with their immediate dependencies by being dynamically linked to them, i.e. recompiled when new versions of those files are saved. It might happen that such a file might not be valid anymore because of changes made to its dependencies: to keep validity we have to make it link statically to the suitable previous version.

To help with this version compatibility issue, we propose a three-colour scheme:

- A file is labelled as *red* (i.e outdated) if it depends statically on an older-than-latest version of another file.
- A file is labelled as *yellow* (i.e tainted) if it depends only on the latest versions of other files, and one or more of those files have a yellow or red status. Yellow status thus tracks the files which are indirectly lagging behind.

– A file is labelled as *green* (i.e. up-to-date) if it depends only on the latest versions of other files, and all those files are also labelled as up-to-date (green status).

The separation between the yellow and red files comes from the fact that red files have to be manually updated to become green again (i.e. by creating a new version of them), whereas yellow files might be fixed by updating the red files that taint them.

The switching to red status can be automated by rewriting `Require` statements on-the-fly to make them refer statically to the last suitable version of the file depended on. This means that fixing a red file can give red status to yellow files that it was tainting, thus pushing the problem upwards in the dependency tree.

If the user wants to export a file together with its dependencies from the repository, a mechanism can be used to convert long file names (with version number) to short ones. The case might arise where a file would refer, directly or by transitivity, to an old version of itself. We can either forbid this or generate fresh file names using standard suffixing techniques.

The procedure of updating the prover itself, although intended to be rare, will be critical. Here a decision will have to be made whether to port all possible versions or only the newest versions of each files and their dependencies. The current system is clearly not yet designed to enable such updates without putting it offline and porting files manually, but such a feature should definitely be designed and implemented.

4   Prototype

4.1   Implementation

To experiment with our idea, we have created a prototype implementation based on off-the-shelf components as much as possible. We chose Coq as our target proof assistant. We used the MediaWiki code-base, the `coqdoc` documentation generator for Coq and PROOFWEB for Coq (Fig. 3). The `coqdoc` tool was modified to generate wiki format rendered pages.

When the user opens a page of our wiki, he/she is presented with a viewing page where the usual contents area is replaced by three subframes. One frame shows the rendered version of the current document, the second one shows the current proof state and the third one displays the Coq error messages.

The user may press the 'up' and 'down' buttons to step over the proof and examine both the proof state and Coq messages. A background coloring scheme allows the user to keep track of the part of the script that was already processed.

The proof is rendered, that is identifiers are colored and linked to their definition, mathematical LATEX comments are rendered, links to internal wiki pages lead to those wiki pages and links to Coq standard library objects lead to the documentation on the Coq website (Fig. 4).

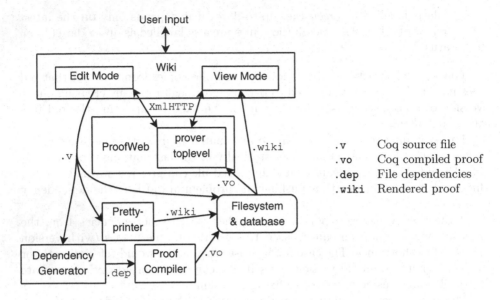

Fig. 3. Data flow diagram for our prototype wiki

The page also includes standard wiki elements, one of which is the 'edit' button. When the user starts editing the page, a similar page is presented, but with the raw proof script (no rendering) in a modifiable text box. The user may modify the script and use the 'up' and 'down' buttons to step over the proof in a similar way as in the view mode (Fig. 5). The processed part of the buffer is frozen.

When satisfied with his work, the user can save the proof. The contents of the buffer are processed in three ways:

- The raw script is saved in the database, to be used by following edits.
- The file is compiled and the corresponding .vo file is stored in the filesystem for processing of files that would include it using **Require** statements.
- A rendered version is generated by **coqdoc** and saved in the database to be displayed in view mode.

The user can see the history of any page as well as display the differences between the sources of any versions, using built-in MediaWiki routines. The textual search mechanism allows to query the source Coq files for any terms.

4.2   Security and Efficiency

The security and efficiency of the server are crucial since unavailability of the proof wiki would make the users not only unable to work, but also unable to access their own files. The security and efficiency of the architecture relies on the security of PROOFWEB, the underlying wiki, the compilation and rendering processes and the communication mechanism.

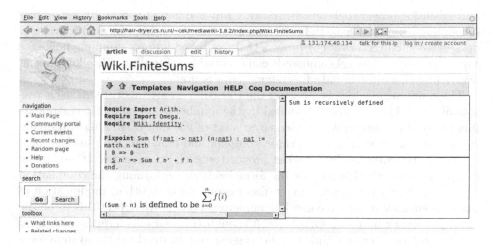

Fig. 4. Screenshot of the prototype showing the rendered version of a a Coq file. The verified part of the edit buffer is colored. The state buffer shows the state of the prover, there are no Coq warnings.

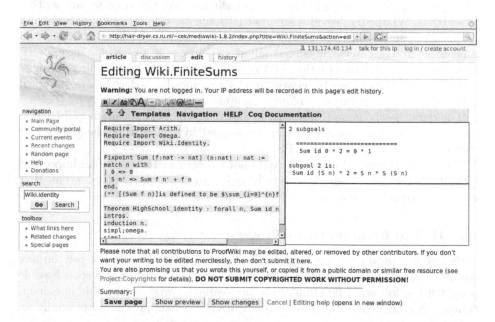

Fig. 5. Screenshot of the prototype showing the editing of the corresponding source file. The verified part of the edit buffer is colored and frozen.

The security and efficiency of PROOFWEB are described in detail in [1], we remind here the most important issues. The solution adopted is *sandboxing*: the PROOFWEB server process is run in a chrooted environment as a non privileged user without network access. The permissions include only reading server files and executing prover toplevels. Provers are run as different users with a modified scheduling policy and have rights to read only the prover libraries and to write in designated subdirectories. For provers that are based on programming language toplevels issuing toplevel commands can be disabled. Finally to disallow storing overly large amounts of data filesystem quota is used.

The sandboxing which is a part of the PROOFWEB architecture makes it reasonably safe. The efficiency is divided by the number of users, but it is straightforward to distribute prover sessions over a set of machines. We are additionally running a Coq dependency generator, compiler and renderer. We run these processes in the same sandbox as the prover toplevel, so we expect them to be comparably secure. However for big formalizations performing the compilation can be costly. Specially when many files depend on each other, modification of one of them may require recompiling numerous proofs. We expect this to be the main bottleneck of a wiki for proofs. Although this can also happen with local proof interfaces, here multiple parallel sessions might overload the system for a longer period.

The proof text verification that PROOFWEB does is independent from page serving performed by the web server and MediaWiki, so we can analyse the latter separately. Wikis are quite secure and efficient. At the time we are writing this article, Wikipedia provides servers that have more than 3 million users and 1.6 million articles without significant efficiency issues. An issue that is often a problem in wikis is vandalism. Disallowing edition by particular users or IP addresses is a common practice, and is already supported in MediaWiki. Discovering vandalism in our framework may sometimes be easier than in standard wikis, since incorrect proofs no longer compile.

The data that is being transferred to and from the wiki is usually public, still the communication mechanism can be secured by configuring the web server that serves the wiki to use HTTPS.

If the wiki is secured properly we do not expect crackers to be an important issue. However the efficiency seems to be quite fragile, in particular it seems that our architecture is quite vulnerable to denial-of-service attacks.

4.3   How to Integrate Other Provers

Although our prototype has been implemented for Coq, we do not rely on any specific Coq feature. We think that extending the wiki to other provers is feasible provided the following functionalities are available: wiki compatible documentation renderer, dependency generator, PROOFWEB support and optionally an index generator.

The renderer does not need to be sophisticated, the only mandatory feature of the renderer is distinction between active proof script from comments. Other features like syntax highlighting and links are not necessary, although they allow a more wiki-like interaction.

The wiki needs to know how to call the dependency generator of the prover, to know what files need to be updated if a particular file is modified. If the prover has a compiler, the wiki needs to know how to compile proofs. The wiki also needs to be able to identify statements that refer to other files during the interactive session.

An optional element is an index generating utility. It is needed for the wiki to distinguish concepts from the new prover's language. This allows not only nice index pages in the wiki, but also searching for particular prover objects, like only definitions or theorems.

Finally PROOFWEB needs to be able to interact with the prover. It already supports some provers. To extend it to a new one, the client part needs to know how to find the ends of complete prover commands and the server part needs to know how to interface with the prover process, in particular it needs to know how to check if commands succeed and how to undo. The details of extending PROOFWEB to a new prover are described in [1].

5 Conclusion and Future Work

5.1 Future Work

The current architecture of the prototype is not satisfying since it relies on a double storage of files: in the database, and on the disk. We are also limited by the way MediaWiki handles its name space. If we adopt the static system where files are never modified, it can be worthwhile to consider moving all the data to the file system, and adopting an architecture where we can have a better control of the name space.

The static naming will require to implement a versioning system for the substitution of `Require` statements and the distributed generation of version numbers, then the three colour scheme will be added. A mechanism for importing and exporting parts of the library will also be necessary, to allow users to have a local copy on which to work without Internet access.

A milestone in this development will be the ability to actually import the Coq standard library and official users contributions to our repository. Only then will we be able to get user feedback and report on the suitability of the repository for Coq users.

The `coqdoc` tool is able to generate index files that contain all constants occurring in the library. We could use such a a feature to generate such wiki pages.

The basic textual search is very limited and proof assistants users often need query types that are far beyond the scope of textual search: find theorems about a given object or do pattern matching on theorem statements. This might be achieved by adapting the Whelp search engine to search our database: it will require a customisation of the indexing technology.

We could also experiment with more advanced rendering tools such as Helm and consider using MathML instead of (currently) HTML with LaTeXimages.

5.2 Conclusion

Although our prototype is still at a very early stage of development, our idea of combining a wiki web site with the PROOFWEB interface looks definitely promising. Surprisingly, we could achieve the current result without many modifications neither to the wiki code-base, nor to PROOFWEB. Most of the work was devoted to database modification and rendering.

We believe that formalising mathematics in a wiki system will foster more cooperation both within prover specific communities and between users of different provers, especially if we can make several provers coexist in the same repository. We also believe that such a project can act as a display of the work on formal proofs for a wider audience.

References

1. Kaliszyk, C.: Web interfaces for proof assistants. In: Autexier, S., Benzmüller, C. (eds.) Proceedings of the FLoCs Workshop on User Interfaces for Theorem Provers (UITP-06), Seattle, pp. 53–64 (To be published in ENTCS) (2006)
2. Paulson, L.D.: Building rich web applications with ajax. Computer 38(10), 14–17 (2005)
3. Nipkow, T., Paulson, L.C., Wenzel, M.: Isabelle/HOL. LNCS, vol. 2283. Springer, Heidelberg (2002)
4. Coq Development Team: The Coq Proof Assistant Reference Manual Version 8.0. INRIA-Rocquencourt (January 2005) URL :http://coq.inria.fr/doc-eng.html
5. Muzalewski, M.: An Outline of PC Mizar. Fondation Philippe le Hodey, Brussels (1993)
6. Bancerek, G., Rudnicki, P.: Information retrieval in MML. In: Asperti, A., Buchberger, B., Davenport, J.H. (eds.) MKM 2003. LNCS, vol. 2594, pp. 119–132. Springer, Heidelberg (2003)
7. Rudnicki, P., Trybulec, A.: On the integrity of a repository of formalized mathematics. In: Asperti, A., Buchberger, B., Davenport, J.H. (eds.) MKM 2003. LNCS, vol. 2594, pp. 162–174. Springer, Heidelberg (2003)
8. Grue, K.: Logiweb - a system for web publication of mathematics. In: Iglesias, A., Takayama, N. (eds.) ICMS 2006. LNCS, vol. 4151, pp. 343–353. Springer, Heidelberg (2006)
9. Asperti, A., Padovani, L., Coen, C.S., Schena, I.: Helm and the semantic math-web. In: Boulton, R.J., Jackson, P.B. (eds.) TPHOLs 2001. LNCS, vol. 2152, pp. 59–74. Springer, Heidelberg (2001) URL: http://helm.cs.unibo.it/smweb.ps.gz
10. Asperti, A., Guidi, F., Coen, C.S., Tassi, E., Zacchiroli, S.: A content based mathematical search engine: Whelp. In: Filliâtre, J.-C., Paulin-Mohring, C., Werner, B. (eds.) TYPES 2004. LNCS, vol. 3839, pp. 17–32. Springer, Heidelberg (2006), URL: http://www.bononia.it/~zack/stuff/whelp.pdf
11. Asperti, A., Coen, C.S., E.T., Zacchiroli, S.: User interaction with the Matita proof assistant. Journal of Automated Reasoning (To appear 2007)
12. Schurmann, C., Pfenning, F., Kohlhase, M., Shankar, N., Owre, S.: Logosphere. A Formal Digital Library Logosphere homepage: http://www.logosphere.org
13. Mizar Development Team: Mizar wiki (2006) URL: http://wiki.mizar.org
14. Niqui, M.: Cocorico: a Coq wiki (2005) URL: http://cocorico.cs.ru.nl/coqwiki
15. Aspinall, D.: Proof General: A generic tool for proof development. In: Schwartzbach, M.I., Graf, S. (eds.) ETAPS 2000 and TACAS 2000. LNCS, vol. 1785, pp. 38–42. Springer, Heidelberg (2000)

Revisions as an Essential Tool to Maintain Mathematical Repositories

Adam Grabowski[1] and Christoph Schwarzweller[2]

[1] Institute of Mathematics, University of Białystok
ul. Akademicka 2, 15-267 Białystok, Poland
`adam@math.uwb.edu.pl`
[2] Department of Computer Science, University of Gdańsk
ul. Wita Stwosza 57, 80-952 Gdańsk, Poland
`schwarzw@math.univ.gda.pl`

Abstract. One major goal of Mathematical Knowledge Management is building extensive repositories, in which the mathematical knowledge has been verified. It appears, however, that maintaining such a repository is as hard as building it – especially for an open collection with a large number of contributors. In this paper we argue that even careful reviewing of contributions cannot cope with the task of keeping a mathematical repository efficient and clearly arranged in the long term. We discuss reasons for revisions of mathematical repositories accomplished by the "core implementors" and illustrate our experiences with revisions of MML, the Mizar Mathematical Library.

1 Introduction

Mathematical knowledge management aims at providing both tools and infrastructure supporting the organization, development, and teaching of mathematics using computers. Large repositories of mathematical knowledge are here of major concern since they provide users with a base of verified mathematical knowledge. We emphasize the fact that a repository should contain verified knowledge only: we believe that (machine-checkable) proofs necessarily belong to each theorem and therefore are an essential part of a repository. For repositories in the sense of Mathematical Knowledge Management community this implies even more: proofs should not only be understandable for the machine, but also – for human users of the repository.

From this follows that mathematical repositories are more than collections of theorems and proofs accomplished by a prover or proof checker. The overall goal is not proving a theorem – though this still is an important and challenging part – but presenting definitions and theorems so that the "natural" mathematical buildup remains visible. Theories and their interconnections must be available, so that further development of the repository can be based upon these. Being not trivial anyway, this becomes even harder to assure for an open repository with a large number of authors.

M. Kauers et al. (Eds.): MKM/Calculemus 2007, LNAI 4573, pp. 235–249, 2007.

So, how to tackle this task? One possibility, of course, is reviewing submissions. Reviewing improves the quality of knowledge and proofs added to the repository, but we shall illustrate that in the long run reviewing cannot ensure that a mathematical repository meets the demands mentioned above. We therefore claim that revisions are an essential part of maintaining mathematical repositories: in order to keep it clean and attractive for users, from time to time a "core team" has to check and improve the organization, quality, and proofs of a mathematical repository.

In the following section we describe and discuss the goals and benefits of revisions compared to a straightforward reviewing process. Then, after a brief introduction to the Mizar system [11], we consider the reviewing process for submissions to MML in Section 3. We describe reviewing criteria and show which insufficiencies can be handled by reviewing. In contrast, Section 4, is devoted to revisions of MML illustrating on the one hand what kind of improvements reviewing cannot perform, on the other hand the role of revisions in maintaining MML. This is the process done mainly by human hand as of now, in the next section we discuss some issues concerned with this activity and describe some traps the developer could meet when enhancing the library. Then we conclude showing some related work and drawing some remarks for future.

2   The Need for Revisions

The goal of a revision is to improve the mathematical repository. In contrast to reviewing submissions, however, here the attention is turned to the repository as a whole, not to a single, new part of it. Consequently, motives for revisions can be for example:

- keeping the repository as small as possible,
- preserving a clear organization of the repository in order to attract authors,
- establishing "elegant" mathematics, that is e.g. using short definitions (without unnecessary properties) or better proofs.

Note that all these points characterize a qualitative repository and can hardly be achieved by reviewing single submissions. Of course there are different possibilities to achieve the points mentioned. Improving the prover e.g. can shorten proofs and hence – simplify the repository. (Re-)organizing a mathematical repository probably demands manipulating the whole file structure, not only the files themselves. Therefore we decided to classify revisions based on their occasion, that is on which kind of insufficiency we want to address. Based on our experiences with the Mizar Mathematical Library we distinguish four major occasions for revisions:

1. improving authors' contributions;
2. improving the underlying prover or proof checker;
3. reorganizing the repository;
4. changing representation of knowledge.

Improving an author's contribution is the classical task of reviewing and is of course to be recommended for mathematical repositories too: nomenclature can be polished up to fit to the yet existing one, definitions can be improved, that is e.g. generalized if appropriate. Proofs are also a matter of interest here, especially keeping them as short as possible, yet still understandable is of major concern. In a large, open repository however, authors sometimes may prove and submit theorems or lemmas not being aware that those are already part of the repository. Similarly, special versions of already included theorems can happen to be "resubmitted". It is doubtful, that this kind of flaws will be detected by ordinary reviewing.

Strenghtening the underlying prover or proof checker has also an impact on the repository. Proofs can be shortened or rewritten in a more clear fashion, both being fundamental properties of attractive mathematical repositories. Even more, theorems in such collection may now be superfluous, because the improved prover accepts and applies them automatically. A typical example here is the additional inclusion of decision procedures.

Reorganizing the repository deals with the fact that a repository is built up by a large number of contributors. For their development authors (should) use already existing theories as a basis. To establish their main results, however, they often have to prove additional theorems or lemmas just because the theory used does not provide them yet. So, these additional facts have to be put in the right place of the repository. Otherwise, it will be hard for other authors to detect them or at least searching the repository becomes less comfortable. In the same direction goes the building of monographs: a frequently used theory should be handled with extra care. Not only should all related theorems be collected in a distinguished place, but also still lacking theorems be complemented, in order to ease working for further authors. These tasks can only be accomplished when considering the repository as whole, that is by revisions.

The last point concerns the development of a repository in the long term. What if after while it turns out that another definition or representation of mathematical objects would serve our purposes better than the one chosen? Should it be changed? Note that a lot of authors already could have used these objects in their proofs, that is changing the definition or representation would imply changing all these proofs – and of course one cannot force authors to redo all their proofs. On the other hand, including both definitions or representations leads to an unbalance: the theory of the new prefered version is much less developed than the one of the old version, so authors hardly will base their developments on the new one. Again, the solution is a revision: In the best case definitions and representations are changed by a "core team", so that ordinary users can furthermore use all theorems without even noticing they changed.

In the following sections we will illustrate these considerations by examples taken from the Mizar Mathematical Library and in particular show how revisions maintain mathematical repositories.

3   Review of MML Submissions

Reviews of submissions to the MML – as reviews of ordinary submissions for conferences or journals – have the overall goal to check whether a submission should be accepted (for inclusion into the MML) and simultaneously improve the quality of a submission. For mathematical repositories, however, the criteria for acceptance and improvements are somewhat different.

Certainly the contents of a submission for a repository should likewise be original and interesting. Original here, of course, means that definitions and theorems presented are not part of the repository, yet. This is easy to check for the main theorem of a submission. For technical lemmas used to establish this main result, however, this task is much more difficult. So, for example, a reviewer will probably neither know nor be willing to check whether a theorem like

```
theorem
  for F,G being FinSequence, k being Nat st
    G = F|(Seg k) & len F = k + 1 holds F = G ^ <*F/.(k+1)*>;
```

is already included in a repository. Even if the textual search via grepping is no longer the only method to find such repetitions since the MML Query by Grzegorz Bancerek [3] is available, even after the volunteer will learn how to use this system, still there is no single automated bunch of tools which removes all repeated theorems effectively. Furthermore, the motivation to check these things in detail will be even decreased, because such a point will not decide between acceptance and rejection.

The question whether a submission is interesting should be handled more liberally. Of course, the usual issues, that is the quality of the main results, apply here also. There is, however, another kind of submissions to repositories: the one that deals with the further development of (basic) theories. This concerns collections of basically simple theorems providing necessary foundations so that more ambitous developments can be easier accomplished. Usually, these are theorems, that easily follow from the definitions, however are so often used, that repeating the proof over and over again is hardly acceptable. Examples here are the theories of complex numbers or polynomials, where among other things we can find the following theorems.

```
theorem
  for a,b,c,d being complex number st
    a + b = c - d holds a + d = c - b;
```

```
theorem
  for n being Ordinal,
      L being add-associative right_complementable add-left-cancelable
          right_zeroed left-distributive (non empty doubleLoopStr),
      p being Series of n,L holds
  0_(n,L) *' p = 0_(n,L);
```

Though hardly interesting from a mathematical point of view, such theorems are important for the development of a repository and should therefore be considered as interesting, too.

Improvements of a submission are a more difficult issue. Firstly, we can consider definitions and notations contained in the submission. Can they be arranged more sparsely, that is can the results be established based on fewer axioms? Is it possible or reasonable (in the actual repository) to generalize the definitions? This also applies to theorems. Note that the theorem from above, though applicable to polynomials, in fact is stated for power series. Again to address these issues a high knowledge of the repository by the reviewers is necessary.

Secondly, when it comes to proofs, there are hardly any guidelines, because proving in particular is a matter of style. We can hardly force an author to change his (finished) proof into another one using completely different proof techniques. What we can do, is to suggest improvements for the presented proof. We can, for example, propose a more accurate use of the proof language to get more elegant or better readable proofs. Or we can give pointers to other theorems in the repository that allow to shorten the proof.

Based on these considerations a reviewing process for Mizar articles, that is for submissions to the Mizar Mathematical Library, has been introduced. Using basically the commonly used scheme accept/revise/reject (and apart from its descriptive grade) the rating of a submission can be[1]

A. accept

requires editorial changes only, which can be done by the editors
B. accept

requires changes by the author to be approved by the editors
C. revise

substantial author's revisions necessary, resubmission for another review
D. decision delayed

revision of MML necessary
E. reject

no hope of getting anything valuable

The most important issue here, of course, is the question whether an article should be included in MML. Note, that there are two grades (A and B) for acceptance. The reason is that accepted articles should be included in the MML as soon as possible to avoid duplication of results during the reviewing phase.[2] So, while submissions rated B or C need feedback from the authors, submissions rated A can be added to MML without further delays.

The most interesting point is D. Note that here already the problem of a revision of the whole repository is addressed. Reviewing can point out that – though the author has proven his main results – the way Mizar and MML support establishing the presented results is not optimal and should be improved.

As the most notable example here, we can cite the newly submitted definition of a kind of a norm for the elements of the real Euclidean plane, which are defined in the Mizar library just as finite sequences of real numbers.

---

[1] There has been and is still going on an email discussion about these options.

[2] There even has been the proposition of making public submissions before reviewing to avoid this problem, but we are not aware of a definite decision concerning this point.

```
definition let n be Nat, f be Element of TOP-REAL n;
  func |. f .| -> Real means
    ex g being FinSequence of REAL st
      g = f & it = |. g .|;
end;
```

where |. g .| is a usual Euclidean norm which was introduced in the MML before as

```
definition let f be FinSequence of REAL;
  func |. f .| -> Real equals
:: EUCLID:def 5
    sqrt Sum sqr f;
end;
```

After the change of the loci type (the submission obtained grade D, of course) from **FinSequence of REAL** into **real-yielding FinSequence** in the EUCLID article the earlier definition was no longer needed which helped to simplify the structure of notions in this new submission.

The decision is not a typical result of majority voting, because referees giving C grade point out possible improvements, so usually the lowest grade counts (luckily, in case of E marks, all three referees agreed).

To summarize the grades for 2006, let us look at Table 1.

Table 1. Number of submissions to the MML and their grades in 2006

	all	A	B	C	D	E
items	39	6	4	20	6	3
% of total	100	15.4	10.2	51.3	15.4	7.7

Basically, all ten submissions graded A and B were included into the MML, and among C and D candidate articles, which were returned to authors, other 15 were accepted; in total there were 25 Mizar articles accepted in 2006, the first year the reviewing procedure as described above was introduced.

All in all we have seen that reviewing MML submissions indeed addresses only the first point mentioned in Section 2. Of course a thorough reviewing process will improve the quality of MML articles and may even pilot authors into the direction of a good style of "Mizar writing". As we can conclude from Table 1, this is the case of the majority of submissions because the authors should enhance the articles according to the referees' suggestions. However there remain situations in which the MML as a whole should be improved; in the long term mere reviewing of submissions cannot avoid this. Here even carefully reviewing of Mizar articles – as already indicated by rate D above – can only help to detect the need for such revisions.

4   MML Enhancing

The Library Committee has been established on November 11, 1989. Its main aim is to collect Mizar articles and to organize them into a repository – the MML. Recently, from this agenda a new additional one was created – the Development Committee, which takes care of the quality of the library as a whole.

4.1   Types of Revisions

For the reasons we tried to point out before, the Mizar Mathematical Library is continuously revised. Roughly speaking, there are different kinds of revisions:

- an authored revision – consists of small changes in some articles in the library when somebody writing a new article notices a theorem or a definition in an old article that can be generalized. This is also the case of D grade as described above. To do this generalization, sometimes it is necessary to change (improve) some older articles that depend on the change. As a rule, a small part of the library is affected.
- an automatic revision – takes place frequently whenever either a new revision software is developed (e.g. software for checking equivalence of theorems, which enables to remove one or two equivalent theorems) or the Mizar verifier is strengthened and existing revision programs can use it to simplify articles.
- a reorganization of the library – although was very rare before, as of now it happens rather frequently. It consists in changing the order of processing articles when the Mizar data base is created. Its main steering force is the division of the MML into concrete and abstract parts.

4.2   MML Versions

Apart of the Mizar version numbering, the MML has also separate indexing scheme. As of the end of 2006, the latest official distribution of the MML has number 4.76.959.

As a rule, the last number, currently 959 shows how many articles are there in the library (this number can be sometimes different because 26 items were removed so far from the MML, but some additional items such as EMM articles, and "Addenda" which do not count as regular submissions, were added). The second number (76) changes if a bigger revision is finished and the version is made official. Although it is relatively small comparing with the age of the library, the changes are much more frequent.

4.3   Some Statistics

The policy of the head of Library Committee – to accept virtually all submissions from the developers and, if needed, enhance it by himself, was then very liberal. For these nearly twenty years there were only three persons taking a chair of a head of the Committee (Edmund Woronowicz, Czesław Byliński, and currently Adam Grabowski); their decisions were usually consulted with the other members of the committee, though.

Such an openness of the repository was justified: in the early years of the Mizar project the policy "to travel to Białystok and to get acquaintance with the system straight from its designer" resulted in the situation all authors knew each other personally, now the situation changed.

The MML evolved from the project, frankly speaking, considered rather an experiment of how to model mathematics to allow many users benefit from a kind of parallel development. Now, when the role of the library is to be much closer to the reality and the MML itself is just one among many mathematical repositories, the situation is significantly different.

Table 2. Submissions to the MML by year

Year	Add.	1989	90	91	92	93	94	95	96	97	98	99	2000	01	02	03	04	05	06
Articles	19	65	136	46	48	33	33	35	57	39	47	65	54	33	42	54	80	48	25

As it can be seen, the first two years were extremely fruitful. No doubt, the first one was most influential, when the fundamentals, such as basic properties of sets, relations, and functions, the arithmetics, and vector spaces were established – to enumerate among many these most important. Some articles from that time were more or less straight translation from those written in older dialect of the Mizar language (Mizar PC, Mizar-2 etc., see [9]). Especially the subsequent year – 1990, when many authors could benefit from introducing the basics, hence they were able to work on various topics in parallel, brought into the Mizar Mathematical Library the bigest number of submissions so far (136 by year), then the number stabilized.

4.4   Towards Concrete and Abstract Mathematics

As it was announced in 2001 [15], the MML will be gradually divided into two parts. As the library is based on the Tarski-Grothendieck set theory, the part devoted to the set theory (and related objects, as relations, functions, etc.) is indispensable. There is, however, huge amount of knowledge for which set theory is essential, but basing on the notion of structure by means of the Mizar language.

There are three parts of the Mizar Mathematical Library:

- concrete, which does not use the notion of structure (here of course comes standard set theory, relations, functions, arithmetic and so on);
- abstract, i.e. STRUCT_0 and its descendants, operating on the level of Mizar structures; both parts are not completely independent – here the concrete part is also reused (abstract algebra, general topology including the proof of the Jordan Curve Theorem, etc.);

– SCM, the part Random Access Turing Machines are modelled, i.e. mathematical model of a computer is described.

This division is reflected in the file mml.lar in the distribution in which is the order of processing of articles when creating Mizar data base is given – the "concrete" articles go first, at the end are those devoted to the SCM series. The process of separating these three parts is very stimulating for the quality of the Mizar library – many lemmas are better clustered as a result of this activity.

As of the beginning of 2007, the division can be summarized in Table 3.

Table 3. Three parts of the MML

Part	Number of articles	% of total
concrete	266	27.70
abstract	640	66.67
SCM	54	5.63
Total	960	100.00

Note that apart of the revisions suggested by the referees when giving D-grades, any user can via TWiki mechanism suggest the change; he may of course also send improved version via email to the Library Committee; as an example, lemmas needed for the Gödel Completeness Theorem were reformulated to provide its better understanding as a result of the external call.

4.5 Library Management

As a first tool of collaborative work on the library we can enumerate Mizar TWiki (wiki.mizar.org) which gradually changes its profile from an experimental – and rarely used – forum into the place where suggestions/experiences with the MML can be described.

As the most important, and probably one of the better known MML tools, we can point out MML Query [3]. It has proven its feasibility when subsequent EMM items were created. Also researchers, when writing their Mizar articles, can find it useful. But usually, typical author does not care too much if his lemma which takes some ten lines of Mizar code is already present in the library. Actually, searching for such auxiliary fact can take much more time than just proving it. This results in many repetitions in the library MML Query cannot cope with. And although the author can feel uncomfortable with multiple hits of the same fact, annotating such situations and reporting it to the MML developers is usually out of his focus.

This is the area where another tool comes in handy. Potentially very useful for the enhancement of the MML as a whole, MoMM (Most of Mizar Matches) developed by Josef Urban was primarily developed to serve as assistant during authoring Mizar articles [18]. It is a fast tool for fetching matching theorems, hence existing duplications can be detected and deleted from the MML (according to [18], more than two percent of main Mizar theorems is subsumed by the others). The work with the elimination of these lemmas is still to be done – many of detected repetitions are useful special cases so their automatic removal is at least questionable.

Another popular software, MML CVS – the usual concurrent version system for the MML was active for quite some time, but then was postponed because the changes were too cryptic for the reader due to the lack of proper marking of items. Actually, one of the most general problems is that there are no absolute names for MML items and the changes are usually too massive to find out what really matters.

5   Traps for the Developer

Usually, the revision process via generalization of notions improves the MML. There are some dangerous issues, however; we address some of them in this section.

5.1   Mind the Gap!

Let us cite an example from the article ABIAN [14]:

```
definition let i be Integer;
  attr i is even means
:: ABIAN:def 1
    ex j being Integer st i = 2*j;
end;
```

These are usual definitions of odd and even integer numbers.

```
definition let i be Integer;
  attr i is odd means
    ex j being Integer st i = 2*j+1;
end;
```

Then, among the others, the theorem stating that all integer numbers are either odd or even, was proven; the proof was very simple, but there was something in it to do, at least both definitions were involved.

```
theorem LEM:
  for i being Integer holds i is odd or i is even;
```

However, after the revision (which was done by the author of ABIAN, after all), the second definition got simplified as follows:

```
notation let i be Integer;
  antonym i is odd for i is even;
end;
```

Still, it seems perfectly correct, introducing antonym we obtain the law of excluded middle automatically, in a sense, and the proof of the above lemma labeled LEM was no longer necessary.

But we made a step too far, as it seems. Because in the definition of even number, the `integer` was not needed (remember the type `Integer` is a shorthand for `integer number`), it was dropped in both – definition of an attribute and its antonym, and the latter got simplified into the form:

```
notation let i be number;
  antonym i is odd for i is even;
end;
```

Unfortunately, e.g. number `Pi` (the Mizar symbol for the usual constant π) can be proven to be odd which can be considered really odd. Observe that any automation of the process of dropping assumption about the types of used loci in the definition of attributes, however possible, could be dangerous.

5.2 Permissive Definitions

There are two unities for vector spaces defined in the MML – with symbol 1. and 1_, and the following definitions:

```
definition let FS be multLoopStr;
  func 1.FS -> Element of FS equals
:: VECTSP_1:def 9
    the unity of FS;
end;
```

```
definition let G be non empty HGrStr such that G is unital;
  func 1_G -> Element of G means
:: GROUP_1:def 5
    for h being Element of G holds h * it = h & it * h = h;
end;
```

where `multLoopStr` is `HGrStr` enriched by an additional selector, namely `unity` and the adjective `unital` in the permissive assumption (after `such that`) assures that the proper neutral element exists.

At first glance, the earlier approach is better – the less complicated a type in a locus, the less problems we have to assure the required type. In the second definition however, the underlying structure has only two selectors instead of three.

5.3 Meaningless Predicates

Suppose we have the following:

```
definition let a,b be natural number;
  assume a <> 0;
  pred a divides b means
    ex x being natural number st b = a * x;
end;
```

Well, we can freely delete the assumption, in Mizar predicate definitions do not require any correctness conditions proven. But, if we forget for a while that within the MML division by zero *is* defined, does it make any mathematical sense for *any* pair of natural numbers? According to the current policy of the Library Committee, we allow for any such underspecification.

5.4 Apparent Generalizations

There are sometimes cases the price for the revision is too high comparing to gains or the enhancing is apparent. If we remind that pathwise connectedness of the topological space T denotes the existence of a function from the unit interval into T which has the values a and b in 0 and 1, respectively, for any pair of points a, b of T, a *path* between two points is just an underlying mapping, if it exists. It is enough however to have an assumption about the existence of appropriate function for just the pair of points currently under considerations.

```
definition let T be TopStruct; let a, b be Point of T;
  assume a, b are_connected;
  mode Path of a, b -> Function of I[01], T means
:: BORSUK_2:def 2
    it is continuous & it.0 = a & it.1 = b;
end;
```

With no doubt, the assumption of the existence of a path not for arbitrary, but just for these two fixed points is more general, the gains from stating every now and then that considered two points can be connected by a path, are at least doubtful. Similarly, we often write `Abelian add-associative right_zeroed right_complementable RealLinearSpace-like (non empty RLSStruct)` dropping an attribute or two to have slightly more general setting instead of using the mode `RealLinearSpace` which is equivalent to that complicated string above.

6   Related Work

Contemporary standards of the publishing process open some new possibilities – there are many journals online, Springer also announces his books/proceedings at their webpage. Paper-printed editions have some obvious limitations, vanishing for electronically stored and managed repositories of knowledge. We can notice, as an example, new functionalities of [10] in comparison to (even online) version of Abramowitz and Stegun [1].

Of course, what is published on paper, is fixed. We can mind some real-life situations – rough sets as an example of obtaining new results via a kind of revision process (originally considered to be classes of abstraction with respect to some equivalnce relation, then some of its attributes were dropped to generalize the notion – see [6], [7]); also Robbins algebras and related axiomatizations of algebras are a good example, when a classical problem could be rewritten and reused when solved. In the aforementioned examples these were subjects for another papers, within the computerized repository the enhancement (the generalization of results) can be obtained via revision process.

As a rule, building an extensive encyclopedia of knowledge needs some investment; on the one hand, it can be considered by purely financial means as "information wants to be free, people want to be paid" [2]. That is the way Wolfram MathWorld [19] has been raised, as a collection closed in style, in fact authored by one person, Eric Weisstein.

But right after this service has been closed due to the court injunction, it soon appeared that the need to bridge this gap is that strong – many volunteers were working to develop a concurrent service to that of Wolfram's, but of the more open type, based on the mechanism similar to Wiki.

The effort of PlanetMath is now a kind of Wikipedia for mathematics (in fact they even cooperate closely); with its content somewhat questionable because virtually anyone can contribute, but frankly speaking, also nobody really asks about the verification of other, even commercial resources. Although we believe the use of proof checkers could enable the automatic verification of the proofs; still the correctness of the definitions, i.e. how the encoded version reflect real mathematical objects, is under question only human can answer to a full extent.

But even in the projects of GNU type, people want to get their payment in another form: at least the added annotation such as "This article is owned by...", as in PlanetMath, which can also be considered a kind of motivation to keep higher standards of the encyclopedia since the authorship is not fully anonymous.

In the MML the authorship is somewhat fixed, there were however, especially recently, cases when the parts of submissions moved between them so that the authorship actually exchanged (as for example, with the formalization of the Zorn Lemma, originally created by Grzegorz Bancerek, and now, after the changes concerned with the move of this article to the concrete part, attributed to Wojciech Trybulec). In a sense, the Mizar library is much closer to Planet-Math, but the official distributions are created by the Library Committee which decides about the acceptance of revisions.

7   Conclusions

To meet the expectations of researchers being potential users of repositories of mathematical knowledge, such collections cannot be frozen. The availability of the contemporary electronic media open new directions of the development of the new encyclopedias yet unavailable for their paper counterparts. The need of the enhancement stems not only of the fact there may be some obvious mistakes in the source; the reasons are far more complex.

In the paper, we tried to point out some of the issues connected with the mechanism of revisions performed on the Mizar Mathematical Library, large repository of computer-verified mathematical knowledge. The dependencies between its items and the environment declaration (notation and especially, constructors) are as of now too complex to freely move a single definition or a theorem between separate articles.

In our opinion, the current itemization of the MML into articles does not fit the needs we expect from the feasible repository of mathematical facts; if we try to keep authors' rights unchanged, there is an emerging need to have some other, smaller items which guarantee the developer's authorship rights, a kind of ownership similar to that used in the PlanetMath project.

Also the better automation of the MML revision process is strongly desirable. However possible, at least to some extent, but due to some difficulties which can be met as we pointed out, the human supervision of such automatic changes will probably always be needed.

Acknowledgments

We are grateful to Andrzej Trybulec and Artur Korniłowicz for their continuous cooperation on the enhancement of the MML. The first author acknowledges the support of the EU FP6 IST grant TYPES (Types for Proofs and Programs) No. 510996. We acknowledge also anonymous referees for their useful suggestions, unfortunately due to space constraints not all of them were reflected in this final version.

References

1. Abramowitz, M., Stegun, I.A.: Handbook of Mathematical Functions; National Bureau of Standards, Applied Mathematics Series No. 55, U.S. Government Printing Office, Washington, DC (1964) see also
   http://www.convertit.com/Go/ConvertIt/Reference/AMS55.ASP
2. Adams, A.A., Davenport, J.H.: Copyright issues for MKM. In: Asperti, A., Bancerek, G., Trybulec, A. (eds.) MKM 2004. LNCS, vol. 3119, pp. 1–16. Springer, Heidelberg (2004)
3. Bancerek, G.: Information retrieval and rendering with MML Query. In: Borwein, J.M., Farmer, W.M. (eds.) MKM 2006. LNCS (LNAI), vol. 4108, pp. 65–80. Springer, Heidelberg (2006)
4. de Bruijn, N.G.: The Mathematical Vernacular, A Language for Mathematics with typed sets. In: Dybjer, P., et al. (eds.) Proc. of the Workshop on Programming Languages, Marstrand, Sweden (1987)
5. Davenport, J.H.: MKM from book to computer: A case study. In: Asperti, A., Buchberger, B., Davenport, J.H. (eds.) MKM 2003. LNCS, vol. 2594, pp. 17–29. Springer, Heidelberg (2003)

6. Grabowski, A.: On the computer-assisted reasoning about rough sets. In: Dunin-Kęplicz, B., et al. (ed.) Monitoring, Security, and Rescue Techniques in Multiagent Systems, Advances in Soft Computing, pp. 215–226. Springer, Heidelberg (2005)
7. Grabowski, A., Schwarzweller, Ch.: Rough Concept Analysis – theory development in the Mizar system. In: Asperti, A., Bancerek, G., Trybulec, A. (eds.) MKM 2004. LNCS, vol. 3119, pp. 130–144. Springer, Heidelberg (2004)
8. Kamareddine, F., Nederpelt, R.: A Refinement of de Bruijn's Formal Language of Mathematics. Journal of Logic, Language and Information 13(3), 287–340 (2004)
9. Matuszewski, R., Rudnicki, P.: Mizar: the first 30 years. Mechanized Mathematics and Its. Applications 4(1), 3–24 (2005)
10. Miller, B.R., Youssef, A.: Technical aspects of the Digital Library of Mathematical Functions. Annals of Mathematics and Artificial Intelligence 38, 121–136 (2003)
11. The Mizar Homepage, http://www.mizar.org/
12. Naumowicz, A., Byliński, Cz.: Improving Mizar texts with properties and requirements. In: Asperti, A., Bancerek, G., Trybulec, A. (eds.) MKM 2004. LNCS, vol. 3119, pp. 190–301. Springer, Heidelberg (2004)
13. PlanetMath web page, http://planetmath.org/
14. Rudnicki, P., Trybulec, A.: Abian's fixed point theorem. Formalized Mathematics 6(3), 335–338 (1997)
15. Rudnicki, P., Trybulec, A.: Mathematical Knowledge Management in Mizar. In: Buchberger, B., Caprotti, O. (eds.) Proc. of MKM 2001, Linz, Austria (2001)
16. Rudnicki, P., Trybulec, A.: On the integrity of a repository of formalized mathematics. In: Asperti, A., Buchberger, B., Davenport, J.H. (eds.) MKM 2003. LNCS, vol. 2594, pp. 162–174. Springer, Heidelberg (2003)
17. Sacerdoti, C.: From proof-asistants to distributed knowledge repositories: tips and pitfalls. In: Asperti, A., Buchberger, B., Davenport, J.H. (eds.) MKM 2003. LNCS, vol. 2594, pp. 30–44. Springer, Heidelberg (2003)
18. Urban, J.: MoMM – fast interreduction and retrieval in large libraries of formalized mathematics. International Journal on Artificial Intelligence Tools 15(1), 109–130 (2006)
19. Wolfram Mathworld web page, http://mathworld.wolfram.com/

The Layers of Logiweb

K. Grue

DIKU, Universitestparken 1, DK-2100 Copenhagen
grue@diku.dk
http://logiweb.eu/,
http://www.diku.dk/~grue/

Abstract. Logiweb is an open source, distributed system for publication of machine checked mathematics. It covers all aspects of electronic publishing: high typographical quality, archival, handling of references to previously published results, and publication of refereed volumes. The present paper is itself produced using Logiweb; and the paper is formally correct in the sense that it has been verified by Logiweb. The paper describes the implementation layers of the Logiweb system as seen by the user: the programming layer, the metalogic layer, the tactic layer, and the object proof layer.

1 Introduction

Logiweb [8,9] is a system for authoring, storing, distributing, indexing, checking, and rendering of "Logiweb pages". Logiweb pages may contain mathematical definitions, conjectures, lemmas, proofs, theories, journal papers, computer programs, and proof checkers.

The main features of Logiweb are that Logiweb offers authors unlimited notational freedom, offers unlimited choice of what kind of logic authors may use, allows authors to reference definitions and lemmas web-published by other authors, and still is able to verify the formal correctness of papers.

From the point of view of an author, Logiweb resembles Mizar [19,16] and TEX [12]: One prepares a source text `mypaper.pyk` with a text editor and runs a command like `pyk mypaper` to get it checked, rendered, and published.

One difference from Mizar is that, on Logiweb, one has to see the result in a web browser (even during the writing of papers where one typically does not web-publish each iteration). Papers are typically rendered in PDF using TEX. Another difference is that one may reference any Logiweb paper published on the Internet. A third difference is that Logiweb has no central authority: anybody can publish anything on the system as easily as one can publish pages on the World Wide Web. A fourth difference is that Logiweb is more resource demanding than Mizar and requires a modern PC to run smoothly.

Yet another difference from Mizar is that Mizar defines a syntax for source files whereas Logiweb leaves the definition of that to the user. This allows authors to write in a spoken math style like "limit as x tends to infinity of 1 over x" as in EzMath [17]. Actually, the name "pyk" of the authoring tool was constructed by

M. Kauers et al. (Eds.): MKM/Calculemus 2007, LNAI 4573, pp. 250–264, 2007.

removing "Vola" from "Volapyk" where "Vola" means "world" and "pyk" means "speak". Spoken mathematics like "limit as x tends to infinity of 1 over x" certainly does look like Volapyk to most people. Furthermore, the extension ".pyk" was not used by anybody. The liberty to define the source language also allows an author to choose e.g. `Lim x -> infty : 1 / x` for denoting a limit. The source language of Logiweb has been used for proof checking at DIKU since 1985.

Having mentioned the differences with Mizar it should also be noted that Mizar .miz and Logiweb .pyk files are not all that different, and the possibility of a miz2pyk filter is being investigated.

1.1 Internet Problems

If one author proves Zorns lemma in ZFC and another proves Hahn-Banachs theorem from Zorns lemma, then the correctness of the second proof depends on the correctness of the former. If the proofs were published as papers on Logiweb, then the paper proving Hahn-Banach would reference the one proving Zorn. When Logiweb checks the Hahn-Banach paper, it fetches the Zorn paper from the Internet and verifies that as well.

Now consider what happens if the author of the Zorn paper makes a slight change of the formulation of Zorns lemma. In that case, the latter paper may end up being incorrect.

Logiweb solves that using *immutability*. Immutability is normal in pure functional programming: if one computes $2::3$ then one gets a pair whose head is 2 and whose tail is 3. After that, one cannot change the head and tail of the pair. The pair remains immutable until it is garbage collected. If one wants a pair whose tail is 4 one has to compute $2::4$ which gives rise to a new pair without affecting the $2::3$ pair.

Logiweb pages are also immutable: once published, they cannot change. So now the author of the Hahn-Banach paper can be sure that even if a new version of the Zorn paper is produced, the Hahn-Banach paper will still reference the old Zorn paper.

But there still is a problem, since the author of the Zorn paper may garbage collect the old version. To counter for that, the author of the Hahn-Banach paper may mirror the Zorn paper. The Hahn-Banach author does so by writing `pyk HahnBanach -mirror` which makes the authoring tool resubmit all transitively referenced papers. Resubmitted papers are stored on the computer of the author, out of reach of other peoples garbage collectors.

The Logiweb software suite contains a demon named "logiweb" which must run on all sites that publish Logiweb papers. All logiweb demons in the world cooperate on indexing all Logiweb papers in the world, and no-one will notice them as long as they do what they are supposed to do. The Logiweb demon is the heart of the system. The pyk compiler is just one possible authoring tool (a wysiwyg authoring tool has been made but is not currently maintained).

The Logiweb demons ensure that, as long as at least one copy of the old Zorn paper remains accessible, no-one will notice that the author of the paper deleted the original.

1.2   Notational Problems

If the Zorn author for some reason uses $x \oplus y$ for the disjoint union of x and y, and if the Hahn-Banach author wants to use $x \uplus y$ instead, then the Hahn-Banach author may define $[x \uplus y \doteq x \oplus y]$ which makes Logiweb macro expand $x \uplus y$ to $x \oplus y$. This allows the Hahn-Banach author to write $x \uplus y$, but Logiweb will macro expand each $x \uplus y$ to $x \oplus y$ before proof checking.

If the Hahn-Banach author then wants to use $x \oplus y$ for some other operation, then the Hahn-Banach author may define his own $x \oplus y$ and define it to his liking. To Logiweb, each operator is identified by two natural numbers, a *reference* and an *index*. The reference is a big number which identifies the home paper of the operator (the paper which introduced the construct) and the index is a small number which distinguishes operators introduced by the same paper. Logiweb will not care about the two $\oplus$ operators having the same rendering. An author may fool himself and others by mistaking the two operators, but to Logiweb the two $\oplus$ operators are as different as any two other operators.

A malicious author could abuse notation to make readers think he has a checked proof of some lemma, but the fraud would be detected the moment somebody else try to use the lemma for further work.

1.3   Foundational Problems

Logiweb was constructed for supporting Map Theory (MT) [5,1] which is alien to "normal" logic: MT builds on lambda calculus [2], has ZFC-power, unlimited lambda abstraction, and equality, and it can e.g. well-order any set by recursive use of Hilberts epsilon operator. But it does not have logical connectives; they have to be defined. A large MT proof [7] has been formalized in Isabelle [18], but existing systems are not very good at supporting MT.

Instead of making a taylor-made system for MT, Logiweb was constructed such that it can support any theory equally well. Furthermore, Logiweb was constructed with human readability in mind. As an example, the text book [6] for teaching logic in computer science was written in a "Logiweb style" before Logiweb itself was designed. That allowed to collect the requirements needed to formalize a complete math book without sacrificing readability.

But the ability to support different kinds of logic raises new problems. Different authors are likely to use different axiomatic systems, and a single paper may use more than one axiomatic system. As an example, the present paper defines both propositional calculus Prop and first order logic FOL. To avoid conflicts, each lemma must state which axiomatic system it is relative to.

It may be non-trivial to use lemmas proved in one axiomatic system in proofs that are relative to another system. Using Prop lemmas in FOL is easy since, as we shall see, FOL builds on top of Prop. It is more cumbersome to prove ZFC lemmas in MT even though every ZFC lemma is provable in MT. If one has to use ZFC lemmas in MT, Logiweb offers two possibilities: The easy one is to state as an axiom scheme that any proved ZFC lemma automatically becomes an axiom of MT. The more cumbersome one is to define a proof tactic which, given a ZFC proof, translates the proof to an MT proof.

1.4 Rendering Problems

Logiweb renders pages using TEX and PDF. From August 11 to November 23, 2004, Logiweb also had support for MathML [15]. Support was removed, however, for several reasons: The automatically generated MathML could easily bring the users browser to the knees, TEX had lots of facilities for bibliographies, indexes, and tables of contents, and the very purpose of MathML is not particularly compatible with Logiweb. MathML provides a nice and easy way to get mathematics on the web and it is easy to write and modify by hand. But Logiweb has no need for a human friendly intermediate format here. Furthermore, PDF and TEX source can be fed directly from Logiweb into EasyChair or even a publishing house. The latter was done in [9].

It should be noted that the rendering machinery of Logiweb is not tied to TEX and PDF. TEX and PDF are only supported in the sense that TEX, BIBTEX, makeindex, and dvipdfm are external programs which Logiweb provides access to during rendering. All other rendering must be done by Logiwebs own machinery. As an example, Logiweb also generates html, but that is done internally.

So any user who wants Logiweb to render pages in MathML may define their own "renderer" and publish it as a Logiweb paper. That would allow anybody else to reference that renderer and themselves write papers which were rendered in MathML.

The OMDoc format [13] is not supported either. But, again, one may add support without opening the code of Logiweb itself. Like MathML, OMDoc seems best suited for information in flux, i.e. information which the author may change now and then without a need for keeping the old version. OMDoc is suited as an output format for Logiweb, i.e. as a format in which papers are rendered. But OMDoc invites inclusion of URLs inside documents. And one should avoid relying on URLs in Logiweb papers since the immutability of Logiweb papers prevents authors from repairing broken links.

Like OMDoc, Logiweb supports linear, tree-like, and DAG-like proofs. At the inter-proof level, DAG-like structures are supported by the ability to use lemmas in a non-circular manner. At the intra-proof level, DAG-like proofs are supported by the ability to use arbitrary elements from the set of premises for further reasoning. Tree-like proofs are supported by the cut sequent operation, and linear proofs are supported by lists formed using the cut sequent operation to bind lines together.

Contrary to OMDoc, Logiweb proofs are either formal (meaning they are checked mechanically) or informal (meaning they are not checked). A partly formal proof of a lemma L, however, may be represented defining a construct P to denote the conjunction of the unproved statements used in the proof and then stating and proving $P \vdash L$ instead of L. The proof of $P \vdash L$ can then be checked formally, and L will be easy to prove if the proof obligation P is proved later on on another Logiweb page.

One particular kind of rendering is rendering of executables. As an example, one could define the Unix "ls" command in a Logiweb paper. Rendering of the paper would then result in an executable named "ls" which does what Unix "ls" is supposed to do provided the definition of "ls" in the paper is correct.

1.5   Availability

Logiweb is available from http://logiweb.eu/. First beta release (version 0.1.1) was released on December 27, 2006. The current version is 0.1.5. Starting from version 0.1.1 it is the intension to make all updates backward compatible. All of Logiweb except its logo is available under the GNU public license.

At the time of writing, Logiweb runs at http://logiweb.eu, http://logiweb. imm.dtu.dk, and http://topps.diku.dk/logiweb. The two first sites support a Logiwiki each, allowing anyone to publish on Logiweb without installing local software. All three sites provide a tutorial.

At the time of writing, the present paper is available at http://logiweb.eu/ grue/pages/Logiweb+layers/latest/. Note that a Logiweb paper has one and only one Logiweb reference but may be available under many URLs. This may happen if a paper is mirrored. Furthermore, the URLs of a paper may change over time, but the Logiweb reference remains fixed.

The Logiweb reference of the present paper is BUpCgdix9lwGZKohkESIwWk vdwwIhPa08-fhoigBB and is based on a RIPEMD-160 hash key [3]. It is the task of the afore-mentioned Logiweb demons to keep track of the relationship between Logiweb references and URLs. Logiweb demons listen at particular URLs such as http://logiweb.eu/logiweb/server/relay/ called "relays". To look up a paper with a given Logiweb reference, append the URL of an arbitrary relay with the string "64/" if the Logiweb reference is expressed base 64, the Logiweb reference itself, and, optionally, a relative address such as "/2/body/tex/page.pdf" in case one wants a rendering rather than raw bytes.

At the time of writing, both http://logiweb.eu/grue/pages/Logiweb+layers/ latest/ and http://logiweb.eu/logiweb/server/relay/64/BUpCgdi x9lwGZKohkESIwWkvdwwIhPa08-fhoigBB/2/ point to the "front page" of the present version of the present paper. But the former may point to some other version in the future while the latter will continue to point to the present version as long as at least one copy of the present version remains on Logiweb. The 2/ at the end of the latter reference means "two levels above the raw bytes", i.e. at address ../../ relative to the raw bytes of the page.

1.6   Overview

The following sections present the programming, metalogic, tactic, and proof layers of Logiweb, respectively.

2   Programming Layer

2.1   Combinations

As an example of the programming and rendering facilities of Logiweb, we define
the binomial coefficient
$$\binom{n}{k} = \frac{n!}{k!(n-k)!}$$
A recursive definition reads:

$$[\binom{n}{k} \doteq \textbf{if } k = 0 \textbf{ then } 1 \textbf{ else } \binom{n-1}{k-1} \cdot n \text{ div } k]$$

As an example, we have $[\binom{4}{2} = 6]\dot{}$.

2.2   Syntax

The source text of Section 2.1 reads:

```
\subsection{Combinations}\label{sec:Combinations}

As an example of the programming and
rendering facilities of Logiweb, we
define the binomial coefficient

"[[[ (( n , k )) = (( n factorial /
k factorial,( n - k ) factorial )) ]]]"

\noindent A recursive definition reads:

"[[[ value define (( n , k )) as
if k = 0 then 1 else (( n - 1 , k - 1 ))
* n div k end define ]]]"

\noindent As an example, we have
"[[ ttst (( 4 , 2 )) = 6 end test ]]".
```

Section 2.1 contains an informal formula $((n,k)) = \ldots$, a value definition
(value define ...), and a test case (ttst ...). The formulas are mixed with TEX
source with formulas delimited by "[...]" (double and triple brackets are like
single brackets but also change to TEX math and display math mode, respectively).

The value definition is what actually defines the binomial coefficient to Logiweb. Logiweb uses that definition when verifying the test case. Logiweb simply ignores the informal formula. The test case in Section 2.1 reveals to
thorough readers that it is formally checked by an almost invisible dot
superscript.

2.3   Aspects

The present paper has an electronic appendix [10] which contains the following:

```
"[[ tex define (( n , k )) as "
\left( \begin{array}{l} "[ n ]"
\\ "[ k ]"
\end{array}\right)" end define ]]"
```

That defines how to render the binomial coefficient. Logiweb uses named parameters "[n]" and "[k]" where TEX uses positional parameters #1 and #2.

The two definitions of the binomial coefficient stated so far define the value and tex aspects of the coefficient, respectively. Logiweb allows to assign an arbitrary number of aspects to any construct and allows users to invent new aspects.

2.4   Headers

The source text of the present paper starts thus:

```
PAGE Logiweb layers

BIBLIOGRAPHY
"check" "http:check/latest/vector/page.lgw",
"base" "http:base/latest/vector/page.lgw"

PREASSOCIATIVE
"check" check
"base" base
""  (( " , " ))
""  (( " / " ))

PREASSOCIATIVE
"base" +"
```

The first line gives the page a local name. The next three lines reference two previously published Logiweb pages (the references given are relative to a "current directory", one may also give full URLs or Logiweb references). The bibliography provides access to all constructs defined on the two mentioned pages plus all their transitively referenced pages.

The next five lines import the "check" construct from the check page, the "base" construct from the base page, and defines two new constructs named $((*, *))$ and $((*/*))$. The new constructs become constructs of the Logiweb layers page. Lines that begin with the empty string "" introduce new constructs. Lines beginning with a page name import a construct from that page. Inside defined and imported constructs, the double quote " servers as a parameter placeholder.

The next two lines import a "gluing" prefix plus. It is gluing in the sense that no space is allowed between the plus sign and its argument. A gluing plus is used for expressing numerals like $^+117$. Importing the gluing plus implicitly imports all constructs with the same priority as the gluing plus, which happens to include a gluing minus. For that reason, one may write $^-117$ even though there is no explicit mention of a gluing minus in the header (for completeness: the base page also defines a unary minus -117 which changes the sign of its argument; the gluing minus only works with numerals and is handled at macro expansion time whereas the unary minus can be applied to anything and is handled at run time).

Constructs mentioned in pre- and postassociativity sections are left and right associative, respectively, when used in text written left to right. Constructs in early associativity sections have higher priority than those in later sections. Constructs in the same associativity section have the same priority. As an example,

```
PREASSOCIATIVE
"check" all " : "
POSTASSOCIATIVE
"check" " imply "
```

makes y = 0 imply all x : x >= 0 imply all x : x >= y have the following tacit parentheses:

$$y = 0 \Rightarrow ((\forall x : x \geq 0) \Rightarrow (\forall x : x \geq y))$$

The input syntax is up to the user. As an example, the present author likes to write imply rather than => for implication, but other authors may have other preferences. Logiweb supports Unicode, so one may even use a $\Rightarrow$-character for implication.

3   Metalogic Layer

3.1   Sequent Calculus

A proof checker is no more than a big, recursive function when expressed in Logiwebs programming language. The check page referenced by the present paper defines one, particular proof checker which we shall refer to as the Logiweb sequent checker. All Logiweb pages which reference the check page as their first reference and which do not define their own proof checker are checked by the sequent checker.

Any user can define their own proof checker, but the sequent checker is general enough that there should be no need for doing so. The sequent checker implements a variant of sequent calculus [4,11]. The calculus allows to formulate axiomatic theories, lemmas, and proofs.

3.2   Terms

The following axioms of FOL [14] illustrate what a term of the Logiweb sequent calculus may look like:

Axiom A1: Πx, y: x ⇒ y ⇒ x   □

Axiom A2: Πx, y, z: (x ⇒ y ⇒ z) ⇒ (x ⇒ y) ⇒ x ⇒ z   □

Axiom A3: Πx, y: (¬y ⇒ ¬x) ⇒ (¬y ⇒ x) ⇒ y   □

Axiom A4: Πt, x, a, b: sub (b , a , x , t) ⊪ ∀x: a ⇒ b   □

Axiom A5: Πx, a, b: x avoid a ⊪ ∀x: (a ⇒ b) ⇒ a ⇒ ∀x: b   □

Rule MP: Πx, y: x ⇒ y ⊢ x ⊢ y   □

Rule Gen: Πu, x: x ⊢ ∀u: x   □

Theory Prop: MP ⊕ A1 ⊕ A2 ⊕ A3   □

Theory FOL: Prop ⊕ Gen ⊕ A4 ⊕ A5   □

Terms of the Logiweb sequent calculus are built up from Πx: y (y is provable for all terms x), x ⊢ y (y is provable if x is provable), x ⊪ y (y is provable if x evaluates to T), and x ⊕ y (both x and y are provable).

Constructs like $x \Rightarrow y$ and $\neg x$ live at the object level and do not mean anything to the Logiweb sequent calculus. The construct x avoid y is defined such that it evaluates to T (truth) if x is an object variable not free in y and sub (b , a , x , t) is T if b is alpha-equivalent to a where x is replaced by t (see the check page for details).

A statement like

Axiom A1: Πx, y: x ⇒ y ⇒ x   □

defines the "statement" aspect of A1 to be Πx, y: x ⇒ y ⇒ x. The Logiweb sequent calculus does not distinguish between axioms, rules, and theories, so all of the declarations leading up to the declaration of FOL above just macro expand to statement definitions. Even lemmas translate into statement definitions. What distinguishes lemmas from other statements is that they also have a "proof" aspect which proves the lemma. Axioms, rules, and theories are statements that do not have proofs.

3.3   Sequents

A Logiweb sequent is a triple $\langle p, s, c \rangle$ where c is a sequent term and p and s are sets of sequent terms. We shall refer to c as the conclusion, to elements of p as premises, and to elements of s as side conditions. A Logiweb sequent represents the statement that if all elements of p are true and all elements of s evaluate to T then c is true.

3.4 Sequent Operations

Logiweb sequent calculus has thirteen sequent operations:

$$a^{\mathrm{I}} \qquad\qquad\qquad\quad \rightarrow \langle\{a\},\emptyset,a\rangle \qquad\qquad \text{Init}$$

$$a \vdash \langle p,s,c\rangle \qquad\qquad \rightarrow \langle p\setminus\{a\},s,a\vdash c\rangle \qquad \text{Infer}$$

$$a \Vdash \langle p,s,c\rangle \qquad\qquad \rightarrow \langle p,s\setminus\{a\},a\Vdash c\rangle \qquad \text{Endorse}$$

$$\Pi x\colon \langle p,s,c\rangle \qquad\qquad \rightarrow \langle p,s,\Pi x\colon c\rangle^{1} \qquad\quad \text{Generalize}$$

$$\langle p,s,a\vdash c\rangle^{\triangleright} \qquad\quad \rightarrow \langle p\cup\{a\},s,c\rangle \qquad\quad \text{Ponens}$$

$$\langle p,s,a\Vdash c\rangle^{\triangleright} \qquad\quad \rightarrow \langle p,s\cup\{a\},c\rangle \qquad\quad \text{Probans}$$

$$\langle p,s,\Pi x\colon c\rangle \,@\, a \quad \rightarrow \langle p,s,\langle c|x:=a\rangle\rangle^{2} \qquad \text{Instantiate}$$

$$\langle p,s,a\Vdash c\rangle^{*} \qquad\quad \rightarrow \langle p,s,c\rangle^{3} \qquad\qquad \text{Verify}$$

$$\langle p,s,a\vdash b\vdash c\rangle^{\mathrm{U}} \quad \rightarrow \langle p,s,(a\oplus b)\vdash c\rangle \quad \text{Uncurry}$$

$$\langle p,s,(a\oplus b)\vdash c\rangle^{\mathrm{C}} \rightarrow \langle p,s,a\vdash b\vdash c\rangle \quad \text{Curry}$$

$$\langle p,s,c\rangle \text{ ie } n \qquad\quad \rightarrow \langle p,s,n\rangle^{4} \qquad\qquad \text{Reference}$$

$$\langle p,s,n\rangle^{\mathrm{D}} \qquad\qquad \rightarrow \langle p,s,c\rangle^{4} \qquad\qquad \text{Dereference}$$

$$\langle p,s,c\rangle;\langle p',s',c'\rangle \rightarrow \langle (p'\setminus\{c\})\cup p,s\cup s',c'\rangle \quad \text{Cut}$$

As an example, $\text{A1}\vdash \text{A1}^{\mathrm{ID}}\,@\,x=0\,@\,y=0$ sequent evaluates as follows:

$$\text{A1}\vdash \text{A1}^{\mathrm{ID}}\,@\,x=0\,@\,y=0 \qquad\qquad\qquad\qquad \rightarrow$$
$$\text{A1}\vdash \langle\{\text{A1}\},\emptyset,\text{A1}\rangle^{\mathrm{D}}\,@\,x=0\,@\,y=0 \qquad\qquad \rightarrow$$
$$\text{A1}\vdash \langle\{\text{A1}\},\emptyset,\Pi x,y\colon x\Rightarrow y\Rightarrow x\rangle\,@\,x=0\,@\,y=0 \rightarrow$$
$$\text{A1}\vdash \langle\{\text{A1}\},\emptyset,\Pi y\colon x=0\Rightarrow y\Rightarrow x=0\rangle\,@\,y=0 \quad \rightarrow$$
$$\text{A1}\vdash \langle\{\text{A1}\},\emptyset,\Pi y\colon x=0\Rightarrow y=0\Rightarrow x=0\rangle \qquad\quad \rightarrow$$
$$\langle\emptyset,\emptyset,\text{A1}\vdash x=0\Rightarrow y=0\Rightarrow x=0\rangle$$

The resulting sequent has no premises, no side conditions, and conclusion $\text{A1}\vdash x=0\Rightarrow y=0\Rightarrow x=0$. For that reason, $\text{A1}\vdash \text{A1}^{\mathrm{ID}}\,@\,x=0\,@\,y=0$ is said to prove $\text{A1}\vdash x=0\Rightarrow y=0\Rightarrow x=0$.

Statements that can be proved in the Logiweb sequent calculus are intuitionistically valid. As an example, if ZFC is Zermelo-Fraenkel set theory with the Axiom of Choice and Zorn is Zorns lemma, then $\text{ZFC}\vdash \text{Zorn}$ is the intuitionistically valid statement that Zorn follows from ZFC. With suitable definitions of ZFC and Zorn, that statement would be provable in the sequent calculus.

4 Tactic Layer

4.1 Proof Construction

It is difficult to express proofs directly in the Logiweb sequent calculus. For that reason, the proof checker defined on the check page defines a number of macros and "tactics" which perform some of the trivial work in constructing proofs.

---

[1] If x is not free in any element of p or s.

[2] If a is free for x in c.

[3] If a evaluates to T.

[4] If the statement aspect of n is c.

A Logiweb page has a "body" which is the parse tree of the source text. Logiweb pages are rendered on basis of the body using tex definitions.

In many situations it is convenient to use shorthand notation. Logiweb supports that by having a macro expansion facility. Among other, the macro expansion facility is responsible for translating axiom declarations into statement definitions. Evaluation of test cases and proof checking is done after macro expansion.

As a further convenience, proofs are "tactic" expanded before they are checked. In principle, tactic expansion is much like macro expansion. But tactic expansion is only applied to proofs, not to test cases. Furthermore, the sequent proof that results from tactic expansion is discarded as soon as it is verified whereas the outcome of macro expansion is kept in the memory of the proof checker. Hence, when the Logiweb proof checker verified the present paper it had access to the macro expanded version of the check page but not to the tactic expanded version. Furthermore, the tactic expanded version of each proof is discarded before the proof checker goes on to the next proof.

4.2   Definitions of Tactics

A tactic is defined by defining the "tactic" aspect of a construct, and the tactic is used by including that construct in a proof. When the proof checker verifies a proof, it invokes the tactics and then sequent evaluates the outcome.

When the proof checker invokes a tactic, it applies the tactic to a "cache". Each page has a "cache" which is a functional array of information. Among other, the cache contains all definitions on the page itself and on all transitively referenced pages. Hence, the tactics have access to all definitions. Since lemmas and proofs macro expand into definitions, one would be able to write a tactic which searches all transitively referenced pages for suitable lemmas. The tactics defined on the check page are more modest, however.

The cache also contains a lot of other information such as the compiled versions of all value definitions, the bodies and macro expanded versions of all transitively referenced pages, and "diagnoses" of all referenced pages. Logiweb allows users to publish incorrect pages. And it may even make sense to reference incorrect pages. But incorrect pages are easy to distinguish from other pages in that they have a non-empty diagnose. That diagnose, when rendered, is supposed to tell what is wrong with the given page.

4.3   Tactic Levels

The system of tactics defined on the check page defines three levels of tactics: A top level tactic which expands entire proofs, medium level tactics which take care of proof constructors, and low level tactics which take care of individual lemmas and rules.

The top level tactic is an evaluator which invokes all the medium level tactics in a proof. The top level tactic is installed in the root of each proof during macro expansion.

Most of the medium level tactics collect information and pass it on. As an example, the sequent operator $x \vdash y$ has an associated tactic which ensures that tactics inside y are aware that they can assume that x holds.

The check page also defines a somewhat more complex "unification" tactic $x \gg y$ which is a medium level tactic. The unification tactic takes two arguments x and y where x is an incomplete proof and y is the desired conclusion. The unification tactic tries to adapt x such that x proves y. It mainly does so by adding instantiation operators $u @ v$ where it guesses v using unification.

Low level tactics may occur inside the x argument of $x \gg y$. Among other, there is a low level tactic for proving axioms like A1. From a human point of view, axioms are assumed rather than proved. But in the sequent calculus, axioms do have to be proved, and the proof depends on context. As an example, if the proof is relative to Prop then the tactic expansion of A1 reads Prop $\vdash \cdots$ and A1 is proved by dereferencing Prop into $MP \oplus A1 \oplus A2 \oplus A3$ and then picking the second element using suitable sequent operations. If the proof is relative to FOL then a proof of A1 is slightly different.

5   Proof Layer

5.1   A Lemma and a Proof

As a simple example, consider the following lemma and proof:

Prop **lemma** Taut: $\Pi x\!: x \Rightarrow x$ □

Prop **proof of** Taut:

L01:	Arbitrary $\gg$	x	;
L02:	A2 $\gg$	$(x \Rightarrow (y \Rightarrow x) \Rightarrow x) \Rightarrow$	
		$(x \Rightarrow y \Rightarrow x) \Rightarrow x \Rightarrow x$	;
L03:	A1 $\gg$	$x \Rightarrow (y \Rightarrow x) \Rightarrow x$	;
L04:	MP $\triangleright$ L02 $\triangleright$ L03 $\gg$	$(x \Rightarrow y \Rightarrow x) \Rightarrow x \Rightarrow x$	;
L05:	A1 $\gg$	$x \Rightarrow y \Rightarrow x$	;
L06:	MP $\triangleright$ L04 $\triangleright$ L05 $\gg$	$x \Rightarrow x$	□

The first line above (the one before line L01) defines the proof aspect of Taut to be Prop $\vdash \cdots$. Effectively that makes all axioms and inference rules of propositional calculus available.

Line L01 translate into $\Pi x\!: \cdots$. That produces the Π in the lemma.

Line L02 macro expands into a local macro definition which defines L02 as shorthand for $(x \Rightarrow (y \Rightarrow x) \Rightarrow x) \Rightarrow (x \Rightarrow y \Rightarrow x) \Rightarrow x \Rightarrow x$ in the lines following Line L02. That is utilized in Line L04. Furthermore, Line L02 macro expands into a cut operation whose left argument is A2 $\gg (x \Rightarrow (y \Rightarrow x) \Rightarrow x) \Rightarrow (x \Rightarrow y \Rightarrow x) \Rightarrow x \Rightarrow x$ and whose right argument is the rest of the proof.

Later on, A2 is tactic expanded to a proof P of A2 by a low level tactic, and the unification tactic massages $P \gg (x \Rightarrow (y \Rightarrow x) \Rightarrow x) \Rightarrow (x \Rightarrow y \Rightarrow x) \Rightarrow x \Rightarrow x$ into $P @ x @ y \Rightarrow x @ x$.

After macro and tactic expansion, the proof has the form of a rather unreadable sequent proof which is verified and then discarded.

5.2   Source

The source of Section 5.1 starts thus:

As a simple example, consider the following lemma and proof:

"[Prop lemma Taut : All #x : #x imply #x end lemma]"

"[Prop proof of Taut :

line L01 : Arbitrary >> #x ;

line L02 : A2 >>
  (#x imply (#y imply #x) imply #x) imply newline
  (#x imply #y imply #x) imply #x imply #x ;

line L03 : A1 >> #x imply (#y imply #x) imply #x ;

line L04 : MP ponens L02 ponens L03 >>
  (#x imply #y imply #x) imply #x imply #x ;

line L05 : A1 >> #x imply #y imply #x ;

line L06 : MP ponens L04 ponens L05 >> #x imply #x qed]"

In the source, the present author has chosen to use #x and #y for meta variables
and to use All for the meta quantifier as opposed to all for an object quantifier.
The newline in Line L02 tells TEX that the given place is a good place for
breaking the line. The blank line between lemma and proof is important to TEX
as it puts TEX in vertical mode before the proof. The blank lines inside the pyk
source of the proof are merely aesthetic.

5.3   Further Examples

As a simple example of a macro, one may define

$$[x \trianglerighteq y \doteq \mathrm{MP} \triangleright x \triangleright y]$$

and then replace e.g. MP ▷ L02 ▷ L03 by L02 ⊵ L03 in line L04 of the proof above.
This illustrates the simplest kind of macro definition available in Logiweb. The
general macro facility is Turing complete.

Further proofs may be found at http://logiweb.eu/.

One topic which has not been touched upon is the handling of the deduction
theorem. Deduction can be handled by macro expansion or tactic expansion or
by including the deduction rule as an inference rule of FOL. At the time of
writing, deduction is treated as an inference rule, but a solution using tactics is
being implemented.

6   Conclusion and Further Work

The macro, tactic, and rendering facilities of Logiweb together with the Logiweb sequent calculus and the proof checker makes it easy to define axiomatic theories and state lemmas and proofs in a readable style close to that of e.g. [14]. The rendering and programming facilities also make it easy to define and render constructs like the binomial coefficient in a style close to that of ordinary mathematics. The present paper only presents very few examples, and it is left to the reader to extrapolate. Larger proofs may be found at logiweb.eu. The reader may construct proofs more complex the that given here in less than half an hour by following the tutorial at logiweb.eu.

A particularly important problem not covered here is the handling of deduction. Deduction is currently being ported from being an inference rule to being a tactic. Proofs that use a deduction inference rule may be found at logiweb.eu and in [9]. Deduction is different in FOL and MT, so it has to be implemented twice. Furthermore, it is planned to investigate the possibility of translating Mizar .miz files to Logiweb .pyk files, and to verify [7] and [6] using Logiweb. Finally, it is the intension to continue doing bug fixes and backward compatible enhancements of the Logiweb system.

References

1. Berline, C., Grue, K.: A κ-denotational semantics for Map Theory in ZFC+SI. TCS 179(1–2), 137–202 (1997)
2. Church, A.: The Calculi of Lambda-Conversion. Princeton University Press, Princeton (1941)
3. Dobbertin, H., Bosselaers, A., Preneel, B.: RIPEMD-160: A strengthened version of RIPEMD. In: Fast Software Encryption, pp. 71–82 (1996) http://citeseer.nj.nec.com/dobbertin96ripemd.html
4. Genzen, G.: The Collected Papers of Gerhard Gentzen. In: Szabo, M. E. (ed.) North-Holland (1969)
5. Grue, K.: Map theory. Theoretical Computer Science 102(1), 1–133 (1992)
6. Grue, K.: Mathematics and Computation, DIKU, Universitetsparken 1, DK-2100 Copenhagen, 7 edn. vol. 1(3) (2001)
7. Grue, K.: Map theory with classical maps. Technical Report 02/21, DIKU (2002) http://www.diku.dk/publikationer/tekniske.rapporter/2002/
8. Grue, K.: Logiweb. In: Kamareddine, F. (ed.) Mathematical Knowledge Management Symposium 2003. Electronic Notes in Theoretical Computer Science, vol. 93, pp. 70–101. Elsevier, New York (2004)
9. Grue, K.: Logiweb - a system for web publication of mathematics. In: Iglesias, A., Takayama, N. (eds.) ICMS 2006. LNCS, vol. 4151, pp. 343–353. Springer, Heidelberg (2006)
10. Grue,K.: The layers of logiweb - appendix. Technical report, Logiweb (2007), http://logiweb.eu/logiweb/server/relay/64/BUpCgdix91wGZKohkESIwWkvdww IhPa08-fhoigBB/2/body/tex/appendix.pdf
11. Kleene, S.C.: Introduction to Metamathematics. Bibliotheca Mathematica, N-H, vol. 1(1964)

12. Knuth, D.: The TeXbook. Addison-Wesley, London (1983)
13. Kohlhase, M.: OMDoc: An open markup format for mathematical documents (version 1.2) (March 16, 2005) http://www.mathweb.org/omdoc/index.html
14. Mendelson, E.: Introduction to Mathematical Logic. Wadsworth and Brooks, 3. edn. (1987)
15. Miner, R., Schaeffer, J.: A gentle introduction to MathML (2001) http://www.dessci.com/en/support/tutorials/mathml/default.htm
16. Muzalewski, M.: An Outline of PC Mizar. Foundation of Logic, Mathematics and Informatics, Mizar User Group, Brussels (1993)
17. Raggett, D., Batsalle, D.: Adding math to Web pages with EzMath. j-COMP-NET-ISDN 30(1–7), 679–681 (1998)
18. Skalberg, S.C.: An Interactive Proof System for Map Theory. PhD thesis, University of Copenhagen (October 2002) http://www.mangust.dk/skalberg/phd/
19. Trybulec, A., Blair, H.: Computer assisted reasoning with MIZAR. In: Joshi, A. (ed.) Proceedings of the 9th International Joint Conference on Artificial Intelligence, Los Angeles, CA, pp. 26–28. Morgan Kaufmann, San Francisco (1985) http://www.mizar.org/

Formal Representation of Mathematics in a Dependently Typed Set Theory

Feryal Fulya Horozal and Chad E. Brown

Universität des Saarlandes, Saarbrücken, Germany
{fulya,cebrown}@ags.uni-sb.de

Abstract. We have formalized material from an introductory real analysis textbook in the proof assistant Scunak. Scunak is a system based on set theory encoded in a dependent type theory. We use the formalized material to illustrate some interesting aspects of the relationship between informal presentations of mathematics and their formal representation. We focus especially on a representative example proved using the system.

1 Introduction

In recent decades, a large amount of mathematics has been formalized in different logical systems using various computer programs. Still, the mathematics that has been formally represented and verified using computers is a tiny percent of the mathematics that has been informally written and published as books and papers. If this informally presented mathematical knowledge is to be transformed into formal versions in mechanized systems, then we must better understand the relationships between the two versions. In order to study the gap between informal and formal representations, we have formalized some material from Bartle and Sherbert's introductory textbook on real analysis [2]. This particular textbook has been studied in the context of formalized mathematics already. In particular, a linguistic analysis of portions of [2] is given in [3]. Likewise, an example from the first chapter of [2] is considered in [1] and [8]. We formalized the material in Scunak [5,6], a system based on set theory encoded into a dependent type theory.

In the next section, we will introduce the type theory of Scunak. In Section 3, we discuss the formalization of material from the textbook [2]. In Section 4, we focus on one small example from [2] (the limit of the sequence $\frac{1}{\sqrt{n}}$ is 0). We have encoded this example in ISABELLE-HOL [12] and MIZAR [14], and we briefly compare the encodings.

2 Preliminaries

We present the mathematical proof assistant Scunak [5,6]. Scunak is based on set theory formalized in a logical framework with dependent types and proof terms. The system is relatively new and has been under development since 2005.

Scunak offers several functionalities to its users.

M. Kauers et al. (Eds.): MKM/Calculemus 2007, LNAI 4573, pp. 265–279, 2007.

- It provides an environment in which one can formalize mathematics from a set-theoretical foundation.
- Like several other proof assistants such as Coq [4], Scunak allows its users to interactively construct proofs (technically "proof terms") using the Scunak Interactive Prover (Scip).
- Scunak can be used as a tutor (Scutor, for a demonstration see [7]) that gives feedback to a user on his proof attempts in an arbitrary state of the proof.
- Scunak can be used for verifying textbook proofs through a process of translating the LATEX representation of informal proofs into a proof term that can be understood and checked by Scunak. A detailed description of the process is given in [8].

In this paper, we focus on the relationship between informal mathematical texts and the formal versions in Scunak. Functionalities of Scunak like Scutor and Scunak's verification component for textbook proofs are beyond the scope of this paper.

2.1 The Scunak Type Theory

The type theory of Scunak is dependent type theory with proof irrelevance. A general frame in which proof irrelevance is discussed can be found in [13]. We briefly give the syntax for the type theory of Scunak.

We assume a countably infinite set of variables and use x to range over this set. In addition to variables, we also assume a countably infinite set of names and use c to range over this set. Names will be used to declare constants, abbreviations and claims in a signature.

The set of terms and types are given inductively as follows:

$$\textbf{Terms } p, r, s, t, \phi, \rho \ldots := x|c|(\lambda x.s)|(s\,t)|\langle s, \rho\rangle|\pi_1(t)|\pi_2(t)$$
$$\textbf{Types } S, T, S_1, T_1 \ldots := \textbf{obj}|\textbf{prop}|(\textbf{pf } p)|(\textbf{class } \phi)|(\varPi x : S.\,T)$$

We often omit parenthesis if it is clear in context where they are missing. A context $\varGamma$ is an ordered list of variables associated with types. We sometimes speak of a term t having a type T (in a context $\varGamma$), and we write $t : T$ (or $\varGamma \vdash t : T$). For the definitions of these notions see [5,9].

Often, we will be discussing a particular object of mathematical discourse such as a set A or a sequence X. During such a discussion we may use a corresponding term $\mathbf{A}$ and $\mathbf{X}$. In each such case, the implicit assumption is that the term (e.g., $\mathbf{A}$) corresponds to the object of discourse (e.g., A).

We describe the types of Scunak below.

- **obj** is the type of all mathematical objects. In set theory it is very common to consider any mathematical object as a set. Scunak reflects this idea by having a synonym **set** for the basic type **obj**.
- **prop** is the type of all propositions.
- Proof types: **pf** p is the type of (all) proofs of the proposition p. Note that **pf** p is empty if p is unprovable. Proof types are a form of dependent types since they depend on propositions.

- Class types: These are types that correspond to the class $\{x|\phi(x)\}$ where $\phi(x)$ is a proposition depending on a mathematical object x. Such types are called *class types* and they depend on predicates ϕ. An inhabitant of a class type **class** ϕ is a pair term $\langle s, \rho \rangle$, where $\phi : \mathbf{obj} \rightarrow \mathbf{prop}$ is a predicate, $s : \mathbf{obj}$ is a mathematical object and $\rho : \mathbf{pf}\ (\phi\ s)$ is the proof of the proposition $(\phi\ s)$. Note that without proof irrelevance, there could be more than one proof of $(\phi\ s)$, and hence more than one representative of type $\mathbf{pf}\ (\phi\ s)$.
- Π-types: The remaining types in Scunak are the dependent Π-types, which are generalizations of simple function types. We write $S \rightarrow T$ for $\Pi x : S.\,T$ if x does not occur in the output type T.

One represents a formal mathematical theory in Scunak by giving a *signature* Σ which is a list of constants, abbreviations and claims. We describe each of these below.

- A *constant* is specified by a name c and a type S. A constant corresponds to a basic constructor or axiom of the theory.
- An *abbreviation* is specified by a name c, a type S and a term t. An abbreviation corresponds to a defined constructor or a proved theorem.
- A *claim* is specified by a name c and a type S. A claim corresponds to a constructor we intend to define, or a proposition we intend to prove. In essence, claims are constants which should become abbreviations in a later version of a signature.

Scunak uses type checking to ensure signatures are declared in a valid manner.

2.2 Mac Lane Set Theory in Scunak

The current version of Scunak provides a variety of set theories, including a theory of hereditarily finite sets (see [6]), forms of Mac Lane set theory (see [5]) and a form of Zermelo-Fraenkel set theory with axiom of choice (**ZFC**, see [9]). Each set theory is given as a signature in the type theory which can be loaded as a "kernel." The user can choose the appropriate set theory by loading the corresponding kernel.

We have worked in a set theory that is a form of Mac Lane set theory with universes, the axiom of choice and foundation (**MACU**). One of the first set theories implemented in Scunak was Mac Lane set theory with universes, but without choice or foundation (**MU**). The signature corresponding to **MU** is given in [5]. Aside from the fact that **MACU** includes choice and foundation (adding two constants), the formulation of universes in the two theories are different (removing two constants). Both the signature for **MU** and **MACU** consist of 29 constants. A description of Mac Lane set theory can be found in [11].

We briefly mention the constants used to construct propositions and set theoretical concepts.

There are three constants in the signature for propositions. There is a constant for the logical connective $\neg$, which is the only logical connective represented by a constant. There are two constants for the basic relations $=$ and $\in$ in set theory.

Six constants are defined in the signature for constructors corresponding to the following axioms of **MACU**.

- **Axiom of empty set:** There is a set $\emptyset$ containing no elements.
- **Axiom of separation:** Given any set A and any property ϕ, there is a set of elements x of A (a subset of A) for which $\phi(x)$ holds.
- **Axiom of power set:** For any set A, there is a set $\mathcal{P}(A)$ (the power set of A) such that the elements of $\mathcal{P}(A)$ are exactly the subsets of A.
- **Axiom of union:** For any set A, there is a set $\bigcup A$ such that if $x \in \bigcup A$, then there is an element $y \in A$ such that $x \in y$.
- **Set adjoin:** For two sets A and B, $\{A\} \cup B$ is a set.
- **Universes:** For any set A, there is a set $Univ(A)$ which contains A, is transitive, and is closed under power set. (Note that $Univ(\emptyset)$ must be an infinite set.)

The remaining constants correspond to deduction rules for the basic set theory [5], as well as choice and foundation.

The other logical connectives and set theoretical notions are given as abbreviations from the constants mentioned above (see [6] for a presentation of the derivations).

2.3 A Modular Treatment of the Real Numbers

If one is formalizing mathematics within a foundational framework, then one must face the question of how to treat the real numbers. Fundamentally, the question is whether the real numbers should be constructed or axiomatized. In Scunak, we want all our mathematical content to be reduced to the basic foundational axioms. To obtain this goal, we could construct a signature with three sections:

1. **Set Theory Intro:** Axioms of set theory and basic set theoretic constructions
2. **Constructing the Reals:** A construction of the reals
3. **Real Analysis Intro:** Results from real analysis

Such a signature would guarantee that all our results can be traced back to the original axioms of set theory.

Our primary goal, however, was to follow the textbook [2], we note carefully how the authors introduce the reals in the first paragraph of Chapter 2 of [2]:

> In this chapter we shall discuss the essential properties of the real number system R. Although it is possible to give a formal construction of this system on the basis of a more primitive set (such as the set N of natural numbers or the set Q of rational numbers), we have chosen not to do so. Instead, we exhibit a list of fundamental properties associated with the real numbers and show how further properties can be deduced from them.

Bartle and Sherbert are quite explicit that they are not constructing a set of reals. However, they also refer to "the" real number system R, indicating that they have a real number system R already. How should this be reflected in the formalized version?

We decided to construct a signature of the following form:

1. **Set Theory Intro:** Axioms of set theory and basic set theoretic constructions
2. **Claiming the Reals:** Claims corresponding to the real number system
3. **Real Analysis Intro:** Results from real analysis

The claims in the second section would behave like basic constants and axioms, but could be later given definitions. The idea was that using claims would force the real analysis section to be independent of the construction of the reals (as in the textbook). This approach enforces a level of modularity between sections. The second section could be replaced by different constructions of the reals so long as the types of the claims corresponding the the real number system are the same.

In addition to the independence of the real analysis section on the construction of the reals, we found that the real analysis section was largely independent of the underlying set theory as well. In particular, while we choose to use **MACU** as the underlying set theory, after encoding the mathematical content it became clear that the axioms of choice and foundation were never used.[1] In fact, since we are working with a claimed set of reals, we do not even need an axiom of universes (or any axiom of infinity). Without difficulty, one can change the underlying set theory to be **MU**, **MACU**, **ZFC** or even a theory of hereditarily finite sets. Note that if the underlying set theory is a theory of hereditarily finite sets, then there is no hope of constructing the real numbers; they must remain open claims in this case.

3   How Does the Scunak Type Theory Reflect Informal Mathematics?

Informal presentations of mathematical knowledge in textbooks are untyped, but their formal versions in most mechanized systems for mathematics correspond to typed representations. We illustrate how the informal presentation of mathematics we have taken from [2] is reflected formally in Scunak by identifying several properties we observe in the formal version as consequences of the Scunak type theory.

3.1   Syntax

We briefly mention the concrete syntax employed for the examples we present in this paper. The conrete syntax used for terms and types is PAM (Pseudo-Automath) syntax [6]. The PAM syntax provides human-readable forms of notation to denote several mathematical operators using a combination of infix notation and special binder notation. We will use the `typewriter` font to present material formalized in PAM syntax.

The symbol `::` is an infix notation for the constant **in** that represents the membership relation $\in$ of sets.

The PAM syntax for the λ-binder is `\`. For convenience we also include some special forms for binders encoded as constants or abbreviations. The PAM syntax for the proof type **pf** p is `|- p`.

Given a set A and a property ϕ, we have the constant

$$\textbf{dsetconstr} : \varPi\textbf{A} : \textbf{obj}. \varPi\phi : (\textbf{class}\,(\textbf{in}\,\textbf{A}) \rightarrow \textbf{prop}).\textbf{obj}$$

corresponding to the Axiom of Separation. Given a set A and a proposition $P(x)$ which depends on an element x of A, the term $(\textbf{dsetconstr}\,\textbf{A}\,(\lambda\textbf{x}.\textbf{P}))$ corresponds to the

---

[1] We should note, however, that some lemmas were left as open claims. It is possible, though unlikely, that some of these lemmas might require choice or foundation.

subset $\{x \in A | P(x)\}$ of A. One can write this as (dsetconstr A (\x.P)) in PAM syntax. PAM syntax also includes the syntactic sugar {x:A|P} for such a term.

Quantifiers are handled in a similar way. The quantifiers derived in the kernel of **MACU** are bounded quantifiers. That is, they are bounded to certain domains (sets) and have the form $\forall x \in A. P(x)$, $\exists x \in A. P(x)$ for a set A and a property $P(x)$ depending on an element x in A. The abbreviations corresponding to the bounded universal and the existential quantifiers have the names **dall** and **dex** and have the type $\Pi\mathbf{A} : \mathbf{obj}. (\mathbf{class}\,(\mathbf{in\,A}) \rightarrow \mathbf{prop}). \mathbf{prop}$.

In PAM syntax, one can write (forall x:A . P) and (exists x:A . P) as syntactic sugar for $(\mathbf{dall\,A}\,(\lambda\mathbf{x.P}))$ and $(\mathbf{dex\,A}\,(\lambda\mathbf{x.P}))$ for a term **P** with type **prop**.

After claiming the set of reals and defining the ordering relation on the reals, the symbol > is given as infix notation for the formal version of the 'greater than' operator.

In order to aid readability, we will sometimes mix notations in the discussion below. Also, we sometimes mention a "type" and give PAM syntax, by which we mean the type specified by the given PAM syntax.

We add variables to a context by giving the variable name, a colon, and a PAM specification of the type, all surrounded by brackets. For example, if we want to introduce a set A into the context, we actually introduce a variable **A** of type **set** into the context using the PAM syntax [A:set].

3.2 Sets as Types

In the Scunak type theory, the notion of set is represented by the basic type **set**, which is a synonym for the basic type **obj** of all mathematical objects.

When we formalize mathematics in Scunak we quite often use sets as types of certain terms, in particular, when we work with elements of sets. For example, consider a set A and an element x of A. We can represent these objects in Scunak by declaring variables **A** and **x** to have certain types in a context. We declare this in PAM syntax as follows: [A:set][x:A].

The type of **x** is **class** (**in A**). Intuitively, this corresponds to the fact that x belongs to the class of objects that are in the set A. The class type **class** (**in A**) is often written as A in PAM syntax leaving out **class** and **in**. This allows any set to be used as the "type" of its elements.

Note that the above representation of $x \in A$ uses dependent types. The type of **x** depends on the variable **A**. This representation is quite compact compared to the representations of simple type systems, since the information $x \in A$ is contained in the type of **x**. In simply typed systems one is required to either assume **A** is the simple type of **x** or add the information $(x \subset A)$ into the formalizations usually as the antecedent of an implication $((x \in A) \Rightarrow \dots)$. This means, one carries $x \in A$ as an extra information in the formalizations.

We now discuss some examples that demonstrate the use of sets as types in the material we have formalized in Scunak.

Fig. 1 shows the definition of the notion of a *lower bound* of a set of real numbers taken from [2] and its corresponding formal representation in Scunak in PAM syntax.

A real number w is represented as an object that is in the set $\mathbb{R}$ of real numbers as [w:R], where R denotes the set of real numbers we have claimed in Scunak.

A subset S of $\mathbb{R}$ is represented as an object that is an element of the power set of $\mathbb{R}$, in PAM syntax as [S:(powerset R)]. Here powerset is the PAM version of the constant named **powerset** with type **obj** → **obj** in the kernel of Scunak corresponding to the axiom of power set.

realLowerBoundOf and realLeq are the formal versions of a *lower bound* of a set and the relation $\leq$. They inhabit the types (powerset R) -> R -> prop and R -> R -> prop, respectively.

Note that the bound variable s in (forall s:S . (realLeq w s)) has the type S, whereas realLeq expects two arguments of type R. Scunak uses a special type conversion mechanism to type-check the application of realLeq w to s. We discuss the mechanism in Section 3.3.

Type Refinement using the Axiom of Separation. As we mentioned earlier, the Axiom of Separation is encoded in Scunak. Here we show how one can use the encoding to give refined types. Given a set A and a property ϕ, we can form the set $\{x \in A | \phi(x)\}$ and use this as a refined version of the type corresponding to A.

A : **set**

ϕ : (**class** (**in A**)) → **prop**

x : **class** (**in** $\{x \in A | (\phi x)\}$)

This style of type refinement corresponds to linguistic specifications in informal mathematical texts. For example, "a **lower bound** w" as stated in Fig. 2 taken from the definition of an infimum of a set of real numbers in [2] is a linguistic specification of a real number that has the property of being a lower bound of a set S of real numbers. The formal version in Fig. 2, uses separation to reflect the informal specification by refining the type R of real numbers with the relation realLowerBoundOf. The resulting type {x:R | (realLowerBoundOf S x)} is the type a variable representing a real number w that is a lower bound of a set S of real numbers.

In MIZAR, there is an alternative type refinement mechanism that uses MIZAR "attributes" (see [15]) to represent such linguistic specifications in textbooks.

In ISABELLE-HOL, one can employ a type definition mechanism rather than type refinements for the presentation of these specifications. For example, a simple type α and a closed, nonempty predicate ϕ on α can be used to define the type, say γ, of terms for which the property ϕ holds. Along with the definition, there is usually a function that serves the purpose of an explicit type conversion between α and γ.

Type refinement using the Axiom of Separation does not require a function for the explicit conversion of type **class** (**in** $\{x \in A | (\phi x)\}$) to type **class** (**in A**). The conversion is performed implicitly by means of certain inference rules in the kernel of

---

Definition. Let S be a subset of $\mathbb{R}$. A number $w \in \mathbb{R}$ is said to be a **lower bound** of S if $w \leq s$ for all $s \in S$.

```
[S:(powerset R)]
[w:R]
(realLowerBoundOf S w):prop=(forall s:S . (realLeq w s)).
```

Fig. 1. An Example of Using Sets as Types

Scunak. By means of this implicit type conversion, the application of the infix operator >, which expects two arguments of type R, to the terms w and v, which have the refined type {x:R| (realLowerBoundOf S x)}, type-checks.

3.3 Type Conversions

From everyday programming languages like C/C++, we are familiar with the notion of implicit type conversions, also known as **coercions**, used for converting numeric types (like the type int of integers and float of floating numbers). The general idea of type conversions is that a variable of a certain type is forced to behave as if it has another type. This means, if a type S is coerced to another type T, then any term that expects a member of T can accept an argument that is a member of S.

Scunak does not have numeric types. Numbers are members of class types. For instance, N and R are PAM notation for the sets $\mathbb{N}$ of natural numbers and $\mathbb{R}$ of real numbers, respectively. Hence N and R can be used as the types of natural numbers and real numbers. This means, there are distinct types for numbers in Scunak. Nevertheless, one can naturally expect a natural number to behave as a real number (since $\mathbb{N} \subseteq \mathbb{R}$ and thus a natural number is a real number). In other words, one technically expects a term with type N to behave as if it has type R. Scunak has a type conversion mechanism for subsets of sets (like $\mathbb{N}$ of $\mathbb{R}$). We describe the mechanism below.

Suppose there are two sets A and B with the property $B \subseteq A$. Given two corresponding terms **A** and **B**, the type **class** (in **B**) can be converted to type **class** (in **A**) if there is a proof of the property $B \subseteq A$ (i.e, if there is a term ρ with type **pf** (**B** $\subseteq$ **A**)). The conversion requires an explicit statement in the presence of a proof of $B \subseteq A$ in the formalizations. The statement needs to be declared only once. Then, in any future formalization, any term that expects arguments corresponding to elements of A can be applied to terms corresponding to elements of B without violating type checking. One should note that the type conversion do not affect the resulting type of an application.

Arithmetical operators such as addition, subtraction, multiplication are given for real numbers and their application to elements of subsets of $\mathbb{R}$ (like naturals, integers, rationals, etc.) is handled through converting the types of elements of subsets of $\mathbb{R}$ to the type of real numbers. The definitions of arithmetical operators are not overloaded for each distinct type of numbers.

Definition. Let S be a subset of $\mathbb{R}$. If S is bounded below, then a lower bound w is said to be an **infimum** (or a **greatest lower bound**) of S if no number larger than w is a lower bound of S.

```
[S:(powerset R)]
[w:{x:R| (realLowerBoundOf S x)}]
(realInfimum S w):prop=
  (not (exists v:{x:R| (realLowerBoundOf S x)} .
    (v > w))).
```

Fig. 2. An Example of Separation in Scunak

3.4 Pair Terms

Pair terms are inhabitants of class types. In the formalizations, pair terms are frequently used to address type checking issues in the case of no available type conversion procedures. The type conversion procedure we have discussed in Section 3.3 is a special case used to convert class types induced by the predicate $(\mathbf{in\,A})$ for a set A. Currently, the only general way to convert a term of a class type or of type $\mathbf{obj}$ to another class type is by using pairs as we will describe below.

Suppose a term $\mathbf{x}$ of type $\mathbf{class}\,\phi$ is expected to behave as a member of type $\mathbf{class}\,\psi$ for predicates ϕ and ψ. If one can prove that the term $\mathbf{x}$ (as an object) satisfies $(\psi\,\mathbf{x})$, then the proof can be used to construct a pair term of type $\mathbf{class}\,\psi$, which can be given as an argument to a term $\mathbf{t}$ that expects a member of the latter type.

Technically, $\mathbf{x}\,:\,\mathbf{class}\,\phi$ is (judgmentally) the same as a pair term $\langle\pi_1(\mathbf{x}),\pi_2(\mathbf{x})\rangle$ with $\pi_1(\mathbf{x})\,:\,\mathbf{obj}$ and $\pi_2(\mathbf{x})\,:\,\mathbf{pf}\,(\phi\,\pi_1(\mathbf{x}))$. The first projection $\pi_1(\mathbf{x})$ of the pair is the object representation of $\mathbf{x}$. If one can prove that $(\psi\,\pi_1(\mathbf{x}))$ holds, then the proof $\rho\,:\,\mathbf{pf}\,(\psi\,\pi_1(\mathbf{x}))$ can be paired together with $\pi_1(\mathbf{x})$ and the resulting pair term has the type expected by $\mathbf{t}$. In PAM syntax, π_1 and π_2 are not written down explicitly.

If a term $\mathbf{x}$ with type $\mathbf{obj}$ is expected to behave as a member of a class type, say $\mathbf{class}\,\phi$ for a predicate ϕ, then $\mathbf{x}$ is paired together with the proof of the proposition $(\phi\,\mathbf{x})$.

An instance of using pair terms in the formalizations in Scunak is the case, where a member of type $\mathbf{class}\,(\mathbf{in\,B})$ is expected to behave as if it has type $\mathbf{class}\,(\mathbf{in\,A})$, but we do not have a proof that $B \subseteq A$ holds. In this case, the explicit conversion of types in Scunak (as we have mentioned in Section 3.3) cannot be applied, since the conversion is specific to sets A and B for which $B \subseteq A$ holds. We use pair terms like in the general case above. The proof we are looking for is that the object representation of the term with type $\mathbf{class}\,(\mathbf{in\,B})$ is an element of the set A.

3.5 Representation of Functions

In Scunak, functions are represented as objects that are functional binary relations on arbitrary sets. This means, an element of the relation's domain is associated with a unique element of the relation's range. The encoding of this representation is presented in [6] by introducing the kernel constants **func**, **ap** and **lam** with their formal definitions. They respectively serve the purpose of declaring functions, applying functions to their arguments and specifying functions.

We briefly mention how these constants are used. A function f from the set A to the set B can be represented as a member of the class of objects that are functions from A to B. For an element a of A the function application $f(a)$ is represented as $(\mathbf{ap\,A\,B\,f\,a})$, where $\mathbf{A}$, $\mathbf{B}$ have type $\mathbf{set}$, $\mathbf{f}$ has type $\mathbf{class}\,(\mathbf{func\,A\,B})$ and $\mathbf{a}$ has type $\mathbf{class}\,(\mathbf{in\,A})$. The type of $(\mathbf{ap\,A\,B\,f\,a})$ is $\mathbf{class}\,(\mathbf{in\,B})$. If $\mathbf{t}$ is a term which has type $\mathbf{class}\,(\mathbf{in\,B})$ when $\mathbf{x}$ is a declared variable with type $\mathbf{class}\,(\mathbf{in\,A})$, then $(\mathbf{lam\,A\,B}\,(\lambda\mathbf{x}.\mathbf{t}))$ has type $\mathbf{class}\,(\mathbf{func\,A\,B})$ and represents the λ-abstraction that takes an element of the set A and returns an element of the set B.

An alternative way to work with functions in Scunak is to use the notion of a *set of functions* represented by the kernel constant $\mathbf{funcSet}\,:\,\mathbf{obj}\,\rightarrow\,\mathbf{obj}\,\rightarrow\,\mathbf{obj}$, which takes two objects (sets) A and B, and returns the set of functions from A to B.

Given two sets A and B, we can represent a function f from A to B using **funcSet** by declaring variables **A**, **B** and **f** as follows:

A : **set**
B : **set**
f : **class** (**in** (**funcSet A B**))

where (**in** (**funcSet A B**)) is a predicate that takes a term and checks whether it is in the set of functions from A to B. In PAM syntax, we write [f:(funcSet A B)].

For declared variables **A** and **B** with type **set**, the semantic interpretation of both a term with type **class** (**func A B**) and a term with type **class** (**in** (**funcSet A B**)) is the same: A function from the set A to the set B.

The corresponding function application and λ-abstraction operators for **funcSet** are **ap2** and **lam2** with the following types respectively.

Π**A** : **set**.Π**B** : **set**.**class** (**in** (**funcSet A B**)) $\rightarrow$ **class** (**in A**) $\rightarrow$ **class** (**in B**)
Π**A** : **set**.Π**B** : **set**.(**class** (**in A**) $\rightarrow$ **class** (**in B**)) $\rightarrow$ **class** (**in** (**funcSet A B**))

The use of **ap2** and **lam2** is similar to that of **ap** and **lam**. For terms **A** : **set**, **B** : **set**, **f** : **class** (**in** (**funcSet A B**)), **a** : **class** (**in A**) and **x** : **class** (**in A**),

- (**ap2 A B f a**) represents an element $f(a) \in B$ for $a \in A$,
- (**lam2 A B** (λ**x.t**)) represents a function f determined by $f(x) = t$ for $x \in A$.

Sequences As a special case of working with functions in Scunak, we present the formalization of the notion of sequences. Fig. 3 shows the informal definition of a sequence of real numbers taken from [2] and its formal representation in Scunak.

We first formalize a general notion of sequences. Given an arbitrary set A, a sequence in the set A is a function from the set $\mathbb{N}$ of natural numbers to A. We define a set constructor called sequenceIn that takes a term corresponding to a set A and returns the set of functions from $\mathbb{N}$ to A using the constant **funcSet**.

Definition. A **sequence of real numbers** (or a **sequence in** $\mathbb{R}$) is a function on the set $\mathbb{N}$ of natural numbers whose range is contained in the set $\mathbb{R}$ of real numbers.

```
[A:set]
(sequenceIn A):set=(funcSet N A).
notation RSeq (sequenceIn R).
[X:RSeq]
```

Fig. 3. Sequences

We instantiate sequenceIn with R to yield the set of functions from $\mathbb{N}$ to $\mathbb{R}$, which we denote as RSeq. Since a sequence in $\mathbb{R}$ is a member of the set represented by RSeq, we can use RSeq as the type of a sequence in $\mathbb{R}$ as [X:RSeq] in PAM syntax.

We define the value of a sequence at index $n \in \mathbb{N}$ using **ap2**. The value of a sequence at index n is the value obtained when the sequence, as a function, is applied to n.

```
[X:(sequenceIn A)]
[n:N]
notation XinfuncSetNA (sequenceIn#U A (\x.(X::x)) X).
(valueAt A X n):A=(ap2 N A <X,XinfuncSetNA> n).
notation subA (valueAt A).
notation sub (valueAt R).
```

The term valueAt takes terms representing a set A, a sequence in A and a natural number n, and returns a term representing the value of the sequence at index n. The pair term in the definition is to ensure that (ap2 N A) is applied to an argument with the expected type (funcSet N A). The type of a sequence in $\mathbb{R}$ is in PAM syntax (sequenceIn R), which is not the type expected by (ap2 N A). The PAM term (sequenceIn#U A (\x.(X::x)) X) is the proof that the context variable X is in the set of functions from $\mathbb{N}$ to A. We obtain the proof by unfolding the definition of sequenceIn. For readability, we declare XinfuncSetNA as notation for this proof.

The last component we need in order to be able to work with sequences is a sequence constructor and we define it using **lam2**.

```
[A:set]
[f:N -> A]
notation lam2NAf
         (sequenceIn#F A (\x.((lam2 N A f)::x)) (lam2 N A f)).
(sequenceconstr A f):(sequenceIn A)=<(lam2 N A f),lam2NAf>.
```

The abbreviation sequenceconstr takes a term representing a set A and a meta-level function with type N -> A and gives back a term corresponding to a sequence in A that is determined by the meta-level function. Note (lam2 N A f) has the type (funcSet N A). lam2NAf is a notation that stands for the proof that the object-level λ-abstraction (lam2 N A f) is a sequence in A. We use lam2NAf to obtain a term of type (sequenceIn A).

4   A Case Study

After presenting the Monotone Convergence Theorem, Bartle and Sherbert give a number of examples which use the Monotone Convergence Theorem. We present the formal version of the first of these examples: $\lim \left(\frac{1}{\sqrt{n}}\right) = 0$. The proof in [2] essentially consists of one sentence. The statement and short proof from [2] are shown in Fig. 4.

Example. $lim \left(\frac{1}{\sqrt{n}}\right) = 0$.

Proof. Clearly, 0 is a lower bound for the set $\{\frac{1}{\sqrt{n}} : n \in N\}$, and it is not difficult to show that 0 is the infimum of the set $\{\frac{1}{\sqrt{n}} : n \in N\}$; hence $0 = lim \left(\frac{1}{\sqrt{n}}\right)$.

Fig. 4. An Example on Sequences

The formalization of the example is divided into the following parts:

- Formalization of necessary notions and theorems the example uses in its statement and proof in a PAM document
- An analysis of the informal proof to generate underspecified lemmata and their formalization
- Formalization of the mathematical statement of the example in a PAM file
- Formalization of the proof interactively in Scip

The underlying notions used in the example are the notions of a lower bound and an infimum of a set (of real numbers), sequences, the limit of a sequence, decreasing sequences, the square root function, the underlying set of a sequence, and the Monotone Convergence Theorem. Once these preliminaries are given, we give claims corresponding to the steps of the proof and then the final result.

The notions of a lower bound and an infimum of a set (of real numbers), and sequences are introduced in Sections 3.2 and 3.5. We have formalized the notion of the limit of a sequence and decreasing sequences as a term `lim:RSeq -> R -> prop` that takes a sequence of real numbers and a real number, and checks whether the proposed number is the limit of the given sequence, and `decreasing:RSeq -> prop` that takes a sequence of real numbers and checks whether the given sequence is decreasing. Whenever `X` is of type `RSeq`, then `(RSeqSet X)` (of type `set`) is defined to be the underlying set of the sequence `X`. The term `RSeqSetSubsetReals` abbreviates a proof that for any `X` of type `RSeq`, the underlying set `(RSeqSet X)` is in the power set of the reals.

Given the notions of limit and decreasing, we can represent the Monotone Convergence Theorem in PAM syntax as shown in Fig. 5. We explain Fig. 5 by giving the same information in natural language:

`x:` Let X be a sequence of reals.
`v:` Assume X is decreasing.
`a:` Let a be a real number.
`apf:` Assume a is a lower bound of the underlying set of X.
`w:` Assume a is an infimum of the underlying set of X. (Note that we cannot assert that a is an infimum unless we know it is a lower bound.)
`monotoneConvTheo-b-2:` The Monotone Convergence Theorem implies the limit of X is a.

The example we will consider is shown in PAM syntax in Fig. 6. We begin by explaining the notation. The symbol `x` is declared as notation for the sequence $\frac{1}{\sqrt{n}}$. Note that n is bound in this expression. This is reflected by the fact that n is λ-bound in the term `(\n.(1/ <(sqrt n),(sqrtNatInR-0 n)>))` which has type `N -> R` (in PAM syntax). `seqconstr` takes this term of function types and creates a term of type `RSeq`. We next declare `s` to be notation for the underlying set of `x`. We declare `SPR` as notation for a term proving `s` is in the power set of the reals. Finally, we declare notation `LB` for the set of lower bounds of the set `s` of reals.

Using this notation, we can represent the facts asserted in the proof of the example. Two facts are stated explicitly in the proof: 0 is a lower bound and 0 is an infimum. These two facts are represented as the claims `bs-example-3-3-3a-1` and

```
[X:RSeq]
[v:|- (decreasing X)]
[a:R]
[apf:|- (a::{x:R|(realLowerBoundOf
                  <(RSeqSet X),(RSeqSetSubsetReals X)> x)})]
[w:|- (realInfimum
         <(RSeqSet X),(RSeqSetSubsetReals X)>
           <a,apf>)]
(monotoneConvTheo-b-2 X v a apf w):|- (lim X a)?
```

Fig. 5. Formalization of Monotone Convergence in Scunak

bs-example-3-3-3a-2 in Fig. 6. These two claims are essentially lemmas we commit to proving at some later time. Note that since the definition of infimum requires knowing that the element is a lower bound, the fact that 0 is a lower bound (as witnessed by the claim bs-example-3-3-3a-1) is used in the type of bs-example-3-3-3a-2. One of the premises of the Monotone Convergence Theorem is that the sequence is monotone (in this case, decreasing). While the text does not explicitly say the sequence $\frac{1}{\sqrt{n}}$ is decreasing, we include this as a third claimed lemma bs-example-3-3-3a-3. Finally, we declare a claim bs-example-3-3-3a corresponding to the main result.

```
notation X
(sequenceconstr R (\n.(1/ <(sqrt n),(sqrtNatInR-0 n)>))).

notation S (RSeqSet X).
notation SPR (RSeqSetSubsetReals X).
notation LB {x:R|(realLowerBoundOf <S,SPR> x)}.

bs-example-3-3-3a-1:|- (0::LB)?
bs-example-3-3-3a-2:
  |- (realInfimum <S,SPR> <0,bs-example-3-3-3a-1>)?
bs-example-3-3-3a-3:|- (decreasing X)?
bs-example-3-3-3a:|- (lim X 0)?
```

Fig. 6. Formalization of the Example in Scunak

After Scunak has read the PAM file containing the information in Figs. 5 and 6, our goal changes to obtaining a proof term for the main result bs-example-3-3-3a. One way to give the proof is simply as a proof term. Since we have given names to the steps of the proof, such a proof term is small (but not enlightening):

```
(monotoneConvTheo-b-2 X bs-example-3-3-3a-3 0
    bs-example-3-3-3a-1 bs-example-3-3-3a-2)
```

Another way to give the proof is to construct it in Scip. A Scip session which constructs the proof is given in Fig. 7. This corresponds more closely to the text. We begin the Scip session with a "use" which lists the known facts we can use in the proof. In our case, we list the Monotone Convergence Theorem along with the claimed steps of the proof. Now we can construct the proof by giving three "facts." First, 0 is a lower bound of $\{\frac{1}{\sqrt{n}} : n \in N\}$. Second, 0 is an infimum of $\{\frac{1}{\sqrt{n}} : n \in N\}$. Note that these two statements correspond directly to the statements given in the textbook proof in Fig. 4 and to the claims given in Fig. 6. The third fact is that the sequence is decreasing. We end the proof by giving the Scip command d, indicating that the proof is done. Essentially, we have stated all the steps in the proof in the PAM file and we have then used Scip to appropriately combine them into a proof term.

```
prove bs-example-3-3-3a
use bs-example-3-3-3a-1 bs-example-3-3-3a-2
    bs-example-3-3-3a-3 monotoneConvTheo-b-2
fact (0::LB)
fact (realInfimum <S,SPR> <0,fact0>)
fact (decreasing X)
d
```

Fig. 7. Construction of the Proof in Scip

We have also experimented with this example in Isabelle-HOL [12] and Mizar [14]. We mention two interesting points.

The first point regards the use of dependent types to state definitions and theorems in a manner as close as possible to the text. In particular, we used the (dependent) type of lower bounds of S in the definition of infimum. In Isabelle-HOL, the restriction to simple types prevented us from using types. Instead, one must ignore such restrictions on arguments when defining concepts such as infimum and include the restrictions as premises when formulating theorems. In Mizar one can define such dependent types, but only if they are nonempty. The most satisfying way we found to define such a type in Mizar was to assume S is a "bounded below subset of reals" when defining the type of lower bounds of S.

The second point regards the binding mechanism for the n in the sequence $\frac{1}{\sqrt{n}}$. One can easily give the sequence as a λ-term in Isabelle-HOL. We had difficulty trying to find an appropriate binding mechanism in Mizar.

In Mizar's library, the notion of a sequence of reals is represented by the mode `Real_Sequence` which is defined in terms of functions from naturals to reals [10]. One can easily use Mizar's `func` definition mechanism to define a unary constructor named `seq333a` which expects a natural number n and returns a real number $\frac{1}{\sqrt{n}}$. However, this does not yield the desired member of `Real_Sequence`. In the end, we formulated the example in Mizar by stating that if X is a real sequence and for all n, X_n is $\frac{1}{\sqrt{n}}$, then the limit of X is 0. Essentially this uses the universal quantifier as the binder, but leaves implicit the fact that the hypothesis determines a unique sequence X. Later, Krzysztof Retel pointed out that we could have used the `func` definition mechanism to define a nullary constructor named `seq333a` which has type `Real_Sequence`.

5 Conclusion

We have demonstrated that mathematical content informally represented in a textbook can be given a precise formal representation in Scunak. Especially useful aspects of Scunak include using sets as types, type conversions for subsets, and the handling of binding constructors (e.g., for binding n in the sequence $\frac{1}{\sqrt{n}}$). However, some aspects of the formal versions in Scunak were problematic. First, sometimes we needed to explicitly include proof objects in terms (as the second part of a pair of class type) for the purposes of type checking. A mechanism allowing users to leave out such proof objects (by looking them up somehow, not by performing proof search) would be helpful. Second, writing proofs as proof terms does not give a very human-readable (or "natural") representation of proofs. A MIZAR-style of proof presentation would be preferable. Essentially one would need a "compiler" which translates MIZAR-style proofs into Scunak proof terms. We leave such improvements as future work.

References

1. Autexier, S., Fiedler, A.: Textbook proofs meet formal logic - the problem of underspecification and granularity. In: Kohlhase, M. (ed.) MKM 2005. LNCS (LNAI), vol. 3863, pp. 96–110. Springer, Heidelberg (2006)
2. Bartle, R.G., Sherbert, D.R.: Introduction to Real Analysis. John Wiley and Sons, New York (1982)
3. Baur, J.: Syntax und semantik mathematischer texte. Diploma thesis, Saarland University, Saarbrücken, Germany (1999)
4. Bertot, Y., Castéran, P.: Interactive Theorem Proving and Program Development. Coq'Art: The Calculus of Inductive Constructions. In: Texts in Theoretical Computer Science. An EATCS Series, Springer, Heidelberg (2004)
5. Chad, E.: Combining Type Theory and Untyped Set Theory. In: Furbach, U., Shankar, N. (eds.) IJCAR 2006. LNCS (LNAI), vol. 4130, pp. 205–219. Springer, Heidelberg (2006)
6. Brown, C.E.: Encoding functional relations in Scunak. In: LFMTP'2006 (September 2006)
7. C.E. Brown. Scunak users manual (2006)
   http://gtps.math.cmu.edu/cebrown/manual.ps
8. Brown, C.E.: Verifying and invalidating textbook proofs using scunak. In: Borwein, J.M., Farmer, W.M. (eds.) MKM 2006. LNCS (LNAI), vol. 4108, pp. 110–123. Springer, Heidelberg (2006)
9. Brown, C.E.: Dependently Typed Set Theory. In: SEKI-Working-Paper SWP–2006–03, SEKI Publications, Saarland Univ (2006) ISSN 1860–5931
10. Kotowicz, J.: Real sequences and basic operations on them. Journal of Formalized Mathematics, 1 (1989)
11. Mac, S.: Mathematics, Form, and Function. Springer, Heidelberg (1986)
12. Nipkow, T., Paulson, L.C., Wenzel, M.: Isabelle/HOL. LNCS, vol. 2283. Springer, Heidelberg (2002)
13. Reed, J.: Proof irrelevance and strict definitions in a logical framework. Technical Report 02-153, School of Computer Science, Carnegie Mellon University (2002)
14. Rudnicki, P.: An overview of the mizar project. In: Workshop on Types for Proofs and Programs, pp. 311–332 (1992)
15. Wiedijk, F.: Mizar: An impression.
    http://www.cs.kun.nl/~freek/mizar/mizarmanual.ps.gz

Restoring Natural Language as a Computerised Mathematics Input Method

Fairouz Kamareddine, Robert Lamar, Manuel Maarek, and J.B. Wells

ULTRA group, Heriot-Watt University
http://www.macs.hw.ac.uk/ultra/

Abstract. Methods for computerised mathematics have found little appeal among mathematicians because they call for additional skills which are not available to the typical mathematician. We herein propose to reconcile computerised mathematics to mathematicians by restoring natural language as the primary medium for mathematical authoring. Our method associates portions of text with grammatical argumentation roles and computerises the informal mathematical style of the mathematician. Typical abbreviations like the aggregation of equations $a = b > c$, are not usually accepted as input to computerised languages. We propose specific annotations to explicate the morphology of such natural language style, to accept input in this style, and to expand this input in the computer to obtain the intended representation (i.e., $a = b$ and $b > c$). We have named this method *syntax souring* in contrast to the usual *syntax sugaring*. All results have been implemented in a prototype editor developed on top of T$_{E}$X$_{MACS}$ as a GUI for the core grammatical aspect of MathLang, a framework developed by the ULTRA group to computerise and formalise mathematics.

1   Introduction

Over several millennia, the mathematical community has developed a prodigious mass of knowledge. Effective communication of this knowledge has been essential to its dissemination. As various results have arisen and circulated, patterns and conventions have been developed for their sound and acceptable communication, leading to a *de facto* style of recording mathematical concepts in natural language. This style is sufficiently standardised to effectively communicate the most esoteric of ideas, while being flexible enough to record the variety of mathematical topics which have been explored in the academy of yesteryear and today.

1.1   State of the Field

Since the advent of computer-aided proof in the 1960s, mathematicians and computer scientists have been seeking effective ways to encode mathematical concepts in languages of varying structure. Some theorem provers are highly rigid and distant from natural language, while others such as Mizar and Isar have a syntax similar to the mathematician's style. Each prover has its proponents and

M. Kauers et al. (Eds.): MKM/Calculemus 2007, LNAI 4573, pp. 280–295, 2007.

favoured applications, but they are all stark and restrictive when compared with the fluidity of natural language. None currently has an infrastructure to provide a direct mapping from a typical natural language mathematical text to its own language but they all have methodologies to offer natural language integration. We group these methodologies into four categories.

1. **Proof code with embedded natural language.** In a typical formal proof language—such as Isabelle [1] or Coq [2]—there are facilities to incorporate natural language alongside formal definitions and proofs. Natural language text parts are treated as commentary in a literate proof document and omitted by the verification. This method uses *structured comments*, akin to programming languages, for generating documentation out of programming code. In a similar fashion, recent developments of intuitive text editors have permitted plugin-interfacing with theorem provers [3,4,5].
2. **Syntax *à la natural language*.** Formal languages often suffer from rough syntax and strict grammar. To soften the use of formal languages some efforts have been made to adapt these syntaxis and grammars to mathematicians' habits. Some developments have gone far in this direction to obtain formal proof documents that *look like* natural language texts. The main examples are Mizar [6] and Isar [7], but more recently some calculi [8,9] were developed pursuing the idea of a formal representation for pseudo-natural language.
3. **Semantic Web data model.** Mathematical natural language is a vague and imprecise language which is unfriendly to computation. Web technologies offer a compromise in the way they encapsulate natural language and extend it with semantic tagging and hyperlinking. OMDoc [10] is a precursor in this domain.
4. **Natural language generator.** If the starting point is a formally defined language then a natural language representation of the formal content can be produced. The proof assistant HE*A*M [11] has this capability. Furthermore, [12] and [13] provide facilities to personalise the natural language generated.

We conclude that the primary input for a theorem prover is generally a formal language and that the natural language of a theorem prover's document is a formalisation side effect. In case 2 the document is written in an altered and restricted natural language while in case 4 the generated (natural) text is only available *after* providing the input through a significantly restricted language. These pseudo-natural languages are by no means the only legitimate representation of mathematics. Recent work—not pertaining to any particular system—has explored the more general issue of comparing various formal representations [14], demonstrating the importance of flexibility in establishing formal models and providing concrete examples such as the formalisation of matrices [15].

1.2   Contributions

From the above it may be seen that for a semantically helpful computerisation of mathematical knowledge, today's systems require the use of a formal language which differs in some way from the common, natural, mathematical language. This paper proposes a method to **restore natural language as the primary**

input for computerised mathematics. The motivation is to provide mathematicians with straightforward tools they can employ to use computers in their everyday work. Efforts towards this goal fall into several categories.

1. *An integrated system for natural-language text input and grammatical categorisation.* A new approach to authoring natural language texts is presented in Section 2. As the natural language text is composed, each word or phrase is placed into a certain grammatical category as enumerated in Table 1. This is achieved by annotating the original natural language text either during or after its composition. A typical work pattern is presented in Section 2.4.

2. *Tools for reconciling complex expressions to simple grammatical categories.* In Section 3 we give several transformations a user may apply to plain text in order to cause the expression to cleanly fit a grammatical classification. These tools are built on top of the aforementioned authoring approach and work to reconcile varying natural writing styles to the stricter grammatical rules. The effect is to duplicate, shuffle, and unfold natural language text so that it is expressed in an explicit manner and strict order. These rewriting rules constitute a "dual" of syntax sugaring which we call *syntax souring*.

3. *An abstract framework to assert the foundational reliability of the proposed system.* The narrative in Section 4 presents an operational system which provides a rigorous framework upon which the denotational meaning can rest. It provides a data structure for mathematical documents, incorporating the grammatical categorisation, syntax souring notions, and a set of rewriting rules which achieve the souring functionality presented in the earlier sections.

Throughout the paper, we motivate our proposal on a supplement example of an excerpt from a textbook [16, Ch. 12] which, due to space limitation, is available as a supplement to this paper at this paper's authors' respective web pages.

1.3   Background: MathLang

In the development of computer proof aids, a major goal is to establish a correspondence between natural language mathematics and some core language (e.g., Automath, Coq, Mizar). The MathLang proposal [17] is to analyse the text in terms of various *aspects* exhibited by the document. [18] outlined the *core grammatical* aspect (CGa). CGa is concerned first with terminology, entities, and modifiers which express the knowledge and moreover their relationships to one another. Table 1 lists in **bold face**, the grammatical categories used

Table 1. MathLang's grammatical categories

term common mathematical objects like "$a + b$" or "an additive identity 0".
set Sets of mathematical objects such as "$\mathbb{N}$".
noun families of **terms** such as "ring".
adjective defines new **nouns** from old ones. E.g., "Abelian" is an **adjective** which modifies the **noun** "ring" to create the new **noun** "Abelian ring".
statement Expressions like "$a + 0 = a$" which describe mathematical properties.
declaration the type signature of a new **term**, **set**, **noun**, **adjective**, or **statement**.
definition defines new symbols in mathematical texts.
step A group of mathematical assertions.
context preliminary assertions prior to a **step**.

at the CGa aspect of MathLang together with the colour coding. In the current paper we focus on a *text and symbol* aspect (TSa) of mathematical knowledge which is able to flexibly represent natural language mathematics.

2   Box Annotation, an Explicit Typing of Expressions

We propose an authoring technique in which the mathematical text is input to the computer exactly as it is written on paper by the mathematician. As an author composes a document, it is desirable to truly derive any formal or symbolic version from this original document. We propose to decorate the original text with extra information. This extra content has to be more precise, complete and computation-friendly than natural language. With such extra information intermingled with the original text, it is possible to ensure that subsequent translations are consistent with and faithful to the natural language text.

2.1   Box Annotation

The approach of this paper augments the original natural language text with supplementary information. We do so by wrapping (at the screen), pieces of text with *annotation boxes*. The background colour of an annotation box informs about the MathLang-grammatical role of the wrapped text (following the colour coding of Table 1). Notice that once we remove these annotation boxes we find the text completely unchanged. Take from our supplement example the sentence "There is an element 0 in R such that $a + 0 = a$". The grammatical information (in terms of MathLang's grammatical constructions) can be easily inferred from the original text as shown by the following annotation boxes. The boxes surrounding "an element 0", "a", "0" and "$a + 0$" indicate that these expressions are **terms**. "R" is wrapped in a **set** box and "an element 0 in R" in a **declaration** box. The box surrounding "$a + 0 = a$" indicates that this equation is a **statement**. The whole sentence is put in a **step** box.

$$\boxed{\text{There is } \boxed{\text{an element } \boxed{0}} \text{ in } \boxed{R} \text{ such that } \boxed{\boxed{a} + \boxed{0}} = \boxed{a}}$$

This expression would correspond to the pseudo-logic code `eq(plus(a,0),a)` which differs from its box-annotated natural language equivalent by its namespaces. The symbol "$+$" corresponds to the identifier `plus`. Accordingly, one might argue that the symbols = and + could have been used with infix notation and relevant symbols' precedence. We would have obtained an expression `a+0=a` which is similar to the natural language sentence's equation. But imagine a situation where, instead of stating the equality between $a + 0$ and a by an equation, the verb "equal" is used: $\boxed{\boxed{a} + \boxed{0}} \text{ equals } \boxed{a}$. The sentence would be printed differently but would still mean that `a+0=a`. An equation and its natural language equivalent should reflect the same meaning (`a+0=a` in our example). Similarly, a natural language sentence and its equivalent formula should get similar box annotations. Our sentence could look like: $\boxed{\boxed{0} \in \boxed{R}}, \boxed{\boxed{a} + \boxed{0}} = \boxed{a}$.

2.2 Interpretation

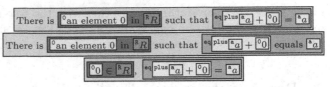

To establish the meaning of the text contained in each annotation box, we attribute to each box its interpretation in our grammar (see Table 1). The boxes surrounding "an element 0" and "0" get 0 as interpretation attribute. The box surrounding "R", "a", "$a + 0$" and the equation get as interpretation attributes R, a, plus and eq, respectively. Each interpretation attribute is printed in a typewriter typeface on the left hand side of the annotation box.

With these examples we see that MathLang's grammar is not a natural language grammar but a mathematical *justifications* grammar (following de Bruijn [19]).

2.3 Nested Annotations

In our example we see also that some boxes are inside other boxes. In the case of our equation, each inner box is interpreted as an argument for its surrounding box. The nesting of boxes indicates that some annotated expressions are sub-expressions of others. It is a straightforward automatic process to create a MathLang grammatical expression out of a text with box annotations.

We show here the MathLang grammatical expression corresponding to our box-annotated text. This expression is written us-

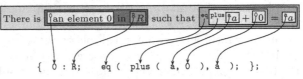

ing the abstract syntax we presented in [18]. Note that this syntax is not meant to be used by the end-user of MathLang, it is only designed for theoretical discussion on MathLang's grammar. The MathLang end-user edits his natural language text with annotation boxes, as shown in Section 2.4 and in the supplement example. The internal syntax used in our implementations follows XML recommendations.

2.4 Automatic Grammatical Analysis

This authoring method with annotation boxes was implemented as a plugin for the scientific text editor $\text{T}_{\text{E}}\text{X}_{\text{MACS}}$. During or after the editing of a natural language text, an author is asked to wrap relevant pieces of text in MathLang's annotation boxes. Customised views are provided within the MathLang plugin to toggle the display of several features of the document including the coloured boxes resulting in this wrapping and the interpretations introduced in Section 2.2. The user easily obtains the following views (once with annotation boxes printed as coloured boxes and then with these boxes hidden):

There is an element 0 in R such that $a + 0 = a$ There is an element 0 in R such that $a + 0 = a$

The MathLang plugin communicates the content of the document to the MathLang grammar checker given in [18], employing $\text{T}_{\text{E}}\text{X}_{\text{MACS}}$ as an integrated graphical environment for natural language input, annotation, and grammar checking.

Continuing with our sentence-example, let us assume that R, 0, $=$ and $+$ were properly introduced in the larger document. When the user is satisfied with his annotation of the sentence, the $\text{T}_{\text{E}}\text{X}_{\text{MACS}}$ plugin is instructed to send the entire document to the type checker. The checker analyses the grammatical structure of the MathLang document and finds out that a has not been properly introduced in our sentence. A set of errors[1] with their locations in the $\text{T}_{\text{E}}\text{X}_{\text{MACS}}$ document are sent back to the plugin to be shown to the user. Here is a view of the text with annotation boxes and their interpretations printed in between angle brackets < and >, and errors' labels printed in between stars *.

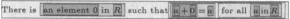

```
Error (e-1): Anticipated instance of "a"
Error (e-2): Categories mismatch, Unspecified expected, not term.
Error (e-3): Types mismatch for "plus", (term,term):term expected, not (Unspecified,term):term.
Error (e-4): Categories mismatch, Unspecified expected, not term.
Error (e-5): Categories mismatch, Unspecified expected, not term.
Error (e-6): Types mismatch for "equal", (term,term):stat expected,
not (term,Unspecified):stat.
```

To fix these errors we simply define a as it was done in the original text (see the supplement example). The extra "for all a in R" text is wrapped in a **context** box annotation which indicates that it forms the context of the equation.

There is $\boxed{\text{an element } 0 \text{ in } R}$ such that $\boxed{a+0=a}$ for all $\boxed{a}$ in $\boxed{R}$

3 Souring Annotation

The grammatical box annotations of Section 2 are guided by the style in which the original natural-language sentences were written. Mathematical writing styles are *uneven* and do not always fit such simplistic annotations. To adapt to any style, we need additional box annotations which help interpret the author's style. We believe it is necessary to separate grammatical and style annotations.

3.1 Syntax Souring

Mathematicians use mathematical natural language as a medium for communicating mathematical knowledge, but this language is highly automation-unfriendly for computer software. We showed in [12] that MathLang has constructions that correspond to the way common mathematical justifications are structured. MathLang is automation-friendly and mimics the mathematical natural language structure of justification. Therefore MathLang authoring does not require the user to alter or translate the document's knowledge for computerisation, although there is a need to adjust the writing style when encoding text directly into the core MathLang language. Because we regard our starting language, natural language, to be the *sweetest* for human readers, we call this modification *syntax souring*.

---

[1] The high number of errors is due to the fact that the checking of the document does not stop after one error is found but analyses the entire document. An error may point at several locations in the document, this to cover all expression involved in a typing error.

This term describes the process of transforming natural language into syntactically formalised language (the core grammatical MathLang of [18]). The additives needed to describe how to perform a transformation of natural language to a core formalised language are known as *souring annotation*.

Syntax sugaring. The notion of *syntax sugaring* is well known by programmers. Syntactic sugar is added to the syntax of programming languages to make it easier to use by humans. Syntax sugaring lightens the syntax without affecting expressiveness.

Programming language
$+$
Syntactic sugar
de-sugaring $\downarrow$
Core programming language

Souring: dual of de-sugaring. Syntactic sugar is usually an additive for the syntax of formal language. *De-sugaring* is the process of getting rid of the sugared bits by replacing them with proper core syntax expressions.

In our case the primary input is the mathematician's natural language which we want to extend for computer software use. *Souring* unfolds the sour bits to produce a *sour document*, i.e. a document which is formal enough to be understood by computer software. The

Natural language
$+$
Grammatical annotations
$+$
Syntactic sour bits
souring $\downarrow$
Core sour language

original document and the sour one do not belong to the same type of document.

The *duality* between syntax sugaring and syntax souring resides in the fact that both are methods to humanise the authoring of rigid languages but have a different starting point (i.e., programming language for syntax sugaring and natural language for syntax souring). De-sugaring adapts rigid languages for human consumption. Souring rigidifies natural language for software use.

3.2   Denotational Representation

We give here the denotational representation which is formalised in Section 4.1.

Document. Our starting point is the mathematician's text (as he wrote it on paper) which is composed by a mixture of natural language text and formulae formed by symbols. This primary input corresponds to $\mathcal{D}_{\mathcal{F}}$ (formed by $\mathcal{F}$ individuals) in the abstract syntax of Section 4.1. We add to this primary input, grammatical and souring annotations that wrap portions of the text. We already saw in Section 2 how we represent grammatical annotations. In this section we explain how we represent the souring annotations discussed in Section 3.1. We denote by T a portion of text which may include formulae, grammatical annotations and souring annotations. We denote by A an arbitrary annotation.

Grammatical annotations. A grammatical annotation is an instance of one of the grammatical categories **term, set, noun, adjective, statement, declaration, definition, context,** or **step** (see Table 1). Each instance of a grammatical annotation may get an attribute which corresponds to the grammatical annotation's interpretation given in Section 2. We represent grammatical annotations by a box whose background colour—according to the colour coding of Table 1—informs the grammatical category and whose interpretation is printed on the upper left-hand side of the box using `courier` typeface. Here is for instance the term a annotated with a **term**-box with "a" as interpretation: $\boxed{^{\mathtt{a}}a}$. We use G, G', G_1, etc., to range over grammatical interpretations. Grammatical annotations correspond to $\mathcal{G}$ labels in the formal system presented in Section 4.1.

Souring annotations. Sour bits correspond to souring annotations. We denote them by a distinguishable font colour and a thicker box for the annotation they describe (i.e., $\boxed{^{\mathtt{list}}a, b, c}$). We define in the rest of this paper the following syntax souring annotations (which correspond to the elements souring labels $\mathcal{S}_u$ of Section 4.1): `position` i, `fold-right`, `fold-left`, `base`, `list`, `hook`, `loop`, `shared` and `map` (where i is a natural number).

Patterns. To describe the souring rules, we need to reason about the annotation boxes contained in a text. To do so, we add parameters to a text T to identify the text patterns that could be transformed. We use two different notations for these parametrised texts: the *in-order* notation where arguments should appear in T in the same order as they appear in the pattern and the *un-ordered* notation where the order of arguments is unimportant. We denote such parametrised notation, with $\boxed{^{A_1}T_1}$, $\ldots$, $\boxed{^{A_k}T_k}$ being the arguments for T, as in the accompanying diagram. Sometimes, optional names

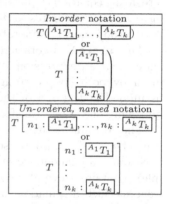

$n_1, \ldots, n_k$ are used as markers to determine the argument's location in the text. The behaviour of parametrised text is reflected in the de-formatting function (see Definition 5) and compatibility property (see Definition 6) stated in Section 4.

3.3 Souring Transformations

In this section we indicate how to use our souring annotations and describe the result of a souring transformation where the souring notation is unfolded to obtain a text where grammatical annotations are similar to those of Section 2. Such a document could then be checked according to the MathLang grammatical checker of [18] discussed briefly in Section 2.4.

Re-ordering. $\boxed{\text{position } i}$ When dealing with a natural language mathematical text, one regularly faces situations where two expressions holding similar knowledge are ordered differently. The re-ordering transformation corresponds to $\rightarrow_{pos}$ of Section 4.2. Considering the expression "a in R" from our supplement example, one can easily

$$T \begin{bmatrix} \boxed{\text{position } 1}T_1 \\ \vdots \\ \boxed{\text{position } n}T_n \end{bmatrix} \xrightarrow{souring} T(T_1, \ldots, T_n)$$

imagine the author using "R contains a" instead. The `position` souring annotation is meant for reordering inner-annotations. The souring rewriting function reorders the elements according to their position indices.

The expressions "a in R" and "R contains 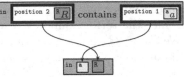 a" should both be interpreted as `in(a,R)` if `in` is the set membership relation. To indicate in the second expression that the order of the argument is not the "reading" order, we annotate R and a with `position 2` and `position 1`, respectively. It is common for binary symbols like $\subset$ to have a mirror twin like $\supset$. The `position` souring annotation usefully gives the same interpretation to twin symbols.

Sharing/chaining. `shared` `hook` `loop` Mathematicians have the habit of aggregating equations which follow one another. This creates reading difficulties for novices yet contributes to the aesthetic of mathematical writing. The **shared** and `hook/loop` souring annotations are solutions which elucidate such expressions.

The **shared** annotation indicates that an expression is to be 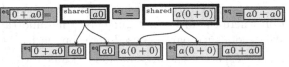 used by both its preceding and following expressions. The shared expression is inlined at the end of the preceding expression and at the beginning of the following one. This transformation corresponds to $\twoheadrightarrow_{share}$ of Section 4.2.

The document example we chose to computerise (see our supplement example) contains several sentences which are made easier to computerise by the use of sharing. The multiple

The full interpretation of this expression being:
```
eq(plus(0,times(a,0)),times(a,0));
eq(times(a,0),times(a,plus(0,0)));
eq(times(a,plus(0,0)),plus(times(a,0),times(a,0)))
```

equation "$0+a0 = a0 = a(0+0) = a0+a0$" is certainly the best example as it requires the use of two **shared** annotations. We can see that $a0$ and $a(0+0)$ are shared by two equations each. We annotate them as being shared to obtain an unfolded result equivalent to "$0+a0 = a0$, $a0 = a(0+0)$, $a(0+0) = a0+a0$".

The tuple of souring annotations `hook/loop` indicates the expression contained in the hook should be repeated in the loop. We named this $T\begin{pmatrix} \text{hook}\,T' \\ \text{loop} \end{pmatrix} \xrightarrow{souring} T\begin{pmatrix} T' \\ T' \end{pmatrix}$ concept chaining because it permits the separation of two expressions which are effectively printed as one in a natural language text. Chaining provides results similar to sharing (any sharing could be expressed in terms of chaining), but is more expressive. This transformation corresponds to $\twoheadrightarrow_{chain}$ of Section 4.2.

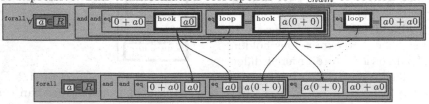

The full interpretation of this expression being:
```
forall(a:R, and( and( eq(plus(0,times(a,0)),times(a,0)),
                      eq(times(a,0),times(a,plus(0,0))) ),
                 eq(times(a,plus(0,0)),plus(times(a,0),times(a,0))) ))
```

Let us see an example where a **shared** souring annotation could not have been used. If we consider the equation we used in the sharing example and decide to quantify this equation over a, we would obtain "$\forall a \in R$, $0+a0 = a0 = a(0+0) = a0+a0$" which is effectively a shortcut for "$\forall a \in R$, $0+a0=a0 \wedge a0=a(0+0) \wedge a(0+0)=a0+a0$". We can see that in this example the individual equations are combined using two binary operators **and**, the combination of whose annotation boxes disallows the use of **shared**.

List manipulations. `fold-right base list` `fold-left base list` `map list` The list souring annotations indicate how lists of expressions have to be unfolded into MathLang interpretations. We define two list folding annotations, **fold-right** and **fold-left**, and a mapping annotation, **map**.

$$\boxed{\text{fold-right } T_f \left[\begin{array}{l} b : \boxed{\text{base }} T_b \\ l : \boxed{\text{list }} T_1 \dots T_k \end{array} \right]} \xrightarrow{\text{souring}} T_f \left[\begin{array}{l} b : T_f \left[b : T_f \left[\dots T_f \left[\begin{array}{l} b : T_b \\ l : T_k \end{array} \right] \dots \right] \right] \\ l : T_2 \\ l : T_1 \end{array} \right]$$

The **fold-right** souring annotation defines a pattern which is repeated for each element of the list argument. For each repeated pattern, the **list** inner annotation is replaced by one element of the list and the **base** inner annotation is replaced by the pattern with the next element of the list. **fold-left** works similarly but starts with the last element of the list. These transformations correspond to $\rightarrow_{fold}$ of Section 4.2.

A major use of the **fold-right** souring annotation is to handle quantification over multiple variables. Considering the sentence "for all a, b, c in R $[\dots]$ $(a+b)+c = a+(b+c)$", we would like to use one single **forall** instance for each variable a, b and c. We simply annotate the list of variables as such and the base equation as **base** and the souring unfolding creates a fully expanded interpretation on our behalf.

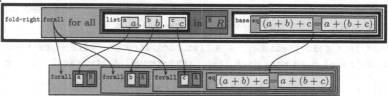

The full interpretation of this expression being:
```
forall(a:R, forall(b:R, forall(c:R,eq(plus(plus(a,b),c),plus(a,plus(b,c)))) ) ) )
```

The **map** souring annotation also defines a pattern but with only one argument being **list**. $$\boxed{\text{map } T_f \left(\boxed{\text{list } T_1 \dots T_n} \right)} \xrightarrow{\text{souring}} T_f(T_1) \dots T_f(T_n)$$ This pattern is also repeated for each element of the list. The resulting expression is a sequence. It corresponds to $\rightarrow_{map}$ defined in Section 4.2.

Similarly to folding, this souring annotation is useful for declarations, definitions or statements over several things. In the case of the sentence "Let a and b belong

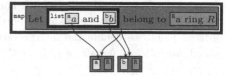

to a ring R" taken from our supplement example, the variables a and b are declared simultaneously.

4   Operational System

Having presented our method in an intuitive, denotational style, we now give the formal system behind it and the foundation for MathLang documents.

4.1   Abstract Syntax

Let $\mathbb{N}$ denote the natural numbers, use $(-;-)$ to denote ordered pairs, and let functions be sets φ of ordered pairs with a domain $\mathrm{dom}(\varphi) = \{a \mid \exists b$ such that $(a;b) \in \varphi\}$. A sequence is a function s for which $\mathrm{dom}(s) = \{n \mid 0 \le n < k\}$ for some $k \in \mathbb{N}$. We write $[]$ for the empty sequence and $[x_0, x_1, \ldots, x_n]$ for the sequence s such that $s(i) = x_i$ for each $i \in \mathrm{dom}(s) = \{0, \ldots, n\}$. Upon that sequence is defined the metric $|s| = n + 1$. We define s_1, s_2, the concatenation of sequences s_1 and s_2, as the new sequence s such that $\mathrm{dom}(s_1, s_2) = \{0, \ldots, |s_1| + |s_2| - 1\}$, $s(i) = s_1(i)$ for $i \in \mathrm{dom}(s_1)$ and $s(i) = s_2(i)$ for $i - |s_1| \in \mathrm{dom}(s_2)$. Concatenation is associative. Moreover, $[], s = s$ and $s, [] = s$.

Let $\mathcal{L} = \mathcal{F} \cup \mathcal{G} \cup \mathcal{S}$ to be the set of labels over which ℓ ranges where elements of $\mathcal{F}$, resp. $\mathcal{G}$, resp. $\mathcal{S}$, are formatting, resp. grammatical, resp. souring labels.

$\mathcal{F}$ (over which f ranges, cf. Definition 4) consists of *formatting instructions* and varies according to the typesetting system used.

$\mathcal{G} = \mathcal{C} \times \mathcal{I}$ where $\mathcal{C} = \{\textbf{term}, \textbf{set}, \textbf{noun}, \textbf{adj}, \textbf{stat}, \textbf{decl}, \textbf{defn}, \textbf{step}, \textbf{cont}\}$, and contains identifiers for the primitive grammatical categories of Table 1. The set $\mathcal{I}$ consists of strings used for identifying abstract interpretations (e.g., 0, R, eq, plus and a are the interpretation strings used in the examples throughout Section 2). We let g, c and i range respectively over $\mathcal{G}$, $\mathcal{C}$ and $\mathcal{I}$.

We let s range over $\mathcal{S} = \mathcal{S}_u \cup \mathcal{S}_i$ where $\mathcal{S}_u$ contains *souring identifiers* to be employed directly by the user while $\mathcal{S}_i$ holds several identifiers used internally for rewriting. $\mathcal{S}_u$ and $\mathcal{S}_i$ are disjoint and are as follows:

$\mathcal{S}_u = \{\texttt{fold-left}, \texttt{fold-right}, \texttt{map}, \texttt{base}, \texttt{list}, \texttt{hook}, \texttt{loop}, \texttt{shared}\} \cup (\{\textbf{position}\} \times \mathbb{N})$

$\mathcal{S}_i = \{\texttt{hook-travel}, \texttt{head}, \texttt{tail}, \texttt{daeh}, \texttt{liat}, \texttt{right-travel}, \texttt{left-travel}\} \cup (\{\textbf{cursor}\} \times \mathbb{N})$

Definition 1 (Document). *Let $\mathcal{D}$ be the smallest set such that:*
1. *$[] \in \mathcal{D}$,*
2. *if $d \in \mathcal{D}$ and $\ell \in \mathcal{L}$ then $[(\ell; d)] \in \mathcal{D}$, and*
3. *if both d_1 and d_2 are elements of $\mathcal{D}$ then $(d_1, d_2) \in \mathcal{D}$.*
A MathLang document is an element of the set $\mathcal{D}$. In addition, we denote by $\mathcal{D}_\mathcal{F}$, $\mathcal{D}_\mathcal{G}$, $\mathcal{D}_{\mathcal{F} \cup \mathcal{G}}$, $\mathcal{D}_{\mathcal{G} \cup \mathcal{S}}$ and $\mathcal{D}_{\mathcal{F} \cup \mathcal{G} \cup \mathcal{S}}$ the sets of documents for which labels are in $\mathcal{F}$, $\mathcal{G}$, $\mathcal{F} \cup \mathcal{G}$, $\mathcal{G} \cup \mathcal{S}$ and $\mathcal{F} \cup \mathcal{G} \cup \mathcal{S}$, respectively.

Remark 1 (Notational convention). For convenience, $[(\ell; d)]$ abbreviates to $\ell\langle d\rangle$. Furthermore, when not ambiguous $\ell\langle[]\rangle$ abbreviates to ℓ. In the case of grammatical labels ordered pairs, we denote the interpretation (second element of the pair) by an adjoined superscript. A pair $(c; i)$ from $\mathcal{G}$ is denoted by c^i. Similarly, an ordered pair from $\{\texttt{position}\} \times \mathbb{N}$ (respectively $\{\texttt{cursor}\} \times \mathbb{N}$) is denoted by a superscript number (second element of the pair) adjoined to $\texttt{position}$ (respectively $\texttt{cursor}$).

Definition 2 (Sub-document). *We define* sub-document, *and we denote by* $\sqsubset_{\mathcal{G}}$ *, the binary relation between documents such that:*

$$d \sqsubset_{\mathcal{G}} d \qquad\qquad (\text{SUB1})$$

$$d \sqsubset_{\mathcal{G}} g\langle d_1 \rangle \ \ if \ d \sqsubset_{\mathcal{G}} d_1 \qquad\qquad (\text{SUB2})$$

$$d \sqsubset_{\mathcal{G}} (d_1, d_2) \ \ if \ d \sqsubset_{\mathcal{G}} d_1 \ or \ d \sqsubset_{\mathcal{G}} d_2 \qquad\qquad (\text{SUB3})$$

Remark 2. It is important to notice that our sub-document property (SUB2) is restricted to grammatical labels which means that for any label $\ell \notin \mathcal{G}$ and any documents d_1 and d_2 such that $d_1 \sqsubset_{\mathcal{G}} d_2$, we have that $d_1 \not\sqsubset_{\mathcal{G}} \ell\langle d_2 \rangle$.

Definition 3 (Label inclusion). *We define* label inclusion, *and we denote by* $\widetilde{\in}_{\mathcal{G}}$ *, the binary relation between a label and a document such that:*

$$\ell \ \widetilde{\in}_{\mathcal{G}} \ \ell\langle d \rangle \qquad\qquad (\text{INC1})$$

$$\ell \ \widetilde{\in}_{\mathcal{G}} \ g\langle d \rangle \ \ if \ \ell \neq g \ and \ \ell \ \widetilde{\in}_{\mathcal{G}} \ d \qquad\qquad (\text{INC2})$$

$$\ell \ \widetilde{\in}_{\mathcal{G}} \ (d_1, d_2) \ \ if \ \ell \ \widetilde{\in}_{\mathcal{G}} \ d_1 \ or \ \ell \ \widetilde{\in}_{\mathcal{G}} \ d_2 \qquad\qquad (\text{INC3})$$

Remark 3. Note that our label inclusion property (INC2) is restricted to grammatical labels, which means that for any labels $\ell_1 \notin \mathcal{G}$ and $\ell_2 \in \mathcal{L}$ such that $\ell_1 \neq \ell_2$, and any document d such that $\ell_2 \ \widetilde{\in}_{\mathcal{G}} \ d$, we have that $\ell_2 \ \widetilde{\notin}_{\mathcal{G}} \ \ell_1\langle d \rangle$.

Definition 4 (Rendering functions). *Let* $\mathrm{f} : \mathcal{D} \to \mathcal{D}_{\mathcal{F}}$ *be a function where:*

$$\mathrm{f}([\,]) = [\,] \qquad\qquad (\text{FORM1})$$

$$\mathrm{f}(\ell\langle d \rangle) = \begin{cases} \ell\langle \mathrm{f}(d) \rangle & if \ \ell \in \mathcal{F} \\ \mathrm{f}(d) & otherwise \end{cases} \qquad\qquad (\text{FORM2})$$

$$\mathrm{f}(d_1, d_2) = \mathrm{f}(d_1), \mathrm{f}(d_2) \qquad\qquad (\text{FORM3})$$

Thus, f *flattens a given document* d *at any label from* $\mathcal{G}$ *or* $\mathcal{S}$*, removing all such labels. Once this is achieved, it will be possible to use* $\mathrm{r} : \mathcal{D}_{\mathcal{F}} \to \mathcal{F}$*, where:*

$$\mathrm{r}([\,]) = \varepsilon \qquad\qquad (\text{REN1})$$

$$\mathrm{r}(f\langle d \rangle) = \mathrm{fill}(f, [\mathrm{r}(d(0)), \ldots, \mathrm{r}(d(|d|-1))]) \qquad\qquad (\text{REN2})$$

$$\mathrm{r}(d_1, d_2) = \mathrm{r}(d_1) \bullet \mathrm{r}(d_2) \qquad\qquad (\text{REN3})$$

Where, in a specific typesetting system, ε *is the blank formatting instruction,* $\bullet$ *is the composition operator and* fill *is a formatting-system-specific function. The function* fill *interprets a formatting instruction (first argument) with a sequence of rendered documents passed as argument. One can imagine a formatting instruction to be a template with holes and* fill *to simply fill these holes. The number of vacancies exhibited by the first argument of* fill *should be equal to the length of the sequence, which is the second argument of* fill*. The function* fill *returns an element of the set* $\mathcal{F}$ *which is a formatting instruction requiring no argument.*

Definition 5 (De-formatting function). *To prepare a document for souring, we strip it of all formatting elements using the function* $\mathrm{df} : \mathcal{D} \to \mathcal{D}_{\mathcal{G}\cup\mathcal{S}}$ *where:*

$$\mathrm{df}([]) = [] \tag{DF1}$$

$$\mathrm{df}(\ell\langle d\rangle) = \begin{cases} d & \textit{if } \ell \in \mathcal{F} \\ \ell\langle\mathrm{df}(d)\rangle & \textit{otherwise} \end{cases} \tag{DF2}$$

$$\mathrm{df}(d_1, d_2) = \mathrm{df}(d_1), \mathrm{df}(d_2) \tag{DF3}$$

4.2 Souring Rewriting Rules

Definition 6 (Compatibility, Reflexive transitive closure, Normal form).
We define the following compatibility property for a rewriting rule $\to_n$.

$$d_1, d, d_2 \to_n d_1, d', d_2 \textit{ if } d \to_n d' \tag{COMP1}$$

$$g\langle d\rangle \to g\langle d'\rangle \textit{ if } d \to_n d' \tag{COMP2}$$

We denote by $\twoheadrightarrow_n$ *the reflexive transitive closure of* $\to_n$.
We define the n-normal form relatively to $\to_n$ *and we denote by* NF_n *the property on a document d such that no* $\twoheadrightarrow_n$ *rewriting can be applied to d.*
Note that our compatibility rule (COMP2) *is restricted to grammatical labels.*
Below are the formal rewriting rules for souring transformations from Section 3.1.

$\mathtt{head}\langle d_1, d_2\rangle \to_{list} d_1, \mathtt{head}\langle d_2\rangle$ where list $\widetilde{\notin}_{\mathcal{G}} d_1$
$\mathtt{tail}\langle d_1, d_2\rangle \to_{list} d_1, \mathtt{tail}\langle d_2\rangle$ where list $\widetilde{\notin}_{\mathcal{G}} d_1$
$\mathtt{daeh}\langle d_1, d_2\rangle \to_{list} d_1, \mathtt{daeh}\langle d_2\rangle$ where list $\widetilde{\notin}_{\mathcal{G}} d_1$
$\mathtt{liat}\langle d_1, d_2\rangle \to_{list} d_1, \mathtt{liat}\langle d_2\rangle$ where list $\widetilde{\notin}_{\mathcal{G}} d_1$
$\mathtt{head}\langle g\langle d_1\rangle, d_2\rangle \to_{list} g\langle\mathtt{head}\langle d_1\rangle\rangle, d_2$
$\qquad\qquad\qquad$ where list $\widetilde{\in}_{\mathcal{G}} d_1$
$\mathtt{tail}\langle g\langle d_1\rangle, d_2\rangle \to_{list} g\langle\mathtt{tail}\langle d_1\rangle\rangle, d_2$
$\qquad\qquad\qquad$ where list $\widetilde{\in}_{\mathcal{G}} d_1$
$\mathtt{daeh}\langle g\langle d_1\rangle, d_2\rangle \to_{list} g\langle\mathtt{daeh}\langle d_1\rangle\rangle, d_2$
$\qquad\qquad\qquad$ where list $\widetilde{\in}_{\mathcal{G}} d_1$
$\mathtt{liat}\langle g\langle d_1\rangle, d_2\rangle \to_{list} g\langle\mathtt{liat}\langle d_1\rangle\rangle, d_2$
$\qquad\qquad\qquad$ where list $\widetilde{\in}_{\mathcal{G}} d_1$
$\mathtt{head}\langle\mathtt{list}\langle g\langle d_1\rangle, d_2\rangle, d_3\rangle \to_{list} g\langle d_1\rangle, d_3$
$\mathtt{tail}\langle\mathtt{list}\langle g\langle d_1\rangle, d_2\rangle, d_3\rangle \to_{list} d_2, d_3$
$\mathtt{daeh}\langle\mathtt{list}\langle d_1, g\langle d_2\rangle\rangle, d_3\rangle \to_{list} g\langle d_2\rangle, d_3$
$\mathtt{liat}\langle\mathtt{list}\langle d_1, g\langle d_2\rangle\rangle, d_3\rangle \to_{list} d_1, d_3$

$g_1\langle d_1\rangle, \mathtt{shared}\langle d\rangle, g_2\langle d_2\rangle \to_{share} g_1\langle d_1, d\rangle, g_2\langle d, d_2\rangle$

$\mathtt{hook}\langle d\rangle \to_{chain} d, \mathtt{hook\text{-}travel}\langle d\rangle$
$\mathtt{hook\text{-}travel}\langle d\rangle, \mathtt{loop} \to_{chain} d$
$\mathtt{hook\text{-}travel}\langle d_0\rangle, d_1, d_2 \to_{chain} d_1, \mathtt{hook\text{-}travel}\langle d_0\rangle, d_2$
$\qquad\qquad$ where loop $\widetilde{\notin}_{\mathcal{G}} d_1$
$\mathtt{hook\text{-}travel}\langle d_0\rangle, g\langle d_1\rangle \to_{chain} g\langle\mathtt{hook\text{-}travel}\langle d_0\rangle, d_1\rangle$
$g\langle d_1, \mathtt{hook\text{-}travel}\langle d_0\rangle\rangle \to_{chain} g\langle d_1\rangle, \mathtt{hook\text{-}travel}\langle d_0\rangle$

$\mathtt{position}^i\langle d_1\rangle, \mathtt{position}^j\langle d_2\rangle \to_{pos}$
$\qquad\qquad \mathtt{position}^j\langle d_2\rangle, \mathtt{position}^i\langle d_1\rangle$
$\qquad\qquad\qquad$ where $j < i$
$\ell\langle\mathtt{position}^1\langle d_1\rangle, d_2\rangle \to_{pos} \ell\langle d_1, \mathtt{cursor}^1\langle d_2\rangle$
$\mathtt{cursor}^i, \mathtt{position}^{i+1}\langle d_1\rangle, d_2 \to_{pos} d_1, \mathtt{cursor}^{i+1}, d_2$
$\ell\langle d, \mathtt{cursor}^i\rangle \to_{pos} \ell\langle d\rangle$

$\mathtt{fold\text{-}right}\langle d_0\rangle \to_{fold} \mathtt{right\text{-}travel}\langle d_2\rangle, d_1$
$\qquad$ where $d_0 \twoheadrightarrow_{souring} d_0'$, $\mathtt{head}\langle d_0'\rangle \twoheadrightarrow_{list} d_1$
$\qquad\qquad$ and $\mathtt{tail}\langle d_0'\rangle \twoheadrightarrow_{list} d_2$
$\mathtt{right\text{-}travel}\langle d_1, d_2\rangle \to_{fold} d_1, \mathtt{right\text{-}travel}\langle d_2\rangle$
$\qquad\qquad\qquad$ where base $\widetilde{\notin}_{\mathcal{G}} d_1$
$\mathtt{right\text{-}travel}\langle g\langle d_1\rangle, d_2\rangle \to_{fold}$
$\qquad\qquad\qquad g\langle\mathtt{right\text{-}travel}\langle d_1\rangle\rangle, d_2$
$\qquad\qquad$ where $g \neq$ base and base $\widetilde{\in}_{\mathcal{G}} d_1$
$\mathtt{right\text{-}travel}\langle d_1\rangle, \mathtt{base}\langle d_2\rangle \to_{fold}$
$\qquad\qquad\qquad d_2, \mathtt{right\text{-}travel}\langle d_2\rangle$
$\qquad\qquad\qquad$ where list $\sqsubseteq_{\mathcal{G}} d_1$
$\mathtt{right\text{-}travel}\langle d_1\rangle, \mathtt{base}\langle d_2\rangle \to_{fold} \mathtt{fold\text{-}right}\langle d_1\rangle$

$\mathtt{fold\text{-}left}\langle d_0\rangle \to_{fold} \mathtt{left\text{-}travel}\langle d_2\rangle, d_1$
$\qquad$ where $d_0 \twoheadrightarrow_{souring} d_0'$, $\mathtt{daeh}\langle d_0'\rangle \twoheadrightarrow_{list} d_1$
$\qquad\qquad$ and $\mathtt{liat}\langle d_0'\rangle \twoheadrightarrow_{list} d_2$
$\mathtt{left\text{-}travel}\langle d_1, d_2\rangle \to_{fold} d_1, \mathtt{left\text{-}travel}\langle d_2\rangle$
$\qquad\qquad\qquad$ where base $\widetilde{\notin}_{\mathcal{G}} d_1$
$\mathtt{left\text{-}travel}\langle g\langle d_1\rangle, d_2\rangle \to_{fold} g\langle\mathtt{left\text{-}travel}\langle d_1\rangle\rangle, d_2$
$\qquad\qquad$ where $g \neq$ base and base $\widetilde{\in}_{\mathcal{G}} d_1$
$\mathtt{left\text{-}travel}\langle d_1\rangle, \mathtt{base}\langle d_2\rangle \to_{fold} d_2, \mathtt{left\text{-}travel}\langle d_2\rangle$
$\qquad\qquad\qquad$ where list $\sqsubseteq_{\mathcal{G}} d_1$
$\mathtt{left\text{-}travel}\langle d_1\rangle, \mathtt{base}\langle d_2\rangle \to_{fold} \mathtt{fold\text{-}left}\langle d_1\rangle$

$\mathtt{map}\langle d\rangle \to_{map} []$ where list $\sqsubseteq_{\mathcal{G}} d$
$\mathtt{map}\langle d_0\rangle \to_{map} d_1, \mathtt{map}\langle d_2\rangle$
$\qquad$ where $d_0 \twoheadrightarrow_{souring} d_0'$, $\mathtt{head}\langle d_0'\rangle \twoheadrightarrow_{list} d_1$
$\qquad\qquad$ and $\mathtt{tail}\langle d_0'\rangle \twoheadrightarrow_{list} d_2$

Definition 7 (Souring rewriting rule). *The souring rewriting rule, denoted by $\rightarrow_{souring}$ is defined as $d_0 \rightarrow_{souring} d_4$ where $d_0 \twoheadrightarrow_{share} d_1$ (d_1 being in a NF_{share}), $d_1 \twoheadrightarrow_{chain} d_2$ (d_2 being in a NF_{chain}), $d_2 \twoheadrightarrow_{pos} d_3$ (d_3 being in a NF_{pos}), $d_3 \twoheadrightarrow_{lists} d_4$ (d_4 being in a NF_{lists}).*

The souring a document is the application of $\twoheadrightarrow_{souring}$ until $NF_{souring}$ is reached.

5 Related Work

The natural-to-abstract work pattern which has been presented in this paper will be useful in a wide variety of settings. One possible area of application is work being done with optical character recognition of mathematics. In the work of the Infty Project [20,21], for example, it is desirable to automate the process of extracting information from printed material. As MathLang becomes capable of being automated, it will provide further aid to extracting semantic information from a document with as little hand-translation as possible.

The $T_{E}X_{MACS}$ plugin environment and method of editing causes MathLang to be a *visual language*. Using visual languages for knowledge representation is becoming more popular, and its benefits are obvious. By displaying and editing the logical structure of a mathematical document, the categorisation of various portions of text is made more clear and the structure more lucid. This could certainly lead to a new generation of literate programming [22].

In Computational linguistics, transformational grammars [23, Ch.5] provide a morphism method similar to souring. Nevertheless they are a natural language grammar and do not provide this separation between the original human-medium (natural language) and the software-medium (MathLang core language).

6 Conclusion and Future Work

We demonstrate in this paper the feasibility of restoring natural language as the primary input for mathematical authoring on computers. This method will benefit mathematicians as it permits the use of computer-assisted authoring without requiring skills in computer-based formalisation. Thus, this method will benefit the mathematical knowledge community as it makes the bridge between traditional and computerised mathematics. Since the souring rewriting rules are defined on top of a generic document format, it should be straightforward to adapt the rules to some specific formatting system and core "sour" language.

The work in [12] established a language and a system for encoding mathematical texts with transformations which permitted viewing the document in various useful forms, especially natural language. The current development improves on this by allowing the user to work directly in natural language while making use of a number of automated features to do the rest. Even so, the set of souring rules with which the system has been augmented is almost certainly incomplete. There are some known mathematical constructs (mentioned below) for which a satisfactory annotation has not yet been found, and there are surely others which have not yet come to the attention of this development team.

Current known shortcomings in the system include good handling of expressions with omitted terms, such as $\overbrace{x + \ldots + x}^{n \text{ times}}$, $2^{2^{\cdot^{\cdot^{\cdot^2}}}}$, and $\frac{1}{1+\frac{1}{1+\cdots}}$. Proper treatment of proof by induction is under active investigation, as well as satisfactory treatment of relations such as modular equivalence e.g., $-1 \equiv 2(\mod 3)$. Other priorities are developing/integrating methods to automate the annotation process, which at present can be redundant and tedious, and gathering data on use and work patterns by mathematicians to guide further tool and interface development. Anecdotal tests have indicated that the system is very easy to use, but further investigation will be necessary to ensure that the present—or any future—implementation provides maximal assistance to the mathematical user.

References

1. Nipkow, T., Paulson, L.C., Wenzel, M.: Isabelle/HOL. LNCS, vol. 2283. Springer, Heidelberg (2002)
2. LogiCal Project, INRIA Rocquencourt, France: The Coq Proof Assistant Reference Manual – Version 8.0 (2004) At ftp://ftp.inria.fr/INRIA/coq/V8.0/doc/
3. Audebaud, P., Rideau, L.: TeXmacs as authoring tool for publication and dissemination of formal developments. ENTCS, Rome, vol. 103, pp. 27–48 (2003)
4. Autexier, S., Benzmüller, C., Fiedler, A., Lesourd, H.: Integrating proof assistants as reasoning and verification tools into a scientific WYSIWIG (ed.). In: User Interfaces for Theorem Provers (UITP '05) [Workshop], Edinburgh (2005)
5. Mamane, L.E., Geuvers, H.: A document-oriented Coq plugin for TeXmacs. In: Mathematical User-Interfaces Workshop 2006 [Workshop], Workingham (2006)
6. Rudnicki, P.: An overview of the Mizar project. In: Proceedings of the 1992 Workshop on Types for Proofs and Programs (1992)
7. Wenzel, M.: Isar – a generic interpretative approach to readable formal proof documents. In: Bertot, Y., Dowek, G., Hirschowitz, A., Paulin, C., Théry, L. (eds.) TPHOLs 1999. LNCS, vol. 1690, pp. 167–184. Springer, Heidelberg (1999)
8. Autexier, S., Sacerdoti Coen, C.: A formal correspondence between OMDoc with alternative proofs and the $\bar{\lambda}\mu\tilde{\mu}$-calculus. [24],pp. 67–81
9. Brown, C.E.: Verifying and invalidating textbook proofs using Scunak. [24], pp. 110–123
10. Kohlhase, M.: OMDoc – An Open Markup Format for Mathematical Documents [version 1.2]. LNCS (LNAI), vol. 4180. Springer, Heidelberg (2006)
11. Asperti, A., Padovani, L., Sacerdoti Coen, C., Schena, I.: HELM and the semantic math-web. In: Boulton, R.J., Jackson, P.B. (eds.) TPHOLs 2001. LNCS, vol. 2152, pp. 59–74. Springer, Heidelberg (2001)
12. Kamareddine, F., Maarek, M., Wells, J.B.: Flexible encoding of mathematics on the computer. In: Asperti, A., Bancerek, G., Trybulec, A. (eds.) MKM 2004. LNCS, vol. 3119, pp. 160–174. Springer, Heidelberg (2004)
13. Padovani, L., Zacchiroli, S.: From notation to semantics: There and back again. [24], pp. 194–207
14. Kerber, M., Pollet, M.: A tough nut for mathematical knowledge management. In: Kohlhase, M. (ed.) MKM 2005. LNCS (LNAI), vol. 3863, pp. 81–95. Springer, Heidelberg (2006)

15. Pollet, M., Sorge, V., Kerber, M.: Intuitive and formal representations: The case of matrices. In: Asperti, A., Bancerek, G., Trybulec, A. (eds.) MKM 2004. LNCS, vol. 3119, pp. 19–21. Springer, Heidelberg (2004)
16. Gallian, J.A.: Contemporary Abstract Algebra. 5th edn. Houghton Mifflin Company (2002)
17. Kamareddine, F., Wells, J.: MathLang: A new language for mathematics, logic, and proof checking. A research proposal to UK funding body (2001)
18. Kamareddine, F., Maarek, M., Wells, J.B.: Toward an object-oriented structure for mathematical text. In: Kohlhase, M. (ed.) MKM 2005. LNCS (LNAI), vol. 3863, pp. 217–233. Springer, Heidelberg (2006)
19. de Bruijn, N.G.: Checking mathematics with computer assistance. Notices of the American Mathematical Society 38, 8–15 (1991)
20. Kanahori, T., Sexton, A., Sorge, V., Suzuki, M.: Capturing abstract matrices from paper. [24], pp. 124–138
21. Raja, A., Rayner, M., Sexton, A., Sorge, V.: Towards a parser for mathematical formula recognition. [24], pp. 139–151
22. Knuth, D.E.: Literate programming. The. Computer Journal 27, 97–111 (1984)
23. Farrell, P.: Grammatical Relations. Oxford Surveys in Syntax and Morphology. Oxford Linguistics (2005)
24. Borwein, J.M., Farmer, W.M. (eds.): MKM 2006. LNCS (LNAI), vol. 4108. Springer, Heidelberg (2006)

Narrative Structure of Mathematical Texts

Fairouz Kamareddine, Manuel Maarek, Krzysztof Retel, and J.B. Wells

ULTRA group, Heriot-Watt University
http://www.macs.hw.ac.uk/ultra/

Abstract. There are many styles for the narrative structure of a mathematical document. Each mathematician has its own conventions and traditions about labeling portions of texts (e.g., *chapter, section, theorem* or *proof*) and identifying statements according to their logical importance (e.g., *theorem* is more important than *lemma*). Such narrative/structuring labels guide the reader's navigation of the text and form the key components in the reasoning structure of the theory reflected in the text. We present in this paper a method to computerise the narrative structure of a text which includes the relationships between labeled text entities. These labels and relations are input by the user on top of their natural language text. This narrative structure is then automatically analysed to check its consistency. This automatic analysis consists of two phases: (1) checking the correct usage of labels and relations (i.e., that a "proof" justifies a "theorem" but cannot justify an "axiom") and (2) checking that the logical precedences in the document are self-consistent. The development of this method was driven by the experience of computerising a number of mathematical documents (covering different authoring styles). We illustrate how such computerised narrative structure could be used for further manipulations, i.e. to build a skeleton of a formal document in a formal system like Mizar, Coq or Isabelle.

1 Introduction

The past forty years have seen a sharp increase in the use of the computer by the mathematician for his work purposes. Such use covers communication, authoring, processing, and checking/verifying mathematical knowledge. There exists already a number of flexible computer tools that allow producing aesthetic presentations of a mathematical document. This presentation, among others, comprises of a clear structure of a document, and usage of a 'fancy' and easy to read fonts and symbols. However, the presentation of a document and its structure also depends on the style of the mathematician and is usually expressed in terms of structural components (e.g., chapter or section) and mathematical components (e.g., lemma or proof). Moreover, a clear appearance of such components as well as explicitly specified relations between such components enhances the readability of the document and makes the navigation of a text more enjoyable.

Different styles of writing mathematics. The presentation of a mathematical document is a matter of writing style and involves among other things, a

M. Kauers et al. (Eds.): MKM/Calculemus 2007, LNAI 4573, pp. 296–312, 2007.

narrative structure of the document. This narrative structure plays an important narration role throughout the theory presented. Clearly expressed relations between mathematical components show logical dependencies which help the reader recognize the theory structure of a paper before reading the details.

The reader could find his way while reading the document depending on how the structure and dependencies are expressed. One could produce a clear structure of a document by specifying explicitly where the important parts (e.g., sections, definitions, etc.) start and end, and also where the dependencies are clearly expressed. In such case, the reader has a clear view of the theory in the document (see Figure 7). Otherwise, if the mathematician writing style is newspaper-like, the reader will have a difficult task finding his way in the document (see the example on the right).

> We prove that *two congruences can be added or subtracted from each other provided both have the same modulus.*
>
> Let
> $$a \equiv b \pmod{m} \text{ and } c \equiv d \pmod{m}. \quad (2)$$
> In order to prove that $a + c \equiv b + d \pmod{m}$ and $a - c \equiv b - d \pmod{m}$ it is sufficient to apply the identities
> $$a + c - (b + d) = (a - b) + (c - d)$$
> $$\text{and } (a - c) - (b - d) = (a - b) - (c - d).$$
> Similarly, using the identity
> $$ac - bd = (a - b)c + (c - d)b,$$
> we prove that congruences (2) imply the congruence $\quad ac \equiv bd \pmod{m}.$
>
> Consequently, we see that *two congruences having the same modulus can be multiplied by each other.* [...]
>
> It follows from the theorem on the multiplication of congruences that *a congruence can always be multiplied throughout by any integer and that each side of a congruence can be raised to the same natural power.* [...]
>
> W.Sierpiński [17, Chapter V, §1]

Motivations. In this paper we concentrate on the computerisation of the narrative structure of a document. Our main motivations are as follow:

1. *To handle the structure of a mathematical document as it appears on paper and at the same time to allow further computerisation and analysis.* Our proposed annotation system can deal with different styles of writing mathematics.
2. *To allow the presentation of a text with different layouts.* Currently the presentation of the structure of a documents is rather linear and it is not clear which parts (chunks of text) of the document depend on which (which theorem depends on which lemma or definition etc.). Ideally the presentation of a document should be flexible, and should allow the full automatic generation of different views of the structure of a document: *dependency graph*, graph of logical precedences, skeleton of the document in a chosen formal system, etc.
3. *To allow further formalisation.* Capturing the narrative structure of a document is not only for computerisation purposes, but also for further formalisation. The automatically generated views of the narrative structure of a text are very important to generate further forms of the text including a more formalised version (in a chosen formal system) as we illustrate in this article.

Contributions. Our contributions can be summarised as follows:

1. *A Document Rhetorical aspect (DRa) ontology and a related annotation system.* We present an ontology and an associated markup system, that offers a way to make explicit the traditional components of a mathematical text

(such as chapters, sections, proofs) and the dependencies between them. The ontology is very easy for the mathematician to use and requires no extra skills.

2. *Automatic processing of the narrative structure of a text.* Automated programs take the mathematician's DRa annotated mathematical text and build a number of internal representations and screen views of the narrative structure of the text. This includes: a *dependency graph* that represents relations between annotated parts of text, and its graph of logical precedences. The internal representations are used for further consistency analysis and formalisation while the screen views show the reader the narrative structure.

3. *Reuse of the narrative structure of a document.* We show how the automatically generated representations of the narrative structure of a document lead to a skeleton of a formal document in the Mizar Language [16]. Similar steps lead to formal skeletons in other formal systems.

Outline. In Section 1.1 we present the MathLang roadmap. Section 2 describes our approach to annotating the structure of mathematical documents and gives the DRa ontology used in the annotation system. Section 3 presents automatic transformations of the document's narrative structure into different views. We also present a formal mathematical model describing those automatically generated views. In Section 4 we present the analysis process of the dependency graph generated from the annotation. In Section 5 we express how the structure and its different views are used to build a skeleton of a part of a Mizar article. Finally, in Section 6 we describe related work, conclude and discuss future work.

1.1   The MathLang Project Roadmap

Since 2001, the MathLang project [3], has developed a number of prototypes for computerising mathematics. MathLang aims to give *alternative* techniques for capturing the mathematical knowledge of a mathematical text in a way that permits the transformation of this knowledge into new computerised and/or formalised versions while accommodating different degrees of formalisation, different mathematical editing/checking tools and different proof checkers. We started from de Bruijn's Mathematical Vernacular [1] (MV), and Nederpelt's Weak Type Theory (WTT) whose proof theory was developed by Kamareddine [7] and were faced with the huge challenge of how to really create a path from original mathematical texts into fully formalised ones and how would this path differ for different choices of texts, text editors, logical frameworks, and proof checkers.

Extensive computerisations of different mathematical texts (some taken fully from natural language to different levels of computerisation and finally to full Mizar), continue to shape the MathLang language. Its expressiveness has been increased in comparison with MV and WTT. MathLang adopted to decompose the computerisation process by means of knowledge components called **aspects**. In the current development of MathLang we have formalised and implemented three aspects: CGa and TSa (see below), and DRa (the subject of this article).

The Core Grammatical aspect (CGa) is a formal language derived from MV and WTT which specifies the grammatical role played by the elements of a mathematical text. CGa has a finite set of grammatical categories: **Terms, sets, nouns, adjectives, phrases, statements, declarations** and **contexts/local-scoping, definitions, steps, blocks**. The MathLang automated type system [6] checks whether the reasoning parts of a document are coherently built.

The Text and Symbol aspect (TSa) builds the bridge between a mathematical text and its grammatical interpretation and adjoins to each CGa expression a string of words and/or symbols which aims to act as its representation. We added information on how each CGa element should be printed on paper or on screen. This makes MathLang's encoding of mathematical texts faithful to traditional mathematical authoring [5]. TSa adds on top of a mathematical text a new dimension to the document where colored boxes represent the grammatical categories of the CGa. We implemented TSa in a *plugin* for the scientific text editor TEXmacs (http://www.texmacs.org/).

2 Annotating the Narrative Structure of a Document

This section gives our approach to annotate mathematical documents. The mathematical text on the left hand column of Figure 7 is used as our main example.

2.1 What Does the Mathematician Have to Do?

To annotate a mathematical text, the mathematician follows three easy steps:

1. He wraps chunks of text with unique boxes and names each box. Unicity allows avoiding problems when stating relations between some boxes. For our example of Figure 7, the names are: $S2, D1, D2, T1, PT1, T2, L1, PL1, PT2$.
2. He assigns to each (name of a) box, structural or/and mathematical rhetorical roles which this box may play. He can either use the structural/mathematical roles listed in Table 1, or specify his own. For our example of Figure 7, we assigned the roles stated in the left hand column in the table below.
3. He makes explicit the relations between wrapped chunks of texts using the relation names of Table 1. For our example of Figure 7, the relations are presented in the right hand side of the table below.

We use RDF triples [10] to represent the relationships between the boxes annotated by the mathematician. Each triple is expressed by a *subject-predicate-object* triple,

Assigned rhetorical roles	Relations
$(S2,$ hasStructuralRhetoricalRole, section$)$	
$(D1,$ hasMathematicalRhetoricalRole, definition$)$	$(PT1,$ justifies, $T1)$
$(D2,$ hasMathematicalRhetoricalRole, definition$)$	$(PT2,$ justifies, $T2)$
$(T1,$ hasMathematicalRhetoricalRole, theorem$)$	$(PL1,$ justifies, $L1)$
$(PT1,$ hasMathematicalRhetoricalRole, proof$)$	$(PT1,$ uses, $D1)$
$(T2,$ hasMathematicalRhetoricalRole, theorem$)$	$(PT2,$ uses, $L1)$
$(L1,$ hasMathematicalRhetoricalRole, lemma$)$	$(PL1,$ uses, $T1)$
$(PL1,$ hasMathematicalRhetoricalRole, proof$)$	$(PL1,$ uses, $D1)$
$(PT2,$ hasMathematicalRhetoricalRole, proof$)$	

where a *predicate* (i.e., a property) denotes a relationship. The order in a triple between *subject* and *object* is significant, and when transformed into a *dependency graph* the direction of the arc the triple makes, always points toward the *object*.

2.2   The Annotation System Ontology

Looking at different styles of mathematical knowledge representation we can distinguish two kinds of document structural units: *division elements* and *mathematical units*. *Division elements* express a textual structure (e.g., chapter or section) of a mathematical text. *Mathematical units*, are usually expressed in mathematical textbooks and papers in terms of theorem, lemma or remark. Some *mathematical units*, for instance "proof", are more or less hinted by the authors' style of writing (see the example in the introduction). The human reader is able to recognise and infer them only by looking carefully at the original text.

We express and tag these structural units (*division elements* and *mathematical units*) explicitly. By making explicit annotations of structure units we refine the content of the already captured original text, and at the same time we give a wider possibility for (semi)automatic text manipulation (see Sections 3 and 5).

Ontology. An ontology is a representation of terms with their relationships in a specific domain. An ontology describes: (1) individuals/instances of a class: the basic objects (e.g., "Bach" is an instance of class "Person" (`http://www.foaf-project.org/`); (2) classes/abstract groups: sets, or collections of objects (e.g., "Person"); (3) relations/properties between objects, e.g., the relation *childOf* (`http://vocab.org/relationship/`) in ("Sebastian Bach", *childOf*, "Ambrosius Bach").

DRa ontology in a nutshell. To model our DRa ontology we used the OWL-DL Web Ontology Language, which is the OWL sub-language so-named due to its correspondence with *description logics* [11].

Following OWL, our DRa ontology makes explicit the formal description of classes (whose names start with capital letter, e.g., StructuredUnit), individuals (e.g., section) and properties/relations (whose names start with small letter, e.g., justifies or hasMathematicalRhetoricalRole) in a domain of the DRa.

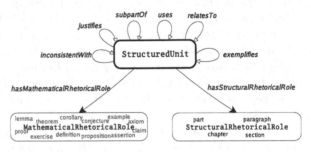

Fig. 1. Part of the DRa annotation system ontology

The DRa concepts are given as three disjoint OWL *classes* [18] (see Figure 1):
1. StructuredUnit.
2. MathematicalRhetoricalRole whose instances are lemma, proof, etc.
3. StructuralRhetoricalRole whose instances are chapter, section, etc.

Relations between various instances are given as OWL *object properties* [18]:
1. The ownership relation between structural units and the roles played in a text, i.e. hasMathematicalRhetoricalRole and hasStructuralRhetoricalRole. E.g., in Figure 7, (*D*1, hasMathematicalRhetoricalRole, definition).

2. The relations between instances of the class StructuredUnit:
   (a) relatesTo, justifies, subpartOf, uses, exemplifies, inconsistentWith.
The relations of the first kind (item 1) are modeled as *object properties* (i.e., link individuals of one class to individuals of another class). The relations presented in item (2) are modeled as *subproperties* of the generic *object property* – specifies, i.e. (A, specifies, B), where A, B are instances of class StructuredUnit.

Relations between instances of the classes MathematicalRhetoricalRole or StructuralRhetoricalRole and the XML schema datatype (xsd:string) are given as OWL *datatype properties* [18] (i.e., they link individuals of a class to the XML Schema datatypes [2]): hasOtherMathematicalRhetoricalRole and hasOtherStructuralRhetoricalRole. The existence of these relations gives the freedom if one wants to provide a new label not appearing in Table 1, this is possible through the usage of a variant property called hasOtherStructuralRhetoricalRole for *division elements* and hasOtherMathematicalRhetoricalRole for *mathematical units*. The range of values of such properties is restricted to the XML Schema datatype "string", e.g., (A, hasOtherMathematicalRhetoricalRole, *discussion*).

Since both *division elements* and *mathematical units* express the boundaries of chunks of text, we included them into one class (StructuredUnit). The two disjoint classes: StructuralRhetoricalRole and MathematicalRhetoricalRole allow to represent the different roles played by *division elements* and *mathematical units*. Instances of the first class are conventional names for *division elements* which might at the same time express the hierarchical level of a document structure, i.e., chapter, section, etc. Instances of the class MathematicalRhetoricalRole are common labels and names for the *mathematical units*, i.e., theorem, corollary, etc. All instances of the classes StructuralRhetoricalRole and MathematicalRhetoricalRole, are fixed conventional labels used to annotate mathematical documents.

Table 1. DRa annotations

Description
Instances for the hasStructuralRhetoricalRole *property:* preamble, part, chapter, section, paragraph, *etc.*
Instances for the hasMathematicalRhetoricalRole *property:* lemma, corollary, theorem, conjecture, definition, axiom, claim, proposition, assertion, proof, exercise, example, problem, solution, *etc.*

Relation
Types of relations: relatesTo, justifies, subpartOf, uses, exemplifies, inconsistentWith

The DRa ontology allows to relate a particular instance of the class StructuredUnit with any instance of StructuralRhetoricalRole and MathematicalRhetoricalRole via the properties hasStructuralRhetoricalRole and hasMathematicalRhetoricalRole respectively. We allow the use of both properties when relating to an instance of a class StructuredUnit. This enables to specify, for instance, that a chunk of text plays the structural role "section" and concurrently plays the mathematical role "theorem". By stating two properties simultaneously in a document annotation we allow to encode different styles of writing mathematics.

While annotating the narrative feature of a document, we make explicit correlations between recognised chunks of text. For this, within the DRa ontology, we introduced other properties which describe relations between instances of the class StructuredUnit and represent dependencies between *mathematical units* and/or *division elements*. Our DRa ontology clarifies important relationships in a text. The properties used to represent relations between chunks of text, have human readable names: relatesTo, justifies, subpartOf, uses, inconsistentWith, exemplifies. In a formal system, some of these properties have formal meanings:

1. $(v_1, \text{justifies}, v_2)$ – v_1 describes a proof object that proves the formula v_2.
2. (v_1, uses, v_2) – (1) All/some variables under the general quantifiers that have been applied in a formula v_2 have been instantiated in formula v_1 which could be proved via simple reasoning where v_2 appears among references needed to prove v_1. (2) The formula v_2 has been unfolded or folded in the formula v_1.
3. $(v_1, \text{subpartOf}, v_2)$ – (1) if v_2 is a formula, then v_1 is an inseparable part of that formula; (2) if v_2 is a proof object, then v_1 is part of that proof object.
4. $(v_1, \text{inconsistentWith}, v_2)$ – if v_1 and v_2 are proof objects of one formula, then the environment in which these proof objects were achieved is inconsistent.

3 Automatic Transformation of a DRa Annotated Text

In this section we show how to use the DRa annotated text to automatically create a number of views of the text including the *dependency graph* (that represents relations between annotated parts of text) and the graph of logical precedences.

$G = (V, A, E)$ where $A \subseteq V \times (MR \cup SR)$, $E \subseteq V \times L_d \times V$
$V = \{v \mid v = nodeId\}$ – set of vertices
$A = \{a \mid a = (v, r) \ \wedge \ r \in MR \cup SR \ \wedge \ MR \cap SR = \emptyset\}$ – set of vertices attributes
$E = \{e \mid e = (v_{src}, \alpha, v_{anch}) \ \wedge \ v_{src}, v_{anch} \in V \ \wedge \ \alpha \in L_d\}$ – set of edges

where
$L_d = \{\text{relatesTo, justifies, subpartOf, uses, inconsistentWith, exemplifies}\}$ – the set of allowed labels in a dependency graph
MR – the set of MathematicalRhetoricalRoles, cf. Table 1
SR – the set of StructuralRhetoricalRoles, cf. Table 1
$nodeId$ – a unique name/identifier given by the user while wrapping the text with boxes

---

Fig. 2. Formal presentation of a *dependency graph*

3.1 The Automatically Generated Dependency Graph of a Document

A document's dependency graph is a directed labeled graph with attributes assigned to the vertices (see Figure 2). The vertices (resp. attributes resp. edges) of such graph are the names of boxes (resp. mathematical or structural rhetorical roles resp. relations) specified by the user during the first (resp. second resp. third) step of the annotation of the document described in Section 2.1.

Figure 3 (and the right hand side of Figure 7) presents the dependency graph of our particular example. This graph consists of (1) relations between parts of the text which are represented by visible arrows, and (2) graph nodes which have specified (but not visible) mathematical or/and structural rhetorical roles. Dependencies between the annotated chunks of text play an important role in mathematical knowledge representation. Thanks to those dependencies, the reader finds his own way while reading the text without the need to understand all its subtleties. Moreover, we will show in the next sections that these dependencies

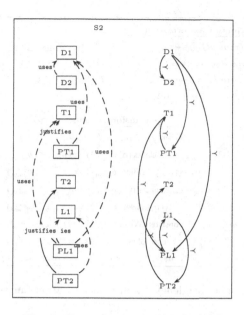

Fig. 3. *Dependency graph* and *GoLP.*

On the left hand side we have the automatically generated presentation of the *dependency graph* constructed from the input of the mathematician in Section 2.1 for our main example of Figure 7. The right hand side of the figure presents automatically generated *GoLP* from the *dependency graph.*

allow one to present other views on a document, and to structure the skeleton of a document in the formal language Mizar. Dependencies graphs (and their views as in Figure 3) are found automatically from the mathematicians' input in Section 2.1.

3.2 Logical Precedences of Mathematical Relations

The annotation identifies and makes explicit different parts of the text, stores either the mathematical or structural or both roles of each chunk of text, and annotates the relations between recognised chunks of text (see Section 2.1). The usage of the DRa system allows us to express relations explicitly in the computerised version of the original text. This explicit representation of relations allows to build a graph of logical precedences between different chunks of the text.

The ***logical precedence*** between two chunks of text indicates the relative positions of the chunks in a sequence of reasoning steps. These (and other) steps, contribute to the analysis of the logical correctness of the original text. *Logical precedence* is independent of the sequential appearance of the chunks of text in a document. For instance, in Figure 7, the "Proof" (node $PT1$) is stated after "Theorem 1.19" (node $T1$). However, the *logical precedence* between $PT1$ and $T1$ is the other way round (see the direction of the arrow established between both nodes in Figure 3). In such a case, we say that $PT1$ logically precedes $T1$.

Graph transformation

$Trans : G_{DG} \rightarrow G'_{GoLP}$

Vertex transformation

$Trans((v, a, e)) = (v', e')$

$Trans_V : V_{DG} \rightarrow V'_{GoLP}$

(where $v' = Trans_V(v)$

$Trans_V(v) = v$

and $e' = Trans_E(e)$)

Edge transformation

$Trans_E : E_{DG} \rightarrow E'_{GoLP}$

$Trans_E((v_{src}, \mathsf{relatesTo}, v_{anch})) = (v'_{src}, \backsimeq, v'_{anch})$

$Trans_E((v_{src}, \mathsf{justifies}, v_{anch})) = (v'_{src}, \prec, v'_{anch})$

$Trans_E((v_{src}, \mathsf{subpartOf}, v_{anch})) = (v'_{src}, \prec, v'_{anch})$

$Trans_E((v_{src}, \mathsf{uses}, v_{anch})) = (v'_{anch}, \prec, v'_{src})$

$Trans_E((v_{src}, \mathsf{inconsistentWith}, v_{anch})) = (v'_{anch}, \prec, v'_{src})$

$Trans_E((v_{src}, \mathsf{exemplifies}, v_{anch})) = (v'_{anch}, \prec, v'_{src})$

(where $v'_{src} = Trans_V(v_{src})$ and $v'_{anch} = Trans_V(v_{anch})$)

Fig. 4. Dependency graph transformation function

We assume two kinds of *logical precedences*: *strong logical precedence* $\prec$ and *not-specified logical precedence* $\backsimeq$. In Section 2.2 we gave a DRa ontology which allows specifying the relations between recognised StructuredUnits in a document. Each such stated relation expresses its own *logical precedence* (see Figure 4).

3.3 The Automatically Generated Graph of Logical Precedences: GoLP

Using the *logical precedence* of each relation (see Figure 4), one can automatically build for a mathematical text, a *graph of logical precedences* (GoLP). Figure 3 gives the automatically generated GoLP for our main example. GoLP is a directed graph with labeled edges, achieved by the automatic transformation of the dependency graph using the transformation function *Trans* (see Figure 4). In a GoLP, the direction of an edge together with a label of that edge expresses the *logical precedence* corresponding to the relation in a dependency graph from which the edge (in the GoLP) was achieved. Figure 5 gives the formal definition of a *graph of logical precedences*. If G is the *dependency graph* (DG) and G' is the *graph of logical precedences* (*GoLP*) shown in Figure 3, using the transformation function *Trans* shown in Figure 4, we can automatically transform G into G'.

$G' = (V', E')$ where $E' \subseteq V' \times L_p \times V'$

$V' = \{v' \mid v' = nodeId\}$ – set of vertices

$E' = \{e' \mid e' = (v'_{src}, \alpha', v'_{anch}) \wedge v'_{src}, v'_{anch} \in V' \wedge \alpha' \in L_p\}$ – set of edges

where $L_p = \{\backsimeq, \prec\}$ – the set of *logical precedences* in GoLP

Fig. 5. Formal presentation of a *graph of logical precedences* (GoLP)

4   Automatic Analysis of the Dependency Graph and GoLP

This section explains the checking of the DRa annotation done in two phases:
1. Checking the annotation of distinct roles of recognised fragments of text and the correct usage of labels and relations.
2. Checking that the logical precedences in the GoLP are self-consistent.

Pre-analysis of the dependency graph. The first phase of checking catches some inconsistencies while representing the different roles of recognised chunks of text and the stated dependencies between them. E.g., if two chunks of text were annotated as "proof" resp. "axiom", and if a relation justifies is stated between them (i.e. $(proof, \text{justifies}, axiom)$), the first validation stage returns two warnings: one on the relation type – which might/should be different, and another on the role of each chunk of text – which was mistakenly specified.

This checking captures other cases. Assume that one has specified simultaneously two MathematicalRhetoricalRoles for a chunk of text, for instance "axiom" and "proposition". In such a case the analysis returns a warning stating that "axiom" cannot be provable, whereas "proposition" can. Similarly, if one simultaneously states two different StructuralRhetoricalRoles for one chunk of text (e.g., "chapter" and "subsection"), the analysis will return a warning. The difference between a "chapter" and a "subsection" is that the background knowledge of a "chapter" is something like an external library for the following sections and subsections, whereas for "subsection" the context is more specific and composed of small chunks of text from the previous sections or chapters, although both "chapter" and "subsection" may use the external knowledge.

Checking the consistency of labels in a GoLP. To allow the analysis of a GoLP we have identified a number of common relational properties for *logical precedences* (see Table 2). These properties are used while checking the labeling consistency in a GoLP – see the following section.

Table 2. Relational properties of *logical precedences*

Relational properties	Not-specified logical precedence	Strong logical precedences
	$C \simeq C' \implies C \prec C' \lor C' \prec C$	
irreflexivity	$\neg(C \simeq C)$	$\neg(C \prec C)$
symmetry	$C \simeq C' \implies C' \simeq C$	
asymmetry		$C \prec C' \implies \neg\,(C' \prec C)$
transitivity		$A \prec B \land B \prec C \implies A \prec C$

We build a *transitive closure* of a GoLP (using for example Roy-Warshall's algorithm [15,19]) from a dependency graph of the original document. Furthermore we check if such built graph is the graph of a *strict partial order* (i.e., that no edge in the *transitive closure* graph has its reflexive image in the GoLP), where the *strict partial order* relation is the *strong logical precedence*.

We illustrate the analysis of consistent labeling on the GoLP based on our example. Take the nodes $D1$ and $PL1$, and the edge $(D1, \alpha', PL1)$, where $\alpha' \in L_p$ (see Figure 3). In the transitive closure of our GoLP we have two paths that form the edge $(D1, \alpha', PL1)$: (1) a direct path $\pi_{D1,PL1}^{d,\alpha'} = \{(D1, \prec, PL1)\}$, and (2) an indirect path $\pi_{D1,PL1}^{ind,\alpha'} = \{(D1, \prec, PT1), (PT1, \prec, T1), (T1, \prec, PL1)\}$. The direct path is labeled with a *strong logical precedence* symbol $\prec$, denoted as $\pi_{D1,PL1}^{d,\prec}$. When evaluating the label of the indirect path $\pi_{D1,PL1}^{ind,\alpha'}$, we have to take into account the relational properties of the *logical precedences* of Table 2. In our case, we use the transitivity of *strong logical precedence* $\prec$ between the three edges of the path $\pi_{D1,PL1}^{ind,\alpha'}$. From this, we obtain the labelled indirect path $\pi_{D1,PL1}^{ind,\prec}$, which has the same label as the direct path $\pi_{D1,PL1}^{d,\prec}$. We conclude that the edge $(D1, \alpha', PL1)$ in the graph of logical precedences (GoLP) is labeled consistently.

Labeling consistency validation is performed on each existing edge in the *transitive closure* of GoLP built from a dependency graph of the original document. Once we go through the whole checking of the graph we can say that the GoLP is valid according to the consistent labeling.

5 From the Document Narrative Structure to the Formal Document Skeleton in Formal Systems

So far, the mathematician's DRa annotations of his text in Section 2.1 have been used to automatically produce the dependency graph and the GoLP of the text which explicit the narrative, structural and logical features of the text. In this section, we explain how the automatically generated dependency graph and GoLP are used for further processing and formalisation of the text. In particular, we express how the dependency graph together with the GoLP are used to build a skeleton of a part of a Mizar article – *Text-Proper*. We do not go into the technical details. Instead, we present roughly the transformation hints based on our main example resulting in the Mizar *Text-Proper* skeleton of Figure 6. For extensive details on the passage from the dependency graph and the GoLP into Mizar *Text-Proper*, Mizar formal proof sketch FPS and full Mizar, see [4].

In Section 2.1, the mathematician specified that a big box named $S2$ is an entire section in the document. In Mizar the *Text-Proper* part of a document could be divided into a sequence of *Sections*, where each *Section* starts with **begin** and consists of a sequence of theorems and definitions together with their proofs. The division of the *Text-Proper* into *Sections* has no impact on the correctness of the Mizar document. Hence, the whole box is indicated to be a section by explicitly specifying **begin** at the very top of the right hand side of Figure 6. It also consists of two lines ::Section and ::Title ... which are treated as Mizar comments, and are solely oriented for the Mizar user consumption, or the reader of the Mizar file. Inside ::Title ... it is a good practice (in the Mizar community) to specify the title of this *Section* of the Mizar document.

Since the mathematician specified for the box $\boxed{D1}$ the MathematicalRhetorical-Role definition, then it is transformed into Mizar syntax as: **definition** :DEF1:

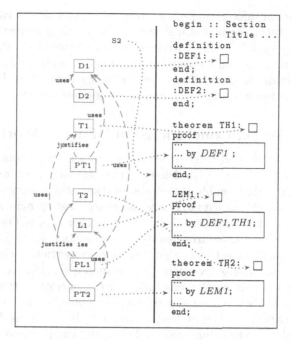

Fig. 6. Transformation into Mizar skeleton

The left hand side reproduces the dependency graph of our example (Figure 7). On the right hand side we show the Mizar *Text-Proper* skeleton of the same example. The arrows from left to right show how the dependency graph is used to build the Mizar *Text-Proper* skeleton. □ stands for holes (incomplete proofs).

$\boxed{D1}$ `end;` (see Figure 6). In Mizar we introduce the label `DEF1` for this definition to be able to refer to it in further reasoning steps.

Since the mathematician specified for the box $\boxed{T1}$ the MathematicalRhetoricalRole theorem, then it is transformed into Mizar syntax as: `theorem` $\boxed{T1}$. Moreover, since the box $\boxed{PT1}$ has the MathematicalRhetoricalRole proof, then we transform it into: `proof` $\boxed{1}$ `end;`. Moreover, since a block of steps having the mathematical role proof is related by justifies to a single statement, we can say that this is a *Justification* in Mizar, which is transformed into a specific form. See the corresponding transformation arrows in Figure 6.

In the dependency graph of our main example we also specified that some blocks of text use other blocks. For instance a block of text named $\boxed{PT2}$ uses statement $\boxed{L1}$. Here, we transform $\boxed{PT2}$ into a specific Mizar *Proof* block, which contains an expression with *Straightforward-Justification* to statement $\boxed{L1}$, where in Mizar it is reused by referring to a label (i.e., `LEM1`) that was assigned to a statement $\boxed{L1}$ during the transformation into the Mizar syntax.

During the transformation of the dependency graph, we use the GoLP of our main example to be able to put annotated and named chunks of text into a proper Mizar order inside the Mizar skeleton.

The above transformation process leads to a part of a Mizar *Text-Proper* skeleton of a Mizar document (given in Figure 6 for our main example).

The grammatical information of the original text, which is captured by the CGa aspect of MathLang and stored in the MathLang document, can be then used to fill more details in the current skeleton of the Mizar document. This better filled document could be transformed later into a proper Mizar document. The work describing these transformation and usage of the MathLang document for the migration process into the Mizar language, has been described in [4].

6   Related Work, Conclusions and Future Work

Many studies have been carried on the structure of documents. For example, the Text Encoding Initiative Guidelines (`http://www.tei-c.org/`) are international standards that enable the representation of a variety of literary and linguistic texts. DocBook (`http://www.docbook.org`), provides a system for writing a structured document using XML. Another tool is OMDoc ([8] - see below). These systems allow to separate/divide a document into a number of structural components (sections or mathematical assertions) which can be annotated in the computerised version. Our proposed markup system is simpler and is concentrated only on the annotation of the narrative structure of mathematical documents, whereas others are more oriented towards capturing other documents subtleties. We believe that separating the concerns during computerisations can play a very helpful role in developing computer tools that can aid various levels of computerisation/formalisation.

OMDoc vs DRa – A short comparison. OMDoc presents mathematical knowledge on three levels: the object and formula level, the statement level, and the theory level. What is made explicit by the DRa markup, is similar to the statement level and partly to the theory level in the OMDoc system. The OMDoc markup distinguishes the knowledge elements of a theory into constitutive ones like symbols, axioms, and definitions (which present the essence of the annotated theory) and non-constitutive ones such as assertions, their proofs, examples (which illustrate properties and attributes of mathematical objects determined by the constitutive statements). This shows a different approach to annotating the same knowledge. The aim of introducing the DRa is to be able to catch and store the narrative structure of the text, and simultaneously allow to stay as close as possible to the original document and the style it was written in. Therefore, on the DRa markup we do not distinguish constitutive or non-constitutive statements. We recognize only one class of elements, called StructuredUnit, and we distinguish the roles they play in mathematical knowledge representation.

Therefore, the purposes/aims of OMDoc and DRa are different. All instances of the class MathematicalRhetoricalRole in the DRa ontology, are presented as disjoint classes in the OMDoc ontology [9]. "axiom" is an ontology class in the OMDoc ontology, whereas in the DRa it is expressed as an instance (individual) of the class MathematicalRhetoricalRole. This particular name "axiom" expresses a role of the text labeled by that name, and hence in the DRa ontology we annotate it by stating the property hasMathematicalRhetoricalRole whose range value is an appropriate instance (i.e., "axiom") of a class MathematicalRhetoricalRole.

Both annotation systems OMDoc and DRa allow to markup dependencies between statements. In the OMDoc file format they are implemented by means of the `for`

```
<definition xml:id="node-D1.def">
    <CMP>A subset $A \subset R$ is inductive if [...]
<assertion xml:id="thm-T1" type="theorem">
    <CMP>Let $J$ be a subset of Z+ [...]
<proof xml:id="proof-PT1" for="#thm-T1">
    <CMP>$J$ is inductive so $J$ contains [...]
```

attribute to OMDoc's elements (e.g., `<proof for="#id-of-assertion">`). A possible encoding of a part of our main example shown in Figure 7 in OMDoc is sketched[1] above. Within the DRa system we annotate the relation as an RDF triple, and it might be expressed in the MathLang internal file using any kind of XML-RDF recommendations.

The other and main advantage of DRa over OMDoc is a possible analysis of the dependency graph and the GoLP, which are automatically built from the performed annotation. This analysis allows to check the annotation of the narrative/rhetorical aspect of a document (see Section 4). Although OMDoc gives a lot of elements and constructions that can be used to structure mathematical documents, these allow the user some software compatibility but no validation yet. The DRa annotation system gives the user a validation tool making it possible to analyse the well-formedness/encoding of the rhetorical aspect of a document.

Other works. [13,14] present a method to express the logical structure of a document and hyperlinks between chunks of mathematical text that enhance the readability of a document and the navigation throughout the text. That method detects the logical structure of a text and several types of hyperlinks from printed mathematical documents. Our approach differs in the sense that we propose an annotation system that allows to express such logical structure and hyperlinks/relations while authoring a document. Moreover, we use the dependency graph achieved from the annotation to build a formal document skeleton (as we have done in Mizar and can be done in other systems).

Conclusions. We have presented in this paper our approach to computerise the narrative aspect of mathematical texts. We built a DRa ontology which described formally the domain of narrative/structural representations of mathematical knowledge in a document. The ontology allows to share a common understanding of the structure of the represented knowledge among other people and software agents. The ontology separates a domain knowledge (DRa) from the operational knowledge – the actual annotation. By using the ontology we annotated/marked up our main example shown in Figure 7.

We presented the meaning behind the DRa annotation and gave automated tools which generate different representations of the document structure. We showed how the encoded Document Rhetorical aspect annotation could be validated for checking the well-formedness of the annotation. We also expressed which mistakes made during annotation we are able to automatically catch.

---

[1] For readability and brevity, we show only the opening tag of each XML element for most elements; we use indentation to express nesting.

Finally we demonstrated how the dependency graph and the graph of logical precedences are used to build the skeleton of Mizar *Text-Proper*.

Future work. The DRa encoding system is a part of the ongoing MathLang project. As future work, we need to concentrate on the evaluation and improvement of the DRa system, to finish the implementation of the DRa validation rules and to test them on bigger examples. We also need to work further on the DRa ontology and to refine the instances of the class StructuralRhetoricalRole. Namely, we need to separate the depth level of structural units labels from the actual meaning of a unit. For instance "section" and "subsection", for the representation purposes, differ only in the embedded relation. Therefore we have to investigate how the depth level can be incorporated within the DRa ontology.

We also need to investigate how a mathematician could add his own intended relation to the DRa system. For instance, he might want to add the explanationOf relation which could be used to express that (*example*, explanationOf, *definition*). We have to incorporate this kind of possibilities within the DRa markup system.

Another advantage is that we do not provide yet another concrete syntax for mathematical encoding. Instead, we incorporate the markup of the narrative aspect of a mathematical text into the existing encoded document. We believe that a clear separation between different aspects of mathematical knowledge and their markup brings a clear guidance for non expert authors. This guidance mainly helps to extract from the original text, different aspects of mathematical knowledge at different phases of its computerisation.

References

1. de Bruijn, N.G.: The mathematical vernacular, a language for mathematics with typed sets. In: Workshop on Programming Logic, Sweden (1987)
2. Biron, P.V., Malhotra, A.: XML Schema Part 2: Datatypes. W3C Recommendation (2001)
3. Kamareddine, F., Wells, J.: MathLang: A new language for mathematics, logic, and proof checking. A research proposal to UK funding body (2001)
4. Kamareddine, F., Maarek, M., Retel, K., Wells, J.B.: Gradual computerisation/-formalisation of mathematical texts into mizar (2007)
   http://www.macs.hw.ac.uk/~retel/
5. Kamareddine, F., Lamar, R., Maarek, M., Wells, J.B.: Restoring Natural Language as a Computerised Mathematics Input Method (2007)
   http://www.macs.hw.ac.uk/~mm20/
6. Kamareddine, F., Maarek, M., Wells, J.B.: Toward an object-oriented structure for mathematical text. In: Kohlhase, M. (ed.) MKM 2005. LNCS (LNAI), vol. 3863, Springer, Heidelberg (2006)
7. Kamareddine, F., Nederpelt, R.: A refinement of de Bruijn's formal language of mathematics. J. Logic Lang. Inform. 13(3), 287–340 (2004)
8. Kohlhase, M.: OMDoc – An Open Markup Format for Mathematical Documents [version 1.2]. LNCS (LNAI), vol. 4180. Springer, Heidelberg (2006)
9. Lange, Ch.: SWiM – A Semantic Wiki for Mathematical Knowledge Management. Technical Report (December 2006)

10. Lassila, O., Swick, R.R.: Resource Description Framework (RDF) Model and Syntax Specification. W3C Recommendation (1999)
11. McGuinness, D.L., van Harmelen, F.: OWL Web Ontology Language Overview. W3C Recommendation (2004)
12. Moller, J.M.: General topology. Authors' notes, last visit 2007-02-25. Available at http://www.math.ku.dk/~moller/e03/3gt/notes/gtnotes.pdf
13. Nakagawa, K., Suzuki, M.: Mathematical knowledge browser with automatic hyperlink detection. In: Kohlhase, M. (ed.) MKM 2005. LNCS (LNAI), vol. 3863, Springer, Heidelberg (2006)
14. Nakagawa, K., Nomura, A., Suzuki, M.: Extraction of Logical Structure from Articles in Mathematics. In: Kohlhase, M. (ed.) MKM 2005. LNCS (LNAI), vol. 3863, Springer, Heidelberg (2006)
15. Roy, B.: Transitivité et connexité. C. R. Acad. Sci. Paris 249, 216–218 (1959)
16. Rudnicki, P., Trybulec, A.: On equivalents of well-foundedness. An experiment in MIZAR. Journal of Automated Reasoning 23(3–4), 197–234 (1999)
17. Sierpiński, W.: Elementary Theory of Numbers. PWN, Warszawa (1964)
18. Smith, M.K., Welty, Ch., McGuinness, D.L.: OWL Web Ontology Language Guide. W3C Recommendation (2004)
19. Warshall, S.: A theorem on boolean matrices. J. ACM 9(1), 11–12 (1962)

A   Original and DRa-Annotated Text of Our Example

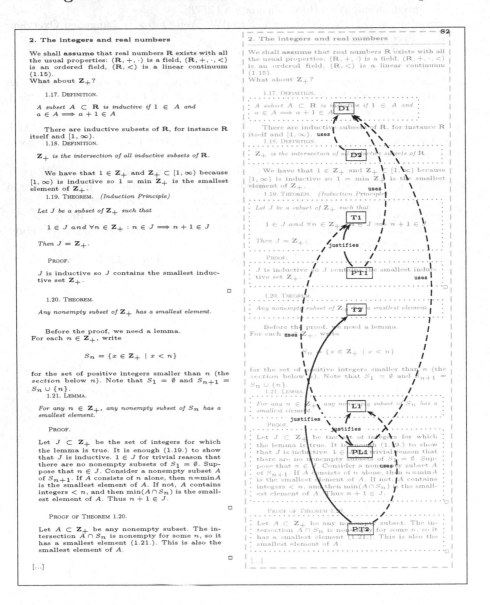

Fig. 7. Fragment of text without and with dependency graph

The original text [12, Chapter III, §2] of the given example is taken from J.M. Moller's notes [12] regarding general topology and is reproduced on the left hand side of the figure. The right hand side of the figure shows the automatically generated dependency graph for the text where relations between parts of the text are represented by visible arrows and graph nodes have specified (but not visible) mathematical or structural rhetorical roles.

Reexamining the MKM Value Proposition: From Math Web Search to Math Web ReSearch

Andrea Kohlhase and Michael Kohlhase

Computer Science, Jacobs University Bremen
{a,m}.kohlhase@iu-bremen.de

Abstract. The interest of the field of Mathematical Knowledge Management is predicated on the assumption that by investing into markup or formalization of mathematical knowledge, we can reap benefits in managing (creating, classifying, reusing, verifying, and finding) mathematical theories, statements, and objects. This global value proposition has been used to motivate the pursuit of technologies that can add machine support to these knowledge management tasks. But this (rather naive) technology-centered motivation takes a view merely from the global (macro) perspective, and almost totally disregards the user's point of view and motivations for using it, the local (micro) perspective.

In this paper we go a first step into a more principled analysis of the MKM value proposition by focusing on motivations for mathematical search engines from the micro perspective. We will use a table-based method called the "Added-Value Analysis" (AVA) developed by one of the authors. Even though we apply the AVA only to mathematical search engines, the method quickly leads to value considerations that are relevant for the whole field of MKM.

1 Introduction

Mathematical knowledge management (MKM) is a field at the intersection of document management, knowledge representation, and meta-mathematics. Like the first (and unlike the third), it is driven by practical motivations, i.e. by the desire to create technologies that help researchers, scholars, students, and engineers in dealing with mathematical knowledge.

The field has been inspired by experiences in *Formal Methods*: formalizing the intended behavior of programs in logic-based specification languages and employing semi-automated deduction technologies to verify programs against these by using formal proofs can indeed lead to safety and security assurances that are considered so valuable in some high-risk domains that they justify the extremely high costs involved. However, most applications of mathematics are less safety-critical, and with peer-review the mathematical community has a sufficient (much less cost-intensive) instrument for verifying their results. Therefore, MKM concentrates on management tasks for mathematical knowledge that involve higher volumes of information, but possibly less formal depth of representation. The intuition is that mathematical practices like notational adaptation,

M. Kauers et al. (Eds.): MKM/Calculemus 2007, LNAI 4573, pp. 313–326, 2007.

document aggregation and translation, semantic search and navigation, classification of mathematical objects, refactoring of mathematical theories, or error spotting can be based on lightweight formal annotations and content markup techniques for mathematics.

In this respect, the MKM approach is similar to the much-hyped *Semantic Web*, and suffers from the same problem: before the inference-based techniques of the field can pay off, a large volume of data (mathematical documents for MKM and web content for the Semantic Web) must be semantically annotated. Both fields also agree on how this should be achieved: rather than waiting for artificial intelligence methods that can automate this, we rely on content authors or volunteers to supply the annotations. Here, we have a very important difference to the case of Formal Methods, where the connection of the sacrifices incurred in program verification are directly linked to the expected benefits, usually a radically reduced risk of liability or reduced insurance payments. In MKM and the Semantic Web, the benefits lie with the "readers", while the sacrifices remain with the "authors" — creating what we call the *"authoring dilemma of MKM"* in [KK04] or is referred to as MKM's "chicken-and-egg problem"[1].

In [Koh06a] the authoring dilemma was traced back to differing perspectives on the problem: the **micro perspective**, a (local) view from within, and the **macro perspective**, a (global) view from without. The direct link between benefits and sacrifices for a user in Formal Methods consists in the alignment of the micro and macro perspectives, whereas the differentiation between authors and readers in MKM and Semantic Web is an expression of its drifting apart. From a macro perspective authors and readers are just roles of a single user, who will change dynamically between them. In contrast, from the micro perspective of that user, the benefits lie in the far future. MKM technology like most other designed systems usually takes the macro standpoint and almost totally disregards the user's point of view and motivations for using it in the micro perspective.

The MKM community sometimes calls for a "business plan" for MKM, but to our knowledge it has never seriously been attempted as a business plan must take the designer's macro perspective *and* the users' micro perspectives into consideration, so that great visionary technology gets actually used. In particular, we believe that MKM's value proposition is lop-sided and must be *re*examined, e.g. marketing experts like Normann and Ramirez argue: *"Like a portrait which over time dictates to those who see it how the person portrayed actually looked, models also tend to transform what they model, constraining that reality within the limits of the model's logic"* [NR98, xvi].

In this paper we go a first step into a more principled analysis of the MKM value proposition. We will use a table-based method called the "**Added-Value Analysis (AVA)**" [KM07] developed by one of the authors to investigate the value constellations (core problems, their solutions, expected benefits, incurred sacrifices, and added values) that may activate people into users of

---

[1] It is often lamented that MKM technologies will only become useful, once we have a large corpus of semantically enhanced background material, but we can only realistically develop one once we have the MKM technologies.

mathematical search engines. Our motivation for this is twofold: first, we are interested in understanding the potential of our own search engine MATHWEB-SEARH [KŞ06, KŞ07] more thoroughly and secondly, we want to introduce the MKM community to the AVA method.

If we want to realize our dream of creating a viable technology and a semantic resource with universal coverage in mathematics [Far05], then we will have to convince many authors that there is value for them in doing so.

2   Added-Value Analysis and Easy-to-Fall-For Catches

The "Added-Value Analysis (AVA)" [KM07] was developed as an interaction design method, that allows and supports a view from micro perspectives, i.e. a theoretically unlimited number of micro perspectives. It differs from well-known interaction design methods because of its focus on the "here-and-now" of a user in the *process of using* the software.

The micro perspective influences a user's evaluation of a concrete context and thereby determines her action. With the Added-Value Analysis this contextual evaluation process can be (re)constructed. It makes use of the **double relativity of added-values**. On the one hand "value" is always relative to the individual (and not to the object under consideration), on the other hand only if this value is fixed, we can add *more* value. This implies that value *can* be manipulated. Moreover, we cannot think any longer in simple value chains, but have to consider rather complex value constellations [NR98], which implies that an object's value may change in the process of using it. In particular, there is no such thing like an added-value per se; its relativities create the basis for understanding taking-action based on underlying value constellations from a micro perspective.

How to Do an Added-Value Analysis? The starting point for the AVA is the understanding the term "Added-Value" based on [Grö97]. In particular, before we can fix the meaning of **"Added-Value"** of a software package, we have to state the *core problem* it tackles. Then we can speak of its (core) value, by evaluating the core problem with respect to the given solution. Afterwards — and only afterwards — we can determine how this value can be enhanced by adding other values. In order to get a handle on a value, the AVA method suggests to systematically split it into benefits received and sacrifices incurred [dChR00] *according to the core problem with the given solution.*

Concretely, we build up a table-like list with the AVA, that contains columns for the core problem, the considered solution, and its evaluation listed in form of lists of benefits and sacrifices, see the following table structure:

#	Pot	Trigger	Core Problem	Solution	Benefits	Sacrifices
0					•	•

The first step consists in determining the *"initial"* problem and a solution for it, then we make state the benefits and/or sacrifices incurred by them. Special care has to be taken that these benefits/sacrifices are such with respect to the core problem *and* the considered solution. Note that the quality of the AVA depends strongly on the precision of all formulations. For instance, if a general problem is addressed with a very specific solution, then a general benefit typically indicates the inappropriateness of the solution level, whereas a very specific benefit indicates the inappropriateness of the problem level.

Now, we **iterate this step** by using either any component of the previous line or any other association at this level as a trigger for a new core problem. For example, if a sacrifice is incurred on the user in Step #1, this can be taken as another core problem for which either a solution exists or is needed. If former, then we can consider it as added-value, if latter, then it becomes a *potential* added-value to the previous problem, which is marked as such in the second column of the table. If a row is triggered by another row in the AVA table, we label this process either by color or by a reference. We use both below, in particular a reader can use a **reference like** "⤳ 5" to follow the use process top-down, and the **color** to follow it bottom-up. Moreover, we explicitly note the trigger so that the association chain is enhanced and hence not easily broken. These streams can be reinterpreted as unfolding use processes, in which utility problems are not present up to a certain depth of already using the software. Naturally such cognitions can be exploited in interaction as well as interface design, but this is not the focus of this paper. For convenience, we use the abbreviation **"NN"** ("nomen nescio" = "I don't know") to indicate, that we either couldn't think of what to write down or that we intentionally do not follow this line of thought (for now).

Added-Value Analysis' Related Work. To fortify our intuition of the AVA method, and especially for its reinterpretation as a value proposition analysis tool, we will compare it to related interaction design methods first and highlight its distinguishing features: In 1996 Batya Friedman coined the term *"Value-Sensitive Design"* [Fri96] with which human values like "autonomy" reenter interaction design. This concept meant a shift from a technology-centric approach towards a user-centric one. The AVA incorporates the value-sensitive design method by looking at the subjective values for the decision-taking for action or non-action from the micro and not only from the macro perspective. Within the *"Humanistic Research Strategy"* [Oul04] human values are considered as well, but the underlying reasoning for action is restricted to experience and culture, whereas the AVA takes the processuality of action into account. Again we might say, that the micro perspective distinguishes both design approaches from the AVA. Finally, we like to point out that the *"Contextual Design"* method [BH99] differs not by objective but by implementation as their method is based on contextual inquiry, ours is based on thought trajectories.

With these differentiations in mind, we will look now at possible applications for the Added-Value Analysis method.

What Is an Added-Value Analysis For? Note that the AVA can be used for any technology, that involves design and use processes. It is not relevant, whether the design process is done *before* the actual use process like in a typical software product or *during* it like writing a diary (in which what is written inspires what will be written and vice versa). But in the latter the processes are much more interwoven and therefore, we don't go into this spider's nest. In former, we can distinguish the time-line of a use of AVA from the designer's standpoint: it can either be applied in, inbetween, or after the design process as the view of micro perspectives is helpful in either one.

- When the AVA is used *in* a software design process, it functions as a *system analysis* tool with a focus on the evolving interaction design.
- *Within* the design process, the AVA helps to find potential added-value services and enables the designer to reevaluate the anticipated utility of the software component in question. As it tends to discover unforeseen user needs, in this phase, we can consider it as a *usability testing tool* as well as a *creativity tool*.
- If AVA is applied to an *already existent* product (which we do not consider extensible for the moment), we can reconstruct its interaction design as well as evaluating its strengths and weaknesses. If we start with a general initial problem in a specific field and are informed participants, we can *reconstruct* the underlying discourse dynamics and hence its *value proposition*.

Overall, the Added-Value Analysis modifies the understanding of the processes that take place when the designed object is actually used. Therefore, we can also see opportunities for using it from the user's perspective, for example for a new manual design or a user's transparency needs.

But the application of AVA holds some easy-to-fall-for catches, that influence many designs and which we will try to spell out in the following subsections.

Catch 0: The Quest for Objectivity. We are used to the fact that an analysis yields objective results. But as the AVA makes heavy use of thought trajectories, which are of subjective nature, and not "objective" logical deductions, its results are "valid" with respect to the distinguished worked-out AVA list — which can hardly be called objective. Contrary to expectations, the wanted "view of micro perspectives" for a better understanding of the processuality of use actions is an organized *quest for subjectivity* (and thereby more often than not exceeding expectations). In particular, the AVA method cannot be used for an objective evaluation of a software product but for a systematic exploration of the "subjective" micro perspectives: it uncovers value constellations; these are values from a micro perspective.

Catch 1: Knowing the Answer Before the Question. Downsizing from the macro perspective to the micro perspective is not easily achieved. First, the setting of the initial core problem is rather difficult, for an example look at the discussion further below. From a macro perspective we can and actually frequently start with rather fuzzy problems like "saving the world" with software "xyz". Then the list of benefits in the evaluation process is rather awkward like "xyz makes people happy". Therefore the hard question is what core problem that "xyz" might actually solve. Note that most people tend to know the explicit answer but not the explicit problem.[2] The underlying reason for this difficulty is, that a user's value considerations are not distributed any longer over a straight road, not even a curvy road: it is a *value landscape* with several dimensions.

Catch 2: Solutions ARE the Benefits. In the process of the AVA, it often seems as if the solutions are already the benefits: they solve the problem and that alone feels like a benefit. This is usually a sign that the AVA analyst is not sufficiently independent from the software author's motivation (from the macro-perspective). On the one hand, if the user considers the solution among others from a micro-perspective, then her evaluation is independent of this perceived benefit as it doesn't deliver an advantage over another solution. On the other hand, if it is the only available solution, then the evaluation isn't necessary for taking action.

Catch 3: Problems Are Rhetoric Questions. A similar problem appears if the core problem is formulated in a way that the solution is not only evident, but the only reasonable possible solution. In that case, we cannot think of any benefits or sacrifices as their is no choice, therefore, the AVA would run into a dead end. Probably though, the problem can be formulated in a more general way, so the phrasing of the problem should be reconsidered in the AVA; this alone is an added value of the AVA at the conceptual level.

Catch 4: Values on Too Low Levels The evaluation takes place with respect to a concrete core problem and one concrete solution for it. Often benefits and sacrifices from a higher level are still considered on a lower level. Typically this happens as soon as benefits or sacrifices are repeatedly phrased very similarly in one process chain. Then the AV analyst better remembers that a user wouldn't be at that point (i.e. on that level) if she hadn't already taken into account all preliminary values. This exactly forms the specific value constellation from a micro perspective. Therefore a very precise evaluation is required and in the AVA the correct level has to be determined for an argument.

Catch 5: The Unfinishedness of AVA. With an AVA one cannot decide what is right and what is wrong in a design, but rather what its consequences may be. Each core problem by itself can be considered as a starting point for another AVA. It is dramatically subjective what the reason for using a software or not. Therefore, in our experience the AVA always feels unfinished, some termed it "the AVA infinity". This is a feature and not a bug of the AVA: if we as designers know of many of these micro perspectives, we might be able to offer manifold solutions e.g. by abstracting in the right direction. Moreover, the realization that the AVA is not a tree, but a graph, enables a complex interplay between several components.

---

[2] Think of the fame of Douglas Adams' number 42. This also resonates with the fact that MKM is a technology-driven community always on the search of a "killer application" much like a person with a hammer desperately looking for a nail.

3 Math Web *R*esearch: An Added-Value Analysis of MATHWEBSEARCH

In the application of the AVA to the MATHWEBSEARCH (MWS) [KŞ06, KŞ07] system, we started with assuming that the initial problem for search engines for mathematics is *"finding content representations of formulae"* (see row 8 in the table below), and immediately fell victim to a combination of Catch 1 and Catch 3 mentioned above, i.e. "knowing the answers before the question" and "problems as rhetoric questions". In fact, finding formulae is just what a formula search engine does, and relying on content markup is the specific approach our system takes. Remember that we are doing this analysis, because we want to get a view of micro perspectives and not to simply reconfirm our design decisions so far. Therefore, we have to take the standpoint of a potential user. But which real user *starts* with the question "finding content representations of formulae"? It seems obvious, that such a user is probably a researcher of math web search in the first place. Hence, this level of argumentation is more at the level of the introduction of a research paper than it represents a starting point for a value analysis.

Suspiciously, this initial problem does not even cover competing designs of mathematical search engines, e.g. [MY03, LM06, MM06]. We interpret this as a sign, that we started on a rather deep level of the AVA and therefore should rephrase the initial problem in a *much* more general way. Hence, when we put ourselves in the user's shoes, we realize that we might rather want to find occurrences of mathematical objects (irrespective of their representations in mathematical formulae; see row 13, 12, and 5), and what we really want are answers to *"mathematical questions"*. Therefore, this is what we will take as the initial question (row 1) of the AVA in the table below.

#	Pot	Trigger	Core Problem	Solution	Benefits	Sacrifices
1			math questions	finding math answers in documents	• documents as reified knowledge ⤳ 3 • standing on the shoulders of giants	• answer space restricted to available documents • judging document credibility ⤳ 4
2			— dito —	ask experts	• get more than you asked for	• finding people who know ⤳ 20 • get more than you asked for
3		reified math knowledge	kinds of math knowledge	math practice: classification into formulae, statements, theories	• provenience of math assertions • epistemic status of document fragments	• three-level distinction enough? ⤳ 5
4	P	judging document credibility		NN	• NN	• NN

Note in particular, the identification of the underlying initial question allows us to see design alternatives of our system from a higher-level perspective (e.g. the MKM one). For instance, if we want answers to mathematical questions, we may ask people who know — a time-tried method in mathematics, which immediately begs the question of how to identify experts.

The other hidden assumption unearthed here, is that MKM search restricts itself to search in documents (in the widest sense) as sources for reified knowledge, and that (of course) we can only find answers in mathematical documents, if they are machine-readable and accessible to the search engine. Once we have found a suitable document, we have to analyze the provenience of the information, i.e. the credibility of a searched document has to be judged. If the documents discern e.g. the three levels of mathematical practice (formulae, statements, and theories), that is if they are suitably marked up (e.g. in a document format like OMDoc [Koh06b] with explicit structural annotations), then the reader can obtain machine-support in principle. In particular, then we can be interested in whether a found or looked-for math statement like an assertion is assumed, conjectured, or proven, and if so, what assumptions the proof is based on, as its provenience is known as well as the epistemic status of this document fragment.

Now, there are various ways of finding relevant mathematical documents, including using metadata like the Mathematical Subject Classification (MSC), by author or keyword search using standard search engines, or (again) by asking experts (see Catch 5). Here, we will only concentrate on formula search (see row 5), as we use the three-level distinction as a trigger to only look for formulae.

#	Pot	Trigger	Core Problem	Solution	Benefits	Sacrifices
5		three-level distinction enough?	finding math formulae	formula search engine	• finding *references* to formulae	• thinking in formulae • crafting queries ↝ 6, ↝ 7 • no differentiation between function and form ↝ 8, ↝ 12 • media change ↝ 7
6		crafting queries	typing correct queries	structural query editor	• GUI-like	• restricted query subset
7	P	crafting queries, media change	typing correct queries	invasive integration	• no learning curve • no media change	• development costs

The main hidden assumption uncovered here is that it is a prerequisite of math search engines that the user has to think in formulae to use them. Obtaining the formula to search for may be a big part of the problem — indeed we teach this to our children as "math word problems" or "algebra story problems". Currently, the consensus in the MKM field seems to be that the pre-formulaic stage should not or cannot be supported by our methods.

Now that we have drilled down and motivated formula search as a (derived) core problem, we can look at the sacrifices incurred: obviously we have to learn how to craft queries, and we have to decide whether we want to find formulae by their form or their function. We consider the latter a sacrifice, since it requires a decision and mental activity by the user and therefore constitutes a hurdle for using MKM technologies. Furthermore, almost all math search engines require a change of medium: they require the user to enter a formula representation into the input form of a special web page (Mihai Şucan's browser plugin for MATHWEBSEARCH [Mat07c] being a partial exception). To support the user in crafting correct queries, we see two alleys, one is supplying a structural input editor like MATHWEBSEARCH [Mat07b] or MATHDEX [Mat07a] do, or to integrate the search functionality into the editor that users are familiar with and use for math document development anyways (see [KK04] for an introduction to invasive authoring).

The querying problem is much simpler to deal with, if we want to search formulae by their form since presentation formats for formulae are in much wider use. Quite generally, there seems to be an *equivalence principle between queries and intended results*: We type keywords into Google to find pages containing these words, and we issue formulae as queries to math search engines. The next block of the AVA table will be concerned with the problem of finding formulae by their form:

#	Pot	Trigger	Core Problem	Solution	Benefits	Sacrifices
8		content / form differentiation	finding formulae by form	formula presentation search engine	• finding formulae by visual cues	• thinking in formula layout ↝ 9,↝ 10
9		thinking in layout	prioritizing visual cues ↝ 11	weighting by glyph composition	• higher visual recall & precision	• assumption: uniform glyph meaning
10	P	— dito —	low mathematical recall	semantic query expansion	• higher math. recall	• assumption: uniform glyph meaning
11	P	prioritizing visual cues	low-priority sub-layouts	glyph similarity search	• targeted fuzzy search • higher visual recall	• lower mathematical precision

It turns out that the main sacrifice about finding formulae by their form is, that we have to think about them by their layout, i.e. we have to know what they look like, which tends to miss relevant occurrences if we completely specify this. Therefore, some presentation formula engines offer some ways of broadening the search by weighting formula parts (which partly recovers a semantical flavor). We can see this as a measure of prioritizing some visual cues over others, and we can (potentially) take this to its logical conclusion by deemphasizing some low-priority parts of the layout to allow for similarity search on these. Further semantics (i.e. presentation-independence) could be added to search by adapting a technique from information retrieval: Query expansion adds semantically similar concepts (e.g. alternative presentations) to a query to obtain better coverage (see e.g. [QF93] for details).

For the problem of finding formulae by function (finally coming to the initial starting point of our AVA), we can state the general prerequisite that users must be able to think about formulae in terms of their function. If they do, and are able to formulate their problem as an instance problem (this is the only query type that MATHWEBSEARCH can answer, i.e. given a query formula

q find an occurrence of a subterm t, such that $\sigma(q) = t$ for some substitution σ), then the system can efficiently find formula occurrences, retrieve and display them. One intended application here is to remember forgotten (i.e. only partially remembered) formulae [KŞ06] (see also row 14), so MATHWEBSEARCH can act as a memory trigger, leaving the user with the task of judging whether the formulae returned are adequate. Here, system support is given by displaying the formula and associated substitutions.

#	Pot	Trigger	Core Problem	Solution	Benefits	Sacrifices
12		content / form differentiation	finding formulae by function	formula content search engine	• finding formulae independent of presentation ⤳ 13	• thinking in formula function
13		finding formulae independent of presentation	finding content formulae	MWS	• finding URL of OM/MathML instance • Retrieve URL	• formulate as instance problem ⤳ 14
14		formulate as instance problem	finding forgotten formulae	MWS	• memory trigger	• judge precision

Another (largely unexplored) application is that we can use instance queries to find applications of general mathematical results (which can be expressed as universally quantified formulae): here we can just use the quantifier-free formula body as a query term (see row 15 and note Catch 2 "benefits are the solutions"). Similarly, we can find counter-examples by negating the body (see 16 and note Catch 4 "values on too low levels"). We do not pursue the encountered catches here, but they imply that there is more work to do. If we apply these two techniques aggressively (and speculatively) over large bodies of mathematics, then we can conceivably even extend this to a (weak) form of formula induction, where we conjecture theorems from known results.

#	Pot	Trigger	Core Problem	Solution	Benefits	Sacrifices
15	P		finding applications	MWS	• finding applications	• formulate input (theorem body) as formula
16	P		finding counterexamples	MWS	• finding counterexamples	• formulate input (theorem body) as formula, then negate
17	P		formula induction		• NN	• NN

Note that all of these applications rely on MATHWEBSEARCH being able to answer instance queries efficiently, using term indexing techniques developed for automated deduction. But there are other indexing methods that allow to index for other kinds of queries: for instance generalization queries. Here, the query consists of a query formula q, and a generalization search engine, that efficiently finds occurrences t of subformulae such that $\sigma(t) = q$. With this kind of query, we can find theorems applicable to a given formula (e.g. when we are stuck in a proof). Another immediate application is to use this technology to determine novelty of research results or to settle priority claims.

#	Pot	Trigger	Core Problem	Solution	Benefits	Sacrifices
18	P		finding applicable theorems	MWS(G)	• finding universal generalization • document as context • creative trigger	• judge context compatibility
19	P		determining novelty/priority		• NN	• NN

We can use (presentation or content) formula search engines to find experts — a leftover task from the beginning of our AVA, when we were examining ways to obtain math answers — by their publications. So, if we can find documents dealing with mathematical objects we are interested in, we can usually find experts for these by looking up their authors, acknowledgments, or citations. Finally, we can use formula search engines to search for data: If we can fit the data (e.g. time series) by closed-form expressions (i.e. formulae), then we can search them via formula search engines. The advantage here is that this approach avoids having to develop direct indexing techniques for data, and we have automatic similarity search via the approximation during formula fitting.

#	Pot	Trigger	Core Problem	Solution	Benefits	Sacrifices
20	P	finding people	finding people	by math objects in publications ⤳ 5	• authors are experts	• bibliographic metadata
21	P		finding similar data	finding fitted formulae	• prescribed	• finding fitting formulae ⤳ 22 • loss of precision by approximation
22	P	finding fitting formulae	fitting formulae to data		• NN	• NN

4 Conclusion

We have presented the Added-Value Analysis as an associative method for exploring a user's value constellations from a micro perspective and applied it to an emerging subfield of MKM: formula search engines. Even though we initially applied the AVA to the specific mathematical search engine MATHWEB-SEARCH, the method quickly led to value considerations that are relevant for the whole field of MKM and to new potential services that could be explored in the future, e.g. semantic query expansion or a glyph similarity search service (see rows 10, 11).

In particular, consider the value propositions of "thinking in formulae" and the pros and cons for "precision/recall of mathematical queries". The first indicates a discrepancy between intuitions about anticipated and intended users for the math search solutions. On the one hand, formula search engines assume the user's capability of thinking in formulae to use them, on the other hand, they dream of either layman users (that want to pose math questions) or professional mathematicians. For the first a GUI (row 6) is developed, but for the latter who typically still prefer pen and paper above editors, OCR methods would be best as input technology — an opportunity for [SSWX05]?

We were also reminded that math web search — as it is handled now — *requires* "formulae competence" as a quality of the user. In particular, a user gets references as results and has to decide whether the URL is relevant herself. The underlying assumptions again imply math-knowledgeable users, for instance mathematicians. One problem might be the above mentioned equivalence principle between query and search result: the query has to be made associative, lightweight, and *not* verifiable, whereas the search result needs to be trustworthy and verifiable.

Another problem we make out consists in the prediction, that the above anticipations for math web search users indicate, that these users are also the ones we are aiming for as authors for data in our systems. If we assume that, then more work has to be done to support such users within their natural "habitat", which we believe strengthens our case for semantic, invasive editors like [Koh05a, Koh05b] or [Koh06c].

Other parts of MKM can also profit from such an analysis, and as the AVA is subjective in the choices made, even a re-analysis of math search might lead to additional insights. Note that the AVA does not directly lead to a recipe for motivating authors to take action and contribute semantically enhanced materials, but it does help to uncover added-value situations, i.e. such situations in which users can obtain value without incurring sacrifices additional to those already amortized by solving their core problem. Such AVA tables not only provide "cheat sheets" for a real MKM business plan, they enhance added-values that can help tip the scale towards using them.

References

[BH99]   Beyer, H., Holtzblatt, K.: Contextual Design. ACM interactions 6(1), 32–42 (1999)

[dChR00] de Chernatony, L., Harris, F., Riley, F.D.: Added value: Its nature, roles and sustainability. European Journal of marketing 34(1/2), 39–56 (2000)

[Far05]   Farmer, W.M.: Mathematical Knowledge Management. In: Schwartz, D.G.(ed.) Mathematical Knowledge Management, pp. 599–604. Idea Group Reference (2005)

[Fri96]   Friedman, B.: Value-Sensitive Design. interactions, ACM, pp. 16–23 (November and December 1996)

[Grö97]   Grönross, C.: Value-driven relational marketing: from products to resources and competencies. Journal of Marketing Management 13, 407–419 (1997)

[KK04]   Kohlhase, A., Kohlhase, M.: CPoint. In: Asperti, A., Bancerek, G., Trybulec, A. (eds.) MKM 2004. LNCS, vol. 3119, pp. 175–189. Springer, Heidelberg (2004)

[KM07]   Kohlhase, A., Müller, N.: Added-Value: Getting People into Semantic Work Environments (submitted 2007)

[Koh05a]   Kohlhase, A. (2005) `http://kwarc.eecs.iu-bremen.de/projects/CPoint/`

[Koh05b]   Kohlhase, A.: Overcoming Proprietary Hurdles: CPoint as Invasive Editor. In: de Vries, F., Attwell, G., Elferink, R., Tödt, A. (eds.) Open Source for Education in Europe: Research and Practise, pp. 51–56. Open Universiteit of the Netherlands, Heerlen (2005)

[Koh06a]   Kohlhase, A.: The User as Prisoner: How the Dilemma Might Dissolve. In: Memmel, M., Ras, E., Weibelzahl, S. (eds.) 2nd Workshop on Learner Oriented Knowledge Management & KM Oriented e-Learning. pp. 26–31 (2006) Online Proceedings at `http://cnm.open.ac.uk/projects/ectel06/pdfs/ECTEL06WS68d.pdf`

[Koh06b]   Kohlhase, M.: OMDoc – An Open Markup Format for Mathematical Documents [version 1.2]. LNCS (LNAI), vol. 4180. Springer, Heidelberg (2006)

[Koh06c]   Kohlhase, M.: stex: Semantic markup in T_EX / L^AT_EX Self-documenting L^AT_EX pacakge (2006) available at `https://svn.kwarc.info/repos/stex/sty/stex.pdf`

[KŞ06]   Kohlhase, M., Şucan, I.: A search engine for mathematical formulae. In: Calmet, J., Ida, T., Wang, D. (eds.) AISC 2006. LNCS (LNAI), vol. 4120, pp. 241–253. Springer, Heidelberg (2006)

[KŞ07]   Kohlhase, M., Şucan, I.: System Description: MathWebSearch 0.3, A Semantic Search Engine. submitted to CADE 21 (2007)

[LM06]   Libbrecht, P., Melis, E.: Methods for Access and Retrieval of Mathematical Content in ActiveMath. In: Iglesias, A., Takayama, N. (eds.) ICMS 2006. LNCS, vol. 4151, Springer, Heidelberg (2006), `http://www.activemath.org/publications/Libbrecht-Melis-Access-and-Retrieval-ActiveMath-ICMS-2006.pdf`

[Mat07a]   Mathdex (seen Mar 2007) web page at `http://www.mathdex.com`

[Mat07b]   Math Web Search (seen April 2007) web page at `http://kwarc.info/projects/mws/`

[Mat07c]   Math Web Search Plugin. (seen April 2007) `http://kwarc.info/projects/mws/plugin.html`

[MM06]   Munavalli, R., Miner, R.: MathFind: a Math-Aware Search Engine. In: SIGIR '06: Proceedings of the 29th annual international ACM SIGIR conference on Research and development in information retrieval, pp. 735–735. ACM Press, New York (2006)

[MY03]   Miller, B.R., Youssef, A.: Technical Aspects of the Digital Library of Mathematical Functions. Annals of Mathematics and Artificial Intelligence 38(1-3), 121–136 (2003)

[NR98]   Normann, R., Ramirez, R.: Designing Interactive Strategy. From Value Chain to Value Constellation. Wiley and Sons, Chichester (1998)

[Oul04]   Oulasvirta, A.: Finding Meaningful Uses for Context-Aware Technologies: The Humanistic Research Strategy. In: Late Breaking Result Papers (April 2004) ISBN 1-58113-703-6

[QF93]   Qiu, Y., Frei, H.-P.: Concept based query expansion. In: Korfhage, R., Rasmussen, E.M., Willett, P. (eds.) Proceedings of the 16th Annual International ACM-SIGIR Conference on Research and Development in Information Retrieval, ACM, pp. 160–169 (1993)

[SSWX05]   Smirnova, E., So, C., Watt, S., Xie, X.: Components for pen-based mathematical interfaces. In: ACA 2005 Special Session on Pen-Based Mathematical Computing (2005)

Alternative Aggregates in MIZAR

Gilbert Lee[1,*] and Piotr Rudnicki[2,**]

[1] Dept. of Computer Science, University of Victoria
gilbert.c.j.lee@gmail.com.
[2] Dept. of Computing Science, University of Alberta
piotr@cs.ualberta.ca

Abstract. MIZAR provides built-in support for defining structures (aggregates) like the familiar algebraic systems of groups or vector spaces. When trying to employ these structures for formalizing graph algorithms we ran into substantial problems stemming from the fact that fields in MIZAR structures are not first class objects. We decided that a different approach would be more suitable for the task at hand. Starting from scratch, we modeled structures as functions. In our approach, fields in structures are first class objects and just this one factor made working with graph algorithms much more convenient. We report on our experience and argue that our approach to aggregates is more suitable for a proof assistant like MIZAR.

1 Introduction

We frequently deal with objects which are composed of a number of other entities. Examples of such cases include the familiar algebraic systems of groups, rings and vector spaces. All mechanized proof assistants known to us offer some support for such structuring. E.g. see the treatment of constructive algebraic hierarchy in COQ[1], the locales of ISABELLE[2], a draft comparison of available modularization tools in several systems[3], or the development of dependently typed records for logical frameworks [4]. For lack of space, we will not be comparing our proposed approach to other systems as we offer a modest solution to a specific problem at hand. We restrict our attention to some aspects of aggregates in MIZAR. These aggregates resemble *records* from ISABELLE/ISAR[5]. In both MIZAR structures and ISABELLE records, the field names are externalized and they cannot be accessed as first-class values.

We propose another "implementation" of aggregates in MIZAR where fields are first class objects. This change combined with the machinery of MIZAR adjectives offers substantial convenience in dealing with graph algorithms.

The general syntax of structure definitions is given in MIZAR grammar[1].

---

* Supported in part by NSERC PGS, iCORE and AIF.
** Supported in part by NSERC PGP 9207.
[1] http://mizar.org/language

M. Kauers et al. (Eds.): MKM/Calculemus 2007, LNAI 4573, pp. 327–341, 2007.
© Springer-Verlag Berlin Heidelberg 2007

```
struct [ ( Ancestors ) ] Structure-Symbol [ over Loci ] (#
       Selector-Symbol -> Type-Expression ,
       ...
       Selector-Symbol -> Type-Expression
#);
```

MIZAR Mathematical Library (MML) contains various structures that are extensively used in developing abstract algebra. These structures form an inheritance hierarchy where e.g. double loop structure (the backbone for rings and fields) is derived from a multiplicative loop structure with unity and an additive loop structure with zero. See [6] for some details of MIZAR algebraic structures and the inheritance mechanism. The first exposition of MIZAR structures was given by A. Trybulec in [7].

2   Built-In Structures and Graphs

When starting our work with graph algorithms in 2002 we tried to use whatever was available about graphs. At that time, MML contained some treatment of graphs starting with the following structure definition from GRAPH_1[8].

```
definition
  struct MultiGraphStruct (#
    Vertices, Edges -> set,
    Source, Target -> Function of the Edges, the Vertices
  #);
end;
```

Every structure definition introduces a number of constructors (cf. [7]):

- A *structure mode* MultiGraphStruct that is used to construct structure types. A *structure type* is used to type objects. An object of a structure type is called a *structure*. E.g. G introduced in let G be MultiGraphStruct is a structure.
- An aggregating functor MultiGraphStruct(#V, E, S, T#) for constructing objects of structure types. In this example, both S and T, must be functions from E to V.
- Four selector functors
  - the Vertices of G
  - the Edges of G
  - the Source of G
  - the Target of G
  where G must be of type (that widens to) MultiGraphStruct.
  The intended meaning of these fields should be obvious from their names. Functions Source and Target map directed edges to their respective endpoints.
- A forgetful functor the MultiGraphStruct of G where G must be of a type that widens to MultiGraphStruct.

– A strict attribute applicable to type MultiGraphStruct such that if G is
a strict MultiGraphStruct then the MultiGraphStruct of G = G. Note that
when we only know that G is a MultiGraphStruct then the above conclusion is
not guaranteed as G can be of some other, wider type than MultiGraphStruct
and have other fields besides the four fields of MultiGraphStruct.

The mode Graph is defined in [8] as a MultiGraphStruct with an additional condition.

mode Graph is Graph-like MultiGraphStruct

where the attribute Graph-like imposes non emptiness of Vertices and was introduced as follows.

```
let G be MultiGraphStruct;
attr G is Graph-like means                          :: GRAPH_1:def 1
    the Vertices of G is non empty set;
```

Using the above as a backbone, one can define new modes for structures with
additional fields, for instance, the mode of weighted graphs.

```
let WL be set;
struct (MultiGraphStruct) WGraphStruct over WL (#
    Vertices, Edges -> set,
    Source, Target -> Function of the Edges, the Vertices,
    Weight -> Function of the Edges, WL
#);
```

The new mode is prefixed by MultiGraphStruct and inherits all the selectors
of this ancestor. The inherited selectors have to be repeated as their order is
important for the new aggregating functor.

The new mode is parameterized by a set which is intended to be used as the
weight for edges. The parameter is just a set at this stage of defining graphs. The
mode introduces a new selector Weight to be used as the Weight of G whenever
G is a structure whose type is derived from WGraphStruct over some set.

Interested readers should consult A. Trybulec in [7] for some information
about the intended semantics of MIZAR structures.

The graphs as defined above have been further developed in the following
MML articles.

– GRAPH_1 [8]: basic definitions for multigraphs.
– GRAPH_2 [9]: vertex sequences of chains.
– GRAPH_3 [10]: Euler circuits and paths.
– GRAPH_4 [11]: essentially a copy of GRAPH_2 for the case of directed graphs.
– GRAPH_5 [12] and GRAPHSP [13]: a rather ad hoc attempt at proving the cor-
  rectness of Dijkstra's algorithm for finding single source shortest path. This
  formalization is based on a specific encoding of weighted graphs as sequences.
– MSSCYC_1 [14], MSSCYC_2 [15]: graph terminology is used for properties of many
  sorted signatures; nothing interesting about graphs as such.

Besides the above, there is also an article **SGRAPH1** [16] in MML dealing with simple graphs based on a completely different formalization of basic notions.

In short, in 2002 MML contained only basic facts about graphs. Some definitions turned out to be inconvenient to work with and had to be changed—the most troubling was the notion of a walk in a graph, originally defined as a chain of edges in [8]. This notion was replaced by a walk as an alternating finite sequence of vertices and edges starting with a vertex. As for graphs with weighted edges, labeled vertices and/or labeled edges, essentially we had to start from scratch. For a while, we tried reusing whatever was available about graphs in MML. Unfortunately, this led to some inconveniences when defining new notions (see below). Some of these inconvenient features are well known within the

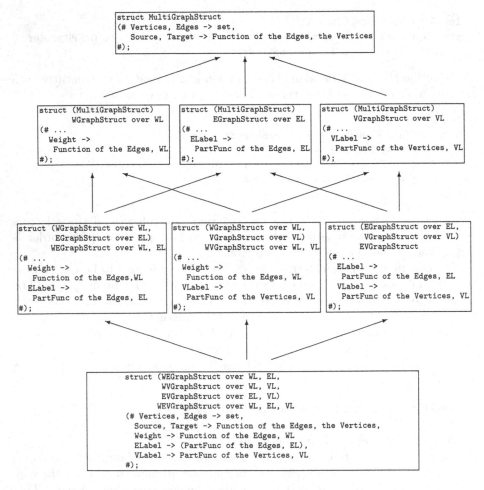

Fig. 1. Possible hierarchy of graph structures. (Note that the ... serve as space savers in this figure. In structure definitions, MIZAR requires listing all the fields of each prefixing structure. In order to save space we use ... instead of repeating the four fields of **MultiGraphStruct**.)

MIZAR community, e.g. defining functors with arguments of a structure type or returning a structure type, see Section 3.

We would like to note that when dealing with graph algorithms, we frequently needed to perform operations on graphs (i.e. structures) that are rarely performed on the familiar algebraic structures. While it happens, defining a new ring by adding an element to the carrier of an existing ring is not a frequent operation. Similarly, one infrequently updates the values of fields in a group. However, update operations on a graph structure are essential in graph algorithms, as at each step of such an algorithm, a graph is transformed into another graph.

We attempted to use the currently available MIZAR structures for our needs. The starting point was the above mentioned `MultiGraphStruct` from which we derived 7 additional graph structures as shown in Figure 1. In order to deal with graph algorithms on weighted graphs, we added three extra fields: weights on edges, labels on vertices, and labels on edges.

3   Two Difficulties

Once the graph structures are defined as above, we need to define a number of functions and modes (type constructors) involving them. This leads to a number of complications. We illustrate them on two examples.

1. Defining a function for labeling vertices.
2. Extending the concept of subgraph to weighted graphs.

3.1   Trouble with Labeling Vertices

When dealing with graph algorithms, it is handy to introduce a collection of helper functions which return graphs, for example a function which labels a given vertex by a given value. In order to guarantee uniqueness, any function returning a structure must return a special, exact form of the structure, expressible by the built-in attribute `strict`. This attribute says that only the specified set of fields is present in the structure. One would expect that one can prove equality of A and B of a structure type S by proving that A and B are equal on S's fields. However, this may succeed only when both A and B are typed as `strict` S.

An object typed as `strict MultiGraphStruct` is a structure with only the `Vertices`, `Edges`, `Source`, and `Target` fields; no other fields are in this structure. The type `strict MultiGraphStruct` cannot be used for objects whose declared type is derived through prefixing from `MultiGraphStruct` as such an object has other fields; however, such an object is a `MultiGraphStruct`. When we have an object of type `MultiGraphStruct`, without knowing whether it is `strict`, then we do not know whether there are any other fields in the object besides the original four: `Vertices`, `Edges`, `Source`, and `Target`. Any object whose type widens to `MultiGraphStruct` can always be considered as a `MultiGraphStruct` despite having additional fields. However, MIZAR offers no tools for proving that an object of a structure type is strict unless the object has been introduced as such.

Let us have a look at the following functor, intended to take a vertex labeled graph G and relabel vertex v by x.

```
definition let G be VGraph, v be Vertex of G, x be set;
   func G.labelVertex(v,x) -> strict VGraph means
      (the MultiGraphStruct of it) = (the MultiGraphStruct of G) &
      (the VLabel of it) = (the VLabel of G) +* (v .--> x);
```

The defined function is named .labelVertex, to be used in infix notation with one left argument and two right ones. The function accepts as its left argument any structure derived from VGraph, e.g. WVGraph. However, the return type of this function is strict VGraph. Thus if the left argument of this function is a WVGraph then the weight information is not "preserved" in the result as only a strict VGraph is returned.

When labeling a vertex in a graph we do not affect any other fields besides VLabel of the graph. Unfortunately, MIZAR does not allow this to be expressed easily as field selectors are not first class objects and we cannot quantify over them. Therefore we cannot say directly that all fields other than VLabel have not been changed.

There are several ways to work around this. We could introduce separate labeling functions, which return different types. For example, for WVGraph which is the type of weighted and labeled graphs we can define the following function.

```
definition let G be WVGraph, v be Vertex of G, x be set;
   func G.labelVertex(v,x) -> strict WVGraph means
      (the WGraphStruct of it) = (the WGraphStruct of G) &
      (the VLabel of it) = (the VLabel of G) +* (v .--> x);
```

Unfortunately, this means that we may need to define many such functions, one for each combination of features we would like to see in a graph. The number of such functions grows linearly with the number of structures we derive, and the latter can grow exponentially with the number of features that we introduce. Add in the customary theorems for each function, and the amount of work required to maintain such a library of helper functions may become immense. Moreover, due to different return types of these functions we would have to maintain quite a number of similar theorems for each of these functions.

There is another solution to reclaiming the information lost when using the original .labelVertex function. One can introduce more helper functions which copy fields from one structure to another as in this example.

```
definition let G be VGraph, G2 be WGraph;
            assume the Edges of G = the Edges of G2;
   func G.copyWeight_WV(G2) -> strict WVGraph means
      the VGraphStruct of it = the VGraphStruct of G &
      the Weight of it = the Weight of G2;
   correctness @proof end;
end;
```

We could then compose the two functions

```
G.labelVertex(v,x).copyWeight_WV(G)
```

to get the WVGraph that we intended. This is somewhat better than the previous approach, but we still need various versions of copy functions in order to deal with all the different types of graphs we would like to consider.

Both these approaches require a lot of extra function definitions which are essentially identical. Ideally, we would like to have a function that takes in some type of a graph, modifies a vertex label, and returns the same type of graph. The function should only modify the vertex labels, and leave all other fields of the "output" graph the same as in the "input" graph. Unfortunately, there simply is no way to specify such a function using MIZAR's built-in structures.

One could consider overcoming this problem by defining all such functions just on the WEVGraph. In doing so we would just postpone the problem as any future prefixing with WEVGraph would bring all the inconveniences back to the surface. In addition, methodologically it seems desirable that when defining functions

- types of arguments are as wide as possible (and not as narrow as convenient);
- return type is as narrow as possible.

3.2   Trouble with Subgraphs

The notion of subgraph is defined in GRAPH_1[8] as follows.

```
definition let G be Graph;
mode Subgraph of G -> Graph means                    :: GRAPH_1:def 17
  the Vertices of it c= the Vertices of G &
  the Edges of it c= the Edges of G &
  for v st v in the Edges of it
    holds (the Source of it).v = (the Source of G).v &
          (the Target of it).v = (the Target of G).v &
          (the Source of G).v in the Vertices of it &
          (the Target of G).v in the Vertices of it;
end;
```

We needed to define the concept of a weighted subgraph in order to talk about a shortest-path subtree when formalizing the Dijkstra shortest-path algorithm. The definition of a weighted subgraph, WSubgraph, could be as follows.

```
definition let G be WGraph;
  mode WSubgraph of G -> WGraph means
    it is Subgraph of G &
    the Weight of it = the Weight of G | the Edges of it
end;
```

(The binary operator | is a restriction on domain of a relation.) A WSubgraph is clearly a subgraph. However, in the definition we give its mother type as WGraph in order to access the Weight selector in the definiens. With this definition

MIZAR automatically processes a WSubgraph as a WGraph but MIZAR does not "automatically" know that it is also a subgraph of G. This is unsatisfactory as then the machinery developed for subgraphs is not available directly; we have to access the machinery by hand. As an example, let us look at the attribute spanning applicable to subgraphs.

```
definition let G be Graph, G2 be Subgraph of G;
  attr G2 is spanning means
    the Vertices of G2 = the Vertices of G;
end;
```

If we have an object w of type WSubgraph of G, we cannot directly say that

```
    w is spanning
```

as WSubgraph of G is not automatically perceived as a Subgraph of G. We can achieve the desired effect but we must use an additional object in order to cast the type.

```
    for w1 being Subgraph of G st w = w1 holds w1 is spanning
```

This type casting is necessary as MIZAR does not see two types of an object simultaneously if one of the types is not derived from the other.

Although none of the above obstacles was fatal, they seemed inconvenient enough that we decided to pursue an alternative treatment of graphs in MIZAR. (Moreover, besides a handful of rudimentary definitions, there was not much material on graphs in MML that we could use when formalizing graph algorithms.)

4   New Implementation of Aggregates

The underlying idea behind our alternate approach to aggregate objects is to have selectors as first class objects and thus to resign completely from using the built-in structures. The MIZAR machinery of attributes (adjective constructors) plays a central role in our approach. Instead of fixing which collection of fields is part of an aggregate, we define what it means for an object to have some particular field. For example, a graph having some associated weight function would have the attribute [Weighted], while a graph having labeled vertices would have the attribute [VLabeled].

We define a GraphStruct[2] as a finite function whose domain is a subset of natural numbers.

```
definition
  mode GraphStruct -> finite Function means          :: GLIB_000:def 1
    dom it c= NAT;
end;
```

---

[2] The name GraphStruct is probably quite misleading at this point, something like Struct would be much better. However, in our experiment we were interested in talking about graphs and not about structures in general.

Although we could have used any fixed set as the domain above, choosing natural numbers seemed more convenient than others. The elements of the domain of GraphStruct play the role of selectors for fields.[3] In the next step, for each selector we are interested in, a functor is defined that returns a unique natural number:

```
definition
  func VertexSelector -> Nat equals 1;
  func EdgeSelector   -> Nat equals 2;
  func SourceSelector -> Nat equals 3;
  func TargetSelector -> Nat equals 4;
  func WeightSelector -> Nat equals 5;
  func ELabelSelector -> Nat equals 6;
  func VLabelSelector -> Nat equals 7;
end;
```

For each selector, we introduce another functor that looks similar to a selector reference used in MIZAR's built-in structures. (We do not have to do this at all but in 2003 it looked like a good idea.)

```
definition let G be GraphStruct;
  func the_Vertices_of G equals G.VertexSelector;
  func the_Edges_of G    equals G.EdgeSelector;
  func the_Source_of G   equals G.SourceSelector;
  func the_Target_of G   equals G.TargetSelector;
end;
```

Now we are in a position to define an attribute that states which properties of a bare-bones GraphStruct make it graph-like. Note that the graph-like attribute expresses properties that earlier were expressible through construction; see the definition of MultiGraphStruct above.

```
definition let G be GraphStruct;
  attr G is [Graph-like] means                        :: GLIB_000:def 11
    VertexSelector in dom G & EdgeSelector in dom G &
    SourceSelector in dom G & TargetSelector in dom G &
    the_Vertices_of G is non empty set &
    the_Source_of G is Function of the_Edges_of G, the_Vertices_of G &
    the_Target_of G is Function of the_Edges_of G, the_Vertices_of G;
end;
```

We define attributes for stating the presence of other fields in a graph.

```
definition let G be GraphStruct;
  attr G is [Weighted] means                          :: GLIB_003:def 4
```

---

[3] A. Trybulec maintains that structure selectors should be a resource that is controlled by the MIZAR processor. Such an approach would avoid possible future conflicts. However, we have conducted the experiment with formalizing graph algorithms without any change of the MIZAR software. The user is entirely responsible for defining and using selectors. A careless user can repeat the natural value given to a selector. We do not see any danger in this while serious inconvenience is certainly possible.

```
WeightSelector in dom G &
G.WeightSelector is ManySortedSet of the_Edges_of G;

attr G is [ELabeled] means                              :: GLIB_003:def 5
ELabelSelector in dom G &
ex f being Function
        st G.ELabelSelector = f & dom f c= the_Edges_of G;

attr G is [VLabeled] means                              :: GLIB_003:def 6
VLabelSelector in dom G &
ex f being Function
        st G.VLabelSelector = f & dom f c= the_Vertices_of G;
end;
```

With the the built-in MIZAR structures, we were constructing new structure
modes with struct $\cdots$ (# $\cdots$ #) and the construction relieved us from proving
the existence of objects of the newly defined types. With our approach, we have
to prove the existence of GraphStructs with specific properties. With the three
additional fields besides the backbone fields of each graph, we have 8 different
kinds of graphs we could possibly deal with. The good news is that we can do
this in one single shot by proving the following existential cluster.

```
registration
  cluster [Graph-like] [Weighted] [ELabeled] [VLabeled] GraphStruct;
  existence proof ... end;
end;
```

After demonstration that such compound objects exist, we assign individual
modes for each subset of features.

```
definition
  mode _Graph is [Graph-like] GraphStruct;
end;
```

```
definition
  mode   WGraph is [Weighted]                           _Graph;
  mode   EGraph is              [ELabeled]               _Graph;
  mode   VGraph is                         [VLabeled] _Graph;
  mode  WEGraph is [Weighted] [ELabeled]               _Graph;
  mode  WVGraph is [Weighted]              [VLabeled] _Graph;
  mode  EVGraph is              [ELabeled] [VLabeled] _Graph;
  mode WEVGraph is [Weighted] [ELabeled] [VLabeled] _Graph;
end;
```

Thanks to the MIZAR attribute system, we get automatic inheritance. For example, MIZAR automatically knows that a WEVGraph is a WGraph, a VGraph and
even a EVGraph, without us having to prove anything.

A feature found in the MIZAR implementation of structures that we have to
emulate by hand is the previously automatic typing of selectors. For example,
in MultiGraphStruct, MIZAR understands that the Source is a function from the

Edges to the **Vertices**. In the new approach, we can get the same by redefining the new "selectors" for various types of arguments in which we make their return more specific.

```
definition let G be _Graph;
  redefine func the_Vertices_of G -> non empty set;

  redefine func the_Source_of G ->
    Function of the_Edges_of G, the_Vertices_of G;

  redefine func the_Target_of G ->
    Function of the_Edges_of G, the_Vertices_of G;
end;
```

5   Overcoming the Difficulties

With our approach we have more control over the fields and we can better address the two limitations we mentioned in Section 3. In the solution, we heavily rely on attributes, i.e. adjective constructors.

Given a selector, we can now easily modify the field associated with that particular selector using the general machinery developed for functions. A functor that accomplishes this task might be as follows.

```
definition let G be GraphStruct, n be Nat, x be set;
  func G.set(n,x) -> GraphStruct equals        :: GLIB_000:def 13
    G +* (n .--> x);
end;
```

where n is a selector. (G +* (n .--> x) is MIZAR lingo, one of many already available, for overwriting G at n by x.)

5.1   Another Way of Labeling Vertices

With the above in hand, we define the primitive graph helper functor for labeling a vertex.

```
definition let G be VGraph, v,x be set;
  func G.labelVertex(v,x) -> VGraph equals      :: GLIB_003:def 22
    G.set(VLabelSelector, the_VLabel_of G +* (v.-->x)) if
    v in the_Vertices_of G otherwise G;
```

If we try to label a WVGraph, the weight information gets preserved. Nonetheless, MIZAR doesn't recognize this automatically and sees the result as only a VGraph. However, since the presence of weight is expressed by an attribute, we can add the missing information using functorial clusters, for a complete set of such clusters see [17].

```
registration let G be WVGraph, v,x be set;
  cluster G.labelVertex(v,x) -> [Weighted];
end;
```

```
registration let G be EVGraph, v,x be set;
  cluster G.labelVertex(v,x) -> [ELabeled];
end;
```

The new labeling functor is much more convenient to use. With MIZAR's built-in treatment of structures, we would have needed many different functions, but with our implementation, we only need one, along with a collection of functorial clusters, one per field. However, we still have to prove several simple theorems stating which selectors have not been affected by the labeling function.

```
theorem                                              :: GLIB_003:46
  for G being WVGraph, v,x being set holds
    the_Weight_of  G = the_Weight_of  G.labelVertex(v,x);
```

```
theorem                                              :: GLIB_003:47
  for G being EVGraph, v,x being set holds
    the_ELabel_of  G = the_ELabel_of  G.labelVertex(v,x);
```

Since we have only one function of updating vertex labels, the number of such theorems will be linear in the number of fields.

5.2   Alternate Subgraphs

A subgraph is now defined as follows.

```
definition let G be _Graph;
  mode Subgraph of G -> _Graph means            :: GLIB_000:def 29
    the_Vertices_of it c= the_Vertices_of G &
    the_Edges_of it c= the_Edges_of G &
    for e being set st e in the_Edges_of it holds
      (the_Source_of it).e = (the_Source_of G).e &
      (the_Target_of it).e = (the_Target_of G).e;
end;
```

Since now the property of having a weight field is an attribute of GraphStruct, we can talk about subgraphs that are weighted, namely [Weighted] Subgraph of G. This is not the same as a WSubgraph in Section 3.2, because the fact that a subgraph of G is weighted does not say that the weights are inherited from G. What we need is another attribute which states this inheritance.

```
definition let G be WGraph, G2 be [Weighted] Subgraph of G;
  attr G2 is weight-inheriting means           :: GLIB_003:def 10
    the_Weight_of  G2 = (the_Weight_of G) | the_Edges_of G2;
end;
```

The mode WSubgraph, analogous to the one with which we had later troubles in Section 3.2 now becomes as follows.

```
definition let G be WGraph;
  mode WSubgraph of G is weight-inheriting ([Weighted] Subgraph of G);
end;
```

Once again, thanks to the MIZAR attribute system, MIZAR automatically understands that a WSubgraph of G is a Subgraph of G, yet also a WGraph because it is both [Weighted] and a _Graph.

Thus, when we deal with a G being a finite real-weighted WGraph and a w being finite real-weighted WSubgraph of G we can write both w is spanning and say that w.cost() <= G.cost(). The .cost functor is defined as follows.

```
definition let G be finite real-weighted WGraph;
  func G.cost() -> Real equals                    :: GLIB_004:def 11
    Sum the_Weight_of G;
end;
```

5.3 Expressing Strictness

MIZAR allows one to extract a strict part out of a structure. For example, for an object G of type WVGraphStruct, we can talk about the WGraphStruct of G which gives us a strict WGraphStruct part of G. With our attribute approach, we can get an analogous effect by restricting the selector set of a graph with a function like this.

```
definition let G be GraphStruct, X be set;
  func G.strict(X) -> GraphStruct equals          :: GLIB_000:def 14
    G | X;
end;
```

As in the case of labeling a vertex, we use functorial clusters to show which features are carried over for a specific argument X.

```
definition
  func WGraphSelectors -> non empty finite Subset of NAT equals
                                                  :: GLIB_004:def 9
    {VertexSelector, EdgeSelector, SourceSelector, TargetSelector,
     WeightSelector};
end;

registration let G be WGraph;
  cluster G.| WGraphSelectors -> [Graph-like] [Weighted];
end;
```

We found it necessary to use this version of strict graph aggregates when defining the set of minimum cost spanning trees for a weighted graph in [18].

6 Final Remarks

The alternative treatment of graphs has been implemented in MIZAR articles [19,20,21,17] by the first author as a part of his MSc thesis, see [22], under supervision of the second author. Then we proved the correctness of the following algorithms.

- Dijkstra's single source shortest path in [18],
- Prim's minimum spanning tree in [18],
- Ford/Fulkerson's maximum network flow algorithm in [23].

All algorithms are proved correct at the level of graph "operations"; we did not venture into formal treatment of implementing priority queues or other data structures.

Following the above work, Broderick Arneson [24] has formalized two algorithms for recognizing chordal graphs from [25] and [26]. These algorithms employ quite complicated labeling of vertices; the formalization is contained in [27] and [28].

In our dealing with graphs, getting past the limitations of built-in MIZAR structures when working with update functions and subgraphs was of paramount importance. We feel that our implementation of aggregates, heavily relying on the MIZAR attributes machinery, has preserved all the benefits of the current MIZAR implementation of structures, yet has given us the extra flexibility to address many issues that we faced. We discussed this alternative implementation of aggregates with Grzegorz Bancerek in mid 1990s; the idea waited for 10 years to be implemented. Now we are considering a design of a systematic way for converting the old style MIZAR structures into the new implementation.

We thank one of the reviewers for valuable comments and suggestions.

References

1. Geuvers, H., Pollack, R., Wiedijk, F.J.Z.: A constructive algebraic hierarchy in Coq. Journal of Symbolic Computation 34(4), 271–286 (2002), See also http://www.cs.ru.nl/~freek/pubs
2. Kammüller, F.: Modular reasoning in Isabelle. In: McAllester, D. (ed.) Automated Deduction - CADE-17. LNCS, vol. 1831, Springer, Heidelberg (2000), See also http://swt.cs.tu-berlin.de/~flokam
3. Kammüller, F.: Comparison of imps, pvs and larch with respect to theory treatment and modularization. Technical report, TU Berlin, Unpublished (1996) see http://swt.cs.tu-berlin.de/~flokam
4. Coquand, T., Pollack, R., Takeyama, M.: A logical framework with dependently typed records. In: Hofmann, M.O. (ed.) TLCA 2003. LNCS, vol. 2701, Springer, Heidelberg (2003)
5. Wenzel, M.: The Isabelle/Isar Reference Manual (2005) See http://isabelle.in.tum.de/doc/isar-ref.pdf
6. Rudnicki, P., Schwarzweller, C., Trybulec, A.: Commutative Algebra in the Mizar System. Journal of Symbolic Computation 32, 143–169 (2001)
7. Trybulec, A.: Some Features of the Mizar Language. Technical report, ESPRIT Workshop, Torino (1993) See also http://mizar.org/project/trybulec93.ps
8. Hryniewiecki, K.: Graphs. Formalized Mathematics 2(3), 365–370 (1990)
9. Nakamura, Y., Rudnicki, P.: Vertex sequences induced by chains. Formalized Mathematics 5(3), 297–304 (1996)
10. Nakamura, Y., Rudnicki, P.: Euler circuits and paths. Formalized Mathematics 6(3), 417–425 (1997)

11. Nakamura, Y., Rudnicki, P.: Oriented chains. Formalized Mathematics 7(2), 189–192 (1998)
12. Nakamura, Y., Chen, J.C.: The underlying principle of Dijkstra's shortest path algorithm. Formalized Mathematics 11(2), 143–152 (2003)
13. Chen, J.C.: Dijkstra's shortest path algorithm. Formalized Mathematics 11(3), 237–248 (2003)
14. Byliński, C., Rudnicki, P.: The correspondence between monotonic many sorted signatures and well-founded graphs. Part I. Journal of Formalized Mathematics, vol. 8 (1996)
15. Byliński, C., Rudnicki, P.: The correspondence between monotonic many sorted signatures and well-founded graphs. Part II. Journal of Formalized Mathematics, vol. 8 (1996)
16. Toda, Y.: The formalization of simple graphs. Journal of Formalized Mathematics, vol. 6 (1994)
17. Lee, G.: Weighted and labeled graphs. Formalized Mathematics 13(2), 279–293 (2005)
18. Lee, G., Rudnicki, P.: Correctness of Dijkstra's shortest path and Prim's minimum spanning tree algorithms. Formalized Mathematics 13(2), 295–304 (2005)
19. Lee, G., Rudnicki, P.: Alternative graph structures. Formalized Mathematics 13(2), 235–252 (2005)
20. Lee, G.: Walks in graphs. Formalized Mathematics 13(2), 253–269 (2005)
21. Lee, G.: Trees and graph components. Formalized Mathematics 13(2), 271–277 (2005)
22. Lee, G.: Verification of Graph Algorithms in Mizar. Master's thesis, University of Alberta, Dept. of Comp. Sci. (2004) See also
    http://www.cs.ualberta.ca/~piotr/Mizar
23. Lee, G.: Correctnesss of Ford-Fulkerson's maximum flow algorithm. Formalized Mathematics 13(2), 305–314 (2005)
24. Arneson, B.: Mizar Verification of Algorithms for Recognizing Chordal Graphs. Master's thesis, University of Alberta, Dept. of Comp. Sci. (2007)
    See also http://www.cs.ualberta.ca/~piotr/Mizar
25. Golumbic, M.C.: Algorithmic Graph Theory and Perfect Graphs. Academic Press, New York (1980)
26. Tarjan, R.E., Yannakakis, M.: Simplie linear-time algorithms to test chordality of graphs, test acyclicity of hypergraphs, and selectively reduce acyclic hypergraphs. SIAM Journal of Computing 13(3), 566–569 (1984)
27. Arneson, B., Rudnicki, P.: Chordal graphs. Formalized Mathematics 14(3), 79–92 (2006)
28. Arneson, B., Rudnicki, P.: Recognizing chordal graphs: Lex BFS and MCS. Formalized Mathematics 14(4), 187–205 (2006)

An Approach to Mathematical Search Through Query Formulation and Data Normalization

Robert Miner and Rajesh Munavalli

Design Science, Inc., St. Paul, MN 55101, USA
robertm@dessci.com, rajvm19@gmail.com
http://www.dessci.com

Abstract. This article describes an approach to searching for mathematical notation. The approach aims at a search system that can be effectively and economically deployed, and that produces good results with a large portion of the mathematical content freely available on the World Wide Web today. The basic concept is to linearize mathematical notation as a sequence of text tokens, which are then indexed by a traditional text search engine. However, naive generalization of the "phrase query" of text search to mathematical expressions performs poorly. For adequate precision and recall in the mathematical context, more complex combinations of atomic queries are required. Our approach is to query for a weighted collection of significant subexpressions, where weights depend on expression complexity, nesting depth, expression length, and special boosting of well-known expressions.

To make this approach perform well with the technical content that is readily obtainable on the World Wide Web, either directly or through conversion, it is necessary to extensively normalize mathematical expression data to eliminate accidently or irrelevant encoding differences. To do this, a multi-pass normalization process is applied. In successive stages, MathML and XML errors are corrected, character data is canonicalized, white space and other insignificant data is removed, and heuristics are applied to disambiguated expressions. Following these preliminary stages, the MathML tree structure is canonicalized via an augmented precedence parsing step. Finally, mathematical synonyms and some variable names are canonicalized.

1 Introduction

This article describes an approach to searching for mathematical notation. The approach aims at a search system that can be effectively and economically deployed, and that produces good results with a large portion of the mathematical content freely available on the World Wide Web today. We have implemented this approach in the Mathdex [10] search service and web site.

Our approach follows the general model for mathematical search developed by Yousef [12]. The basic concept is to linearize mathematical notation as a sequence of text tokens, which are then indexed by a traditional text search engine. The text search engine performs atomic queries for terms in the usual way, computing

M. Kauers et al. (Eds.): MKM/Calculemus 2007, LNAI 4573, pp. 342–355, 2007.

rankings for documents using a standard vector space model based on term frequencies and inverse document frequencies. Conceptually, a query for a more complex mathematical expression then becomes roughly analogous to a phrase query, where the text notion of phrase has somehow been suitably adapted to mathematics.

Apart from the algorithmic problem of devising query types and ranking methods suitable for mathematics from the building blocks of atomic term queries, there is the practical challenge of applying the approach to the technical content actually available on the Web. Text is relatively straightforward to identify and extract in most document formats, and to normalize for searching purposes (via stemming, etc.) By contrast, mathematics is often hard to identify and extract, and is encoded in many different ways, both at the level of markup and notation.

To address this problem, the strategy we have employed is to convert all content to a common format, XHTML+MathML, again leveraging existing third-party tools whenever possible. We then employ a multi-pass normalization algorithm that attempts to produce a canonical MathML representation for equivalent mathematical notations. By notational equivalence of expressions, we mean that a typical user looking at them would judge them to be the same mathematical notation. For example, expressions that differ trivially in spacing or that have markup differences that make no visual difference to the typeset appearance should share the same canonical representation.

There are several advantages of this general approach. A key practical advantage is that it leverages the very considerable amount of effort that has gone into developing effective, highly optimized text search systems and conversion tools. In our case, we have chosen to use the Apache Lucene [1] text search engine and Apache Nutch [2] web crawler, together with a variety of existing tools for conversion to MathML, particularly blahtex [11] and LaTeXML [13].

At a more theoretical level, by favoring notational similarity over mathematical similarity, we believe this approach offers users simpler, more familiar query formulation. This also works well with a much larger class of documents, since in most cases, sufficiently detailed semantics for mathematical manipulation are not adequately specified, and cannot today be inferred programmatically with sufficient reliability. At the same time, in many contexts where mathematical semantics are available, the appeal and promise of semantic search is great. In particular, see [3], [4], [5], [6], and [7] for current work in this area. The search system for Wolfram Research's Functions [17] web site is also an interesting approach to semantic search.

This work is supported in part by the National Science Foundation through the National Science Digital Library program under grant number 0333645.

2   Query Formulation

2.1   Mathematical N-Grams

Users want to formulate short queries, and obtain complete answers. Unfortunately, short queries are generally ambiguous in capturing the user's information need. This fundamental dynamic of information retrieval is magnified in the context of mathematical search. When a user forms a text query by giving one or two keywords, it is likely that the keywords will appear literally in relevant documents. By contrast, when a mathematical expression is given as a query, it is likely that most relevant documents will not literally contain the query expression, but will instead contain expressions that merely share one or more common sub-expressions with query expression. For example, a user querying for $x + y$ may want to match $y + x$ and $x + 2 + y$ with a certain degree of relevance. Thus, the paramount challenge of mathematical search is to identify relevant results by finding expressions that are similar to a query expression while differing in variable names, order and structure in potentially non-trivial ways.

In text retrieval systems, character-based n-grams are effectively used to overcome difficulties such as misspellings and multiple tenses. Instead of indexing individual words, n-grams, (sequences of n consecutive characters appearing in a word) are indexed. At a higher structural level, indexing word-based n-grams (sequences of n consecutive words in a document) is more efficient and effective than searching for literal phrase matches. At both levels, focusing on n-grams instead of literal matches gives a natural and effective way of identifying words and phrases that are similar to a query based on the degree of overlap in the constituent n-grams.

Working by analogy, an obvious approach to quantifying similarity to a mathematical query expression is to build an index of "mathematical n-grams." A mathematical n-gram will consist of a sequence of n consecutive building blocks of some sort, but these building blocks could range from entire expressions, at one end of the spectrum, to individual symbols and characters at the other end of the spectrum, depending on the granularity of the information desired.

We chose to use atomic mathematical notations as the basic unit for mathematical n-grams. These mathematical n-grams range from a single variable to short sub-expressions. For us atomic notations have a one-to-one correspondence with a node or set of adjacent nodes belonging to the same parent in the presentation MathML representation of an expression. This approach constructs mathematical n-grams with retrieval characteristics similar to the text n-grams while still preserving meaningful mathematical structure [14].

The length of an n-gram is defined recursively. All MathML token nodes are considered to be 1-grams. The length of a non-token MathML node is defined as the sum of the lengths of its child nodes, omitting a short list of <mo> nodes containing common operators, such as $+$, $-$ and invisible times. This is analogous to ignoring stop words like "he" "is", "and," etc., in text search retrieval. Also, n-grams starting or ending with these stop word operators, as in $+z$ or $2y+$, are not indexed or searched. These operators appear only inside higher n-grams, to give importance to exact matches. Currently, we only consider n-grams of length up to 5.

N-grams are categorized by notational role, which is typically identified by the parent element of the nodes included in the n-gram. Each category of n-gram is indexed separately. Following Lucene terminology, these categories are called fields in the index. Consider some examples. In the MathML expression, `<msup>` `<mi>x</mi>` `<mn>2</mn>` `</msup>`, the `<mn>` node is placed in the "superscript" field. In the expression `<mfrac><mi>x</mi><mi>y</mi><mfrac>`, both the x and y nodes belong to parent element `<mfrac>`, but in this case, finer control over the notation role is desirable. So here, the first child x is indexed in the "numerator" field whereas y is indexed in the "denominator" filed. Similarly in `<mroot>` `<mi>x</mi>` `<mi>y</mi>` `</mroot>`, x is indexed as "base" and y as "root".

For an example of mathematical n-gram construction, consider the expression

$$\frac{x^2}{2y + z}$$

encoded in presentation MathML as:

```
<math>
   <mfrac>
      <msup> <mi>x</mi> <mn>2</mn> </msup>
      <mrow>
         <mrow> <mn>2</mn> <mo> &it; </mo> <mi>y</mi> </mrow>
         <mo>+</mo>
         <mi>z</mi>
      </mrow>
   </mfrac>
</math>
```

Here, the n-grams structure is:

1 grams: x, 2, 2, Invisible Times, y, +, z
2 grams: x^2 , $2y$
3 grams: $2y + z$
4 grams: none
5 grams: $\frac{x^2}{2y+z}$

The search space is divided into fields to increase the precision of query matching. During query time, the query expression is broken down into n-grams. The index is then searched for each n-gram, both in its own primary field, as well as other extended fields, which are mathematically meaningful in that query context. For example, matches in the "nth root" field of the index for a query sub-expression in the "square root" field have a high likelihood of being relevant, where as matches in the "denominator" field of the index are less likely to be relevant. Note that n-grams are only indexed in their own primary fields to accurately represent the information in the document. Limiting indexing to the primary field helps retrieve documents with high precision whereas querying on both primary and extended fields helps retrieve documents similar to the query thus increasing the recall.

The weighting for a match in an extended field is derived from the weight of the primary field. This weight is proportional to an informal notion of structural similarity between the fields. We have precompiled a matrix of similarity values between different fields that are used to assign weights for the extended fields. As will be discussed below, query terms must exceed a threshold query weight to be considered, and hence extended fields with very low similarity and significance to the entire query sub-expression will be automatically dropped.

2.2   Term Level Query Formulation

Naively searching for all constituent n-grams in a query expression would not only slow down the overall search speed but also retrieve a huge number of non-relevant documents. Merely because a query expression contains x as a 1-gram does not mean that any document containing an x is relevant. To address this problem, it is desirable to choose a small subset of the potential query terms. During the query tree construction, each node in the tree is assigned a weight depending on its structural complexity, n-gram length and depth in the tree. Once the tree is constructed, we select the query terms that satisfy a minimum weight threshold condition, beginning at the root of the tree and traversing the branches all the way to the leaf nodes. Empirically, a good minimum weight threshold is about 25% of weight of the root term.

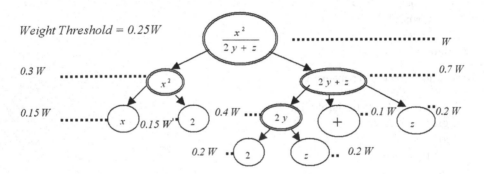

Fig. 1. Weighted Query Tree with a weight of W and minimum threshold $0.25W$

The query term nodes which are selected for inclusion in the final query object are shown in double circled with weight $>= 0.25W$.

The weight of each query term is computed as follows. It is directly proportional to the complexity and length and inversely proportional to the depth of the term node.

    Weight(term) = (Complexity * Length) / Depth + Special Boost

Complexity: Complexity of a term node is computed as the sum of complexities of its child nodes and a predetermined complexity value of the current node. A leaf node typically has complexity of 1. Complex structures like fractions, square roots, matrices etc., get a higher weight thus pruning away the less complex structures which might result in noisy matches.

Length: Length of a term node is computed as the sum of lengths of its child nodes, with the exception of stop word operators as explained above. Longer n-grams have better structural overlap with the query yielding more relevant documents.

Depth: Depth of a term node is equal to the number of branch traversals from the root node to term node under consideration. In general, the higher the depth, the less relevant it is to the query expression.

Special Boost: Certain operators and operands have special meaning in a mathematical expression. For example, common elementary functions such as sin, log, etc., often play a special role in capturing the intention of the user. We have pre-compiled a list of these special operators and operands that are used to boost the weight of the term at query time.

Wild card queries for mathematical expressions are also supported analogously to text wild cards. The scope of the wild card matching is limited to the current mathematical structure. For example, $\frac{x}{*}$ matches any fraction with x in the numerator, but not $x + y + z$. Wild card query terms have less weight compared to their regular counterparts. This is to avoid matching longer ambiguous expressions.

2.3   Expression Level Query Formulation

Query formulation with appropriate weights is vital for effective information retrieval. The challenges of formulating an effective query is emphasized in equation retrieval due to complexity of the mathematical structure and the vagueness in defining mathematical similarity to a given query expression. Meeting the information retrieval need of a diverse user population, from high school math students to research scientists, poses an added challenge.

An effective query ideally should be built around the search task, together with an understanding of the indexing and relevance sorting mechanisms involved. For example, a user with sufficient knowledge to boost the weight of individual query terms and specify specific fields in which to look for them can pose a more intelligent query. Unfortunately, domain expertise, awareness of underlying conceptual search model and its effect on retrieval performance are all the qualities of an expert user. The general user population is far from being willing or able to formulate an intelligent query. Consequently, an expression-level query formulation algorithm, incorporating such expertise, is necessary to attempt to heuristically transform a simple query expression into an advanced query.

We have chosen to optimize our query formulation algorithm to maximize recall to as far as possible without considerably degrading the precision. To this end, our algorithm uses three different logical operations, depending on context, to combine the individual query terms, specifically Boolean, Spanning and Disjunction max operations. The following sections explain how these different logical operations are used and their effect on retrieval performance, using the query expression $\frac{2y+z}{x^2}$ as an example.

Boolean Logic. In general, Boolean IR systems yield high precision but low recall. Terms are either present or absent from the document. By contrast, vector space models result in high recall due to partial matches, and vector space models generally tend to outperform Boolean models due to the fact that precise queries are difficult to construct and Boolean model output has no sense of relevance ranking. For this reason, within the Lucene atomic term queries (in our case queries for individual n-grams) are performed using a vector space model. However, it is still useful to combine the result sets from term queries using the standard Boolean operators AND, OR and NOT[16].

In our algorithm, Boolean combinations of sub-queries are used in situations when the sub-queries have an inherent Boolean logical relationship from the notational structure of the parent query. For example, a query for a fraction naturally decomposes as a query for the numerator and the denominator. For the query example above, the Boolean query "numerator: $2y + z$ AND denominator: x^2 most precisely retrieves documents containing a $2y + z$ in the numerator field and x^2 in the denominator field. Note that precision nonetheless suffers, since there is no guarantee both sub-expressions occur in the same fraction within a document.

Span Logic. Because of commutativity and associativity of common arithmetic operators, some flexibility in the order of terms is mandatory for any mathematical search system. Our algorithm uses span logic to do this. Span logic restricts the search space to only those documents where sub-query objects appear within a given proximity of one another defined by a span scope. In our algorithm, the span scope is computed based on the length of the sub-query objects as well as their weights. To provide mathematically meaningful flexible structure matching, span scopes are limited to consecutive terms belonging to the same index field. For the query example above, a span query "numerator: (2y, z){4}" would match those documents where $2y$ and z appear in the numerator field within an n-gram distance of 4. This would match an expression like $2y + x + z$ in the numerator. Use of span logic generally enhances overall recall.

Disjunction Max Logic. Disjunction max logic is used to select the best match possible among set of terms appearing in different search fields. Unlike the Boolean OR operation, disjunction max operation does not add the search spaces. It rather picks out those search fields that have the maximum possibility of finding relevant documents. This helps to rank the documents in the order of structural preference rather than based on multiple occurrences of a term in the extended fields. Without disjunction max logic, a document with multiple occurrences of x^2, in non-denominator fields would rank higher than a document with only one occurrence x^2 in the denominator field.

Query weights for span and disjunction max logics are significantly higher than the Boolean logic counterparts to achieve higher precision. Empirically we have observed that judicious application of span and disjunction max logic increases precision without undue loss of recall. There is, however, a fair amount of art involved in tuning query weights and logic types. Our algorithm currently uses a precompiled table of weights and types, based on extensive experimentation, for building up expression queries, taking into account the types of n-grams and index fields involved.

The overall weight of the final query object for a mathematical expression must be normalized for the purpose of comparison of different query objects. Normalization is particularly important when the user further restricts the search scope by providing text search terms.

Text queries are formulated in similar, if simpler, fashion. At index time, a document is analyzed to extract title, heading and body content, which is then indexed in separate search fields. During query time, user text query terms are combined and searched in all three fields with different weight levels. The title is given the highest weight followed by heading and body. Text query weights are normalized with respect to math expression query weights. Both math expression query and text query objects are combined with Boolean logic OR in our current implementation.

3 Data Normalization

The approach to query formulation for mathematical search described in section 2 relies heavily on the assumption that notationally equivalent expressions will be indexed the same way. Unfortunately, mathematical expressions encountered in real documents are very different from mathematical expressions in the abstract; they are represented in diverse markup languages, created by specific authors and tools, all of which introduce their own quirks, conventions and errors. In practice, mathematical expressions that, in the abstract, should be identical can appear very differently to a search system.

The search algorithm described above was originally developed and tested against a limited set of carefully controlled test documents. In order to explore the effects of various strategies for addressing the goals of section 2.1 on precision and recall, it was necessary to use test data where it was possible to determine with confidence what the expected results of a given query should be. We expected the algorithm would perform less well with real-world data, but when we began testing it, we were somewhat surprised to discover that artifacts of encoding, conversion, authoring tools and author coding choices completely dominated, rendering the algorithm virtually useless. Consequently, it was clear that an effective mathematical search system would need to combine a rigorous data normalization component as well as a good search algorithm.

By analyzing documents from a variety of sources, we identified seven areas in which data normalization was required. Since our algorithm operates on mathematical expressions encoded in MathML, the starting point of our analysis included both documents containing MathML directly from the publisher as well as document where we converted the mathematics to MathML using 3rd party conversion tools. We attempted to include MathML from most major authoring and/or conversion systems, including Mathematica, MathType, MathFlow, LaTeXML, Hermes, TeX4ht, itexmml and TtM.

3.1   XML and MathML Error Correction

The first and most fundamental data normalization that the documents we examined required was the correction of XML and MathML errors. Documents containing MathML from publishers or produced with MathML tools largely contained valid MathML. However, essentially all converted documents contained XML errors, MathML errors or both, often systematic and in large numbers. Typically XML errors were unclosed elements, but also run-away attribute values, malformed tags, and other serious errors. MathML errors were usually incorrect child counts, e.g. an `<msup>` element with more or less than 2 child elements.

To address these issues, our indexing workflow begins by passing documents through an error correcting parser we developed that inserts missing end elements, quotation marks, etc. in order to produce well-formed XML. It then checks for MathML element counts, either wrapping extra arguments in merror tags or inserting empty merror elements as necessary. It also strips unrecognized elements and attributes, as well as MathML attributes whose values are not legal.

In addition, during this normalization pass, we implemented a number of ad hoc rules via regular expressions to correct particular systematic problems with the output of specific tools.

3.2   Character Normalization

Perhaps the most significant factor contributing to poor recall for our algorithm turned out to be differences in character data that made little or no visual difference. Unicode contains a number of characters that look like a minus sign, an absolute value bar, and so on. Consequently, a second critical normalization phase chose canonical representatives for each equivalence class of such characters.

3.3   Removal of White Space and Other Non-significant Data

Trivial differences in white space and other non-significant data gives rise to problems similar to the characters discrepancies of the preceding section. While white space can carry meaning in mathematics through alignment, etc., we considered it unlikely that a user would be successful in formulating effective queries to find such cases. On the other hand, the vast majority of white space differences we observed caused serious problems in basic searches, e.g. the presence or absence of a small amount of space before the differential in an integral. Consequently, our third normalization pass removes suspect white space constructs, such as `<mtext>` elements containing only whitespace, `<mspace>`, `<mphantom>`, and `<mpadded>` elements.

Though not technically white space, we also remove other similarly troublesome constructs in this pass, such as `<maction>` elements and semantic annotations. We also remove redundant `<mrow>`s, as in this example: `<mrow> <mrow> <mi> x </mi> </mrow></mrow>`.

3.4   Correction of Poor or Ambiguous MathML

The most technically challenging phase of normalization is the application of heuristic rules to improve poor or ambiguous MathML. This has several aspects. The first is fairly straightforward, and consists in the corrections of common errors in MathML coding that produce visually acceptable results using markup that is at odds with the intended mathematical structure.

A well-known example of such an error is attaching a superscript to a parenthesis, instead of the entire base expression to which it applies. Other common problems in converted material include a superscript encoded as an msubsup¿ with a dummy script, or even a <mmultiscript> with a single non-empty script. Another common case is a function where the MathML structure it at odds with the natural grouping of the function with its argument, eg. $f(z) = w$ where the $(z) = w$ is grouped in an <mrow>.

Beyond such relatively clear-cut problems, there are a number of mathematical notations that are commonly encoded in several different ways. The most important issue is the encoding of elementary functions such as sin x. There are at least three common coding conventions:

<mo>sin/mo>mi>x/mi>
<mi>sin/mi>mi>x/mi>
<mi>sin/mi>mo>ApplyFunction/mi>mi>x/mi>

While one may argue one is preferable to another, all are valid, so in these cases, our normalization algorithm merely picks a canonical representative.

Another important case where it is necessary to pick a canonical representation is decimal numbers. Both American and European conventions for the comma and decimal point are common, and one format or the other must be chosen. A second issue especially common in the output of translators is that each digit is separately tagged as an <mn> in numbers such as 123. The comma and decimal are similarly tagged as operators. For negative numbers, the minus sign may be separately tagged as an operator or be included in the CDATA of an <mn>.

Finally, it is very common for expressions to be sub-optimally structured from the point of view of the XML nesting structure reflecting the mathematical structure. One commonly encounters expressions such as as an unstructured row of characters. In other cases, the same expression will be encoded hierarchically with nested mrows¿ grouping arguments and operators. Since the nesting structure can have a large impact on subsequent normalization operations as well as n-gram formation, we apply heuristics to enrich sub-optimal MathML with additional structure. This involves several steps.

First, we attempt to pair fence operators, and refine the MathML structure by adding <mrow>s to group fenced terms. For parentheses, brackets, braces and other fence operators with a left and a right delimiter, a deterministic algorithm is possible, at least once occurrences of fence characters where they do not function as fences (e.g. in the interval notation $(-1, 0]$ and so on) have been noted. However, for notations such as absolute value bars, where the same character functions as both a right and a left fence character, genuinely ambiguous notations arise. In these situations, we apply heuristics to group terms. After fences have been disambiguated, we apply heuristics to distinguish between ambiguous multiplications and function applications, e.g. is a function applied to an argument, or a multiplication of terms.

3.5   Tree Refinement

The preceding normalizations pave the way for a critical tree refinement stage, where the MathML structure is refined so that no mrow¿ contains operators with more than one precedence level. In simple cases, the tree refinement algorithm reduces to standard operator precedence parsing. However, in order to be well defined on all MathML expressions, and to actually produce the canonical MathML structure that is our goal, it must be augmented in several ways.

First, it is necessary to assign precedences to the many hundreds of notations that function as an "operator" in terms of MathML coding, which includes many mathematical notations that are not commonly considered in traditional arithmetic precedence parsing. Second, the algorithm must be extended to accommodate the MathML notion of an embellished operator, typically an operator symbol that has been decorated with scripts or accents, etc. The outline of such an augmented precedence parser was first presented by N. Soiffer and B. Smith of Wolfram Research to the W3C Math Working group in 1996-7.

3.6   Synonyms

The normalization steps described in the preceding sections do a fairly good job of producing a normal form for most commonly encountered encodings of the vast majority of MathML expressions. While severely pathological MathML expressions still cause problems, our empirical results suggest this class of expressions is no more than a few tenths of a percentage of all expressions.

Where notational normalization leaves off, however, mathematical normalization begins, and the line is not always clear. For the practical and theoretical reasons enumerated in the introduction, we chose not to go down the path of attempting to identify mathematically similar expressions in the context of this project, and instead we have focused on notationally similarity for purposes of ranking search results. Nonetheless, after experimenting with our algorithm on a collection of test documents normalized as described above, pragmatism obliged us to make a few concessions toward mathematical equivalence. Consequently, in a sixth and final normalization pass, we select a canonical representative from equivalence classes of mathematical synonyms.

A very basic and widespread type of notational synonym consists in varying conventions for parenthesizing arguments of elementary functions, e.g. vs. . Similarly, the choice of language for elementary function names introduces additional sets of synonyms. Beyond that there are many notational synonyms from relatively simple ones, e.g. competing notations for permutations, to complex ones, such as notational synonyms for differentiation. To date, we have only implemented normalization for a few basic synonym classes, where empirical experiment suggested that recall for searches was perceived as particularly poor to most users without them. This is a very coarse, subjective criterion of course, and more sophisticated handling of synonyms is an obvious area of improvement for the future.

3.7 Variable Names

Another area that empirical experiments indicated as critical to user perception of search recall involves variable names. The question of when a variable name is significant is a subtle one, and involves both psychology and mathematical equivalence. A user looking for is quite likely interested in finding documents containing the expression . At the same time, as user looking for is less likely to want to see results like .

The approach we have taken is to define equivalence classes of variable names that are traditionally used for similar mathematical purposes across many areas of mathematics. Examples would be $\{i, j, k\}$ used as indicies, $\{s, t\}$ used as parameters, $\{f, g\}$ used as function names , and many others. At indexing time, we index an expression first using its original variable names, but then also in a separate index where variable names are abstracted as merely labeled instances of variables from one of these equivalence classes. That is, would be indexed as something like

```
function_var_1 ( parameter_var_1 +  parameter_var_2)
```

This enables use at query time to query for both exact and abstract variable names, with exact names naturally given much greater weight.

While this technique improves perceived search recall, it is definitely a blunt instrument. More sophisticated handling of the variable name problem is an area for future work. In particular, it is probably necessary to give users at least some control over variable name handling.

4 Our Evaluation

An ideal IR model should retrieve all relevant documents and only relevant documents, in the order of their relevance. Performance evaluation of an information retrieval model targeting a diverse user group with, different information needs is challenging. Evaluation methods should credit IR models that can retrieve highly relevant documents in the order of their relevance with appropriate rankings. Almost all the evaluation methods suffer from the problem of defining appropriate relevance levels. Binary relevance does not reflect the way humans judge the relevance of a document. Instead each document has a degree of relevance.

We adopted the method of Tetsuya Sakai's Average Gain Ratio[15] using multiple relevance levels. For all the evaluations, our n-gram model with equal weighting and zero threshold cut-off was used as the base model. Subsequent iterations of improvement in algorithms were informally evaluated against this base model to assess the performance. Queries with varying degree of mathematical complexity and length were selected for this purpose. For each of these queries, our team members assigned multiple relevance levels to the all the documents in the index.

In our informal trials, disjunction max scoring greatly increased the number of relevant documents in the top few as compared to using Boolean scoring. Selecting only those query terms with query weight greater than the minimum threshold filtered spurious documents with no relevance to the query. Also boosting certain query terms over others helped better capture the higher-level mathematical intent of queries. We also found that normalization substantially improved recall. Without normalization, recall was quite poor, due to the variety of ways in which equivalent expressions were encoded. Each of the normalization steps described in section 3 was found to improved recall.

While a careful, quantitative evaluation of precision and recall remains to be done, the following examples are suggestive of the current performance. In a collection of around 40,000 documents (see the next section) a query for the expression $s_1, \ldots, s_n \in \{-1, 1\}$ that is known to be in a document in the collections does not find that page in the top 20 results. No match rates higher than 2 stars on a scale of 0 to 5, with the top hits being lengthy formulas containing $\in \{-1, 1\}$ several times in one case, and $s_1, \ldots, s_n$ in the other. At the same time, a search for $T_{f(x)} L_{f(x)}$ returns the document in which the expression is known to reside as the top hit (it's 2.5 star ranking belying a normalization problem with the star scors). The second document rates half a star, and contains $df : T_x X \to T_{f(x)} Y$. This kind of variation in search performance from query to query is typical, and suggests further research and more careful evaluation is still needed.

5   The Mathdex Search Engine

We have implemented the search algorithm and normalization workflow described above as the Mathdex web application. Users enter mathematics query expressions via a graphical equation editor applet. Additional text query terms are entered via a standard HTML text box. Query results are presented to the user in a list, with a document synopsis and a 'best match' equation for each result document. The synopsis is prepared at indexing time by extracting significant phrases, based on term frequencies and document location, e.g. titles and headings are more likely to appear in the synopsis, as are sentences with using rare words or mathematical expressions that appear repeatedly in the document. The best match equation is selected similarly at indexing time, and displayed to the user via a mouseover area. The best match equation is displayed separately, because it may not appear in the synopsis, since in general the best match equation and surrounding text will not be a good choice for conveying what the document is about to the user. If there are multiple equations with the same 'best match' rank, only the first one is displayed currently.

Subject to copyright restrictions on the original document, users may also be able to view a cached copy of a document in the result set. In cache documents, all matching mathematical expressions are highlighted. This is accomplished by adding JavaScript code to the cache version of the document that searches the document content for expressions matching the query terms. Cache content is prepared in multiple versions, using images and MathML for the mathematics. Content negotiation is used to send highest functionality version supported by the user's browser.

At the time of this writing, Mathdex currently indexes around 25,000 documents from the arXiv[9], 12,000 pages from Wikipedia containing mathematics, approximately 1300 pages from Connexions[8], and around 1000 pages of Wolfram MathWorld[18]. We plan to expand the volume of content indexed significantly in the future.

References

1. Apache Foundation: Lucene Project, http://lucene.apache.org
2. Apache Foundation: Nutch Project, http://lucene.apache.org/nutch
3. Asperti, A., Guidi, F., Coen, C.S., Tassi, E., Zacchiroli, S.: A Content Based Mathematical Search Engine. In: Filliâtre, J.-C., Paulin-Mohring, C., Werner, B. (eds.) TYPES 2004. LNCS, vol. 3839, pp. 17–32. Springer, Heidelberg (2006)
4. Asperti, A., Selmi, M.: Efficient Retrieval of Mathematical Statements. In: Asperti, A., Bancerek, G., Trybulec, A. (eds.) MKM 2004. LNCS, vol. 3119, pp. 17–31. Springer, Heidelberg (2004)
5. Grzegorz, B.: Information Retrieval and Rendering with MML Query. In: Borwein, J.M., Farmer, W.M. (eds.) MKM 2006. LNCS (LNAI), vol. 4108, pp. 266–279. Springer, Heidelberg (2006)
6. Bancerek, G., Rudniki, P.: Information Retrieval in MML. In: Asperti, A., Buchberger, B., Davenport, J.H. (eds.) MKM 2003. LNCS, vol. 2594, pp. 119–132. Springer, Heidelberg (2003)
7. Cairns, P.: Informalising Formal Mathematics: searching the mizar library with Latent Semantics. In: Asperti, A., Bancerek, G., Trybulec, A. (eds.) MKM 2004. LNCS, vol. 3119, pp. 17–31. Springer, Heidelberg (2004)
8. Braniuk, R. et al.: Connexions, http://cnx.org
9. Cornell University Library: The arXiv, http://arxiv.org
10. Design Science, Mathdex, http://www.mathdex.com
11. Harvey, D.: blahtex, http://www.blahtex.org/
12. Miller, B.R., Youssef, A.: Technical Aspects of the Digital Library of Mathematical Functions. In: Annals of Mathematics and Artificial Intelligence, vol. 38(1-3), pp. 121–136. Springer, Netherlands (2003)
13. Miller, B.: DLMF, LaTeXML and some lessons learned. In: The Evolution of Mathematical Communication in the Age of Digital Libraries, IMA "Hot Topic" Workshop (2006) http://www.ima.umn.edu/2006-2007/SW12.8-9.06/abstracts.html#Miller-Bruce
14. Ogilvie, P., Callan, J.: Using Language models for flat text queries in XML retrieval. In: Proceedings of INEX 2003, pp. 12–18 (2003)
15. Tetsuya, S.: Average Gain Ratio: A Simple Retrieval Performance Measure for Evaluation with Multiple Relevance Levels, ACM SIGIR (2003)
16. Salton, G., Fox, E., Wu, H.: Extended Boolean Information Retrieval. Communication of the ACM 26(11), 1022–1036 (1983)
17. Trott, M.: Trott's Corner Mathematical Searching of The Wolfram Functions Site. The Mathematica Journal 9(4), 713–726 (2005)
18. Weisstein, E.: Wolfram MathWorld, http://mathworld.wolfram.com

Extended Formula Normalization for ϵ-Retrieval and Sharing of Mathematical Knowledge

Immanuel Normann and Michael Kohlhase

Computer Science, Jacobs University Bremen
{i.normann,m.kohlhase}@jacobs-university.de

Abstract. Even though only a tiny fraction of mathematical knowledge is available digitally (e.g in theorem prover or computer algebra libraries, in documents with content-markup), our current retrieval methods are already inadequate. With further increase in digitalization of mathematics the situation will get worse without significant advances.

When searching a formula, we often want to find not only structurally identical occurrences, but also all (logically) equivalent ones. Furthermore, we want to retrieve whole mathematical theories (i.e. objects with prescribed properties), and we want to find them irrespective of the nomenclature chosen in the respective formalization.

In this paper, we propose a normalization-based approach to mathematical formula and theory retrieval modulo an equivalence theory and concept renaming, and apply the proposed algorithm to end-user querying and knowledge sharing. We test the implementation by applying it to a first-order translation of the Mizar library.

1 Introduction

The last two decades have seen a slow but steady accumulation of formalizations (or at least content-representation) of mathematical knowledge. We have the MIZAR Mathematical library [Rud92, Miz] with over 40 000 theorems, definitions, and proofs, as well as the the PVS [ORS92, PVS], NuPRL [CAB+86, Nup], and Coq [Tea] libraries of comparable size. But the developments tend to be system-specific, non-interoperable, and redundant. Even inside a single library, it is often simpler to reprove a theorem than finding an equivalent or stronger one to reference. At the level of mathematical theories, the problem is aggravated, since making theories applicable usually involves renaming (or reinterpreting) vocabularies. In this area the field of mathematical knowledge management (MKM) has not yet delivered its initial promise, i.e. that an investment into formalization (or content markup) would yield improvements in automated management. The process of the (human) mathematical community, which is based on peer-review, communication, understanding and reformulating mathematical theories seems to deliver more theory-reuse than the MKM-based counterpart. We believe that to change this, we must solve the *knowledge retrieval* problem at the heart of finding applicable theorems and theory reuse. It is here that our current technology is inadequate. To find a theorem to refer to in a library we should not

M. Kauers et al. (Eds.): MKM/Calculemus 2007, LNAI 4573, pp. 356–370, 2007.

have to know its exact mathematical structure, or even its visual appearance[1].
Consider for instance three of many more possible variants to formalize "f is a
continuous function":

$$\forall \varepsilon.\varepsilon > 0 \Rightarrow \exists \delta.\forall x.\forall y.0 < |x - y| \wedge |x - y| < \delta \Rightarrow |f(x) - f(y)| < \varepsilon$$
$$\forall \varepsilon.\exists \delta.\forall x,y.\varepsilon > 0 \Rightarrow (0 < |x - y| \wedge |x - y| < \delta \Rightarrow |f(x) - f(y)| < \varepsilon) \quad (1)$$
$$\forall \varepsilon.\exists \delta.\forall x,y.\varepsilon > 0 \wedge |x - y| < \delta \wedge 0 < |x - y| \Rightarrow |f(x) - f(y)| < \varepsilon$$

A query for one should also find the others, since they are all equivalent. Given
an equality theory E, we speak of an **E-retrieval** task, if we want to find all
occurrences of formulae t in a collection that are E-equal to a query q. If the
theory E includes logical equivalence, we speak of an ε-retrieval task. Unfortunately semantical equivalence is undecidable in general, so we will have to make
due with search engines that satisfies the query intention approximatively. The
best approximation could be achieved by enlisting automated theorem provers
during search, but performance and termination problems make this approach
intractable, so we have to restrict our notion of equivalence. There is also another
reason to restrict equivalence during retrieval: under full logical equivalence, all
tautologies become equivalent, since they are all valid.

Rather than coming up with a theoretical solution of relaxed equivalence, we
will integrate approximation into the retrieval process itself to ensure efficiency.
We view logical equivalence as an equality theory and relax it by viewing two
formulae as equivalent, iff they can be co-normalized. Given a normalization
function ν, we say that the formulae are **ν-equivalent**, iff their ν normal forms
are syntactically identical. ν-equivalence provides an efficient and useful notion
of equivalence[2], efficiency in **$\nu\varepsilon$-retrieval** (i.e. ε-retrieval modulo ν-equivalence)
is achieved by normalizing the search corpus of the digital library at indexing
time. This moves the cost of ν-retrieval almost completely to the preprocess-
ing/indexing phase: normalizing the query formulae is cheap, since they are
usually rather small.

We will consider two applications of $\nu\varepsilon$-retrieval that need two slightly different forms of normalization in this paper. In $\nu\varepsilon$-retrieval for formula search
— e.g. to extend the MATHWEBSEARCH formula search engine [Kc06, Kc07]
— we are interested in a ν-retrieval of formulae with given constants (e.g. the
absolute value function in the formulae in (1)). If we want to $\nu\varepsilon$-retrieve whole
mathematical theories (i.e. objects with prescribed properties), we want to find
them irrespective of the nomenclature chosen in the respective formalization.
We will concentrate on the latter application in this paper, since it is the more
radical application and indicate where we need to make modifications to the
normalization process for the former.

An application of $\nu\varepsilon$-retrieval, which we will not cover in this paper is library
merging. Currently, formalized mathematical content is scattered across many

---

[1] Which can differ considerably, even if the structure is known. Consider for instance
the presentational variants $\forall x.e^x > 0$ as equivalent to $\forall t.\exp(t) > 0$.

[2] Semantical formula search machines as MBASE and MATHWEBSEARCH are optimized on speed performance by supporting only alphabetical renaming of bound
variables efficiently.

digital libraries which overlap considerably; e.g. one can find a basic Algebra concept like "group" in almost every digital math library. Merging these libraries would obviously improve the knowledge accessibility since redundant searches could be avoided. More importantly, however, knowledge could be enlarged if theory inclusions can be detected.

2   Formula Normalization for Equivalence

This section introduces all normalizers used in this paper. Recall that in term rewriting we say that a formula (or term) is in $\mathcal{R}$-*normal form* iff it is not reducible w.r.t. a given set $\mathcal{R}$ of rewrite rules. $\mathcal{R}$ is called *normalizing* iff every input formula has a $\mathcal{R}$-normal form. Note that normalizing rewrite rules do not necessarily produce unique normal forms as they depend on the application order. Any terminating choice of application order yields unique normal forms and thus a formula transformation mapping. Note furthermore, that any normalizing algorithm f is idempotent, i.e. $f(f(t)) = f(t)$ for any formula t. For the purposes of this paper, we need to generalize: We will call a formula transformation mapping a **normalizer** iff it is idempotent and preserves the semantics of formulae. Normalizer given by a set $\mathcal{R}$ of rewrite rules, are prime examples, but we will also encounter others below.

The methods described in this paper are largely independent of the choice of particular logic, we will presuppose a standard formulation of first-order logic with the quantifiers $\forall$ and $\exists$. We will use a kind of vector notation on quantification binders, where $\overline{x}$ stands for a sequence $x_1, \ldots, x_n$ of variables. Contrary to standard expositions of first-order logic, we will ruthlessly use numbers as variable-, constant-, function- and predicate symbols in the normalization process. These should be considered as purely presentational devices for an infinite, ordered supply of (traditional) symbols of the respective arities. Let us explicitly note that all methods presented in this paper work with typed and higher-order logics with little change. In extensional higher-order logics, where propositions can be embedded into terms more care must be taken to ensure that the embedded propositions are also normalized. Moreover, logics with sequential logical operators as used e.g. in PVS [ORS92, PVS] (where in a formula $A \wedge B$, the subformula B will never be evaluated and might even be ill-typed if A can be determined to be false) will need some adaption.

The most prominent normalizers involved here are the various transformations to negation-, prenex-, and conjunctive/disjunctive-normal form. Their rewrite rules can be found in introductory text books on logic. Next we introduce two straightforward but useful normalizers:

Merge Bindings. The *merge bindings normalizer* ∇_{mb} is defined by the rewrite rule $Q\overline{x}.Q\overline{y}.\varphi \rightarrow Q\overline{xy}.\varphi$ for both quantifiers $Q \in \{\forall, \exists\}$.

Sort Binders. The normalizer *sort binders* (∇_{sb}) is defined as procedure lexicographically sorting all the binders of a formula, e.g. $\nabla_{sb}(\forall z, w.\exists y, x, z.\phi) = \forall w, z.\exists x, y, z.\phi$.

The following sections contain some normalizers that — even though they seem simple — seem not to have been discussed in the literature yet even though AC-normalization must have been considered (and rejected) for AC-unification. Incidentally, neither of the solutions presented here were found in a recent related article about "Semantic Matching for Mathematical Services" [NP05]. At least the *binding-last* strategy is an improvement over the folklore solution.

2.1 α-Normalization

An α-**normalizer** is a formula transformation mapping that replaces each non-logical symbol occurrence in a formula φ by a number where different occurrences of equal variables are substituted by the same number, but different variables by different numbers. Obviously this procedure is idempotent and always terminating. Moreover renaming bound variables does not change the semantics of a formula: input and output formulae are said to be α-**equivalent**. Renaming free variables or constants does not change the semantics of a formula provided that we rename consistently all involved free variables and constants of the context where the formula lives (cf. Section 3). Depending on the application, the notion of what constitutes a context may change, and we assume a representation that makes contexts explicit, e.g. as nested contexts in the Coq libraries [Tea] or in OMDoc documents [Koh06]. For our theory-retrieval application, the context will be a theory.

Let ∇_{as} be the α-normalizer induced by a depth-first but binding-last order. The latter means that the numbering of binding occurrences is executed after the numbering of the formula body is finished. For instance take the formula $\forall x, y. R(f(x, y), x)$ then the procedure is as follows:

1. The depth-first walk through the body of the formula builds the renaming $\sigma = [R \to 1, f \to 2, x \to 3, y \to 4]$ and simultaneously applies it to the body which yields $\forall x, y.\underline{1(2(3, 4), 3)}$.
2. The renaming σ is $\overline{\text{applied to}}$ the binding occurrences: $\forall \underline{3, 4}.1(2(3, 4), 3)$

As a renaming of α-normalization is bijective we can always easily reconstruct the original formula. To make the explicit notation of renaming more convenient we make use of this bijectivity: Instead of $\sigma = [v_1 \to 1, \dots, v_n \to n]$ we will just write $\sigma = [v_1, \dots, v_n]$.

Binding-last is Superior to Binding-first Numbering. Naive α-normalization does binding-first numbering; i.e. the numbering of the variables is determined by the order in binding-occurrence. To see the advantage of the binding-last strategy, consider the following equivalent formulae $t_1, \dots, t_4$:

$$\forall x, y. R(x, y) \approx \forall x, y. R(y, x) \approx \forall y, x. R(x, y) \approx \forall y, x. R(y, x)$$

If we apply binding-first numbering and ∇_{sb} to them, we obtain two normal forms irrespective of the order — e.g.

$$\forall 1, 2.3(1, 2) \text{ for } t_1 \text{ and } t_3 \quad \text{and} \quad \forall 1, 2.3(2, 1) \text{ for } t_2 \text{ and } t_4$$

for ∇_{sb} before numbering. But $\nabla_{sb}(\nabla_{as}(t)) = \forall 2, 3.1(2, 3)$ for all four formulae!

2.2   Normalization for ACI Operators

Consider two applications $f(x_1, \ldots, x_n)$ and $f(y_1, \ldots, y_n)$ where f is an associative, commutative and idempotent (ACI) operator, e.g. the binary logical operators, set union, or set intersection. For ACI-retrieval the task is to find, a permutation π for the arguments $y_1, \ldots, y_n$ such that $x_i = y_{\pi(i)}$ for $i = 1, \ldots, n$. Matching such expressions cause combinatorial explosion if it is done naively. In fact this is one of the weak point of almost all math search engines known to the authors ([DGH96] being a notable exception).

For ACI normalization the respective expression is translated into an object of nested symbol tuples/sets: Let A be an ACI and N be an non-ACI operator; c an atomic and φ_i any kind of expression. Then we translate inductively:

$$c \longmapsto c$$
$$A(\varphi_1, \ldots, \varphi_n) \longmapsto \langle A, \{\varphi_1, \ldots, \varphi_n\}\rangle$$
$$N(\varphi_1, \ldots, \varphi_n) \longmapsto \langle N, \langle\varphi_1, \ldots, \varphi_n\rangle\rangle$$

Obviously every result expression can be uniquely translated back; i.e. the translation is an isomorphic representation of the original formulae. For the following let us call them ACI representations. For instance $(a \wedge b) \vee (b \wedge c) \Rightarrow (a \vee c)$ (in prefix notation $\Rightarrow (\vee(\wedge(a, b), \wedge(b, c)), \vee(a, c)))$ translates to

$$\langle\Rightarrow, \langle\vee, \{\langle\wedge, \{a, b\}\rangle, \langle\wedge, b, c\rangle\}, \langle\vee, \{a, c\}\rangle\rangle\rangle.$$

We will now define an **ACI normalizer** ∇_{aci} that normalizes the ACI representation of a given expression φ. During the ACI normalization an intermediate ACI expression φ_N and a set of renamings Σ is computed with $\sigma(\pi\varphi) = \varphi_N$ for all $\sigma \in \Sigma$ and an appropriate and semantics preserving permutation π of symbols in φ. Our ACI normal forms after all will be each of $\sigma\varphi$. Most importantly this φ_N will be the minimal expression with respect to a given term ordering. In fact the choice of the term ordering is irrelevant. It just needs to be fixed once, so without loss of generality we use to the following, inductively defined term ordering:

- numbers are ordered as usual, but symbols are considered to be equal.
- A number is smaller than a symbol,
- a symbol is smaller than a tuple,
- for two tuples that with fewer arguments is the smaller
- for two tuples of same arity that tuple is smaller whose first component (going from left to right) is smaller than that of the other tuple, e.g. $\langle 1, 2, \{a, b\}\rangle < \langle 1, 3, 4\rangle$ since $2 < 3$ in the second component (the tail of the tuple doesn't matter).
- a tuple is smaller than a set.
- the comparison between set works similar to that of tuples when sets are represented as ordered tuples.

Note, since symbols are considered to be equal the term ordering of ACI expression is not total in general. However, the set of all ACI expressions where symbols are replaced by numbers has a total term ordering (since numbers are totally ordered whereas symbols aren't). This is important, because we are looking for the unique representative ACI expression.

The ACI normalization of $\langle \Rightarrow, \langle \vee, \{\langle \wedge, \{a, b\}\rangle, \langle \wedge, b, c\rangle\}, \langle \vee, \{a, c\}\rangle\rangle\rangle$ from our above example returns the normalized ACI expression

$$\langle 0, \langle 1, \{\langle 2, \{3, 4\}\rangle, \langle 2, \{3, 5\}\rangle\}, \langle 1, \{4, 5\}\rangle\rangle\rangle$$

together with these two renamings

$$[imp, \vee, \wedge, b, a, c], \qquad [imp, \vee, \wedge, b, c, a]$$

We now want to sketch our algorithm of ACI normalization: The basic idea is to replace all symbols of the input expression stepwise by numbers augmenting the renamings in parallel with those symbols just removed from the expression. The key idea for that is to remove the next "minimal" symbol from the expression and replace it by a number determined by the current renaming: If this symbol is a member of a renaming then return the corresponding number (see the convention introduced in Section 2.1) and otherwise append the symbol to the renaming and return the corresponding new number.

Let us fortify our intuition with an example and use *prox* for the procedure which fetches all "minimal" symbols from a given expression. Let s be a symbol, n be a number, φ_i be an arbitrary ACI expression then we define inductively:

$$\text{prox}(s) = \{s\} \quad \text{prox}(n) = \emptyset$$

$$\text{prox}(\langle \varphi_1, \varphi_2 \ldots, \varphi_n\rangle) = \begin{cases} \text{prox}(\varphi_1) & \text{if } \varphi_1 \text{ contains symbols} \\ \text{prox}(\langle \varphi_2 \ldots, \varphi_n\rangle) & \text{otherwise} \end{cases}$$

$$\text{prox}(\{\varphi_1 \ldots, \varphi_n\}) = \text{prox}(\varphi_1) \cup \ldots \cup \text{prox}(\varphi_n)$$

For instance *prox* applied on $\{\langle 1, a\rangle, \langle 1, b\rangle\}$ evaluates to:

$$\text{prox}(\{\langle 1, a\rangle, \langle 1, b\rangle\}) \mapsto \text{prox}(\langle 1, a\rangle) \cup \text{prox}(\langle 1, b\rangle) \mapsto \text{prox}(a) \cup \text{prox}(b) \mapsto \{a, b\}$$

The procedure $\min_\sigma$ takes an expression and a set of renamings and filters all renamings that make this expression minimal if applied. For instance $\min_\sigma$ applied on the expression $\varphi := \langle \{a, b\}, \{b, c\}\rangle$ and the renamings $[a, b]$ and $[a, c]$ would return only $[a, b]$. To understand this we have to apply them on φ: $[a, b]\varphi = \langle \{1, 2\}, \{2, c\}\rangle$ and $[a, c]\varphi = \langle \{1, b\}, \{2, c\}\rangle$. The second result expression is greater than the first one. Hence only $[a, b]$ makes φ minimal.

At last we need the $\oplus$ operator that appends a symbol to a renaming; i.e. $[s_1, \ldots, s_n] \oplus s_{n+1} = [s_1, \ldots, s_n, s_{n+1}]$. We extend its definition to arrays of renamings and arrays of symbols by applying $\oplus$ on those components.

With these procedures we write an ACI algorithm as recursive function ∇_{aci} terminating on a fixpoint:

$$\nabla_{\text{aci}}(\Sigma, \varphi) := \begin{cases} \Sigma\varphi & \text{if } \varphi = \nabla_{\text{aci}}(\Sigma, \varphi) \\ \min_\sigma (\Sigma \oplus \text{prox}(\Sigma\varphi))(\varphi) & \text{otherwise} \end{cases}$$

Hereby φ is the input ACI expression and Σ a set of renamings evolving along the recursion. Initially Σ is the empty set. The algorithm always terminates since the fixpoint is reached as soon as all symbols of the initial expression φ are replaced by numbers and each recursion step replaces at least one symbol by a number.

For a better understanding we demonstrate the algorithm on the example input $\varphi := \langle \{a, b, c\}, \{a, c, d\} \rangle$:

$$\mathrm{prox}(\Sigma\varphi) = \{a, b, c\}$$

$$\Sigma \oplus \mathrm{prox}(\Sigma\varphi) = ([a], [b], [c])$$

$$(\Sigma \oplus \mathrm{prox}(\Sigma\varphi))(\varphi) = \begin{pmatrix} \langle \{1, b, c\}, \{1, c, d\} \rangle \\ \langle \{a, 1, c\}, \{a, c, d\} \rangle \\ \langle \{a, b, 1\}, \{a, 1, d\} \rangle \end{pmatrix}$$

$$\Sigma := \min_{\sigma} (\Sigma \oplus \mathrm{prox}(\Sigma\varphi))(\varphi) = ([a], [c]) \qquad\qquad (1.\text{recursion})$$

$$\mathrm{prox}(\Sigma\varphi) = (\{b, c\}, \{a, b\})$$

$$(\Sigma \oplus \mathrm{prox}(\Sigma\varphi)) = ([a, b], [a, c], [c, a], [c, b])$$

$$(\Sigma \oplus \mathrm{prox}(\Sigma\varphi))(\varphi) = \begin{pmatrix} \langle \{1, 2, c\}, \{1, c, d\} \rangle \\ \langle \{1, b, 2\}, \{1, 2, d\} \rangle \\ \langle \{2, b, 1\}, \{2, 1, d\} \rangle \\ \langle \{a, 2, 1\}, \{a, 1, d\} \rangle \end{pmatrix}$$

$$\Sigma := \min_{\sigma} (\Sigma \oplus \mathrm{prox}(\Sigma\varphi))(\varphi) = ([a, c], [c, a]) \qquad\qquad (2.\text{recursion})$$

$$\vdots$$

$$\Sigma := \min_{\sigma} (\Sigma \oplus \mathrm{prox}(\Sigma\varphi))(\varphi) = ([a, c, b], [c, a, b]) \qquad\qquad (3.\text{recursion})$$

$$\vdots$$

$$\Sigma := \min_{\sigma} (\Sigma \oplus \mathrm{prox}(\Sigma\varphi))(\varphi) = ([a, c, b, d], [c, a, b, d]) \qquad\qquad (4.\text{recursion})$$

So $([a, c, b, d], [c, a, b, d])$ are exactly those renamings which minimize φ (i.e. our φ_N mentioned at the beginning of this section) namely to $\langle \{1, 2, 3\}, \{1, 2, 4\} \rangle$. Hence we have two normal forms: $\langle \{a, c, b\}, \{a, c, d\} \rangle$ and $\langle \{c, a, b\}, \{c, a, d\} \rangle$.

With a slight modification of the ACI normalization algorithm we can also handle AC normalization (i.e. without I). For that we simply have to replace sets in ACI expressions by multisets. Most prominent candidate expressions for AC normalizations are addition and multiplication.

Finally it should be mentioned that AC(I) normalization as introduced here doesn't allow for a more efficient term matching in the worst case than the naive approach does meaning testing all permutations. Such a worst case is an expression where none of its AC(I) subexpressions share common symbols. In practice, however, this is rather the exception as our experiments with the Mizar library shows (s.section 4).

2.3 Concatenating Normalizers to an Overall Normalizer

Now let ∇ be the normalizer obtained by chaining all the normalizers discussed so far in the following order: elimination of $\Rightarrow$ and $\Leftrightarrow$, negation normal form, prenex normal form, and conjunctive normal form, merge binding, ACI normalization, formula abstraction, and sort binding. This order of normalizers is empirically optimal in the sense that each normal form of this overall normalization would represent a maximal equivalence class of formulae; i.e. every other order of normalizers would yield smaller equivalence classes or at most of equal size.

Finally an illustrative example demonstrates the normalization process on some example formula:

	$\forall x.\forall z.R(x,z) \Rightarrow \exists y.(\forall w.Q(y,w) \Rightarrow R(x,y)) \wedge R(y,z))$
$\Rightarrow$-elim	$\forall x.\forall z.\neg R(x,z) \vee \exists y.(\forall w.\neg Q(y,w) \vee R(x,y)) \wedge R(y,z))$
prenex form	$\forall x.\forall z.\exists y.\forall w.\neg R(x,z) \vee ((\neg Q(y,w) \wedge R(x,y)) \vee R(y,z))$
CNF	$\forall x.\forall z.\exists y.\forall w.(\neg R(x,z) \vee \neg Q(y,w) \vee R(x,y)) \wedge (\neg R(x,z) \vee R(y,z))$
merge binding	$\forall x,z.\exists y.\forall w.(\neg R(x,z) \vee \neg Q(y,w) \vee R(x,y)) \wedge (\neg R(x,z) \vee R(y,z))$
∇_{aci}	$\forall x,z.\exists y.\forall w.(R(y,z) \vee \neg R(x,z)) \wedge (R(x,y) \vee \neg R(x,z) \vee \neg Q(y,w))$

The initial and the final formula are equivalent as each normalization step preserves the semantics.

Up to now we have considered equivalence transformations whereby the constants stay the same eventually. For a matching modulo renaming of constants we need a different representation of normal forms gained by the final step which we call **formula abstraction**. This normalization step returns for each input formula φ a pair $(\hat{\varphi}, \overline{p})$ which we call the skeleton and the parameter of φ. The parameter represent the constants of φ in the order from left to right as they occur at first in φ. The skeleton is φ after replacing all its constants by placeholders and a subsequent α-normalization. For instance the formula abstraction of $\forall x.R(x, f(x))$ returns $\forall 1.0(1, 0(1))$ as skeleton and $[R, f]$ as parameter.

Formula abstraction followed by sort binding are finally the last steps of the overall normalizations. In the remainder of this paper we assume a formula being completely normalized through all these steps when we talk of its skeleton and parameter respectively.

3 An Illustrative Example for Knowledge Sharing

Let us now see how the $\nu\epsilon$-retrieval can be used for knowledge sharing and expansion by detection of theory inclusions. The basic idea behind theory inclusion is to find a signature morphism between theories such that the axioms of the translated source theory are theorems in the target theory. For that we normalize all statements of the target theory and all axioms of the source theory. If we can find for each skeleton of the source theory's axioms a syntactically identical statement skeleton from the target theory then we try to find a consistent mapping between the parameter of the source theory and those of the target theory.

If we can find such a mapping between parameters we have found a signature morphism which allows for theory inclusion.

An elementary example from arithmetic should demonstrate how this works in principle. In contrast to the experiments on the MIZAR library reported in section 4, this is only a contrived example aimed at illustrating the process.

For the greatest common divisor (gcd) and the least common multiple (lcm) one finds the properties that both gcd and lcm are associative and commutative. Moreover we have the dual absorption properties:

$$\forall a, b. \gcd(a, \operatorname{lcm}(a, b)) = a \qquad \forall a, b. \operatorname{lcm}(a, \gcd(a, b)) = a$$

In lattice theory we have two operators $\sqcup$, $\sqcap$ and their associativity and commutativity as axioms as well as absorption. Assume that the absorption axioms are formalized by some author as follows:

$$\forall x. \forall y. x = (y \sqcap x) \sqcup x \qquad \forall x. \forall y. x = (y \sqcup x) \sqcap x$$

All absorption formulae, from arithmetic and from lattice theory, would normalize to a single skeleton $\varphi := \forall 1, 4.1 = 2(1, 3(1, 4)))$ due to the commutativity property of all involved operators. The corresponding parameters are:

$$p_1 = [\gcd, \operatorname{lcm}] \quad p_2 = [\operatorname{lcm}, \gcd] \quad p_3 = [\sqcup, \sqcap] \quad p_4 = [\sqcap, \sqcup]$$

A consistent mapping from the parameters of lattice theory to those of arithmetic is for instance:

$$\sigma := p_1 \circ p_3 = p_2 \circ p_4 = [\sqcap \to \gcd, \sqcup \to \operatorname{lcm}]$$

Thus we translated with σ the axioms of the lattice theory into theorems of arithmetic. The gained knowledge expansion is that all theorems from lattice theory are also theorems in arithmetic after applying this translation. Note that this translation is found via normalization that goes beyond simple α-equivalence, as it also takes the ACI properties of the operators into account.

4   Experiments on a Real World Math Library

MIZAR is a representation format for mathematics that is close to mathematical vernacular used in publications and a deduction system for verifying proofs in the MIZAR language. The continual development of the MIZAR system has resulted a centrally maintained library of mathematics (the MIZAR mathematical library MML [Miz]). The MML is a collection of MIZAR articles: text-files that contain theorems and definitions, and proofs. Currently the MML (version 4.76.959) contains 959 articles with 43149 theorems and 8185 definitions. Introductory information on MIZAR and the MML can be found in [RST01, Wie99]. The MIZAR language is based on Tarski-Grothendiek set theory [Try90], it is essentially a first-order logic with an extremely expressive type systems that features dependent types as well as predicate restrictions, see [Ban03] for details.

In our experiments we don't operate on the MML in its original format, but in its translated equivalent in first order logic format [Urb03], which contains 1530811 formulae distributed over 12529 files. The main reason for using this version is that for our experiment we need a simple notion of theory, namely as a set of axioms (for the purposes of this paper we don't differentiate between definitions and axioms), which constitute the theory, and a set of theorems, which were derived from them. In the original MIZAR format this simple notion is hidden behind combinations of very MIZAR specific notions such as "article", "vocabulary", "notation", "cluster", etc. To map these notions into our simple notion of theory would need a quite deep understanding of the MIZAR system. A structurally faithful translation of the MML into the OMDoc format is currently under way [BK07], and we will rerun our experiments on that and compare the results.

The translated MIZAR library is represented in DFG syntax [HKW96], a first order syntax that was designed to be easily parsed. Its concepts map straightforwardly to our needs: The main object of a file in DFG syntax is a "problem" which we interpret as a "theory". A problem contains arbitrarily many formulae being either of the type "axiom" or "conjecture". What is called conjecture there in DFG syntax corresponds to our theorems as we assume that all conjectures of the MML in DFG syntax are already proven and thus can be called theorems.

4.1 MML and Its Export in DFG-Syntax

We will need to review the basics of Josef Urban's translation of the MML to understand the retrieval experiment: Each object of MML gets a context-independent name and all types and properties (e.g. commutativity, transitivity, etc.) are translated into one or more first order formulae. Formulae are relativized with respect to the typed variables occurring in them. For instance the MIZAR expression

```
for x being Real holds x-x = 0
```

translates to

```
forall([x], implies(v1_arytm(x), equal(k3_real_1(x,x),0))).
```

where v1_arytm(x) encodes the type information x being Real. The translation approach leads to specific artifacts worth mentioning for our experiments:

- First of all the "one theorem = one (self-contained) problem" principle induces a considerable amount of redundant axiom repetitions in the assumptions.
- The translation from types and type hierarchies to first order formulae causes another blow up of formulae.
- Even worse many of these type translations result in redundant tautologies like forall([x], implies(and(true,v1_arytm(x)),true)). Since the basic MIZAR type set translates to true all these tautologies result probably from types involving set.
- Some formulae, which have less than 20 subterms in the original MML format, transform to monster formulae with over 800 subterms.

To cope with these we had to adapt our initial normalization procedure:

- Normalizing, in particular to CNF, becomes to expensive for large formulae. We decided to exclude formulae with more than 100 subterms from normalization. Such very large formulae wouldn't occur in handwritten math libraries anyway.
- Formulae where an associative and commutative operator have more than 10 arguments are excluded from AC normalization. Again such formulae are rather unlikely in hand written libraries - even after building the CNF.
- The frequent existence of `true` as part of formulae suggested an additional normalization step to eliminate all these `true` occurrences. For instance a formula like `implies(true,and(true,r(x)))` reduces to `r(x)`. Moreover all axioms which reduce to a single `true` are excluded from the insertion into the database since they don't influence the semantic of a theory.
- The `true` elimination normalization step, however, induced the subsequent artifact of con- and disjunctions with only one argument, which were handled the obvious way; e.g. `and(true,r(x))` reduces to `and(r(x))` and finally to `r(x)`

As a statistical result of the database insertion process we found that the original amount of 1530811 formulae can be reduced to 1416653 formulae due to elimination of 114158 tautologies. 18472 of these 1416653 were of that said large size that they were excluded from normalization. With the exclusion of large formula, normalization ran in about two hours on a contemporary PC, which is acceptable for an indexing-time step.

The most interesting number in this phase, however, is the ratio between the number of original formulae vs the number of skeletons. We consider this as a measure of redundancy or the other way round as indicator for potential theory reuse. It turns out that these 1416653 original formulae are instances of just 18155 skeletons; i.e. about every 80 formulae share the same skeleton. It must be said, however, that this factor significantly relies on the "one theorem = one (self-contained) problem" principle (see above). Hence a deeper analysis is needed to find out which percentage of formulae with common skeletons are not just simple copies generated by that principle.

4.2 THEOSCRUTOR: A Knowledge Base Architecture for Normalized Formulae

THEOSCRUTOR, our implementation of a theory search engine, has a simple architecture given in Figure 1. For indexing (i.e. initialization of the database) all files of the MML in DFG syntax are fed to a parser and then normalized and finally all normalized formulae are inserted into a MySQL database. The parser and the normalizer are implemented in Haskell; a database record contains a *file name* and *line number* referencing the occurrence of the normalized formula, its *skeleton*, and its *parameter*.

To query theory inclusions of a source theory S this theory is fed as file in DFG syntax to the request process which consists of the same parse and

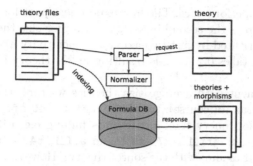

Fig. 1. Architecture of the theory inclusion search engine THEOSCRUTOR

normalization steps as the indexing process. The response of this request is a list of target theories T_i together with a set of signature morphisms $\{\sigma_j\}_{i,j}$. Each pair (i,j) of theory and signature morphism represents a theory inclusions as described in section 3. As additional information to each signature morphism a mapping between source axioms and target axioms or theorems is attached.

4.3 Querying with THEOSCRUTOR

To test the performance of THEOSCRUTOR, we ran various test queries on the database. The first kind of test is a single formula query. This can be considered as special theory inclusion query where the source theory has only one axiom. The test is relevant, because it allows for a performance comparison to the MATHWEBSEARCH even though there are differences: MATHWEBSEARCH supports subterm instantiation queries (i.e given a term t, it returns all subterms t' of formulae with $\sigma(t') = t$), and we support ν-equivalence. Hence the search result sets can't be the same in general. We are actually comparing ν-retrieval using database indexing technology with term matching using term indexing technology in our test. Moreover the MATHWEBSEARCH corpus on which the test runs contains 77000 formulae.

We took as an example query the theorem from the file `aff_1__t40_aff_1` of the DFG syntax version of MML : $\forall.a, b, c.\neg(f(a) \wedge (\neg g(a)(h(a) \wedge l(a)))) \wedge (m(b, u(a)) \wedge m(c, u(a)) \Rightarrow m(r(a, b, c), s(u(a)))$. As query request the theorem is interpreted as axiom. Our search engine needs 200 ms to retrieve from 175 theory inclusions with 88 target theories. Very wide spread formulae are more expensive in search time of course, but still acceptable. As witness query the law of commutativity returns 11502 theory inclusions belonging to 4447 target theories within just 6.4 seconds. An analogous[3] query of commutativity with MATHWEBSEARCH took 0.9 seconds returning.

Theory inclusion queries with multiple formulae is actually a distinctive feature of THEOSCRUTOR, though it can partially be simulated by document-scoped Boolean queries in MATHWEBSEARCH. Some experiments should give an idea

---

[3] To simulate the formula abstraction, we replaced symbols with query variables.

of THEOSCRUTOR performance. The first experiment puts focus on a very small source theory whereas the second investigates average size theories: The theory of monoids is constituted by just two axioms: associativity and neutral element. Querying monoids takes 0.2 seconds resulting in 291 theory inclusions with 141 target theories.

For the experiment with average size theories we took theories from MML itself (in DFG syntax of course): the theory aff_1__t40_aff_1 with 40 axioms. The theory inclusion query took 1.6 seconds finding e.g. different 192 theory inclusions with targets aff_1__t37_aff_1 and aff_1__t47_aff_1 (both theories basically share their axioms with the source theory). However, due to the artifact that every theory contains only one theorem, only three different theorem reuses are gained from these 192 theory inclusions.

5 Conclusion and Future Work

We have proposed a normalization-based approach to mathematical formula and theory retrieval modulo an equivalence theory and concept renaming. Concretely, we have developed a waterfall of normalizers that empirically maximize the ϵ-equivalence classes, while keeping normalization tractable during search index creation. One of the strengths of the normalization-based approach is that we can adopt a flexible notion of scope of constant renaming, allowing to tailor the method not only to $\nu\epsilon$-retrieval, but also to theory retrieval. The former is a user-level task for the working mathematician, where constants should keep their meaning, whereas the latter is a knowledge-engineering task for a library maintainer, where constants must be open to renaming for re-interpretation in different contexts.

With THEOSCRUTOR we have an implementation of the proposed approach. We tested it on a real-life task: the MIZAR library and shown the steps involved to be tractable (after some practical adaptations).

One may object that the theory inclusions found by our system are relatively trivial from a mathematicians perspective. This is not surprising since normalization is essentially based on pure logical equivalence transformation — sophisticated proofs as mathematicians appreciate are not involved. However, this perspective neglects an important aspect of our original goal, namely to improve the accessibility of knowledge in large digital libraries. Whether a theory inclusion is trivial or not from a mathematicians point of view is secondary if our goal is to expand our knowledge base. Moreover what is folklore to one mathematician in one research area is sometimes completely unknown to another mathematician from a different area and certainly to a mathematically interested layman too.

The strength of automated detection of theory inclusion via normalization is the ability of scanning masses of formulae. Mathematicians are unsurpassable in their dedicated field, but machines are good in precision and mass processing - they can discover useful things which are simply overlooked by humans.

We have concentrated on normalization-based ϵ retrieval in this paper, as the normalization properties of logical connectives and quantifiers are given by the base logic. The normalization could be extended by normalizers for constants that are not abstracted over. To enable this, THEOSCRUTOR would have to scan the source theory axioms for e.g. the statements of the ACI properties of addition. Currently, this is beyond the scope of our implementation.

We will re-run our normalization experiments on the structurally faithful OM-Doc translation of the MML currently under way [BK07], and compare the results with the first-order version. It would also be interesting to experiment with other CNF transformations as normalizers, e.g. the very powerful FLOT-TER [WGR96] implementation.

References

[Ban03]   Bancerek, G.: On the structure of Mizar types. Electronic Notes in Theo-
           retical Computer Science, 85(7) (2003)
[BK07]   Bancerek, G., Kohlhase, M.: The mizar mathematical library in omdoc
           (submitted 2007)
[CAB$^+$86] Robert, L., Constable, S., Allen, H., Bromly, W., Cleaveland, J., Cremer,
           R., Harper, D., Howe, T., Knoblock, N., Mendler, P., Panangaden, J.,
           Sasaki, J., Smith, S.: Implementing Mathematics with the Nuprl Proof
           Development System. Prentice-Hall, Englewood Cliffs (1986)
[DGH96]   Dalmas, S., Gaëtano, M., Huchet, C.: A deductive database for mathemat-
           ical formulas. In: Limongelli, C., Calmet, J. (eds.) DISCO 1996. LNCS,
           vol. 1128, pp. 287–296. Springer, Heidelberg (1996)
[HKW96]   Hähnle, R., Kerber, M., Weidenbach, C.: Common syntax of dfg-
           schwerpunktprogramm "deduktion". Interner Bericht 10/96, Universität
           Karlsruhe, Fakultät für Informatik (1996)
[Kc06]   Kohlhase, M., Şucan, I.: A search engine for mathematical formulae. In:
           Calmet, J., Ida, T., Wang, D. (eds.) AISC 2006. LNCS (LNAI), vol. 4120,
           pp. 241–253. Springer, Heidelberg (2006)
[Kc07]   Kohlhase, M., Şucan, I.: System description: MathWebSearch 0.3, a se-
           mantic search engine. submitted to CADE 21 (2007)
[Koh06]   Kohlhase, M.: OMDoc – An Open Markup Format for Mathematical Doc-
           uments [version 1.2]. LNCS (LNAI), vol. 4180. Springer, Heidelberg (2006)
[Miz]
[NP05]   Naylor, W., Padget, J.A.: Semantic matching for mathematical services.
           In: Kohlhase, M. (ed.) MKM 2005. LNCS (LNAI), vol. 3863, pp. 174–189.
           Springer, Heidelberg (2006)
[Nup]   The NuPrl online theory library. Internet interface at
           http://simon.cs.cornell.edu/Info/Projects/NuPrl/Nuprl4.2/
           Libraries/Welc.html
[ORS92]   Owre, S., Rushby, J.M., Shankar, N.: PVS: a prototype verification system.
           In: Kapur, D. (ed.) Automated Deduction - CADE-11. LNCS, vol. 607, pp.
           748–752. Springer, Heidelberg (1992)
[PVS]   Pvs libraries. http://pvs.csl.sri.com/libraries.html
[RST01]   Rudnicki, P., Schwarzweller, C., Trybulec, A.: Commutative algebra in the
           Mizar system. Journal of Symbolic Computation 32, 143–169 (2001)

[Rud92]   Rudnicki, P.: An overview of the mizar project. In: Proceedings of the 1992
            Workshop on Types and Proofs as Programs, pp. 311–332 (1992)
[Tea]   Coq Development Team. The Coq Proof Assistant Reference Manual. IN-
            RIA. see, http://coq.inria.fr/doc/main.html
[Try90]   Trybulec, A.: Tarski Grothendieck set theory. Formalized Mathemat-
            ics 1(1), 9–11 (1990)
[Urb03]   Urban, J.: Translating mizar for first-order theorem provers. In: Asperti,
            A., Buchberger, B., Davenport, J.H. (eds.) MKM 2003. LNCS, vol. 2594,
            pp. 203–215. Springer, Heidelberg (2003)
[WGR96]   Weidenbach, C., Gaede, B., Rock, G.: Spass & flotter, version 0.42. In:
            McRobbie, M.A., Slaney, J.K. (eds.) Automated Deduction - Cade-13.
            LNCS, vol. 1104, Springer, Heidelberg (1996)
[Wie99]   Wiedijk, F.: Mizar: An impression (1999)
            http://www.cs.kun.nl/~freek/notes

Towards Mathematical Knowledge Management for Electrical Engineering

Agnieszka Rowinska-Schwarzweller[1] and Christoph Schwarzweller[2]

[1] Chair of Display Technology, University of Stuttgart
Allmandring 3b, 70569 Stuttgart, Germany
schwarzweller@lfb.uni-stuttgart.de
[2] Department of Computer Science, University of Gdańsk
ul. Wita Stwosza 57, 80-952 Gdańsk, Poland
schwarzw@math.univ.gda.pl

Abstract. We explore mathematical knowledge in the field of electrical engineering and claim that electrical engineering is a suitable area of application for mathematical knowledge management: We show that mathematical knowledge arising in electrical engineering can be successfully handled by existing MKM systems, namely by the Mizar system. To this end we consider in this paper network theory and in particular stability of networks. As an example for mathematical knowledge in electrical engineering we present a Mizar formalization of Schur's theorem. Schur's theorem provides a recursive, easy method to check for BIBO-stability of networks.

1 Introduction

The aim of mathematical knowledge management is to provide both tools and infrastructure supporting the organization, development, and teaching of mathematics with the help of effective up-to-date computer technologies. To achieve this ambitious goal it should be taken into account that the predominant part of potential users will not be professional mathematicians themselves, but rather scientists or teachers that apply mathematics in their area. This point has been adressed lately with the consideration of physics [HKS06] or geo-sciences [Ses07]. In this paper we inspect another application area for mathematical knowledge management: electrical engineering.

The situation of mathematics in electrical engineering is — as in other engineering sciences — twofold. On the one hand there is a number of areas, such as for example network theory, control engineering or filter design, based on clean mathematical fundamentals and results. On the other hand, however, even in these areas the newest developments often do not rely on these results. Electrical engineers essentially use systems like MathLab or Maple providing a convenient environment to accomplish their applications. These systems, however, do not provide mathematical exactness for the verification of results nor include the newest theoretical results from the area. Consequently, knowledge in electrical engineering is often propagated by reusing experimental results that proved to

M. Kauers et al. (Eds.): MKM/Calculemus 2007, LNAI 4573, pp. 371–380, 2007.
© Springer-Verlag Berlin Heidelberg 2007

be successfully. One reason is, that the use of exact mathematical results for these applications is too expensive to be explicitely performed. Furthermore — maybe also as a consequence of the above reason — there are theoretical results that could be advantageously used in applications but are not sufficiently known to electrical engineers.

In this situation mathematical knowledge management can contribute in two ways. Firstly, the widespread use of mathematical knowledge management systems incorporating electrical engineering could lead to a rediscovering and broader use of theoretical results in applications by electrical engineers. Secondly, the support in using these results could help filling the gap between fundamentals and applications in the sense that more new applications are based on mathematical fundamentals.

In this paper we focus on network theory [Unb93], in particular on network stability. Network theory deals with the mathematical description, analysis, and synthesis of electrical (continous and time-discrete) networks. For a realible application such systems have to be stable, that is for an arbitrary (bounded) input the output have to be bounded again. In case of highly-precise filters, however, it turns out that checking for stability is often hard to accomplish numerically. In this situation for example Schur's theorem [Sch21] permits an easy method to decide whether a network is stable by computing a chain of polynomials with decreasing degrees. We shall discuss the mathematical fundamentals and prequisites of Schur's theorem and present a Mizar formalization of this theorem.

The plan of the paper is as follows. In the next section we give a brief introduction to network theory focusing on the stability of networks and Schur's theorem [Sch21]. Then after a short review of the Mizar system [Miz07] we present our formalization of Schur's theorem in section 4. Finally, we discuss our results, draw conclusions for mathematical knowledge management in electrical engineering and give some hints for further work.

2   Networks and Their Stability

As mentiond in the introduction the stability of networks is one of the main issues when dealing with the analysis and design of electrical circuits and systems. In the following we briefly review definitions and properties of electrical systems necessary to understand the application of Schur's theorem to electrical networks. In electrical engineering stability applies to the input/output behaviour of networks (see figure 1). For (time-) continous systems one finds the following definition. For discrete systems an analogous definition is used.

Definition 1. ([Unb93])
A continous system is (BIBO-)[1] stable, if and only if each bounded input signal $x(t)$ results in a bounded output signal $y(t)$.

Physically realizable, linear time-invariant systems (LTI systems) can be described by a set of linear equations [Unb93]. The behaviour of a LTI system

---

[1] BIBO stands for Bounded Input Bounded Output.

then is completely characterized by its impulse response $h(t)$.[2] If the impulse response of auch a system is known, the relation between the input $x(t)$ and the output $y(t)$ is given by the convolution integral

$$y(t) = \int\limits_{-\infty}^{\infty} x(\tau)h(t-\tau)d\tau. \tag{1}$$

Furthermore, a LTI system is stable, if and only if its impulse response $h(t)$ is absolute integrable, that is there exists a constant K such that

$$\int\limits_{-\infty}^{\infty} |h(\tau)|\, d\tau \leq K < \infty. \tag{2}$$

In network and filter analysis and design, however, one commonly employs the frequency domain rather than the time domain. To this end the system is described based on its transfer function $H(s)$. In case the Laplace transformation is used we have[3]

$$H(s) = \int\limits_{-\infty}^{\infty} h(t)e^{-st}dt. \tag{3}$$

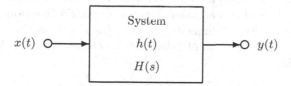

Fig. 1. LTI system with one input $x(t)$ and one output $y(t)$

The evaluation of $H(s)$ for $s = j\omega$ — in case of convergence — enables the qualitative understanding of how the system handles and selects various frequencies, so for example whether the system describes a high-pass filter, low-pass filter, etc. Now the necessary condition to demonstrate the stability of LTI systems in the frequency domain reduces to show, that the $j\omega$-axis lies in the Laplace transformation's region of convergence (ROC).

For physically realizable LTI systems, such as the class of networks with constant and concentrated parameters, $H(s)$ is given in form of a rational function with real coefficients, that is

$$H(s) = \frac{a_n s^n + \ldots + a_0}{b_m s^m + \ldots + b_0}, \quad a_i, b_i \in \mathbb{R}. \tag{4}$$

---

[2] $h(t)$ is the output of the system, when the input is the Dirac delta function $\delta(t)$.
[3] Note that this is a generalization of the continous-time Fourier transformation.

In this case the region of convergence can be described by the roots of the denominator polynomial: If $s_i = \sigma_i + j\omega_i$ for $i = 1, \ldots m$ are the roots of $b_m s^m + \ldots + b_0$, the region of convergence is given by

$$\Re\{s\} > \max\{\sigma_i, \ i = 1, \ldots m\}.$$

To check stability it is therefore sufficient, to show that the real part $\Re\{s\}$ of all poles of $H(s)$ is smaller then 0. The denominator of $H(s)$ is thus a so-called Hurwitz polynomial.

The stability problem for discrete-time signals and systems can be analized with the same approach. For a given discrete-time transfer function $H(z)$ in the Z- domain, it has to be checked whether the unit circle is contained in the region of convergence. Hence for all poles z_i of $H(z)$ we must have $|z_i| < 1$. Using bilinear transformations [OS98]

$$z := \frac{1+s}{1-s}. \tag{5}$$

it is thus sufficient to check whether the denominator of

$$H(z)|_{z:=\frac{1+s}{1-s}} \tag{6}$$

is a Hurwitz polynomial.

The practical examination of stability of highly-precise filters, however, turns out to be very hard. In practical applications the poles of concern are usually close to the axis $s = j\omega$ or the unit circle $|z| = e^{j\omega}$ respectively. Thus numerical determination of the poles is highly error-proning due to its rounding effects. In digital signal processing in addition degrees of transfer functions tend to be very high, for example 128 and higher in communication networks.

It is here that the theorem of Schur [Sch21] comes into play. Using the conjugate polynomial

$$f^*(x) := a_0^* - a_1^* x + a_2^* x^2 - \ldots + (-1)^n a_n^* x^n \tag{7}$$

of a complex polynomial $f(x) = a_0 + a_1 x + a_2 x^2 + \ldots a_n x^n$ a polynomial $g(x)$ of smaller degree is constructed, so that $g(x)$ is a Hurwitz polynomial if and only if $f(x)$ is. The construction itself is fairly easy: it is essentially a division by a linear polynomial.

Theorem 1. ([Sch21])
Let $\Re\{\xi\} < 0$. Then $f(x)$ is a Hurwitz polynomial if and only if $|f(\xi)| < |f^*(\xi)|$ and

$$g(x) := \frac{f^*(\xi)f(x) - f(\xi)f^*(x)}{x - \xi}$$

is a Hurwitz polynomial.

The fact that the degree of $g(x)$ is strictly smaller than the one of $f(x)$ then allows to check stability of networks without explicitly computing roots of polynomials. Note that in addition ξ can always be chosen as -1, so that division can actually be performed by shifting. This however is not widely known in the area of network theory and we are not aware of any system using Schur's theorem for performing stability checks.

3   The Mizar System

The logical basis of Mizar [RT01, Miz07] is classical first order logic extended, however, with so-called schemes. Schemes introduce free second order variables, in this way enabling amongothers the definition of induction schemes. In addition Mizar objects are typed, the types forming a hierarchy with the fundamental type set. The user can introduce new (sub)types describing mathematical objects such as groups, fields, vector spaces or polynomials over rings or fields. To this end the Mizar language provides a powerful typing mechanism based on adjective subtypes [Ban03].

The current development of the Mizar Mathematical Library (MML) relies on Tarski-Grothendieck set theory — a variant of Zermelo Fraenkel set theory using Tarski's axiom on arbitrarily large, strongly inaccessible cardinals [Tar39] which can be used to prove the axiom of choice —, though in principle the Mizar language can be used with other axiom systems also. Mizar proofs are written in natural deduction style as presented in the calculus of [Jaś34]. The rules of the calculus are connected with corresponding (English) natural language phrases so that the Mizar language is close to the one used in mathematical textbooks. The Mizar proof checker verifies the individual proof steps using the notion of obvious inferences [Dav81] to shorten the rather long proofs of pure natural deduction.

The basic theories necessary for Schur's theorem are already contained in MML: Polynomials (over arbitrary rings) have been defined in [Mil01b]. The original goal here was to prove the fundamental theorem of algebra. The complex numbers have been introduced in [Byl90] as objects in their on right. To use the theory of polynomials we need, however, the ring structure of complex numbers. Fortunately, this has been established in [Mil01a]. Consequently, using Mizar we were able to apply — besides the theory of polynomials — both general ring (or field) theorems for complex numbers and special theorems valid for complex numbers only.

4   Mizar Formalization of Schur's Theorem

4.1   Some Preliminaries About Polynomials

Although the theory of polynomials in Mizar is rather well developed, division of polynomials had not been introduced, yet. This, however, can be done (for arbitray fields) in a straightforward way following the well-known literature.[4] We defined two functors div and mod for the quotient and the remainder, respectively. The keyword it denotes the object being defined. Note that Mizar requires an existence and a uniqueness proof for functors. Here, however, these have to be performed for the first definition only, because the definition of mod employs solely arithmetics of polynomials — including the just defined functor div. Therefore existence and uniqueness in this case is automatically derived by the Mizar checker.

---

[4] See for example [GG99].

```
definition
let L be Field;
let p,q be Polynomial of L such that q <> 0_.(L);
func p div q -> Polynomial of L means
  ex r being Polynomial of L st p = it *' q + r & deg r < deg q;
end;

definition
let L be Field;
let p,q be Polynomial of L such that s <> 0_.(L);
func p mod q -> Polynomial of L equals
  p - (p div q) *' q;
end;
```

Divisibility of polynomials can then be introduced by the condition `p mod q = 0._(L)`, where `0._(L)` is the zero polynomial, or by the equivalent condition that there exists a polynomial `h` such that `p = h * q`. For our purposes it is essential that a polynomial $p(x)$ is divisible without remainder by the linear polynomial $x - z$, if z is a root of $p(x)$.[5] It was therefore necessary to show that for every root z of a polynomial $p(x)$ the polynomial $x - z$ is a divisor of $p(x)$. To do so, we introduced the polynomials `rpoly(k,z)` $= x^k - z^k$ and `qpoly(k,z)` $= x^{k-1} + x^{k-2} * z + x^{k-3} * z^2 + ... + x * z^{k-2} + z^{k-1}$. Note that for k > 1 we have `rpoly(1,z) * qpoly(k,z) = rpoly(k,z)`, which allows for the construction of a polynomial `h` such that `r(1,z) * h = p`. We thus get

```
theorem
for L being Field
for p being Polynomial of L
for z being Element of L st z is_a_root_of p holds rpoly(1,z) divides p;
```

Note again, that this property is shown for polynomials over arbitrary fields. In the next section when dealing with Schur's criterium, we shall use the complex number version of this theorem.

4.2   Schur's Theorem

Using the general Mizar theory of polynomials for our purposes, that is for polynomials over the complex numbers, is straightforward. We just instantiate the parameter L describing the coefficient domain with the field of complex numbers `F_Complex` from [Mil01a]. So an object of type

`Polynomial of F_Complex`

combines the theory of polynomials with the one of complex numbers. Hence for such objects we have available both the predicate `is_root_of` defined for polynomials and the functor `Re` giving the real part of a complex number. This allows for the following definition of Hurwitz polynomials.

---

[5] Compare theorem 1 in section 2.

```
definition
let f be Polynomial of F_Complex;
attr f is Hurwitz means
  for z being Element of F_Complex st z is_a_root_of f holds Re(z) < 0;
end;
```

The examination of polynomials with a degree smaller or equal then 1 is rather uncomplex. Constant polynomials are not Hurwitz, except for the zero polynomial which is. A linear polynomial $p(x) = x - z$ obviously is Hurwitz if and only if the real part of z is smaller than 0. This condition carries over to arbitrary polynomials of degree 1. Hence we get the following three theorems for the basic cases.

```
theorem
0_.(F_Complex) is non Hurwitz;
```

```
theorem
for z being Element of F_Complex st z <> 0.F_Complex
holds z * 1_.(F_Complex) is Hurwitz;
```

```
theorem
for z1,z2 being Element of F_Complex st z1 <> 0.F_Complex
holds z1 * rpoly(1,z2) is Hurwitz iff Re(z2) < 0;
```

In addition we proved some other properties of Hurwitz polynomials needed later, so for example that $f * g$ is Hurwitz if and only if f and g are Hurwitz or that for a complex number $z \neq 0$ we have $z * f$ is Hurwitz if and only if f is Hurwitz.

To prove Schur's theorem for the general case we needed to introduce the conjugate of a complex polynomial as given by equation (7). This is accomplished by a Mizar functor *' defining the coefficients of the conjugated polynomial appropriately.[6] For that we use the functor power(G) which describes exponentiation for arbitrary groups G, here again instantiated with F_Complex, the field of complex numbers. Note that after instantiating G with F_Complex the resulting type of the functor power(F_Complex) is automatically accomodated, so that it is no problem multiplying its result with another complex number.

```
definition
let f be Polynomial of F_Complex;
func f*' -> Polynomial of F_Complex means
  for i being Element of NAT holds
  it.i = power(F_Complex).(-1.F_Complex,i) * (f.i)*';
end;
```

Thus prepared we could already state Schur's theorem in Mizar. However, to shorten writings we decided to introduce another functor describing the nominator polynomial of Schur's construction. The functor eval describes evaluation of polynomials.

---

[6] Note that the functor *' is then overloaded, because it also stands for conjugation of complex numbers as can be seen in the following definition.

```
definition
let f be Polynomial of F_Complex;
let z be Element of F_Complex;
func F*(f,z) -> Polynomial of F_Complex equals
  eval(f*',z) * f - eval(f,z) * f*';
end;
```

Taking into account that the Mizar functor |. .| gives the absolute value of complex numbers, we then get the following formulation of Schur's theorem. Note again that rpoly(1,z) is the polynomial $p(x) = x - z$.

```
theorem
for f being Polynomial of F_Complex st deg(f) >= 1
for z being Element of F_Complex
    st Re(z) < 0 & |.eval(f,z).| < |.eval(f*',z).|
holds f is Hurwitz iff F*(f,z) div rpoly(1,z) is Hurwitz;
```

The proof of the theorem relies on a thorough examination of the relation between the real part $\Re(z)$ of a complex number z and the values of $|f(z)|$ and $|f^*(z)|$ in case f is a Hurwitz polynomial. It turns out that whether $\Re(z)$ is smaller or greater than 0 completely determines which value $|f(z)|$ or $|f^*(z)|$ is greater. This allows later to argue about the roots of the nominator polynomial, that is of the polynomial F*(f,z).

```
theorem
for f being Polynomial of F_Complex st deg(f) >= 1 & f is Hurwitz
for z being Element of F_Complex
holds (Re(z) < 0 implies |.eval(f,z).| < |.eval(f*',z).|) &
      (Re(z) > 0 implies |.eval(f,z).| > |.eval(f*',z).|) &
      (Re(z) = 0 implies |.eval(f,z).| = |.eval(f*',z).|);
```

The corresponding proof is rather technical. In Mizar, however, the application of theorems for complex numbers has been automatized in the sense that a number of basic theorems are automatically applied, in this way shortening the proof [NB04]. In addition the Encyclopedia of Mathematics in Mizar (EMM) collecting theorems of a theory — in this case concerning complex numbers — originally spread over the whole repository produced a kindly working environment to accomplish the task.

Note also that this theorem implies that even for polynomials with degree > 1, it is not always necessary to reduce the problem of stability to a basic case: If we find a complex number z with $\Re(z) < 0$ such that $|f(z)| \geq |f^*(z)|$ we immediately get that f is not a Hurwitz polynomial.

```
theorem
for f being Polynomial of F_Complex st deg(f) >= 1
holds (ex z being Element of F_Complex
        st Re(z) < 0 & |.eval(f,z).| >= |.eval(f*',z).|)
implies f is non Hurwitz;
```

The rest of the proof basically applies the theorem from above two times, once for each direction. We first proved the following, more general version of Schur's theorem from [Sch21]: For complex numbers z_1 and z_2 such that $|z_1| > |z_2|$ and a complex polynomial $f(x)$ with degree ≥ 1 holds $f(x)$ is a Hurwitz polynomial if and only if $g(x) = z_1 * f(x) - z_2 * f^*(x)$ is a Hurwitz polynomial: Because of $|z_1| > |z_2|$ we have now $|f(x)| \geq |f^*(x)|$, if $\Re(x) \geq 0$, and hence $|z_1 * f(x)| > |z_2 * f^*(x)|$, which shows the first direction. For the other direction we only note, that $f(x) = z_1' * g(x) - z_2' * g^*(x)$ with

$$z_1' = \frac{z_1^*}{|z_1|^2 - |z_2|^2} \quad \text{and} \quad z_2' = -\frac{z_2}{|z_1|^2 - |z_2|^2},$$

so that $|z_1'| > |z_2'|$ finishes the proof.

From this Schur's theorem easily follows by instantiating z_1 with $f^*(z)$ and z_2 with $f(z)$ giving essentially the functor F*(f,z) from above. Note that we here need in addition that the denominator polynomial $p(x) = x - z$, that is rpoly(1,z), divides the nominator polynomial $f^*(z) * f(x) - f(z) * f^*(x)$, that is F*(f,z). This, however, is ensured by the fact that z is a root of $f^*(z) * f(x) - f(z) * f^*(x)$ and the — automatically available — complex number version of the main theorem of section 4.1.

So this part of the proof requires both arithmetics — including conjugates — and abstract values of complex numbers and arithmetics of polynomials over complex numbers. In Mizar, as already mentioned, this is achieved by instantiating the general theory of polynomials with the field of complex numbers. Then of course the absolute value, defined originally for complex numbers, is available for the coefficients of complex polynomials, also. Consequenly, the just described proof steps could be accomplished based on these two theories without other preparations or additional lemmas.

5   Conclusions

In this paper we have considered electrical engineering as an application area for mathematical knowledge management. We have focused on stability theory of networks and have shown by a Mizar formalization of Schur's theorem that interesting mathematical knowledge in electrical engineering can be successfully handled with mathematical knowledge management systems.

We believe that both electrical engineering and mathematical knowledge management can benefit from a further development of collaboration in the area of mathematical knowledge. The combination of mathematical knowledge managements systems and repositories with analysis and design tools for electrical networks can provide electrical engineers with a thorough mathematical basis for their work. In addition this would lead also to the use of less known theoretical results, such as for example Schur's theorem, in new applications.

For mathematical knowledge management electrical engineering can serve as an additional test bed, in which new developments can be tried out. And, of course, in this way a whole group of potential new users of mathematical knowledge management systems could be addressed.

References

[Byl90]   Byliński, C.: The Complex Numbers. Formalized Mathematics 1(3), 507–513 (1990)

[Ban03]   Bancerek, G.: On the Structure of Mizar Types. In: Geuvers, H., Kamareddine, F. (eds.) Proc. of MLC 2003, ENTCS, vol. 85(7) (2003)

[Dav81]   Davies, M.: Obvious Logical Inferences. In: Proceedings of the 7th International Joint Conference on Artificial Intelligence, pp. 530–531 (1981)

[DeB87]   de Bruijn, N.G.: The Mathematical Vernacular, a language for mathematics with typed sets. In: Dybjer, P., et al. (ed.) Proc. of the Workshop on Programming Languages, Marstrand, Sweden (1987)

[GG99]   von zur Gathen, J., Gerhard, J.: Modern Computer Algebra. Camebridge University Press, Camebridge (1999)

[HKS06]   Hilf, E., Kohlhase, M., Stamerjohanns, H.: Capturing the Content of Physics: Systems, Observables, and Experiments. In: Borwein, J.M., Farmer, W.M. (eds.) MKM 2006. LNCS (LNAI), vol. 4108, pp. 165–178. Springer, Heidelberg (2006)

[Jaś34]   Jaśkowski, S.: On the Rules of Suppositon in Formal Logic. In: Studia Logica, vol. 1 (1934)

[Mil01a]   Milewska, A.J.: The Field of Complex Numbers. Formalized Mathematics 9(2), 265–269 (2001)

[Mil01b]   Milewski, R.: The Ring of Polynomials. Formalized Mathematics 9(2), 339–346 (2001)

[Miz07]   The Mizar Home Page, http://mizar.org

[NB04]   Naumowicz, A., Byliński, C.: Improving Mizar texts with properties and requirements. In: Asperti, A., Bancerek, G., Trybulec, A. (eds.) MKM 2004. LNCS, vol. 3119, pp. 190–301. Springer, Heidelberg (2004)

[OS98]   Oppenheim, A.V., Schafer, R.W.: Discrete-Time Signal Processing, 2nd edn. Prenctice-Hall, New Jersey (1998)

[RT01]   Rudnicki, P., Trybulec, A.: Mathematical Knowledge Management in Mizar. In: Buchberger, B., Caprotti, O. (eds.) Proc. of MKM 2001, Linz, Austria (2001)

[Ses07]   The SESAME Network, http://www.mkm-ig.org/projects/sesame/

[Sch21]   Schur, J.: Über Algebraische Gleichungen, die nur Wurzeln mit negativen Realteilen besitzen. Zeitschrift für angewandte Mathematik und Mechanik 1, 95–110 (1921)

[Tar39]   Tarski, A.: On Well-Ordered Subsets of Any Set. In: Fundamenta Mathematicae, vol. 32, pp. 176–183 (1939)

[Unb93]   Unbehauen, R.: Netzwerk- und Filtersynthese: Grundlagen und Anwendungen (4. Auflage); Oldenbourg-Verlag (1993)

Spurious Disambiguation Error Detection

Claudio Sacerdoti Coen* and Stefano Zacchiroli*

Department of Computer Science, University of Bologna
sacerdot@cs.unibo.it, zacchiro@cs.unibo.it

Abstract. The disambiguation approach to the input of formulae enables the user to type correct formulae in a terse syntax close to the usual ambiguous mathematical notation. When it comes to incorrect formulae we want to present only errors related to the interpretation meant by the user, hiding errors related to other interpretations (*spurious errors*).

We propose a heuristic to recognize spurious errors, which has been integrated with the disambiguation algorithm of [6].

1   Introduction

In [6] we proposed an efficient algorithm for parsing and semantic analysis of ambiguous mathematical formulae. The topic is particularly relevant for the Mathematical Knowledge Management community since every mathematical assistant sooner or later faces the need of letting its user type formulae. When the user is not acquainted with a system or its library—as it happens when using mathematical search engines [1,3,7]—we cannot assume the knowledge of a language other than the usual corpus of ambiguous mathematical notation.

Our algorithm mimics a mathematician behavior of disambiguating a formula by choosing the only possible interpretation that has a meaning in the current context. However when a formula is not correct, every interpretation is "equally" meaningless. Nevertheless, a mathematician seems to be able to understand which interpretation is more likely, spotting the genuine errors in the formula.

Example 1. If f is known to be a real-valued function on vectors, the formula $f(\alpha \cdot x + \beta \cdot y + z) = \alpha \cdot f(x) + \beta \cdot f(y) + z$ is not correct and a mathematician would probably assert that z is not used properly in the right hand side of the equation. Instead, the algorithm of [6] would return several alternative error messages such as: in `"f(`$\alpha \cdot$`x`$\overrightarrow{+}$`...`$\overrightarrow{+}$`z) = ..."`: x is a vector, but is used as a scalar.

A possible way out is designing a disambiguation algorithm able to rate the possible interpretations so that the one expected by a mathematician ranks first. Also in those cases were several possible interpretations are meaningful, this approach is necessary to choose automatically among them or to ask the

---

* Partially supported by the Strategic Project "DAMA: Dimostrazione Assistita per la Matematica e l'Apprendimento" of the University of Bologna.

M. Kauers et al. (Eds.): MKM/Calculemus 2007, LNAI 4573, pp. 381–392, 2007.

user providing a sensible default. In [2] we proposed such an algorithm that was designed to tackle the case of correct formulae with multiple interpretations. In this paper we address the case of formulae for which no correct interpretation can be found.

Consider again Example 1. We need to find a criterion to identify the given error message as spurious, i.e. as an error relative to an interpretation that is not the one expected by the user. Note that a formula can contain more than one genuine error: they are all the errors in the expected interpretation of the formula. The heuristic criterion we propose is the following.

Criterion 1 (Spurious error detection). *An error is* spurious *when it is localized in a sub-formula F such that there is an alternative interpretation of the whole formula such that no error is localized in F.*

Intuitively an error is spurious when no genuine error is spatially co-located with it, i.e. genuine errors are to be found elsewhere. In Example 1 if we interpret all the operators in the left hand side as operations on vectors we do not obtain any error message in the left hand side. Hence the genuine error must be in the right hand side.

The main goal of this paper is the integration of spurious error detection in the efficient algorithm proposed in [6]. We proceed as follows. In Section 2 we formalize the specification of the class of disambiguation algorithms. In Section 3 we provide an improved description of the algorithm proposed in [6], proving that it is a member of the disambiguation algorithm class, while in Section 4 we extend the algorithm with spurious error detection.

2   Disambiguation Algorithm Specification

Traditionally semantic analysis maps an abstract syntax tree (AST) of a formula to a term—its semantics—in some calculus. In an ambiguous setting, semantic analysis rather maps an AST to *a set of* terms; the set can then be rated according to some criterion to identify the best semantics. To represent in a concise way a set of terms sharing a common structure, we use a term containing non linear placeholders in the spirit of [4,5]. We say that a term t' is an *instantiation* of t if it is obtained filling zero or more of its placeholders. For instance $?_1 = ?_2 + ?_2$ represents the set of terms $\{t_1 = t_2 + t_2 \mid t_1, t_2 \text{ terms}\}$; $?_1 = 0 + 0$ and $0 = 0 + 0$ are two instances belonging to that set.

Lemma 1. *If t_1 is an instance of t_2 then the set of instances of t_1 is a subset of the set of instances of t_2.*

Proof. By definition of instantiation.   □

Among all the terms that are semantics of a given AST, we are interested only in those that are well-typed. Thus, we are interested in terms with placeholders only when they denote non-empty sets of well-typed instantiations. We assume the existence of a *refiner* $\mathcal{R}(\cdot)$, that is a function from terms to outcomes. An *outcome* is either the distinguished symbol ✓ or an informative error message. The latter is returned when the set of well-typed instantiations of the input term is (known to be) empty. For instance $\mathcal{R}(f(?_1) = 1) = $ ✓ whereas $\mathcal{R}(f(?_1) = f + 1) = $ "f is a function, but is used as a scalar". In the latter case the error message is relevant to every possible instantiation; in the former there is no guarantee that every possible instantiation is well-typed. Still, the following lemma holds.

Lemma 2. *A term t without placeholders is well-typed iff $\mathcal{R}(t) = $ ✓*

Proof. t is the only instance of itself thus, by definition of $\mathcal{R}(\cdot)$, $\mathcal{R}(t) \neq $ ✓ iff t is not well-typed.   □

We are now ready to describe the specification of a disambiguation algorithm for an AST t. Let $Dom(t)$ be the set of occurrences of overloaded symbols in t. For each $s \in Dom(t)$, let $\mathcal{D}_s$ be the set of possible choices for s.

An *interpretation* ϕ for t is a partial function $Dom(t) \ni s \mapsto u_s \in \mathcal{D}_s$. Intuitively a (partial) interpretation restricts the set of semantics of t resolving the overloading for the occurrences in the domain of ϕ. When an interpretation is a total function a unique semantics is determined. To formalize this intuition we associate to a partial interpretation ϕ a term with placeholders $[\![t]\!]_\phi$, where all (applications of) occurrences of symbols not in the domain of ϕ have been interpreted as fresh placeholders. For instance, when $\phi = [+_1 \mapsto \textit{point-wise sum}]$, $[\![(f+g)(x)=f(x)+g(x)]\!]_\phi$ denotes $(f+g)(x) = ?_1$. Note that the arguments of the second occurrence of plus have been omitted.

We denote with Φ_t the set of all (partial) interpretations for t and with $\hat{\Phi}_t$ the set of all total interpretations. We call $\bot$ the function everywhere undefined and we denote as $\phi[s \mapsto u]$ the function that maps s to u and behaves as ϕ elsewhere. The set of interpretations is ordered by the usual order on partial functions: $\phi_1 \sqsubseteq \phi_2$ iff $\forall s, \phi_1(s) = u \Rightarrow \phi_2(s) = u$. The minimum of Φ according to $\sqsubseteq$ is $\bot$.

Lemma 3. $\phi_1 \sqsubseteq \phi_2$ *iff* $[\![t]\!]_{\phi_2}$ *is an instance of* $[\![t]\!]_{\phi_1}$.

Proof. By structural induction on t and by cases on the definition of $[\![\cdot]\!]$. Since, for the sake of brevity, we omitted its definition, the present lemma can be seen as a required property of $[\![\cdot]\!]$.   □

Together with Lemma 1, Lemma 3 confirms the intuition that the more overloading is resolved, the smaller the set of semantics.

A disambiguation algorithm partitions the set of semantics of an AST into classes of well-typed terms and classes of terms characterized by the same typing error. Since Lemma 2 holds only for placeholder-free terms, all terms in the well-typed class must have no placeholders. We will use the notion of cover to grasp partitions at the interpretation level, and the notion of typing cover to grasp well-typedness.

We say that a set of interpretations S *covers* a set of interpretations T, written $S \rhd T$, when $\forall \phi \in T, \exists! \phi' \in S, \phi' \sqsubseteq \phi$.

Lemma 4. *If $S \rhd T$ then for each $\phi_1 \in T$ there exists an unique $\phi_2 \in S$ such that $[\![t]\!]_{\phi_1}$ is an instance of $[\![t]\!]_{\phi_2}$.*

Proof. By Lemma 3 and the definition of cover.   □

Corollary 1. *If $S \rhd \hat{\Phi}_t$ and $\phi_1, \phi_2 \in S, \phi_1 \neq \phi_2$ then the set of instances of $[\![t]\!]_{\phi_1}$ is disjoint from the set of instances of $[\![t]\!]_{\phi_2}$.*

Proof. Suppose per absurdum that u is an instance of both $[\![t]\!]_{\phi_1}$ and $[\![t]\!]_{\phi_2}$. Let $u' \in \hat{\Phi}_t$ be an instance of u. By Lemma 4 $\phi_1 = \phi_2$, but by hypothesis we know $\phi_1 \neq \phi_2$.   □

Theorem 1. *$S \rhd \hat{\Phi}_t$ iff $\{\{u \mid u$ is an instance of $[\![t]\!]_\phi\} \mid \phi \in S\}$ is a partition of $\{u \mid \exists \phi \in \hat{\Phi}_t, u = [\![t]\!]_\phi\}$ (i.e. the set of all semantics of t).*

Proof. The forward implication is by Lemma 4 and Corollary 1. For the converse implication consider an arbitrary but fixed $\phi \in \hat{\Phi}_t$. By hypothesis there is a unique $\phi' \in S$ such that $u = [\![t]\!]_\phi$ is an instance of $[\![t]\!]_{\phi'}$. Thus $S \rhd \hat{\Phi}_t$.   □

We say that a set of interpretations A' is a *refinement* of a set of interpretations A, written $A \Diamond A'$ when $A \rhd A'$ and for all $u \in \hat{\Phi}_t$ such that there is a $\phi \in A$ such that u is an instance of $[\![t]\!]_\phi$ there exists a unique $\phi' \in A'$ such that u is an instance of $[\![t]\!]_{\phi'}$.

Theorem 2. *If $A \cap B = \emptyset$, $A \cup B \rhd \hat{\Phi}_t$ and $A \Diamond A'$, then $A' \cup B \rhd \hat{\Phi}_t$.*

Proof. By Theorem 1 $\{\{u \mid u$ is an instance of $[\![t]\!]_\phi\} \mid \phi \in A \cup B\}$ partitions the set of all semantics of t. $\{\{u \mid u$ is an instance of $[\![t]\!]_\phi\} \mid \phi \in A' \cup B\}$ partitions the same set by definition of $A \Diamond A'$, where the requirement $A \rhd A'$ is fundamental to avoid interference with B. Hence the thesis by Theorem 1.   □

A set S of interpretations is said to be *typing* when for all $\phi \in S$ if $\mathcal{R}([\![t]\!]_\phi) = \checkmark$ then $\phi \in \hat{\Phi}_t$. In particular a *typing cover* is a cover $S \rhd \hat{\Phi}_t$ that is also typing. Intuitively a disambiguation algorithm returns a typing cover equipped with rating information for its interpretations (that will be called classification).

Theorem 3. *For each typing cover S and for each term u in the set of all semantics of t, u is well-typed iff $\mathcal{R}([\![t]\!]_\phi) = \checkmark$ where ϕ is the only interpretation in S such that u is an instance of $[\![t]\!]_\phi$.*

Proof. If $\mathcal{R}([\![t]\!]_\phi) \neq \checkmark$ by definition of $\mathcal{R}(\cdot)$. Otherwise by Lemma 2 and definition of typing cover.   □

We also expect something more that cannot be grasped formally: if u is not well-typed then the error message for $\mathcal{R}([\![t]\!]_\phi)$ should also be relevant for u. This property is inherited from the refiner.

Lemma 5. $\{\bot\} \rhd \hat{\Phi}_t$. It is typing iff $\mathcal{R}(\llbracket t \rrbracket_\bot) \neq \checkmark$ or $Dom(t) = \emptyset$.

Proof. Trivial by definition of $\hat{\Phi}_t$ and $\mathcal{R}(\cdot)$. $\qquad\qquad\qquad\qquad\qquad\square$

To rate covers, we assume that to each interpretation ϕ is associated a rate $\rho(\phi)$. A rate is an element of a partially ordered set $(A, \preceq)$, such that $\rho(\phi_1) \preceq \rho(\phi_2)$ iff $\llbracket t \rrbracket_{\phi_1}$ is more likely to be the intended meaning of t than $\llbracket t \rrbracket_{\phi_2}$.

Formally, a *disambiguation algorithm* takes as input an AST t and returns a typing and covering classification Σ. A *classification* Σ is a set of tuples $\langle \phi, o, r \rangle$ such that:

1. for all $\langle \phi, o, r \rangle \in \Sigma, o = \mathcal{R}(\llbracket t \rrbracket_\phi)$, and r belongs to some partially ordered set $(B, \trianglelefteq)$;
2. for all $\langle \phi_1, o_1, r_1 \rangle, \langle \phi_2, o_2, r_2 \rangle \in \Sigma$, if $\phi_1 = \phi_2$ then $o_1 = o_2$ and $r_1 = r_2$.

A classification Σ is a *covering classification* if $S_\Sigma = \{\phi \mid \langle \phi, o, r \rangle \in \Sigma\}$ is a cover; it is a *typing classification* when S_Σ is typing.

We choose for B the set $\{\text{\ding{111}}, \text{\ding{111}}, \text{\ding{111}}\} \times A$ ordered lexicographically by the orders: $\text{\ding{111}} \leq \text{\ding{111}} \leq \text{\ding{111}}$ and $\preceq$.

Every classification can be partitioned into the set of (so far) successful and the set of failing interpretations as follows:

$$(\Sigma)^\checkmark = \{\langle \phi, o, r \rangle \in \Sigma \mid o = \checkmark\}$$
$$(\Sigma)^\times = \Sigma \setminus (\Sigma)^\checkmark$$

Example 2 (Naive Disambiguation Algorithm). The *naive disambiguation algorithm* (NDA for short) is the disambiguation algorithm that, when applied to an AST t, computes the typing and covering classification $\Sigma = \{\langle \phi, o, r \rangle \mid \phi \in \hat{\Phi}_t,\ o = \mathcal{R}(\llbracket t \rrbracket_\phi),\ r = \rho'(o, \phi)\}$ where:

$$\rho'(o, \phi) = \begin{cases} \langle \text{\ding{111}}, \rho(\phi) \rangle & \text{if } o = \checkmark \\ \langle \text{\ding{111}}, \rho(\phi) \rangle & \text{otherwise} \end{cases}$$

The rating function $\rho'(\cdot, \cdot)$ gives priority to successes over failures; outcomes being equal, it falls back to the interpretation rating.

We call this algorithm "naive" since its computes the typing cover $S_\Sigma = \hat{\Phi}_t \rhd \hat{\Phi}_t$ of maximum cardinality. Its execution is computationally expensive since it invokes the refiner $|S_\Sigma| = |\hat{\Phi}_t| = \prod_{s \in Dom(t)} |\mathcal{D}_s|$ times.

Example 3 (NDA execution). Consider the (non-typable) AST corresponding to $\mathbf{f}(\alpha \cdot \mathbf{x} + \beta \cdot \mathbf{y} + \mathbf{z}) = \alpha \cdot \mathbf{f}(\mathbf{x}) + \beta \cdot \mathbf{f}(\mathbf{y}) + \mathbf{z}$, where $+$ is left-associative, $\mathbf{x}, \mathbf{y}, \mathbf{z}$ are globally declared as real vectors, α, β are reals, and $\mathbf{f}$ is a real-valued function on vectors. The symbol "$+$" is overloaded on scalar and vector sums; "$\cdot$" is overloaded on scalar and external products.

NDA returns a classification consisting of 2^8 error messages (not necessarily unique), where 2 are the possible choices for each occurrence of overload symbols and 8 is the number of occurrences of "$\cdot$" and "$+$". The "expected" error message

"z is a vector, but is used as a scalar" is drowned in a sea of errors
like (re-ordered here for reader's sake):

- "x is a vector, but is used as a scalar"
- "y is a vector, but is used as a scalar"
- "z is a vector, but is used as a scalar"
- "α·x is a vector, but is used as a scalar"
- "β·y is a vector, but is used as a scalar"
- "α·x + β·y is a vector, but is used as a scalar"
- ...
- "f(x) is a scalar, but is here used as a vector"
- "f(y) is a scalar, but is here used as a vector"
- ...

We can only hope that $\rho(\cdot)$ does a great job ranking first the expected interpretation. In practice we are not aware of any rating function that performs well looking only at the interpretations.

3   An Efficient Disambiguation Algorithm

In terms of efficiency we can do better than NDA. The key observation for improvement is that a single invocation of the refiner on a term with placeholders can rule out the whole set of its instances. More precisely, if the refinement of such a term fails, all of its instances are not well-typed (and will fail in the same way). Thus, it is not necessary to compute the largest typing and covering classification as NDA does: intuitively, the smaller the classification, the more efficient the algorithm.

A typing and covering classification can be built incrementally starting from a covering classification. Indeed if a covering classification Σ is not typing it must contain a partial interpretation $\phi \in S_{(\Sigma)^{\checkmark}}$. A more precise classification can be obtained replacing the interpretation ϕ with a set of more instantiated interpretations S such that $S \triangleright \{\phi\}$. Since $\phi_1 \sqsubseteq \phi$ for each $\phi_i \in S$, the domain of ϕ_1 (a subset of $Dom(t)$) is bigger than the domain of ϕ. Thus the refinement process ends in a finite number of steps since $Dom(t)$ is finite; moreover it yields a typing classification.

To increase efficiency, we can enforce the invariant that all interpretations $\phi \in S_{(\Sigma)^{\checkmark}}$ share a common domain. Thus at each step we have to extend at once the domain shared by all ϕs. Let Σ be a classification such that the interpretations in S_Σ are defined on the same domain and let $s \in Dom(t)$. We define:

$$\Sigma_s = \{\langle \phi, o, r \rangle \mid \exists \phi' \in S_\Sigma, \exists u \in \mathcal{D}_s, \phi = \phi'[s \mapsto u], o = \mathcal{R}(\llbracket t \rrbracket_\phi), r = \rho'(o, \phi)\}$$

Lemma 6. *Let Σ be a classification such that the interpretations in S_Σ are defined on the same domain and let $s \in Dom(t)$. $\Sigma \langle\!\rangle \Sigma_s$.*

Proof. By construction of Σ_s and definition of $\diamondsuit$.   □

The refinement process outlined above can now be formally described. At the n-th step we have the covering (not typing) classification Σ_n. Choosing s outside the domain of the ϕs in $S_{(\Sigma_n)\checkmark}$, we obtain the next covering classification $\Sigma_{n+1} = ((\Sigma_n)')_s \cup (\Sigma_n)^\chi$. Since the functions in $S_{(\Sigma_{n+1})\checkmark}$ are more defined that those in $S_{(\Sigma_n)\checkmark}$ the most natural choice for the initial covering classification is $\Sigma_0 = \{\langle \bot, o, r \rangle \mid o = \mathcal{R}([\![t]\!]_\bot), r = \rho'(o, \bot)\rangle\}$.

Example 4 (Refinement process). Consider the AST of Example 2. Picking occurrences $s \in Dom(t)$ according to the pre-visit order of the AST, the first steps of the refinement process yield the following covering classifications (where for the sake of brevity errors have been substituted by χ):

$$\Sigma_0 = \{\langle \phi_1, \checkmark, \langle \text{\rotatebox{180}{!}}, \rho(\phi_1)\rangle\rangle\} \qquad \text{where } [\![t]\!]_{\phi_1} = f(?_1) = ?_2 \text{ and } \phi_1 = \bot$$

$$\Sigma_1 = \{\langle \phi_{11}, \checkmark, \langle \text{\rotatebox{180}{!}}, \rho(\phi_{11})\rangle\rangle, \qquad [\![t]\!]_{\phi_{11}} = f(?_1 \overrightarrow{\mp} z) = ?_2$$
$$\langle \phi_{12}, \chi, \langle \text{\rotatebox{180}{!}}, \rho(\phi_{12})\rangle\rangle\} \qquad [\![t]\!]_{\phi_{12}} = f(?_1 + z) = ?_2$$

$$\Sigma_2 = \{\langle \phi_{111}, \checkmark, \langle \text{\rotatebox{180}{!}}, \rho(\phi_{111})\rangle\rangle, \qquad [\![t]\!]_{\phi_{111}} = f(?_1 \overrightarrow{\mp} ?_2 \overrightarrow{\mp} z) = ?_3$$
$$\langle \phi_{112}, \chi, \langle \text{\rotatebox{180}{!}}, \rho(\phi_{112})\rangle\rangle, \qquad [\![t]\!]_{\phi_{112}} = f(?_1 + ?_2 \overrightarrow{\mp} z) = ?_3$$
$$\langle \phi_{12}, \chi, \langle \text{\rotatebox{180}{!}}, \rho(\phi_{12})\rangle\rangle\} \qquad [\![t]\!]_{\phi_{12}} = f(?_1 + z) = ?_2$$

$$\Sigma_3 = \{\langle \phi_{1111}, \checkmark, \langle \text{\rotatebox{180}{!}}, \rho(\phi_{1111})\rangle\rangle, \qquad [\![t]\!]_{\phi_{1111}} = f(\alpha \overrightarrow{\cdot} x \overrightarrow{\mp} ?_1 \overrightarrow{\mp} z) = ?_2$$
$$\langle \phi_{1112}, \chi, \langle \text{\rotatebox{180}{!}}, \rho(\phi_{1112})\rangle\rangle, \qquad [\![t]\!]_{\phi_{1112}} = f(\alpha \cdot x \overrightarrow{\mp} ?_1 \overrightarrow{\mp} z) = ?_2$$
$$\langle \phi_{112}, \chi, \langle \text{\rotatebox{180}{!}}, \rho(\phi_{112})\rangle\rangle, \qquad [\![t]\!]_{\phi_{112}} = f(?_1 + ?_2 \overrightarrow{\mp} z) = ?_3$$
$$\langle \phi_{12}, \chi, \langle \text{\rotatebox{180}{!}}, \rho(\phi_{12})\rangle\rangle\} \qquad [\![t]\!]_{\phi_{12}} = f(?_1 + z) = ?_2$$

$\ldots$

Theorem 4 (Correctness of the Refinement Process). *The above refinement process implements a disambiguation algorithm, i.e. for each AST t, $\Sigma_{|Dom(t)|}$ is a covering and typing classification.*

Proof. By induction on $|Dom(t)|$ we prove that $\Sigma_{|Dom(t)|}$ is covering.
Base case. By Lemma 5 Σ_0 is a covering classification.
Inductive case. Let Σ_n be a covering classification per inductive hypothesis. By definition $\Sigma_{n+1} = ((\Sigma_n)')_s \cup (\Sigma_n)^\chi$. By Theorem 2 and Lemma 6, Σ_{n+1} is covering.

To prove that $\Sigma_{|Dom(t)|}$ is typing the reader can prove by induction that all the ϕs in $S_{(\Sigma_n)\checkmark}$ are defined on a subset of $Dom(t)$ of cardinality n. The thesis follows trivially.   □

The above refinement process is parametric in how the next symbol $s \in Dom(t)$ is chosen at each step. In [6] we discussed the implication of such a choice on the computational complexity in terms of numbers of refiner invocations. The best choice corresponds to a pre-visit of the abstract syntax tree t.

We now present the *efficient disambiguation algorithm* (EDA for short) of [6]. It proceeds by recursion on $Dom^{\text{list}}(t)$, which is the list of overloaded symbol occurrences in t obtained in a pre-visit traversal.

$$f(\Sigma, l) = \begin{cases} \Sigma & \text{if } l = [] \\ f((\Sigma_s)^{\prime}, tl) \cup (\Sigma_s)^{\times} & \text{if } l = s :: tl \end{cases}$$

$$EDA(t) = f((\Sigma_0)^{\prime}, Dom^{\text{list}}(t)) \cup (\Sigma_0)^{\times}$$

Theorem 5 (Correctness of EDA). *EDA implements a disambiguation algorithm.*

Proof. By Theorem 4 it is sufficient to prove that the classification returned by EDA is the same returned by the refinement process. We observe that

$$\begin{aligned} \Sigma_n &= ((\Sigma_{n-1})^{\prime})_{s_n} \cup (\Sigma_{n-1})^{\times} \\ &= ((((\Sigma_{n-2})^{\prime})_{s_{n-1}} \cup (\Sigma_{n-2})^{\times})^{\prime})_{s_n} \cup (((\Sigma_{n-2})^{\prime})_{s_{n-1}} \cup (\Sigma_{n-2})^{\times})^{\times} \\ &= ((((\Sigma_{n-2})^{\prime})_{s_{n-1}})^{\prime})_{s_n} \cup (((\Sigma_{n-2})^{\prime})_{s_{n-1}})^{\times} \cup (\Sigma_{n-2})^{\times} \qquad\qquad (\dagger) \\ &= (((((\Sigma_{n-2})^{\prime})_{s_{n-1}})^{\prime})_{s_n})^{\prime} \cup \\ &\quad\; (((((\Sigma_{n-2})^{\prime})_{s_{n-1}})^{\prime})_{s_n})^{\times} \cup (((\Sigma_{n-2})^{\prime})_{s_{n-1}})^{\times} \cup (\Sigma_{n-2})^{\times} \\ &= \ldots \\ &= ((\cdots (((((\Sigma_0)^{\prime})_{s_1})^{\prime})_{s_2})^{\prime} \cdots)_{s_n})^{\prime} \cup \qquad\qquad\qquad\qquad\qquad (\ddagger) \\ &\quad\; ((\cdots (((((\Sigma_0)^{\prime})_{s_1})^{\prime})_{s_2})^{\prime} \cdots)_{s_n})^{\times} \cup \cdots \cup (((\Sigma_0)^{\prime})_{s_1})^{\times} \cup (\Sigma_0)^{\times} \end{aligned}$$

where ($\dagger$) is justified by the two identities $((\Sigma)^{\times})^{\prime} = \emptyset$ and $((\Sigma)^{\times})^{\times} = (\Sigma)^{\times}$. The reader can verify that the pseudo-code of EDA is a recursive formulation of ($\ddagger$) for $n = |Dom(t)|$. $\square$

Example 5 (EDA execution). Consider the AST of Example 2. EDA yields a smaller classification, containing "just" 6 error messages:

1. `"in f(?₁ + z) =?₂: z is a vector, but is used as a scalar"`
2. `"in f(?₁+?₂⃗+z) =?₃: ?₁+?₂ is a scalar, but is used as a vector"`
3. `"in f(α · x⃗+?₁⃗+z) =?₂: x is a vector, but is used as a scalar"`
4. `"in f(α⃗·x⃗+β · y⃗+z) =?₁: y is a vector, but is used as a scalar"`
5. `"in f(α⃗·x⃗+β⃗·y⃗+z) =?₁ + z: z is a vector, but is used as a`
   `scalar"`
6. `"in f(α⃗·x⃗+β⃗·y⃗+z) =?₁⃗+z: ?₁ + z is a vector, but is used as a`
   `scalar"`

where (5) is the expected one, while the other errors are spurious. The rating of errors is unchanged with respect to Example 2.

4   A Humane Disambiguation Algorithm

We look for a restriction of Criterion 1 which can be integrated in EDA. The characteristic of EDA (with respect to the general refinement process) is the pre-visit ordering of $Dom(t)$. This implies that:

a. to interpret an occurrence s, every occurrence s' that precedes s in pre-order must be interpreted too;

b. when an interpretation ϕ yields an error, every occurrence s' that follows in pre-order the last occurrence s added to the domain of ϕ will not be interpreted by any interpretation $\phi' \sqsupseteq \phi$.

Together, (a) and (b) imply that not every sub-formula F will be interpreted in any possible way. Actually, (b) is a consequence of (a). This imposes a non negligible restriction of Criterion 1 for efficiency reasons, yielding:

Criterion 2 (Efficient spurious error detection). *An error message relative to an interpretation ϕ of an AST t is* spurious *iff there exists an occurrence $s \in Dom(t)$ and an interpretation ϕ' such that:*

1. $\phi(s) \neq \phi'(s)$;
2. $\phi'(s') = \phi(s')$ for all s' that precedes s in pre-order;
3. ϕ' is total on the occurrences of overloaded symbols occurring in the sub-tree rooted at s;
4. $\mathcal{R}(\llbracket t \rrbracket_{\phi'}) = \checkmark$.

Dropping (2)—imposed by (a)—from the conditions above we obtain a more formal writing of Criterion 1. We now address the issue of integrating Criterion 2 in EDA.

$f(\Sigma, l)$, the core of EDA, does not work directly on t, but rather on the list l, which is an abstraction of the occurrences of overload symbols in t. In l the tree-structure of t has been lost. As a consequence, without changing its input, we cannot make f recognize spurious errors using Criterion 2. As a solution we could make f work by recursion on t by integrating in f a pre-visit traversal. Still, we prefer to avoid binding f to the data type of AST of formulae and to keep separate the construction of $Dom(t)$ from the actual disambiguation.

Therefore we introduce the new $Dom^{\text{tree}}(t)$ datatype which is a tree representation of $Dom(t)$. $Dom^{\text{tree}}(t)$ is a tree which contains only the nodes $s \in Dom(t)$ and preserves the ancestor-descendent relation of t. As a concrete representation of $Dom^{\text{tree}}(t)$ we adopt the well-known first-child/next-sibling representation. This representation allows to implement straightforwardly a pre-visit of the tree recognizing when all children of a given node have been traversed. Note that the pre-visit order is imposed by the efficiency analysis given in [6] and recognizing the end of children traversal is necessary for Criterion 2.

We call the algorithm that recognizes spurious errors the *humane disambiguation algorithm* (HDA for short). It proceeds by recursion on $Dom^{\text{tree}}(t)$ and, at the end of children traversal, lowers the rate of spurious errors. The pseudo code of HDA is given below:

$$g(\Sigma, t) = \begin{cases} \Sigma & \text{if } t = nil \\ g((\Sigma_1)^\nu, b) \cup p((\Sigma_1)^\nu, (\Sigma_1)^\times \cup (\Sigma_s)^\times) & \text{if } t = \begin{smallmatrix} s \to b \\ \downarrow \\ c \end{smallmatrix} \\ \text{where } \Sigma_1 = g((\Sigma_s)^\nu, c) \end{cases}$$

$$p(\Sigma_{ok}, \Sigma_{err}) = \begin{cases} \Sigma_{err} & \text{if } \Sigma_{ok} = \emptyset \\ \{\langle \phi, o, r \rangle \mid \langle \phi, o, \langle m, p \rangle \rangle \in \Sigma_{err}, r = \langle \text{\textbf{!}}, p \rangle\} & \text{if } \Sigma_{ok} \neq \emptyset \end{cases}$$

$$HDA(t) = (\Sigma')^\nu \cup p((\Sigma')^\nu, (\Sigma')^\times \cup (\Sigma_0)^\times)$$
$$\text{where } \Sigma' = g((\Sigma_0)^\nu, Dom^{\texttt{tree}}(t))$$

g has the same role f had in EDA, while $p(\cdot, \cdot)$ (mnemonic for "prioritize") lowers the rate of spurious errors to **!**, which is the lowest rating.

Theorem 6 (Correctness of HDA)

1. *HDA implements a disambiguation algorithm.*
2. *An error in a classification returned by HDA is spurious according to Criterion 2 iff it is rated $\langle \text{\textbf{!}}, \rho(\phi) \rangle$.*

Proof. We just give a sketch of the proof, which is involved due to the complexity of the code.

(1) By Theorem 5 it is sufficient to prove that the classification returned by HDA is equal to the classification returned by EDA up to rates. Since both algorithms perform a pre-visit of the input tree, we can consider "parallel" executions of them. At the nth step EDA is called on the list $s_n :: tl$ while HDA is called on the tree $\begin{smallmatrix} s_n \to b \\ \downarrow \\ c \end{smallmatrix}$. The nodes that EDA will encounter processing tl are the same (and in the same order) of those HDA will encounter processing c at first and then b. The thesis is reduced to a proof by induction on the length of tl that $f((\Sigma_{s_n})^\nu, tl)$ is equal to $(g((\Sigma_{s_n})^\nu, c))^\times \cup g(g((\Sigma_{s_n})^\nu, c)^\nu, b)$ up to rates.

(2) Recursion is never performed on elements of the current classification corresponding to errors. Thus once an error has been down-rated by $p(\cdot, \cdot)$ its rating will never be raised again.

Suppose that at a given iteration $p(\cdot, \cdot)$ lowers the rating of an error ϵ relative to an interpretation $\phi \in (\Sigma_s)^\times \cup (g((\Sigma_s)^\nu, c))^\times$. We interpret that as ϵ being located in $\begin{smallmatrix} s \\ \downarrow \\ c \end{smallmatrix}$. The set $\tilde{S} = S_{(g((\Sigma_s)^\nu, c))^\nu}$ is not empty since ϵ has been down-rated.

We consider now two cases: either there exists $\phi' \in \tilde{S}$ such that $\phi(s) \neq \phi'(s)$ or not. In the former case s and ϕ' satisfy all the requirements of Criterion 2. In the latter case let $\phi' \in \tilde{S}$. Let $s' \in c$ be the last occurrence that follows s in pre-order such that $\phi(s') \neq \phi'(s')$. Consider now the recursive call on $\begin{smallmatrix} s' \to b' \\ \downarrow \\ c' \end{smallmatrix}$ and iterate the above reasoning. Since this time $\phi(s') \neq \phi'(s')$, ϵ is now properly down-rated according to Criterion 2. When the recursive call on c returns ϵ is still correctly down-rated and $p(\cdot, \cdot)$ leaves its rate unchanged.   □

Example 6 (HDA execution). Consider again the AST of Examples 2 and 5. The first recursive invocation is $g(\Sigma, \tau)$ where: $\Sigma = \{\langle \bot, \checkmark, \langle \text{\ss}, \rho(\bot) \rangle \rangle\}$ and $\tau = \begin{smallmatrix} & +\rightarrow b \\ \downarrow & \\ c & \end{smallmatrix}$.

g computes

$$\Sigma_s = \{\langle \phi_{11}, \checkmark, \langle \text{\ss}, \rho(\phi_{11}) \rangle \rangle, \quad \text{where} \quad [\![t]\!]_{\phi_{11}} = f(?_1 \overline{\mp} z) = ?_2$$
$$\langle \phi_{12}, \boldsymbol{\mathsf{X}}, \langle \text{\ss}, \rho(\phi_{12}) \rangle \rangle\} \qquad\qquad [\![t]\!]_{\phi_{12}} = f(?_1 + z) = ?_2$$

and then calls itself recursively on $(\Sigma_s)^\checkmark$ and c yielding

$$\Sigma_1 = \{\langle \phi_{11111}, \checkmark, \langle \text{\ss}, \rho(\phi_{11111}) \rangle \rangle, \quad \text{where} \quad [\![t]\!]_{\phi_{11111}} = f(\alpha \overrightarrow{\cdot} x \overline{\mp} \beta \overrightarrow{\cdot} y \overline{\mp} z) = ?_1$$
$$\langle \phi_{11112}, \boldsymbol{\mathsf{X}}, \langle \text{\ss}, \rho(\phi_{11112}) \rangle \rangle, \qquad\qquad [\![t]\!]_{\phi_{11112}} = f(\alpha \overrightarrow{\cdot} x \overline{\mp} \beta \cdot y \overline{\mp} z) = ?_1$$
$$\langle \phi_{1112}, \boldsymbol{\mathsf{X}}, \langle \text{\ss}, \rho(\phi_{1112}) \rangle \rangle, \qquad\qquad [\![t]\!]_{\phi_{1112}} = f(\alpha \cdot x \overline{\mp} ?_1 \overline{\mp} z) = ?_2$$
$$\langle \phi_{112}, \boldsymbol{\mathsf{X}}, \langle \text{\ss}, \rho(\phi_{112}) \rangle \rangle\} \qquad\qquad [\![t]\!]_{\phi_{112}} = f(?_1 + ?_2 \overline{\mp} z) = ?_3$$

Since $(\Sigma_1)^\checkmark$ is not empty, all the errors in $(\Sigma_s)^\times$ and $(\Sigma_1)^\times$ are recognized as spurious and their rating is lowered to $\text{\ss}$. In particular the new rating for the error associated to ϕ_{12} will remain the same in the final classification returned by HDA. Errors coming from $(\Sigma_1)^\times$ were already recognized as spurious; this is not always the case.

Eventually HDA yields the same errors of Example 5, but rated differently: the expected one—error (5)—is rated $\langle \text{\ss}, \rho(\phi_5) \rangle$ (ranking first) while the remaining spurious errors are rated $\langle \text{\ss}, \rho(\phi_i) \rangle$.

5 Conclusions

In this paper we proposed a heuristic criterion to detect spurious errors in ambiguous formulae. An error is spurious when it is not relative to the formula interpretation expected by the user. We integrated the criterion in the efficient disambiguation algorithm of [6].

We also believe that the specification of a disambiguation algorithm (Section 2) and the description of our efficient disambiguation algorithm (Section 3) are an improvement over previous descriptions in the literature.

We have implemented the proposed algorithm in the Matita proof assistant [2] and experimented with it in an ongoing formal development of Lebesgue's dominated convergence theorem in an abstract setting. Actually this formalization effort has motivated the study of spurious error identification since in the abstract setting there are plenty of overloaded operators and it was not unusual to be faced with too many error messages to be useful. In the current implementation in Matita we have decided to hide spurious errors from the user, unless explicitly asked for. This choice has decreased dramatically the amount of error messages, but in the general case is still possible to be faced with more than 1 genuine (i.e. not spurious) error. The problem of how effectively present multiple error messages to the user belongs to the user-interface field and will be discussed in a forthcoming paper.

For efficiency reasons, the criterion implemented in Matita is Criterion 2, that is a restriction of Criterion 1. There are cases of undetected spurious errors in Matita that would have been caught by the more general criterion. Consider for instance the right hand side of the formula given in Example 1. According to Criterion 1 the only two genuine errors are:

- "in $?_1 + z$: z is a vector, but is used as a scalar"
- "in $?_1 + ?_2 \overrightarrow{+} z$: $?_1 + ?_2$ is a scalar, but is used as a vector"

however, according to Criterion 2, we also get the errors:

- "in $\alpha \cdot f(x) \overrightarrow{+} ?_1 \overrightarrow{+} z$: $\alpha \cdot f(x)$ is a scalar, but is used as a vector"
- "in $\alpha \overrightarrow{\cdot} f(x) \overrightarrow{+} ?_1 \overrightarrow{+} z$: $f(x)$ is a scalar, but is used as a vector"

Moreover Criterion 1 is debatable itself: are both the above "genuine" errors really genuine? Would mathematicians agree that the second error is spurious since the number of scalars is greater than the number of vectors in the sum? What if there were two scalars and two vectors in the same sum? Or does the order matter? Does the first addend determines the signature of the sum?

Unable to convince ourselves that a general answer to the above questions exists, we claim that Criterion 1 is widely acceptable and never gives false positives. Whether the gap between the two criteria can be reduced without loosing efficiency is an open research direction.

References

1. Asperti, A., Guidi, F., Sacerdoti Coen, C., Tassi, E., Zacchiroli, S.: A content based mathematical search engine: Whelp. In: TYPES 2004. LNCS, vol. 3839, pp. 17–32. Springer, Heidelberg (2004)
2. Asperti, A., Sacerdoti Coen, C., Tassi, E., Zacchiroli, S.: User interaction with the Matita proof assistant. Journal of Automated Reasoning, Special Issue on User Interface for Theorem Proving (To appear 2007)
3. Bancerek, G., Rudnicki, P.: Information retrieval in MML. In: Asperti, A., Buchberger, B., Davenport, J.H. (eds.) MKM 2003. LNCS, vol. 2594, Springer, Heidelberg (2003)
4. Geuvers, H., Jojgov, G.I.: Open proofs and open terms: A basis for interactive logic. In: Bradfield, J.C. (ed.) CSL 2002 and EACSL 2002. LNCS, vol. 2471, pp. 537–552. Springer, Heidelberg (2002)
5. César Muñoz. A Calculus of Substitutions for Incomplete-Proof Representation in Type Theory. PhD thesis, INRIA (November 1997)
6. Sacerdoti Coen, C., Zacchiroli, S.: Efficient ambiguous parsing of mathematical formulae. In: Asperti, A., Bancerek, G., Trybulec, A. (eds.) MKM 2004. LNCS, vol. 3119, pp. 347–362. Springer, Heidelberg (2004)
7. Zentralblatt MATH. http://www.emis.de/ZMATH/

Methods of Relevance Ranking and Hit-Content Generation in Math Search*,**

Abdou S. Youssef

Department of Computer Science
The George Washington University
Washington DC, 20052 USA

Abstract. To be effective and useful, math search systems must not only maximize precision and recall, but also present the query hits in a form that makes it easy for the user to identify quickly the truly relevant hits. To meet that requirement, the search system must sort the hits according to domain-appropriate relevance criteria, and provide with each hit a query-relevant summary of the hit target.

The standard relevance measures in text search, which rely mostly on keyword frequencies and document sizes, turned out to be inadequate in math search. Therefore, alternative relevance measures must be defined, which give more weight to certain types of information than to others and take into account cross-reference statistics. In this paper, new, multi-dimensional relevance metrics are defined for math search, methods for computing and implementing them are discussed, and comparative performance evaluation results are presented.

Query-relevant hit-summary generation is another factor that enables users to quickly determine the relevance of the presented hits. Although the hit title accompanied by a few leading sentences from the target document is simple to produce, this often fails to convey to the user the document's relevant excerpts. This shifts the burden onto the user to pursue many of the hits, and read significant portions of their target documents, to finally locate the wanted documents. Clearly, this task is too time-consuming and should be largely automated. This paper presents query-relevant hit-summary generation methods, outlines implementation strategies, and presents performance evaluation results.

1 Introduction

Digital math libraries consist mostly of equations, graphs, tables, numerous embedded mathematical expressions, and text. Clearly, users will need specialized search systems to find and locate quickly the math information that is most relevant to their needs. A number of search systems have been built and are undergoing further enhancements, such as the NIST DLMF search system [13,17,18] and the Design Science's Mathdex [10].

---

* This work was done in part at the National Institute of Standards and Technology, USA, as part of the DLMF Project.
** This work was supported in part by the National Science Foundation (NSF), USA, under Grant No. 0208818.

M. Kauers et al. (Eds.): MKM/Calculemus 2007, LNAI 4573, pp. 393–406, 2007.
© Springer-Verlag Berlin Heidelberg 2007

For enhanced utility and user-satisfaction, math search systems must not only maximize precision and recall, but also present the query hits in a form that makes it easy for the user to identify quickly the truly relevant hits. To meet that requirement, the search system must sort the hits according to domain-appropriate relevance criteria, and provide with each hit a query-relevant summary of the hit target.

The standard relevance measures in text search, which rely mostly on keyword frequencies and document sizes, turned out to be inadequate in math search. Therefore, alternative relevance measures must be defined, which give more weight to certain types of information than to others, such as definitions, theorems, "standard" functions and operators, and frequently referenced items. In this paper, new, multi-dimensional relevance metrics are defined for math search, methods for computing and implementing them are discussed, and comparative performance evaluation results are presented.

Query-relevant hit-summary generation, or simply *hit packaging*, is another factor that enables users to quickly determine the relevance of the presented hits, and thus determine the most relevant hits. Although the hit title, possibly accompanied by a few leading sentences from the target document, forms a fast and simple way for hit packaging, it often fails to convey to the user the document's relevant excerpts. This shifts the burden onto the user to pursue many of the hits, and read significant portions of their target documents, to finally locate the wanted documents. Clearly, this task is too time-consuming, and should be done by the software on behalf of the user. This paper presents query-relevant hit-summary generation methods, outlines implementation strategies, and shows substantiating illustrations.

2   Background and Related Work

Three types of math search systems have received attention and/or have been built. The first is field-based search systems, which are now widely deployed in several mathematics databases and by many mathematical content providers, such as Zentralblatt's ZMATH and MathDi [19,9], the Jahrbuch Database [6], AMS's MathSCiNet [1], and various professional mathematical socities. Such systems are intended for conventional library search, and are outside the scope of this paper. The second is formal-math search, such as the search systems developed and researched by Guidi et al [4,5], MoWGLI of the Helm project [14], and MIZAR [2]. Formal-math search systems are highly specialized and usually intended for advanced mathematicians, and are thus outside the scope of the paper.

The third type of math-search is math-aware fine-grain search such as the DLMF search system [13,17,18], the Design Science's Mathdex Web search system [10,11], and Mathematica search system [12]. This type of math search is indended for general use by students, educators, researchers, and professionals, in mathematics, physical sciences, and engineering. It is this kind of search that requires further investigation for relevance ranking and hit packaging, which are the focus of this paper.

Relevance scoring has received much research attention in text search for over three decades [16,15]. Although several relevance metrics have been developed and studied, most are elaborations and variations of one central metric, often referred to as the *tf-idf metric* (term frequency inverse document frequency). Essentially, this metric is predicated on the assumptions that (1) the higher the relative frequency of a query keyword in a hit document is, the more relevant the document is, and (2) the more frequent a term is in the whole database, the less important its occurrences are. One implication is that if two documents have the same number of occurrences of the keywords but one is a smaller document than the other, the smaller document ranks higher because its relative term-frequency (i.e., number of keyword occurrences divided by the document size) is larger.

Such traditional considerations are highly inadequate in math search. For example, if the hit targets are equations, a smaller-size equation is not necessarily more relevant or more important to the user. Also, in math, the frequency of occurrence of a term is much less important than the mathematical significance of that term. Finally, the importance of a term is context-dependent, especially in math, as for example in what part of a math structure the term lies, and what other terms the term co-occurs with.

The shortcomings of traditional relevance scores were recognized in Web search, especially by Google. It was realized that the importance, and thus relevance, of a document/page depends more on who publishes it, how many links point to it, how many times it is visited, and such, than the "uninformed" statistics of term frequencies and document sizes.

These same considerations can be utilized in math search, but after significant adaptations and specialization to math contents. For example, the number of times a particular math entity (e.g., equation) is referenced in a document/site can be a very telling indication of the relative importance of that entity. In addition to cross-reference statistics, domain-specific term weighting can be taken into account in relevance scoring, with great expected benefits. For example, if a query includes among its keywords the term "Bessel" and the variable name "x", then intuitively the first term is much weightier than the second term.

The relevance metrics used in the current generation of mostly experimental math search systems are primarily identical to the ones used in conventional text search, that is, the tf-idf metric. In this paper, alternative metrics are developed and shown to yield better results.

The other subject of focus in this paper, which has an equal bearing on helping users find relevant information fast, is hit-description generation. Hit-description generation, or hit packaging, has never been viewed as a major issue in text search, and has thus been done in a rather simple way. Prior to Web search, text search systems often reported each hit as a document title, sometimes accompanied by a few leading sentences in the hit's document. In Web search, such as in Google, the hit package consists of the page title of the hit, accompanied with 2–4 lines of sentences or sentence fragments that contain the keywords of the query, usually highlighted.

As math search is still in its early experimental phases, where more pressing issues have had to be addressed first, the same methods used in text search are used by necessity, until more specialized alternatives are found. In Mathdex of Design Science, the Web page title and the first couple of lines of the Web page contents of the hit are displayed with the hit. As a significant enhancement, a special button is added next to each hit, which when moused over, shows one equation or math expression that made the page match the query. In early experimental versions of DLMF search, two hit-packaging methods were used, depending on the nature of the hit. If the hit target has a small amount of contents, such as equations or even graphs and small-size tables, the entire target content is presented in the hit itself, providing immediacy and directness. If, on the other hand, the hit target is a section of a chapter, the hit description consists of the section title and the chapter title. Mathematica search is somewhat more advanced in providing hit descriptions. Like Google, Mathematica offers with each hit about 2 lines of sentence fragments that contain query keywords. Mathematica's hit packaging may be adequate for Mathematica contents, which tend to be short descriptions of functions or portions of code mixed with some text, but it will not be sufficient for general-purpose math search.

Clearly, much more representative and query-relevant descriptions of hit targets should be generated per math-hit. The reason is that the user will be able to judge faster the value and relevance of the hit without having to pursue many hits and read long passages in them before the valuable and truly relevant information is found. Techniques generating such descriptions/summaries are presnted in this paper.

3   Relevance Ranking in Math Search

Before the new relevance metrics are introduced and related considerations discussed, it is instructive to look at the standard tf-idf metric. For a query q and a hit-target document d in some presumed database DB, the tf-idf relevance metric value is:

$$Relevance_q(d) = \sum_{query\ terms\ t} tf(t, d) \times idf(t)$$

where

$$tf(t, d) = \sqrt{\frac{frequency(t)}{|d|}}$$

and

$$idf(t) = \log \frac{|DB|}{number\ of\ documents\ containing\ t}.$$

($|d|$ is the number of terms in document d, and $|DB|$ is the number of documents in the database.)

Note that the first factor, $tf(t, d)$, represents the frequency of a term in the document, normalized by the document size, and that the second factor, $idf(t)$, represents the inverse of the number of documents containing the term relative to the total number of documents in the database. The square root and the log are meant to attenuate the contributions of those factors to various degrees.

A deeper look into the formula reveals that the first factor attempts to capture the importance (or weight) of the term t with respect to the document d (and thus the relevance of the document relative to the term t), while the second factor attempts to capture the weight of the term t with respect to the database as a whole.

The paper will preserve this paradigm of expressing the relevance of a document to a term in terms of of the weight of the term vis-a-vis the document and the weight of the term vis-a-vis the database. What will change is the way of measuring each of those factors; $tf(t, d)$ will be replaced by a general term-document weight function **Weight**(t, d), and $idf(t)$ will be replaced by a term weight function **Weight**(t), and a math object weight function **Weight**(mo) will be introduced (where a math object can be a full document or some small items such as an equation or even a sentence); all such weight functions will be elaborated later. Furthermore, since various aspects will influence those factors, and some aspects are absolutely more important than others, it will be determined that a multidimensional relevance metric, which is then a *relevance vector*, is a more apt way of measuring relevance and thus of sorting the hits.

As argued earlier, mere frequency and size statistics do not fully capture the importance and relevance of documents. Rather, several other static (i.e., query-independent) and dynamic (i.e., query-dependent) aspects have to be taken into account when computing **Weight**(t, d), **Weight**(t) and **Weight**(d).

Static Weight Information

Many math terms have intrinsic importance due to what they stand for, and some terms have more intrinsic importance than others. For example, special function names stand for much more than a moot variable name. Similarly, certain operators, such as integration ($\int$), exponentiation and division, are more important than variable names. This type of intrinsic importance of terms in themselves is called *categorical importance*. Categorical importance is a primary determinant of the term-weight function **Weight**(t).

Accordinly, the term-weight function **Weight**(t) can be defined as follows:

$$\textbf{Weight}(t) = \textbf{Quantify}(\textbf{Type}(t)),$$

where

- **Type** maps a term to a category based on some typology or taxonomy of terms from a term-importance perspective. For example, the term categories can be "operator", "special-function" and "regular" (for everything else).
- **Quantify** is a mapping that maps a term-type into a positive real number associated with that type, where the more important a type is, the larger

its associated number is. For example, one can have **Quantify**(regular) = 1, **Quantify**(operator) = 2, and **Quantify**(special function) = 4.

Much like terms, math objects (e.g., equations or full documents) have intrinsic importance irrespective of the query. Several aspects feed into that importance:

1. the type of the math object, such as equation, graph, table, bibliographic item, notation item, and so on;
2. the categorical importance of the member terms and other constituent (i.e., subset) objects;
3. the number and possibly types of cross-references made to the object by other objects in the database (or even on the Web). The types of cross-references are taxonomized in two ways. In the first taxonomy, a cross-reference can be *local* or *global*:
   - a local cross-reference is one where the referring object and the referred-to object belong to one and the same division of information, such as one chapter or one Website;
   - a global cross-reference is one where the two objects belong to two different divisions of information.

In the second taxonomy, cross-references can be *definitional cross-references* or *propositional cross-references*:
   - A reference from object A to object B is definitional if both of the following conditions are met:
     - Object B defines some mathematical term/concept c
     - Object A refers explicitly to object B as the object that defines c.
   - A reference from object A to object B is propositional if both of the following conditions are met:
     - Object B states and/or proves some proposition (where the term "proposition" is used in a broad sense, so it encompasses theorems, lemmas, corollaries, "inline" substantiated or stipulated claims, etc.)
     - Object A refers explicitly to object B as the primary location of proposition P.

Accordingly, the math-object weight function **Weight**(mo) can be defined as follows:

$$\mathbf{Weight}(mo) = \mathbf{Combine}(\mathbf{Quantify}(\mathbf{Type}(mo)), \mathit{TW}(mo), \mathbf{CR}(mo)),$$

where

   - **Type** and **Quantify** are like those for terms except here the categories are those of math objects;
   - $\mathit{TW}(mo)$ captures the weight of the terms that make up the math object mo;
   - $\mathbf{CR}(mo)$ captures and quantifies the statistics of cross-reference pointers pointing to mo;

- **Combine** combines the various aspects (i.e, object type, wieght of the constituent terms of the object, and cross-reference information) into either a
  scalar or a vector value, as explained next.

Combining several factors of various degrees of importance into a single
ranking-metric can be done in two ways. The first way is to map the vector
$V = (x_1, x_2, \ldots, x_n)$ of factors into a scalar value S, such as by adding or
multiplying the components, where every component x_i is magnified by some
weight w_i to reflect its relative importance. That is, the scalar value formula
can be $S = \prod_{i=1}^{n} x_i^{w_i}$ or $S = \sum_{i=1}^{n} w_i x_i$, among many possibilities of combining
weighted factors. With scalar metrics, the ranking is done by straightforward
sorting of objects according to their scalar ranking metric.

The other way of combining factors is to map the vector of factors $V =$
$(x_1, x_2, \ldots, x_n)$ into another, carefully ordered vector of factors $V' = (y_1, y_2,$
$\ldots, y_m)$, resulting in a *vector ranking metric* of the same as or smaller dimen-
sionality than that of the original vector V. The first component y_1 corresponds
to the factor of highest weight, y_2 corresponds to the factor of the second highest
weight, and so on. The ranking of objects is then done by lexigraphic sorting of
the vector metric values of the objects.

Vector ranking metrics have several advantages. First, there is no need to concern oneself about how the weights of the various factors should be quantified
and factored into metric formula. Second, and more importantly, vector metrics
and lexicographic sorting stricly enforce the policy that a most important factor should not be overwhelmed by a comination of less important factors. For
example, if an object A has the highest y_1 value among a set of objects, it will
rank ahead of all the other object regardless of the values of the other y_is. In
particular, if definitional types of objects are desired to rank at the tops of hits,
the system can have the first component of V' correspond to object type, and
give the largest value to definition types (compared to other object types such
as propositions, graphs, etc.).

In this paper, the vector ranking metric approach is adopted. To be precise,
the **combine** function used employs a hybrid of scalarization and vectorization
as seen next.

The $\mathbf{TW}(mo)$ function can have scalar or vector values. Specifically, assume
that the types of terms are $\{T_1, T_2, \ldots, T_k\}$, as for example $\{$regular, operator,
special-function$\}$. Then,

$$\mathbf{TW}(mo) = \sum_{i=1}^{k} \mathbf{Quantify}(T_i) \times N_i(mo)$$

or

$$\mathbf{TW}(mo) = (N_1(mo), N_2(mo), \ldots, N_k(mo))$$

where

$$N_i(mo) = \text{number of terms of type } T_i \text{ in the object } mo.$$

The $\mathbf{CR}(mo)$ function maps the cross-reference information into a vector that reflect the number of cross-references of the four possible types identified earlier:

$$\mathbf{CR}(mo) = (GD(mo), GP(mo), LD(mo), LP(mo))$$

where

- $GD(mo)$ = number of global, definitional cross-references to object mo
- $GP(mo)$ = number of global, propositional cross-references to object mo
- $LD(mo)$ = number of local, definitional cross-references to object mo
- $LP(mo)$ = number of local, propositional cross-references to object mo.

Dynamic Weight Information

Dynamic weight information relates to the weight of math object mo relative to the terms t of a query q. That information is incorporated into the function $\mathbf{Weight}(t, mo)$ or generally $\mathbf{Weight}(q, mo)$.

One possible definition of $\mathbf{Weight}(q, mo)$ is the same as $\mathbf{TW}(mo)$ except that the terms will be limited to those that are in the intersection of the object and the query. An elaboration on this definition would be to factor in the number $ND(q, mo)$ of the query keywords that are defined in the object mo. Therefore, assuming that the types of terms are $\{T_1, T_2, \ldots, T_k\}$,

$$\mathbf{Weight}(q, mo) = (ND(q, mo), N_1(q, mo), N_2(q, mo), \ldots, N_k(q, mo))$$

where

$$N_i(q, mo) = |\{t \mid \mathbf{Type}(t) = T_i \text{ and } t \in mo \text{ and } t \text{ is a keyword of the query } q\}|.$$

Overall Relevance Vector Metric

Based on the preceding analysis and discussions, the overall relevance metric is a vector made up of the components of $\mathbf{Weight}(q, mo)$ vector and the $\mathbf{Weight}(mo)$ vector, ordered according to what the system designer's assigned relative importance of each component.

It is the author's judgement, and for the sake of performance evaluation presented later, that a good relevance metric be a vector where the relevance components, ordered from the highest importance to the lowest importance, are:

- $ND(q, mo)$, which is the number of query keywords defined in the object mo,
- $N_i(q, mo)$ for the top one or two most important term types, where $N_i(q, mo)$ is the number of terms of type T_i that occur in both the query and the object,
- $\mathbf{CR}(mo) = (GD(mo), GP(mo), LD(mo), LP(mo))$. It captures the global and local definitional/propositional cross-reference statistics,
- the remaining $N_i(q, mo)$s
- $\mathbf{TW}(mo)$, which is the term-weight of the object mo, expressed either as a vector or a weighted sum,
- (optional) $\mathbf{Quantify}(\mathbf{Type}(mo))$, reflecting preferences for certain document/object types over others,
- tf-idf(q, mo), as a final tie-breaker.

3.1   Speed Performance Evaluation of Hit Ranking

The relevance ranking scheme discussed in this paper has been implemented and tested on the DLMF testbed. Several queries with a range of numbers of hits were tested to measure the overhead of relevance ranking. A sample of the results is presented in Table 1. The table shows the queries, the number of hits per query, the search time for identifying the hits but without ranking, and the time to perform the relevance computation and relevance ranking of the hits. As can be seen, the relevance computation and ranking time is usually higher than the search time, and, naturally, it is higher for larger numbers of hits. Nevertheless, for a standalone database of the size range of DLMF (i.e., about 1000 pages of contents containing over ten thousands equations), the number of hits will usually be in the tens, hundreds, or at most in the thousands, the relevance ranking overhead ranges from a tenth of second to at most a second, which is quite acceptable.

For Web search relevance ranking, where the number of hits could conceivable be in the hundreds of thousands or even millions, the relevance ranking time will be significantly higher. However, the overhead can be managed down to practical ranges. One possibility is to do a two-stage ranking. In the first stage, a coarse relevance metric is applied, which takes into account a carefully selected small subset of the relevance criteria when computing relevance, and instead of sorting all the hits, find the top 100 (or so) hits. In the second stage, a full-fledged relevance evaluation and sorting of those 100 hits is done and the hits are presented to the user in hit-pages, about 10 hits per page. Since the truly relevant hits are very likely to be in the top 100, and most users rarely search down beyond that level, this approach will often be sufficient. In the rare cases where a user wishes to see the hits below rank 100, the 2nd stage is repeated on the next 100 hits, and so on.

It is left to future work to address the important question of determining which subset of relevance criteria makes a good coarse-grained relevance metric to be used in the first stage of the 2-stage Web search relevance ranking process.

Table 1. Speed Performance of Search and Hit Ranking (All time measurements are in milliseconds)

Query	Number of Hits	Search Time	Hit-ranking Time
Ai^2	7	16	15
$\int \sin$	19	15	16
eulerBeta	28	15	32
$\sin^2$	80	15	78
jacobiSN OR Si	94	31	63
eulerGamma	653	31	344
cos	666	16	297
sin	707	15	329
z	2499	16	828

3.2   Outcome Performance Evaluation of Hit Ranking

Outcome performance evaluation of relevance ranking is extremely subjective. A thorough evaluation of this sort will be left to future work, where a statistically significant number of users and a benchmark of queries are identified and used, and a metric of user satisfaction is decided upon and utlized in the collection of user assessments of the search system, including the relevance ranking and the hit-description generation which is discussed in the next section.

For now, suffice it to say that based on the expectations that definitions will be sought after more often and by more users, and based on the valuation scheme where the definitions/equations/plots that are cross-referenced more often are of more weight, the outcome is far superior to the default tf-idf relevance ranking approach. Hundreds of queries were tested. In each and every case, definitions and notations of the query keywords ranked on top, and items of higher cross-reference values ranked higher. Under the tf-idf relevance model, such hits were "buried" in the second, third, or fourth page of hits.

We predict that future evaluation of user satisfaction will confirm the hypothesis that the new relevance metrics are far superior to the tf-idf metric. Of course, further refinements will be suggested by the future subjective evaluation.

4   Hit-Description Generation in Math Search

As mentioned earlier, the rather simple way of putting together the hit-title and a few leading sentences of the hit-target fails to convey to the user why a doument matched and whether the matching parts are indeed relevant. It will be much better to the user if those parts are extracted and provided with the hit so the user can quickly dtetermine whether or not a hit is worth pursuing. Furthermore, of those parts are determined carefully, they may often be all that the user needs from a document, thus saving him/her from extra efforts. This section will provide new methods for determining query-relevant excerpts from math documents. Before starting, it must be noted that if the hit targets are small math objects (e.g., equations or graphs), then such objects should be displayed directly with the hits as they make the best representation of hits. Therefore, for the rest of this section, it will be assumed that the hit targets are relatively sizable objects that cannot be conveniently displayed along with hits, such as sections, chapters, articles, and so on.

The approach to hit-description generation consists of several tasks. Some tasks must be carried out at indexing time, while other tasks must be at search time. One major goal is to minimize the computations that must be done at search time so that query turn-around time is short enough for users.

Index-Time Tasks for Hit Generation

1. Fragment each document in the database into very small units of information, where a unit can be (1) an equation, (2) a sentence, which may contain inline math expressions, (3) a graph, (4) a fragment of a table (in the case

where tables are large), (5) a title of a chapter/section/subsection, (6) a notational item, and so on. This fragmentation will take place when the documents are indexed.

2. Each fragment is then turned into a mini-document with its own ID. The mini-document contains, besides its contents, several fields of information that will facilitate and speed up the hit-description generation at search time. One field is the ID of the document of which the mini-document is a fragment. Other fields contain static information that will be used to measure the relevance vector of the mini-document at search time.

3. Index the fragments (i.e., mini-documents) of all the documents, and store the index information in a separate index structure, termed the *fragment index*. That index is different from the index for the documents. Note that fragment contents and the fields in the fragment are stored verbatim in the fragment index. The reason for this will be explained below.

Search-Time Tasks for Hit Generation

At search time, when the IDs of the hits that match a query have been determined, the hits are presented one page at a time (typically 10 hits per page). For each page of hits, the descriptions of those hits are generated. The following outlines the tasks to be performed:

1. For each hit in a hit-page, identify the ID of the target document, and formulate a *derivative query* made up of the conjunction of the original query and the ID of the target document.

2. Submit the query for search against the *fragment index*. This results in several "sub-hits", each of which is a fragment of the hit target document.

3. If no sub-hits are returned, relax the derivative query so that the keywords in the original query are combined into a disjunctive query (i.e., an OR-query of the keywords), and repeat step 2, resulting this time with one or more sub-hits.

4. The sub-hits are then relevance-ranked using the relevance vectors described in the previous section. Note that the relevance vector values of the fragments can be computed fast because much of the weight information (i.e., the static weight information) is stored in the fragment index, and thus need not be computed from scratch.

5. A few top-scoring sub-hits (i.e., fragments) are selected, retrieved from the fragment index, and combined (in document-order) into a descriptive summary that is presented along with the hit title in the hits page.

Several remarks are in order. First, this hit-description method requires no file IO since all the fragment contents are stored in the fragment index, which is a file that remains open as long as the search system is running. This greatly speeds up the hit-description generation process. Second, the identification of the relevant excerpts (i.e, fragments) is rather fast and straightforward: it is a search-within-search process. Third, the relevance ranking of the matching fragments is also fast since the static weight statistics are computed and stored at indexing time, thus reducing the conputation time for obtaining the relevance vectors of the fragments. Last, hit-description generation requires considerably more disk space to store the fragment index, which is much larger that the document index because the actual fragments are stored in the fragment index. However, since disk space is very inexpensive, the cost overhead is not a serious disvantage.

4.1   Speed Performance Evaluation of Hit-Description Generation

The same performance evaluation was done for hit-description generation as for relevance ranking. A sample of the results is shown in Table 2. The table shows the queries, the time it takes to derive the description of a single hit, and the time to derive the descriptions of 10 hits that make up a hit-page. As can be seen, the time for generating the descriptions for the hits in one page ranges from a few to less than 300 milliseconds.

It is important to note that based on the two Tables 1 and 2, the total wait time for a query to be processed and searched, plus the time to relevance-rank all the hits, plus the time to generate the hit-descriptions per 10-hit page, is about one second or less, making quite feasible the whole approach of math searching, relevance ranking, and hit packaging.

Table 2. Speed Performance of Hit-Description Generation (All time measurements are in milliseconds, and a page has 10 hits)

Query	Hit-packaging Time per Hit	Hit-packaging Time per Page
Ai^2	26	260
$\int \sin$	10.11	101.1
eulerBeta	7	70
$\sin^2$	19.42	194.2
jacobiSN OR Si	4.97	49.7
eulerGamma	6.33	63.3
cos	4.28	42.8
sin	4.16	41.6
z	5.32	53.2

4.2   Outcome Performance Evaluation of Hit-Description Generation

The outcome performance is subjective to some extent. Nevertheless, extensive testing was done on the DLMF testbed on over 100 queries, and the hit-descriptions were examined closely. For each hit, the 5 top-ranking fragments that were identified and presented as the description were found to be truly the most query-relevant and representative of the hit-document. For example, for the query "sin", the top-ranking hit was the one where $\sin z$ is defined, and the description of that hit is:

1. Definitions and Periodicity (in Elementary Functions Chapter)

$\sin z = \frac{e^{iz} - e^{-iz}}{2i}$, ... $e^{\pm iz} = \cos z \pm i \sin z$, ... $\tan z = \frac{\sin z}{\cos z}$, ...

$\csc z = \frac{1}{\sin z}$, ... $\cot z = \frac{\cos z}{\sin z} = \frac{1}{\tan z}$

The hit-document contains other contents involving sin, such as $\sin(z + 2k\pi) = \sin z$, but because the number of fragments per hit-description was limited to 5, some fragments had to be left out. If the description size is set to larger numbers of fragments, more will be included per description. The ideal hit-description size in math search is an aspect that requires further research.

Of course, a thorough subjective evaluation involving a large number of users and a carefully selected benchmarks of queries will have to be conducted in the future.

5 Conclusions

In this paper, new relevance ranking metrics and hit-description generation techniques were presented and analyzed, and their performance was evaluated. It was found that the new relevance metrics are far superior to the conventional tf-idf metric, and the new hit-descriptions are more query-relevant and representative of the hit targets than convential methods of providing the title and some leading sentences of the target document. Furthermore, it was determined that the system response time was about one second or less, which attests to the feasibility of the new approaches working collectively as a system.

Future research will focus on subjective evaluation of the new techniques, with a cross-section of users, using standard testbeds and query benchmarks that the research community will hopefully generate and agree upon. Refinements and extensions of the techniques will undoubtedly have to be carried out as a result of the subjective evaluation and the users' feedback. Also, incorporating highlighting into the hit-descriptions, and turning each fragment in a hit-description into a hot link that would lead the user to the right location in the hit target, are subjects for further research and implementation.

References

1. MathSciNet. American Mathematical Society (AMS).
   http://www.ams.org/mathscinet
2. Bancerek, G.: The 5th International Conference on Mathematical Knowledge Management, Wokingham, UK, pp. 266–279 (August 11-12, 2006)
3. Einwohner, T.H., Fateman, R.: Searching techniques for integral tables. International symposium on Symbolic and algebraic computation, ACM (1995)
   http://torte.cs.berkeley.edu:8010/tilu
4. Guidi, F.: Searching and Retrieving in Content-based Repositories of Formal Mathematical Knowledge. Ph.D. Thesis in Computer Science, University of Bologna, Technical report UBLCS 2003-06 (March 2003)

5. Guidi, F., Schena, I.: A Query Language for a Metadata Framework about Mathematical Resources. In: The 2nd International Conf. Mathematical Knowledge Management, Bertinoro, Italy (February 2003)
6. Jahrbuch Database. http://www.emis.de/MATH/JFM/JFM.html
7. Lozier, D.W.: The DLMF Project: A New Initiative in Classical Special Functions. In: International Workshop on Special Functions - Asymptotics, Harmonic Analysis and Mathematical Physics. Hong Kong (June 21-25, 1999)
8. Lozier, D.W., Miller, B.R., Saunders, B.V.: Design of a Digital Mathematical Library for Science, Technology and Education. In: Proceedings of the IEEE Forum on Research and Technology Advances in Digital Libraries; IEEE ADL '99, Baltimore, Maryland (May 1999)
9. MathDi (Mathematics Didactics Database). http://www.emis.de/MATH/DI.html
10. Mathdex search tool: http://www.mathdex.com:8080/mathfind/search
11. Mathdex description:
    http://www.ima.umn.edu/2006-2007/SW12.8-9.06/activities/Miner-Robert/index.html
12. Mathematica: http://www.mathematica.com
13. Miller, B., Youssef, A.: Technical Aspects of the Digital Library of Mathematical Functions. Annals of Mathematics and Artificial Intelligence 38, 121–136 (2003)
14. MoWGLI: Mathematics on the Web: Get It by Logics and Interfaces.
    http://mowgli.cs.unibo.it/
15. Salton, G., McGill, M.J.: Introduction to Modern Information Retrieval. McGraw Hill, New York (1993)
16. Baeza-Yates, R., Ribeiro-Neto, B.: Modern information retrieval. Addison-Wesley, London (1999)
17. Youssef, A.: Information Search And Retrieval of Mathematical Contents: Issues And Methods. In: The proceedings of the ISCA 14th International Conference on Intelligent and Adaptive Systems and Software Engineering (IASSE-2005), July 20-22, Toronto, Canada (2005)
18. Youssef, A.: Roles of Math Search in Mathematics. In: The 5th International Conference on Mathematical Knowledge Management, Wokingham, UK, pp. 2–16 (August 11-12, 2006)
19. Zentralblatt MATH database at European Mathematical Information Service (EMIS). http://www.emis.de/ZMATH/

Author Index

Lecture Notes in Artificial Intelligence (LNAI)

Vol. 4265: L. Todorovski, N. Lavrač, K.P. Jantke (Eds.), Discovery Science. XIV, 384 pages. 2006.

Vol. 4264: J.L. Balcázar, P.M. Long, F. Stephan (Eds.), Algorithmic Learning Theory. XIII, 393 pages. 2006.

Vol. 4259: S. Greco, Y. Hata, S. Hirano, M. Inuiguchi, S. Miyamoto, H.S. Nguyen, R. Słowiński (Eds.), Rough Sets and Current Trends in Computing. XXII, 951 pages. 2006.

Vol. 4253: B. Gabrys, R.J. Howlett, L.C. Jain (Eds.), Knowledge-Based Intelligent Information and Engineering Systems, Part III. XXXII, 1301 pages. 2006.

Vol. 4252: B. Gabrys, R.J. Howlett, L.C. Jain (Eds.), Knowledge-Based Intelligent Information and Engineering Systems, Part II. XXXIII, 1335 pages. 2006.

Vol. 4251: B. Gabrys, R.J. Howlett, L.C. Jain (Eds.), Knowledge-Based Intelligent Information and Engineering Systems, Part I. LXVI, 1297 pages. 2006.

Vol. 4248: S. Staab, V. Svátek (Eds.), Managing Knowledge in a World of Networks. XIV, 400 pages. 2006.

Vol. 4246: M. Hermann, A. Voronkov (Eds.), Logic for Programming, Artificial Intelligence, and Reasoning. XIII, 588 pages. 2006.

Vol. 4223: L. Wang, L. Jiao, G. Shi, X. Li, J. Liu (Eds.), Fuzzy Systems and Knowledge Discovery. XXVIII, 1335 pages. 2006.

Vol. 4213: J. Fürnkranz, T. Scheffer, M. Spiliopoulou (Eds.), Knowledge Discovery in Databases: PKDD 2006. XXII, 660 pages. 2006.

Vol. 4212: J. Fürnkranz, T. Scheffer, M. Spiliopoulou (Eds.), Machine Learning: ECML 2006. XXIII, 851 pages. 2006.

Vol. 4211: P. Vogt, Y. Sugita, E. Tuci, C.L. Nehaniv (Eds.), Symbol Grounding and Beyond. VIII, 237 pages. 2006.

Vol. 4203: F. Esposito, Z.W. Raś, D. Malerba, G. Semeraro (Eds.), Foundations of Intelligent Systems. XVIII, 767 pages. 2006.

Vol. 4201: Y. Sakakibara, S. Kobayashi, K. Sato, T. Nishino, E. Tomita (Eds.), Grammatical Inference: Algorithms and Applications. XII, 359 pages. 2006.

Vol. 4200: I.F.C. Smith (Ed.), Intelligent Computing in Engineering and Architecture. XIII, 692 pages. 2006.

Vol. 4198: O. Nasraoui, O. Zaïane, M. Spiliopoulou, B. Mobasher, B. Masand, P.S. Yu (Eds.), Advances in Web Mining and Web Usage Analysis. IX, 177 pages. 2006.

Vol. 4196: K. Fischer, I.J. Timm, E. André, N. Zhong (Eds.), Multiagent System Technologies. X, 185 pages. 2006.

Vol. 4188: P. Sojka, I. Kopeček, K. Pala (Eds.), Text, Speech and Dialogue. XV, 721 pages. 2006.

Vol. 4183: J. Euzenat, J. Domingue (Eds.), Artificial Intelligence: Methodology, Systems, and Applications. XIII, 291 pages. 2006.

Vol. 4180: M. Kohlhase, OMDoc – An Open Markup Format for Mathematical Documents [version 1.2]. XIX, 428 pages. 2006.

Vol. 4177: R. Marín, E. Onaindía, A. Bugarín, J. Santos (Eds.), Current Topics in Artificial Intelligence. XV, 482 pages. 2006.

Vol. 4160: M. Fisher, W. van der Hoek, B. Konev, A. Lisitsa (Eds.), Logics in Artificial Intelligence. XII, 516 pages. 2006.

Vol. 4155: O. Stock, M. Schaerf (Eds.), Reasoning, Action and Interaction in AI Theories and Systems. XVIII, 343 pages. 2006.

Vol. 4149: M. Klusch, M. Rovatsos, T.R. Payne (Eds.), Cooperative Information Agents X. XII, 477 pages. 2006.

Vol. 4140: J.S. Sichman, H. Coelho, S.O. Rezende (Eds.), Advances in Artificial Intelligence - IBERAMIA-SBIA 2006. XXIII, 635 pages. 2006.

Vol. 4139: T. Salakoski, F. Ginter, S. Pyysalo, T. Pahikkala (Eds.), Advances in Natural Language Processing. XVI, 771 pages. 2006.

Vol. 4133: J. Gratch, M. Young, R. Aylett, D. Ballin, P. Olivier (Eds.), Intelligent Virtual Agents. XIV, 472 pages. 2006.

Vol. 4130: U. Furbach, N. Shankar (Eds.), Automated Reasoning. XV, 680 pages. 2006.

Vol. 4120: J. Calmet, T. Ida, D. Wang (Eds.), Artificial Intelligence and Symbolic Computation. XIII, 269 pages. 2006.

Vol. 4118: Z. Despotovic, S. Joseph, C. Sartori (Eds.), Agents and Peer-to-Peer Computing. XIV, 173 pages. 2006.

Vol. 4114: D.-S. Huang, K. Li, G.W. Irwin (Eds.), Computational Intelligence, Part II. XXVII, 1337 pages. 2006.

Vol. 4108: J.M. Borwein, W.M. Farmer (Eds.), Mathematical Knowledge Management. VIII, 295 pages. 2006.

Vol. 4106: T.R. Roth-Berghofer, M.H. Göker, H.A. Güvenir (Eds.), Advances in Case-Based Reasoning. XIV, 566 pages. 2006.

Vol. 4099: Q. Yang, G. Webb (Eds.), PRICAI 2006: Trends in Artificial Intelligence. XXVIII, 1263 pages. 2006.

Vol. 4095: S. Nolfi, G. Baldassarre, R. Calabretta, J.C.T. Hallam, D. Marocco, J.-A. Meyer, O. Miglino, D. Parisi (Eds.), From Animals to Animats 9. XV, 869 pages. 2006.

Vol. 4093: X. Li, O.R. Zaïane, Z. Li (Eds.), Advanced Data Mining and Applications. XXI, 1110 pages. 2006.

Vol. 4092: J. Lang, F. Lin, J. Wang (Eds.), Knowledge Science, Engineering and Management. XV, 664 pages. 2006.

Vol. 4088: Z.-Z. Shi, R. Sadananda (Eds.), Agent Computing and Multi-Agent Systems. XVII, 827 pages. 2006.

Vol. 4087: F. Schwenker, S. Marinai (Eds.), Artificial Neural Networks in Pattern Recognition. IX, 299 pages. 2006.

Vol. 4068: H. Schärfe, P. Hitzler, P. Øhrstrøm (Eds.), Conceptual Structures: Inspiration and Application. XI, 455 pages. 2006.

Vol. 4065: P. Perner (Ed.), Advances in Data Mining. XI, 592 pages. 2006.

Vol. 4062: G.-Y. Wang, J.F. Peters, A. Skowron, Y. Yao (Eds.), Rough Sets and Knowledge Technology. XX, 810 pages. 2006.